Conservation Biology

Edited by Misha Bob

SYRAWOOD
PUBLISHING HOUSE
New York

Published by Syrawood Publishing House,
750 Third Avenue, 9th Floor,
New York, NY 10017, USA
www.syrawoodpublishinghouse.com

Conservation Biology
Edited by Misha Bob

International Standard Book Number: 978-1-68286-420-3 (Hardback)

Cataloging-in-publication Data

Conservation biology / edited by Misha Bob.
p. cm.
Includes bibliographical references and index.
ISBN 978-1-68286-420-3
1. Conservation biology. 2. Nature conservation. 3. Ecology. I. Bob, Misha.
QH75 .C66 2017
639.9--dc23

Printed in China.

TABLE OF CONTENTS

PREFACE

Conservation biology is a discipline that integrates several fields in order to study the biodiversity on earth. It also delves into preserving species and their respective habitats from extinction. This book is compiled in such a manner, that it will provide in-depth knowledge about the theory and practice of this discipline. Various techniques and strategies of conservation biology are elucidated in the text like systematic conservation planning, measuring extinction rates, etc. The book aims to shed light on some of the unexplored aspects of this field and the recent researches in this area. This book includes contributions of experts and scientists which will provide innovative insights into this field. It is an essential guide for both academicians and those who wish to pursue conservation biology further.

All of the data presented henceforth, was collaborated in the wake of recent advancements in the field. The aim of this book is to present the diversified developments from across the globe in a comprehensible manner. The opinions expressed in each chapter belong solely to the contributing authors. Their interpretations of the topics are the integral part of this book, which I have carefully compiled for a better understanding of the readers.

At the end, I would like to thank all those who dedicated their time and efforts for the successful completion of this book. I also wish to convey my gratitude towards my friends and family who supported me at every step.

Editor

High Risks of Losing Genetic Diversity in an Endemic Mauritian Gecko: Implications for Conservation

Steeves Buckland[1,2,3]*, Nik C. Cole[1,4,5], Jim J. Groombridge[6], Clemens Küpper[2], Terry Burke[2], Deborah A. Dawson[2], Laura E. Gallagher[3], Stephen Harris[1]

1 School of Biological Sciences, University of Bristol, Bristol, United Kingdom, **2** NERC Biomolecular Analysis Facility, Department of Animal and Plant Sciences, University of Sheffield, Sheffield, United Kingdom, **3** National Parks & Conservation Service, Reduit, Mauritius, **4** Durrell Wildlife Conservation Trust, Jersey, Channel Islands, **5** Mauritian Wildlife Foundation, Vacoas, Mauritius, **6** Durrell Institute of Conservation and Ecology, School of Anthropology and Conservation, University of Kent, Kent, United Kingdom

Abstract

Genetic structure can be a consequence of recent population fragmentation and isolation, or a remnant of historical localised adaptation. This poses a challenge for conservationists since misinterpreting patterns of genetic structure may lead to inappropriate management. Of 17 species of reptile originally found in Mauritius, only five survive on the main island. One of these, *Phelsuma guimbeaui* (lowland forest day gecko), is now restricted to 30 small isolated subpopulations following severe forest fragmentation and isolation due to human colonisation. We used 20 microsatellites in ten subpopulations and two mitochondrial DNA (mtDNA) markers in 13 subpopulations to: (i) assess genetic diversity, population structure and genetic differentiation of subpopulations; (ii) estimate effective population sizes and migration rates of subpopulations; and (iii) examine the phylogenetic relationships of haplotypes found in different subpopulations. Microsatellite data revealed significant population structure with high levels of genetic diversity and isolation by distance, substantial genetic differentiation and no migration between most subpopulations. MtDNA, however, showed no evidence of population structure, indicating that there was once a genetically panmictic population. Effective population sizes of ten subpopulations, based on microsatellite markers, were small, ranging from 44 to 167. Simulations suggested that the chance of survival and allelic diversity of some subpopulations will decrease dramatically over the next 50 years if no migration occurs. Our DNA-based evidence reveals an urgent need for a management plan for the conservation of *P. guimbeaui*. We identified 18 threatened and 12 viable subpopulations and discuss a range of management options that include translocation of threatened subpopulations to retain maximum allelic diversity, and habitat restoration and assisted migration to decrease genetic erosion and inbreeding for the viable subpopulations.

Editor: João Pinto, Instituto de Higiene e Medicina Tropical, Portugal

Funding: This work was funded by the Natural Environment Research Council (NERC). SB was supported by the University of Bristol, a Rufford Small Grant and Interessengruppe Phelsuma sponsors. The funders had no role in study design, data collection and analysis, decision to publish, or preparation of the manuscript.

Competing Interests: The authors have declared that no competing interests exist.

* Email: steeves.buckland@bristol.ac.uk

Introduction

The effects of habitat fragmentation on genetic structure are well documented [1]. Disconnected habitat fragments harbour small, isolated populations [2], which can lead to loss of genetic diversity, inbreeding depression and reduced levels of population-wide fitness [3–5], all factors that can increase the risk of extinction [6]. Reduced genetic diversity following habitat fragmentation [7], combined with low dispersal [8], can also limit the ability of populations to adapt to environmental change [6]. Therefore, a detailed knowledge of how population genetic diversity is structured across fragmented landscapes, and the extent of genetic differentiation, connectivity and effective population sizes (N_e), are key to formulating a conservation strategy that maintains genetic variability and promotes the evolutionary potential of threatened species [9,10].

Understanding patterns of population genetic structure frequently poses a challenge to conservation managers. It is important to determine whether the observed genetic structure is a consequence of recent population fragmentation and isolation, or a remnant of historical localised adaptation. Misdiagnosing the former when the latter is true risks disrupting patterns of local adaptation and outbreeding depression if incompatible populations are mixed. Conversely, interpreting structured patterns to be signatures of local adaptation when they are a consequence of isolation and drift risks inappropriate management to maintain existing genetic patterns, when maximising gene flow between populations might reduce genetic loss and the risk of extinction. Identifying the origins of genetic structure is particularly important when deciding whether to use translocation to reinforce existing populations and/or establish new populations, and to determine how many founding individuals are required to retain the existing genetic diversity. It is also important to interpret genetic patterns alongside ecological factors such as habitat loss when deciding on the most appropriate management option.

In Mauritius, successive occupations by the Dutch (1638–1710), French (1721–1810) and British (1810–1968) have destroyed a large part of the ecosystem through habitat destruction and the introduction of invasive alien species [11]; over this period,

Mauritius has experienced one of the highest known rates of extinction in the world [11]. Of the 17 described species that once formed a rich endemic terrestrial reptile community [12], only five remain on mainland Mauritius (five are extinct, the other seven only survive on small offshore islands). Four of the species still found on the mainland are day geckos (*Phelsuma*), and these support a wide range of ecological functions such as pollination, predator-prey dynamics and seed dispersal [11]. *Phelsuma guimbeaui* (lowland forest day gecko) is the species most vulnerable to extinction. It persists in only 30 small and isolated subpopulations, many of which may be lost within the next decade due to increasing urbanisation, habitat loss, and the impact of invasive species such as *Phelsuma grandis* (giant Madagascar day gecko), introduced to Mauritius in the 1990s (Buckland et al., submitted).

Here, we characterise genetic diversity and structure for *P. guimbeaui* and consider the most appropriate strategy for its long-term conservation. We used a suite of microsatellite markers for *P. guimbeaui* [13] and two mtDNA markers to: (i) quantify levels of genetic diversity, the extent of population structure and genetic differentiation within and between subpopulations; (ii) estimate effective population size for each subpopulation and the degree of gene flow (migration); and (iii) examine the phylogenetic relationships of haplotypes among the different subpopulations. We evaluated whether the levels of genetic structure and diversity reflected recent or ancestral patterns, and interpret our findings in the light of the well-documented chronology of habitat loss recorded in Mauritius by the early European colonists. We then used simulations to: (iv) estimate the probability of survival and retaining rare alleles in a subpopulation; and (v) estimate the number of individuals that should be translocated initially from a subpopulation, and the number of geckos that need to be translocated at timed intervals thereafter, to form a new population without loss of genetic variation. Finally, we make recommendations for the short- and long-term genetic management of *P. guimbeaui*.

Methods

Ethical statement

The capture and tissue sampling were approved by the University of Bristol's Ethical Review Committee (University Investigation Number UB/11/031) and the National Parks and Conservation Service, Ministry of the Agro-Industry, Mauritius.

Study sites and field methods

For the purpose of this study, we defined a subpopulation as inhabiting a piece of wooded habitat separated by barriers such as a major road, large area of agriculture, non-habitable planted/non-planted exotic trees such as *Pinus taeda* (loblolly pine) and *Psidium cattleianum* (strawberry guava), or human habitation. Between 2007 and 2009, we searched the western part of Mauritius and found a total of 30 subpopulations (Figure 1) covering areas ranging in size from 0.006 to 1 km^2. Twelve of the 30 subpopulations were in the Black River mountains in high-quality native forest (>90% native cover).

From 17 January to 2 September 2011, samples were collected from 13 accessible subpopulations for which we could obtain permission. Geckos were captured by hand or with a nylon noose on the end of a telescopic pole. The tail tip (~5 mm) was removed with a sterile scalpel blade, placed in a labelled 1.5 ml microfuge tube containing 94% alcohol and stored at $-20°C$. We only caught one gecko in subpopulation L11 and three each in L12 and L13 (Figure 2); these were not included in the microsatellite analyses. Sample sizes in the other ten subpopulations ranged from 29 to 37. The nearest-neighbour distances among these ten subpopulations ranged from 0.6 to 27.3 km, and the area of each subpopulation ranged from 0.002 to 0.5 km^2 (Figure 2).

DNA extraction and microsatellite amplification

DNA was extracted using an ammonium acetate salt precipitation protocol [14]. The quality and integrity of the DNA was evaluated by gel electrophoresis. DNA concentration was quantified using a Fluostar OPTIMA (Bmg Labtech Ltd, Aylesbury, UK).

We selected 41 polymorphic microsatellite loci with high amplification success, developed specifically for *P. guimbeaui* [13]. We amplified 312 samples in 2 µl multiplex polymerase chain reactions (PCRs) comprising *ca* 10 ng DNA, 1 µl Qiagen Master Mix (Qiagen Ltd, Manchester, UK) [15], 0.2 µM reverse primer, 0.2 µM forward fluorescent primer, and covered by mineral oil. PCRs were conducted with touchdown cycling conditions detailed in [13]. The resulting fragments were sized using an ABI 3730 48-well capillary DNA Analyser (Applied Biosystems, Foster City, CA, USA) and genotypes scored in GENEMAPPER v3.7 (Applied Biosystems, Foster City, CA, USA). The 41 loci were subsequently subjected to an elimination process whereby unsuitable loci were dropped from further analyses (see Results).

Microsatellite analyses

Forty samples were randomly selected, re-PCRed and genotyping error rates per allele [16] estimated using MICROSATELLITE TOOLKIT [17]. Large allelic dropout and stutter-band scoring errors were investigated in MICRO-CHECKER v2.2.3 [18]. We used ML-RELATE [19] to identify 55 full- and half-sib relationships, which were excluded since their inclusion could bias the population structure results [20]. We tested whether microsatellite loci were under selection using LOSITAN [21] and subsequently removed any markers so identified. All loci were tested for deviation from Hardy-Weinberg (HW) equilibrium and all pairs of loci tested for linkage disequilibrium (LD) in each subpopulation using GENEPOP v4.0.10 [22]. False discovery rate [23] was used to correct P values in multiple tests. The null allele frequency per locus was estimated using CERVUS v3.0 [24].

To detect genetic diversity within each subpopulation, the number of different alleles, mean number of different alleles (N_A), allelic richness (A_R) and private allelic richness (P_A) were calculated using a rarefaction approach in the software HP-RARE [25]. The mean observed (H_O) and expected heterozygosities (H_E) for each subpopulation were estimated in CERVUS v3.0 and the mean inbreeding coefficient (F_{IS}) was calculated using GENETIX v4.05 (http://kimura.univ-montp2.fr/genetix/), based on 10,000 permutations. Differences in A_R and H_E were examined using a parametric ANOVA conducted in R 2.15.3 (R Development Core Team 2013).

Three different analyses were used to assess population structure. First, the Bayesian clustering software STRUCTURE v2.3.2 [26] was used. This uses a Markov Chain Monte Carlo (MCMC) approach to assign individuals to their most likely genetic cluster (K) and we used the admixture model with correlated allele frequencies [27]. Ten independent runs of 1,000,000 MCMC iterations, each with a burn-in of 500,000 generations, were explored. No prior information was provided on the geographical locations of samples [26]. The model with the highest log posterior probability Pr (X|K) [26] and highest delta K [28] was examined to identify the most likely value of K. These two parameters were computed using the online version of STRUCTURE HARVESTER [29]. CLUMMP v1.1.2 [30] was then used to infer the global cluster membership (Q) by averaging

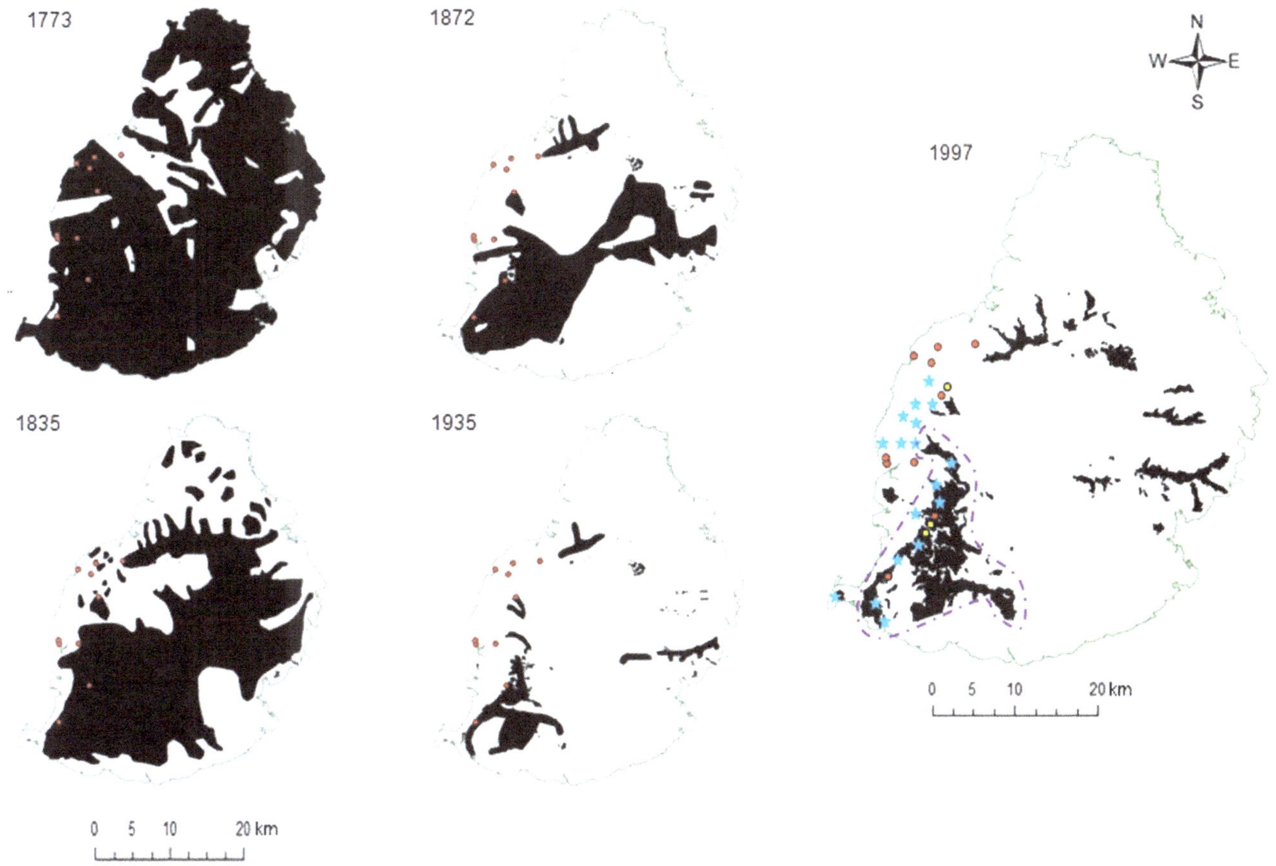

Figure 1. Pattern of deforestation in Mauritius from 1773 to 1997. The red dots indicate the 10 subpopulations for which both microsatellite and mtDNA analyses were conducted, and the yellow dots the three subpopulations for which only mtDNA analyses were carried out. The blue stars mark subpopulations not sampled and the black region within the purple dotted line on the 1997 map shows the Black River mountains. All subpopulation locations were transposed by 1 km.

the results from the independent runs. Clusters were visualised as bar plot charts in DISTRUCT v1.1 [31]. Second, we used TESS [32] to infer the degree of population structure. TESS uses spatial information regarding the location of each individual together with its genotype. For each K value (2–10), an admixture model with ten independent runs of 100,000 sweeps and a burn-in of 20,000 was performed. The lowest deviance information criterion (DIC) was used to choose the best-fitting model. Third, we used the Bayesian assignment method in GENECLASS2 [33] and population assignment was conducted as in [34] with type 1 error set to 0.05, using 1,000 iterations and simulation computation [35].

Weir and Cockerham's F_{st} [36] values were used to assess population genetic differentiation: a pairwise F_{st} matrix was computed and statistical significance estimated with a permutation test of 9,999 replicates in GenAlEx 6.5 [37]. A hierarchical Analysis of Molecular Variance (AMOVA) [38] was used to establish the major sources of genetic variation. The statistical significance of the AMOVA was assessed with 9,999 permutations in GenAlEx 6.5. The software SPAGeDi v1.3 [39] was used to investigate patterns of isolation by distance (IBD) [40]: this was visualised by plotting pairwise genetic distances ($F_{st}/(1\text{-}F_{st})$ against the log-transformed geographical distances among the ten subpopulations.

To determine whether levels of genetic diversity (i.e. A_R and H_E) were related to patterns of habitat loss and isolation, we compiled data on forest cover between 1773 and 1935 from [41] and in 1997 from [42]. The habitat maps were then digitised in ARC GIS 10.1 (ESRI, Redlands, CA, USA). Only the 1997 map was accurately georeferenced as the others were originally hand drawn in the 1940s from historical records of logging and cane production (Figure 1). We then tested whether genetic diversity was related to the time for which subpopulations had been isolated using Pearson's coefficient of correlation in R 2.15.3. We assumed that the loss of native habitats prevented natural migration and eventually reduced genetic diversity. *P. guimbeaui* is a habitat specialist and is mostly reliant on native habitats that have high tree diversity and tall, large trees with cavities (Buckland et al., submitted). The maximum dispersal distance of 28 adult *P. guimbeaui* monitored for a year was <15 m (S. Buckland, unpublished data). There are no data on dispersal in sub-adult and juvenile *P. guimbeaui*, but the maximum recorded dispersal distance for *P. ornata* (ornate day gecko), a similar-sized Mauritian species, was 87 m [43]. While *P. guimbeaui* also persists in the presence of four exotic tree species, i.e. *Eucalyptus tereticornis* (eucalyptus), *Haematoxylum campechianum* (campeche), *Mangifera indica* (mango) and *Terminalia arjuna* (terminalia), these trees are usually scattered and probably act as ecological traps; there are no data on the density of *P. guimbeaui* in these exotic plantations, and it has never been recorded on buildings or in agricultural land.

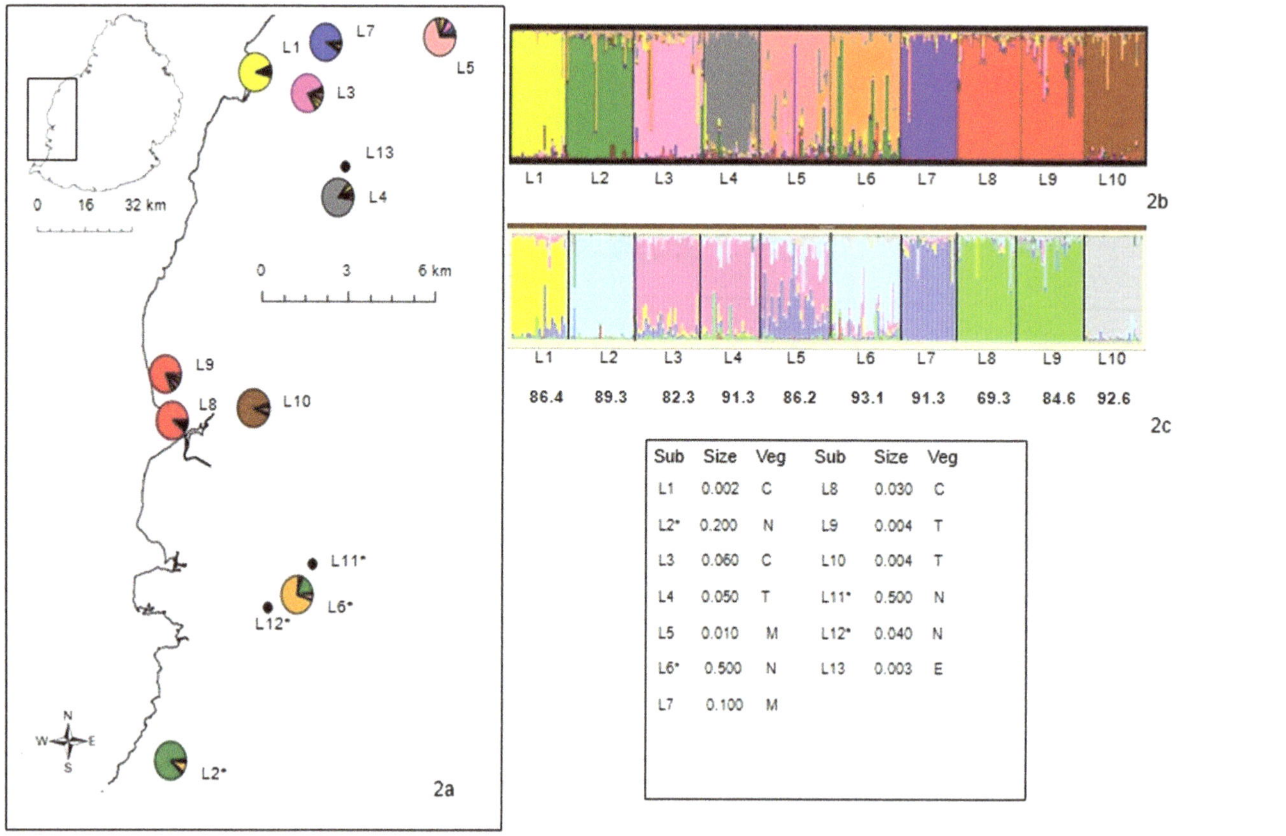

Figure 2. Location and assignment to genetic clusters of ten subpopulations of *Phelsuma guimbeaui*. a Subpopulations used for the microsatellite analyses. Colours in the pie charts indicate the proportion of genetic clusters identified using STRUCTURE 2.3.2. The three subpopulations only used for mtDNA analyses (L11, L12 and L13) are also shown in Figure 2a; subpopulations marked with an asterisk are in the Black River mountains. **b** Bar plots showing the genetic identity of individual samples generated using STRUCTURE 2.3.2. **c** Bar plot output from TESS with each subpopulation's labelling assignment (%) from GENECLASS2 shown below. The box gives details of each subpopulation (Subpop): vegetation (Veg) was classified as exotic campeche forest (C), exotic eucalyptus forest (E), exotic mango orchard (M), native forest (N) and exotic terminalia forest (T); size is the area (km^2) occupied by each subpopulation.

Effective population size and contemporary migration

N_e was estimated using two different techniques. First, we used a point estimate approach using LD in N_e ESTIMATOR [44]. Second, a coalescent computation as implemented in MIGRATE [45] was used, in which the following settings were applied: slice sampling with uniform theta prior boundaries of 0 and 100; a Brownian microsatellite model and constant mutation rate; four static chains (default temperatures); and Watterson estimator theta initially estimated from F_{st}. We carried out four independent runs with 5,000,000 iterations after an initial burn-in of 1,000,000 iterations. The software BOTTLENECK v1.2.02 [46] was also used to detect any recent decline in N_e. We used a general vertebrate multiple step mutation rate ($p_g = 0.22$) [10] and a reptile multiple step mutation rate ($p_g = 0.46$) [47].

Three different techniques were used to estimate contemporary rates of migration between the subpopulations. First, we used a Bayesian approach implemented in BAYESASS v1.3 [48]. The number of iterations was set at 20,000,000, with an initial burn-in of 5,000,000 and a thinning of 2,000 chains. The delta value for allele frequency and inbreeding was kept at the default value of 0.15 and migration was changed to 0.1. Five independent runs were conducted to test for convergence and consistent results. Second, we used GENECLASS2 to detect first-generation migrants. We used the simulation algorithm in [35] and specified L_home as the likelihood criterion, where the number of simulated individuals and type 1 error were set at 10,000 and 0.05, respectively. Third, prior information about the locations of individuals was supplemented using the USEPOPINFO model and MIGRPRIOR was set at 0.001 to output estimates of migration between subpopulations in STRUCTURE v2.3.2 [49].

Estimating risks of extinction and retention of rare alleles

We first used simulations in the R package AlleleRetain [50] to estimate the probability of survival and retaining rare alleles in each subpopulation. We also estimated the number of individuals that would need to be translocated per subpopulation to retain the maximum number of rare alleles. We used the estimated size of each subpopulation, obtained by multiplying N_e by ten [51], and life-history information such as age of maturity and mating system from field and captive data (N. Cole, R. Budzinski and S. Budzinski, unpublished data). The R codes with history information for the simulations are given in Table S1. To evaluate different possible interventions to minimise the risks of extinction and capturing the maximum number of rare alleles in a translocated population, we investigated the following scenarios: (i) the initial number of translocated individuals; (ii) the number of assisted migrants after translocation; and (iii) the frequency at which assisted migrants were translocated, where assisted migrants were geckos translocated in subsequent years after the initial translocation, and the frequency at which assisted migrants were translocated is the number of years after the initial translocation when assisted migrants were translocated. All simulations were conducted across 1,000 replications over a 50-year period.

Mitochondrial sequencing

Partial regions of *cyt b* and *16S rRNA*, two mitochondrial genes frequently used in phylogenetic research [52], were amplified in a 10 µl reaction mixture containing 10 ng of DNA, 5 µl of Mytaq HS DNA Polymerase Mix (Bioline Reagents Ltd., London, UK) and 4 µM of each primer. Primers were designed for mtDNA loci in Primer3 [53] (Table S2). Cycling parameters consisted of an initial denaturation at 95°C for 60 s and 34 cycles starting with denaturation at 95°C for 15 s, annealing at 59°C for 15 s and final extension at 72°C for 10 s. The presence of amplified product was confirmed by visualising a fraction of the product on an agarose gel. PCR products were cleaned up with Exo-SAP-IT (Amersham Biosciences, Piscataway, NJ, USA), precipitated with ethanol and sequenced using the Big Dye Terminator v3.1 Cycle Sequencing Kit on an ABI 3730 DNA Analyser (Applied Biosystems, Foster City, CA, USA).

Mitochondrial DNA analyses

The start and end of sequences were trimmed. A consensus sequence was obtained by aligning the forward and reverse sequence for each individual in CodonCode Aligner (CodonCode Corporation, Dedham, MA, USA) and complementary alignment was conducted in MEGA v5.05 [54] using the ClustalW algorithm. Mitochondrial DNA sequences were concatenated in Geneious (Biomatters Ltd, Auckland, New Zealand). To quantify the genetic variation in the concatenated sequences, haplotype diversity (H) and nucleotide diversity (π) [55] were calculated using DnaSP v5.10.01 [56]. AMOVA and F_{st} were calculated in GenAlEx 6.5; analyses were performed separately per subpopulation (one to eight individuals) and for the entire population.

Phylogenetic trees of different haplotypes were inferred by applying the maximum-likelihood method in MEGA 5.05 [54] and a Bayesian approach in MRBAYES [57]. MODELTEST v3.7 [58] was used to find the best-fit model of evolution according to the corrected Akaike information criterion (AICc): models were GTR + G and HKY + G for *cyt b* and *16S rRNA*, respectively. For maximum likelihood, a phylogenetic tree with 1,000 bootstraps was used to explore the robustness of tree topology. Bayesian analysis was conducted with four chains of 10,000,000 replications, with sample frequency of 2,000 and discarding the first 25% of replicates as burn-in. Two independent runs were conducted to produce a consensus tree that we used to explore relationships between haplotypes. We considered the runs as having converged when their split frequency was <0.01 and the potential scale reduction factor was close to 1 [57]. The tree was rooted with *P. ornata* (Genbank AY221451.1). We also built a statistical parsimony network using TCS v1.7 [59]. Sites with missing data (nucleotides and gaps) were not considered when sequences were collapsed into haplotypes.

Neutrality tests (Tajima's D and F_S) [60,61] and pairwise mismatch distribution were used to detect any signal of recent demographic expansion or increase from a founder population. Geographical regions with negative neutrality indices and unimodal mismatch distributions are expected to show demographic expansion [62]. The raggedness statistic, r, was used to test whether the observed mismatch distribution was significantly different from the expected unimodal distribution [63]. All analyses and tests for significance were performed in DnaSP v5.10.01 with 10,000 coalescent simulations.

To compare the genetic diversity of *P. guimbeaui* with other species of the same genus, we compared π between homologous 464-bp *cyt b* sequences from *P. guimbeaui*, two subspecies of *Phelsuma astriata* (Seychelles small day gecko) and three subspecies of *Phelsuma sundbergi* (Seychelles giant day gecko), all of which are common in the Seychelles [64].

Results

Amplification success and genotyping errors

Amplification success of loci varied from 83.0 to 100.0% (mean ± SD 96.2±3.7%), except for *Pgu 043*, which amplified only 57.6% of the time and so was excluded from subsequent analyses (Table S3). Amplification success (proportion of loci amplifying) varied across samples from 61.7 to 100.0% (mean ± SD 93.7±6.1%). Sampled individuals with an amplification success of less than 70.0% were excluded from the analysis. Loci *Pgu 005* and *Pgu 033* were discarded because of stutter-band scoring errors. Three loci, *Pgu 009*, *Pgu 017* and *Pgu 040*, displayed minor evidence of genotyping errors, with a maximum error rate of 0.033 per allele (Table S3); these loci were retained because of the relatively low mean error rate of 0.004 for all 41 loci. After screening for data quality, a total of 260 geckos genotyped at 38 loci were retained for subsequent testing for HW disequilibrium, LD and the presence of null alleles.

Selection of loci for population genetic analyses

Significant LD was observed in three pairs of loci. However, we retained these loci because the LD was not consistently high across all subpopulations. The number of loci in HW disequilibrium (Table S4) and significant null alleles (proportion >0.1) in the different subpopulations (Table S5) varied from zero to ten. Only loci showing evidence of null alleles and HW disequilibrium ($P < 0.05$) in a maximum of three subpopulations were retained for analysis. Heterozygotes were present in males and females at every locus, indicating no linkage to sex chromosomes. We excluded loci with indications of selection, evidence of alleles differing by 1 base pair, and allele sizes greater than 500 base pairs because the ABI 3730 DNA analyser could not distinguish fragment sizes larger than 500 base pairs accurately. We retained 20 loci for the population genetics analyses (Table S3).

Genetic diversity of subpopulations

The 20 retained markers were all highly polymorphic, with the number of alleles per locus ranging from 11 to 60 (Table S6). Genetic diversity was high at all sites: mean H_E varied from 0.844 to 0.891, H_O from 0.792 to 0.876, N_A from 10.6 to 15.5 and A_R from 8.4 to 10.7. P_A was observed in all the subpopulations and mean P_A varied from 0.400 to 0.960 (Table 1). Significant differences were detected in mean H_E ($F_9 = 9.12$, $P<0.001$) and A_R ($F_9 = 10.17$, $P<0.001$) between the following subpopulations: (i) subpopulation L10 had significantly lower H_E and A_R compared to all other subpopulations except L1 and L7; and (ii) L1 and L7 had significantly lower H_E and A_R compared to subpopulations L2, L3, L4, L5 and L6. There was also a significant positive F_{IS} varying between 0.006 and 0.108 in all subpopulations (Table 1), which indicated inbreeding [65].

Based on the maps of habitat loss (Figure 1), the different subpopulations have been isolated from each other for approximately 0 to 239 years (L2 and L6, 0 years; L4, 77 years; L7 and L10, 140 years; L1, L3, L8 and L9, 177 years; L5, 239 years). We found no correlation between genetic diversity and habitat loss measured as time of isolation (A_R: $r = -0.328$, $t_8 = -0.981$, $P> 0.05$; H_E: $r = -0.183$, $t_8 = -0.526$, $P>0.05$).

Table 1. Population genetic indices (± standard deviation) for ten subpopulations of *Phelsuma guimbeaui*.

Subpop	A	H_E	H_O	N_A	A_R	P_A	F_{IS}
L1	22	0.852±0.018	0.818±0.019	10.7±3.1	8.7±2.2	0.400±0.727	0.042
L2	28	0.877±0.016	0.833±0.016	14.8±4.7	10.4±2.5	0.960±0.791	0.051
L3	29	0.880±0.014	0.817±0.016	13.9±4.8	10.0±2.6	0.800±1.060	0.074
L4	23	0.883±0.012	0.825±0.019	12.3±3.1	9.9±2.1	0.900±1.051	0.067
L5	29	0.889±0.013	0.795±0.017	14.5±.4	10.4±2.6	0.710±0.731	0.108
L6	29	0.891±0.014	0.853±0.015	15.5±5.4	10.7±2.8	0.830±0.894	0.043
L7	23	0.856±0.012	0.807±0.019	11.3±3.3	8.8±.2	0.500±0.611	0.059
L8	26	0.873±0.013	0.833±0.017	12.8±4.3	9.6±2.4	0.680±.964	0.047
L9	26	0.882±0.014	0.876±0.015	14.0±4.3	10.2±2.5	0.660±0.747	0.006
L10	25	0.844±0.016	0.792±0.019	10.6±.8	8.4±1.9	0.630±0.671	0.062

Subpop = subpopulation; A = number of individuals sampled; H_E = mean expected heterozygosity and standard deviation; H_O = mean observed heterozygosity and standard deviation; N_A = mean number of alleles and standard deviation; A_R = mean allelic diversity and standard deviation; P_A = mean private allelic richness and standard deviation; F_{IS} = inbreeding coefficient. N_A, A_R and P_A were based on a minimum sample of 22 geckos per subpopulation.

Population structure, differentiation and IBD

The high membership coefficients to genetic clusters (0.705 to 0.899) found using STRUCTURE supported strong genetic differentiation of subpopulations. The highest delta K and log probability values were observed at K = 9, suggesting that all subpopulations except L8 and L9 were genetically distinct (Figure 2a, b). On average, 86.6% (range 69.3 to 93.1%) of individuals were correctly assigned to their respective subpopulations using the assignment test in GENECLASS2 (Figure 2c). Using TESS, the lowest DIC was obtained for K = 10, indicating that all subpopulations were genetically distinct (Figure 2c). We found a significant pattern of IBD using the microsatellite genotypes ($R^2 = 0.182$, $P = 0.010$) (Figure 3). Pairwise F_{st} comparisons among all subpopulations suggested that there was small to moderate genetic differentiation ($P<0.001$), with F_{st} estimates between 0.016 and 0.072 [66]. The greatest degrees of genetic differentiation were between subpopulations L1 and L2, L1 and L10, L1 and L8, L2 and L7, L2 and L10, L4 and L7, and L7 and L10. The lowest differentiation was found between subpopulations L2 and L6, and L8 and L9 (Table 2). Consistent with the analyses of population structure, the AMOVA confirmed the existence of significant genetic variation at different hierarchical levels, with 4% ($F_9 = 0.042$, $P<0.001$) of variation occurring between subpopulations, 10% among individuals ($F_{250} = 0.108$, $P<0.001$) and 86% ($F_{260} = 0.145$, $P<0.001$) within individuals.

Bottlenecks, migration and effective population size

Using the mutation rate estimate for reptilian microsatellite loci [47], subpopulations L1, L3 and L5 showed evidence of recent bottlenecks, with a significant Wilcoxon's test ($P<0.05$) suggesting an excess of heterozygotes. However, no signs of a bottleneck were observed when the lower general vertebrate mutation rate was used [10]. Using the Bayesian approach in BAYESASS v1.3, the mean probability of no migration occurring per subpopulation in any generation was 0.833 (range 0.675 to 0.992) and the mean migration rate per generation was 0.019 (range 0.000 to 0.121). However, the estimated migration rates between subpopulations were different in each run and so the results were unreliable. No immigrants were detected in the STRUCTURE analysis (USEINFOPOP model). Because the migration results were inconsistent between the different methods, we used the first-generation migrants from GENECLASS2 to calculate the general migration rate [67] and found little evidence of migration across the ten subpopulations (Table S7). Modal estimates of N_e for the different subpopulations were small, ranging from 44 to 167 in N_e ESTIMATOR and 19 to 96 in MIGRATE (Table 3).

Mitochondrial DNA

We obtained partial *cyt b* with 464 base pairs (accession numbers HG779461-HG779540) and *16S rRNA* with 313 base pairs (accession numbers HG518676-HG518755). We analysed concatenated partial *cyt b* and *16S rRNA* sequences (777 base pairs) of 80 individuals from 13 subpopulations (Figure 2a, Table 4). No stop codon was identified, indicating that the true genes rather than nuclear pseudogenes were amplified. We identified 108 sites with missing data (unidentified nucleotides or gaps), 607 monomorphic sites and 62 polymorphic sites (ten singletons and 52 parsimony informative sites). Twenty-five unique haplotypes (Hap_1-25) were recorded in the parsimony network and phylogenetic tree (Figure 4). The number of haplotypes ranged from one to five per subpopulation, with Hap_3 (18.8%) being the most common (Figure 4, Table S8). Seven haplotypes were shared between different subpopulations: (i) Hap_1 and Hap_2 in L1 and L3; (ii) Hap_3 in L3, L4, L5, L6 and L11; (iii) Hap_4 in L2 and L12; (iv)

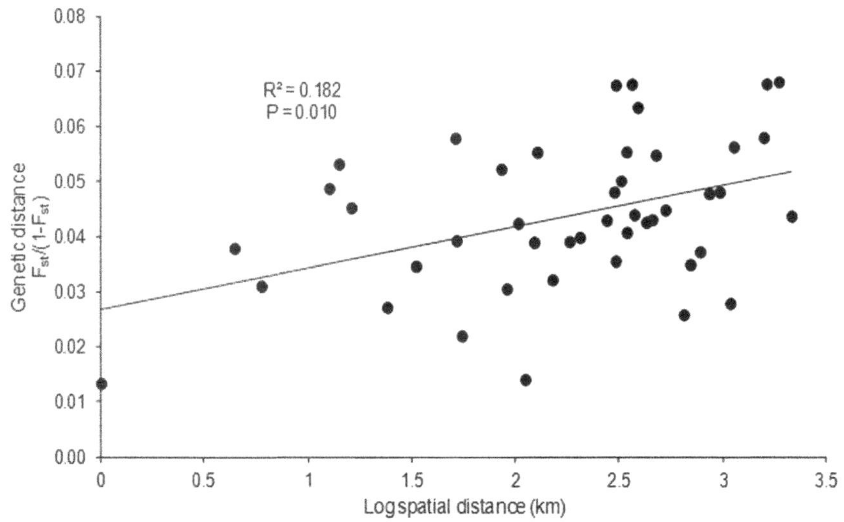

Figure 3. Isolation by distance (IBD) in *Phelsuma guimbeaui* using microsatellite markers. Genetic distance $F_{st}/(1-F_{st})$ is plotted against log spatial distance (km).

Hap_8 in L4 and L13; (v) Hap_9 in L4 and L10; and (vi) Hap_16 in L7 and L10 (Figure 4, Table S8). Unique haplotypes were present in all subpopulations, except some of those where we had a very small sample size, i.e. L11 and L13, but not L12.

Results for mismatch distribution, H and π per subpopulation and for the entire population, are shown in Table 4. Overall, the entire population had negative values for D and F_S, but these were not significant, and a unimodal distribution with $P>0.05$ for the r index suggested demographic expansion. Bayesian and maximum-likelihood phylogenetic tree topologies were similar, and so only the Bayesian results are presented. The Bayesian topology shows clades with low support, with posterior probability <80% (Figure 4). *P. guimbeaui* was three to five times more genetically diverse than the Seychellois geckos, i.e. $\pi = 0.015$ in *P. guimbeaui* compared to $\pi = 0.005$ for *P. a. astriata*, $\pi = 0.005$ for *P. a. semicarinata*, $\pi = 0.003$ for *P. s. ladiguensis*, $\pi = 0.004$ for *P. s. longinsulae* and $\pi = 0.003$ for *P. s. sundbergi*.

Extinction risks and retaining rare alleles through translocation

AlleleRetain simulations averaged across 1,000 replications over 50 years showed that the probability of survival for each subpopulation varied from 0.090 to 0.740, and the probability of retaining rare alleles varied from 0.025 to 0.292 when no migration occurred (Table 5). Under the different scenarios, the simulation results showed that an increase in the number of geckos included in the initial translocation will lead to an increase in the probability of survival and capture of rare alleles, but stabilised at around 20 individuals in most subpopulations (Table S1). With no assisted migration, the probabilities of survival and of retaining rare alleles were nearly zero, but increased significantly when assisted migrants were added to the new translocated population, with a maximum of 0.890 (CI 0.858–0.915) and 0.736 (CI 0.694–0.773) for the probability of survival and retention of rare alleles, respectively (Table S1). The maximum probabilities of survival and of rare allele retention were highest when assisted migration occurred yearly after initial translocation, but gradually decreased to a minimum of 0.206 (CI 0.171–0.244) for probability of survival and 0.126 (CI 0.098–0.159) for retention of rare alleles when assisted migration was only conducted every five years (Table S1).

Discussion

The microsatellite analysis revealed a high degree of population structure and genetic diversity across ten isolated subpopulations. Similarly, the mtDNA data revealed high levels of genetic diversity across the fragmented population despite severe habitat loss and isolation during the last 250 years. Phylogenetic analyses based on mtDNA suggest that the 13 subpopulations were all formerly part of a panmictic population, while the microsatellite analyses indicate that the subpopulations became genetically differentiated through habitat loss and isolation following human colonisation of Mauritius. The N_e was low and simulations suggested that there is a high risk of genetic erosion and extinctions in the next 50 years if subpopulations remain in isolation. While our analyses are based on microsatellite data from ten, and mtDNA data from 13, of the 30 known subpopulations of *P. guimbeaui*, we believe that our data are representative as we covered the known range of *P. guimbeaui* and the non-sampled subpopulations were scattered among those that we did sample.

Population structure, genetic diversity and migration

Since this is the first study to use nuclear markers to examine genetic diversity in the genus *Phelsuma*, comparisons with congeneric species were not possible. The microsatellite measures of genetic diversity were relatively high compared to other reptiles [68,69] despite subpopulations having been isolated for periods up to 239 years. However, we could not find any evidence for a correlation between genetic diversity and time of isolation. Similarly, the high level of nucleotide diversity in *P. guimbeaui* compared to common species of *Phelsuma* in the Seychelles suggests that the various subpopulations of *P. guimbeaui* are still genetically diverse. The negative D and F_S, while not significant, and their unimodal distributions imply that *P. guimbeaui* has experienced a recent population expansion. This could be due to an unexpectedly high proportion of rare alleles originating from a founder effect [70], possibly because *P. guimbeaui* has colonised expanding habitats dominated by *Eucalyptus tereticornis*, *Haematoxylum campechianum*, *Mangifera indica* and *Terminalia arjuna*. All the subpopulations we sampled, except L2, L6, L11 and L12, were in exotic plantations. However, since these plantations are small and widely scattered, natural migration between them is unlikely and they do

Table 2. Pairwise F_{st} values among the 10 subpopulations of *Phelsuma guimbeaui*; all values were significant at P<0.001 based on 999 permutations.

Subpop	L1	L2	L3	L4	L5	L6	L7	L8	L9	L10
L1										
L2	0.070									
L3	0.029	0.060								
L4	0.050	0.055	0.036							
L5	0.045	0.050	0.024	0.031						
L6	0.047	0.019	0.034	0.045	0.026					
L7	0.046	0.067	0.035	0.063	0.037	0.045				
L8	0.061	0.051	0.044	0.039	0.033	0.037	0.056			
L9	0.045	0.044	0.032	0.043	0.026	0.030	0.039	0.016		
L10	0.072	0.067	0.044	0.050	0.040	0.049	0.063	0.049	0.045	

Subpop = subpopulation.

not provide suitable alternative habitats for the long-term conservation of *P. guimbeaui*.

Our analyses suggest that either nine (STRUCTURE) or all ten (TESS and GENECLASS2) of the sampled subpopulations were genetically distinct. There was a relatively moderate or high level of admixture between subpopulations L2 and L6, and between L8 and L9, which may be due to the relative proximity of these paired sites or connectivity. While most sites were separated by large expanses of urbanisation or agriculture, subpopulations L2 and L6 were in a continuous forest within the Black River mountains but isolated by forest dominated by invasive plants, predominantly *Psidium cattleianum*. However, individual *P. guimbeaui* were seen on scattered native trees in areas of invaded forest, suggesting that low levels of migration may be occurring. While subpopulations L8 and L9 were separated by 0.6 km of unsuitable habitat, including a 70 m wide strip of bare land, the admixture results suggest that migration could still be occurring between them, even though the other analyses tend to suggest little or no migration. AMOVA and F_{st} analyses identified genetic structure with small to moderate levels of genetic differentiation between the geographically isolated subpopulations. A small F_{st} can be a sign of historical differentiation and recent gene flow [71] or a signal of shared descent [72]. The relatively low F_{st} values between subpopulations L2 and L6 and L8 and L9 may also be due to the high level of admixture between these pairs of subpopulations, even though TESS and GENECLASS2 predicted that they were genetically different. These results suggest that habitat fragmentation (or lack of connectivity) has had an impact on genetic differentiation in many subpopulations by limiting migration. There are two plausible non-mutually exclusive explanations that may account for the level of genetic differentiation observed among the subpopulations. First, restricted dispersal can produce a pattern of IBD [73]. According to this model, gene flow will decline linearly with geographical distance. Second, isolation due to barriers roads and agricultural areas may have enhanced genetic differentiation. More than 95% of Mauritius' native forest has been lost since human colonisation in 1638 [74]. Habitat destruction coincided with planting of the four types of exotic plantation that provide alternative habitats for *P. guimbeaui*. The initial founder effect of these colonisations, together with the cumulative effects of genetic drift and low effective population size, may have promoted genetic differentiation.

Effective population size and bottlenecks

Values for N_e ESTIMATOR were consistently higher than those obtained with MIGRATE, and the former LD-based method may provide a better estimate of N_e because it records recent N_e (up to two generations) and so is less likely to be affected by historical events [75]. We consider our N_e estimates to be small and so these subpopulations are at high risk of extinction [76]. We had expected that many subpopulations would have been subjected to a genetic bottleneck. However, only subpopulations L1, L3 and L5 showed evidence of a recent bottleneck when the reptilian multiple-step mutation rate was used for demographic models. This could be because the reptilian mutation rate was inappropriate, since it was estimated from just one species of skink [47]. Additionally, this method will only detect recent bottlenecks and is sensitive to small sample sizes [10], and the effects of gene flow [77], so our bottleneck results for *P. guimbeaui* should be interpreted with caution.

Phylogenetic relationships of haplotypes

Phylogenetic analyses provided strong evidence that the spatial genetic structure and differentiation were absent in *P. guimbeaui*

Table 3. Mean effective population sizes with 95% confidence intervals in parentheses for ten subpopulations of *Phelsuma guimbeaui* estimated using N_e ESTIMATOR and MIGRATE.

Subpopulation	N_e ESTIMATOR	MIGRATE
L1	123 (90–191)	19 (4–38)
L2	121 (100–154)	54 (33–77)
L3	145 (116–193)	57 (35–80)
L4	72 (60–91)	60 (34–88)
L5	125 (102–159)	96 (64–135)
L6	167 (133–224)	93 (51–125)
L7	106 (82–148)	26 (4–46)
L8	76 (64–92)	42 (19–71)
L9	121 (98–157)	75 (44–121)
L10	44 (39–51)	24 (6–43)

prior to habitat loss. Shared haplotypes among the subpopulations suggest that they originally formed part of a panmictic population. For example, Hap_3 was shared by five subpopulations, among which L3 and L11 were more than 20 km apart. The absence of genetic structure within the mtDNA data is also supported by historical vegetation maps, which show that most of the western part of Mauritius had continuous forest cover, and *P. guimbeaui* may have occurred throughout these lowland dry forests prior to human colonisation. This also suggests that *P. guimbeaui* was widely dispersed when the forest was not fragmented.

Contrasting signals of genetic structure between nuclear and mtDNA data have been observed in other species [78]. There are several alternative explanations, including differences in mutation rate and sex-biased dispersal. MtDNA has a lower mutation rate than microsatellites [47,79] and mtDNA lineages may not yet have become differentiated following the onset of habitat fragmentation [7]. The population structure revealed by the microsatellite

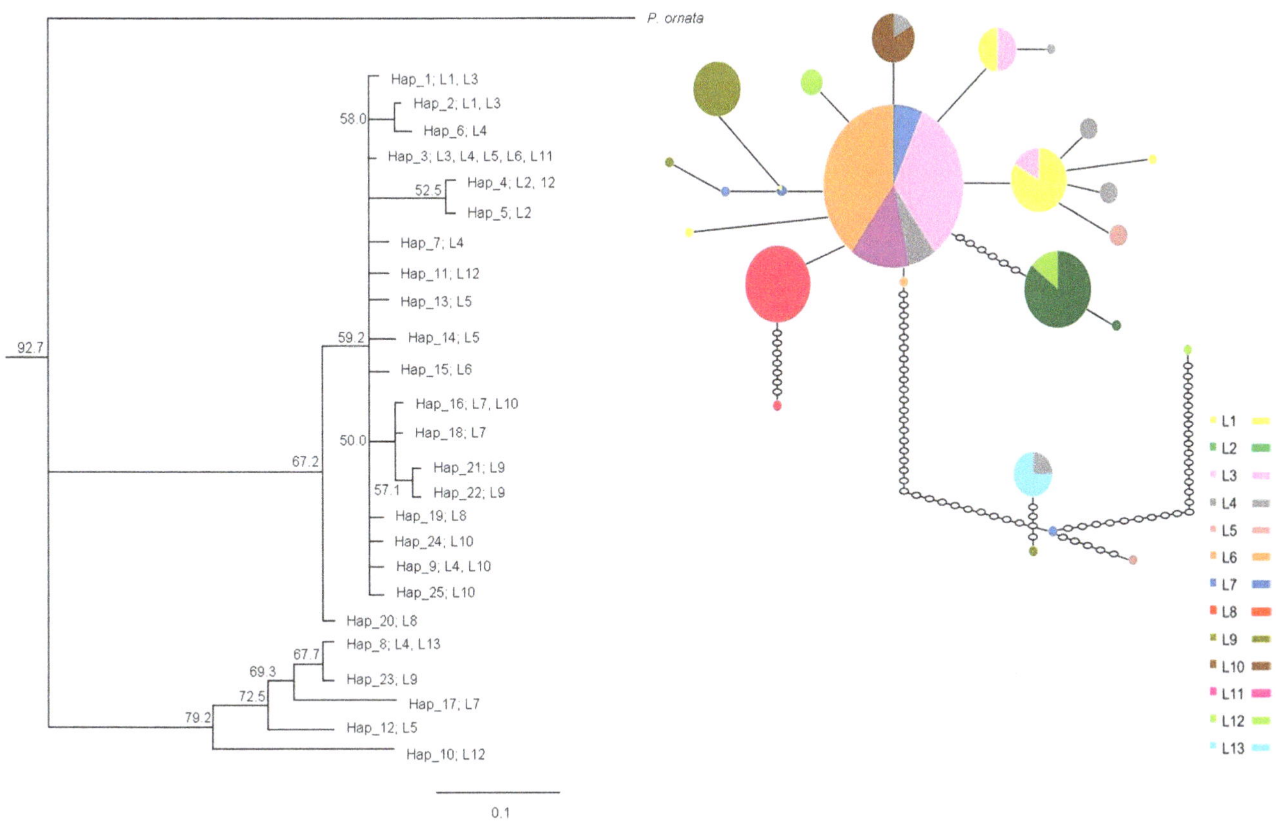

Figure 4. Phylogenetic relationships of mtDNA haplotypes in *Phelsuma guimbeaui*. The Bayesian tree was produced in MRBAYES with subpopulation identity (L1 to L13) shown at the end of each branch. In the parsimony network, the circles represent different haplotypes, with their size proportional to the number of geckos. Open circles represent predicted but missing or unsampled haplotypes.

Table 4. Population statistics of the concatenated mtDNA sequences for 13 subpopulations of *Phelsuma guimbeaui*.

Subpop	Area	A	H_N	Haplotype	H	π	D	Fs	Mismatch distribution	r
L1	0.002	7	2	Hap_1,2	0.476	0.001	0.687	1.741	Multimodal	0.042*
L2	0.200	7	2	Hap_4,5	0.286	<0.001	−1.021	−0.095	Multimodal	0.042*
L3	0.060	8	3	Hap_1,2,3	0.607	0.001	−0.084	0.044	Multimodal	0.042*
L4	0.050	6	5	Hap_3,6,7,8,9	0.933	0.022	−1.423	1.723	Multimodal	0.042
L5	0.010	8	4	Hap_3,12,13,14	0.857	0.027	0.558	6.934	Multimodal	0.042
L6	0.500	7	2	Hap_3,15	0.286	<0.001	−1.023	−0.095	Unimodal	0.042*
L7	0.010	7	3	Hap_16,17,18	0.524	0.013	−1.821*	5.811	Bimodal	0.042*
L8	0.030	8	2	Hap_19,20	0.250	0.003	−1.700*	4.213	Multimodal	0.042*
L9	0.004	7	3	Hap_21,22,23	0.714	0.017	−1.721*	4.221	Multimodal	0.042*
L10	0.004	8	3	Hap_9,16,24,25	0.643	0.002	−0.222	−0.470	Bimodal	0.042*
L11	0.500	1	-	Hap_3	-	-	-	-	-	-
L12	0.040	3	3	Hap_4,10,11	1.000	0.044	-	2.422	Multimodal	0.042
L13	0.003	3	1	Hap_8	0.667	0.002	-	1.133	Multimodal	0.042*
Overall		80	25		0.931	0.016	−0.856	−0.752	Multimodal	0.042

* denotes significant at $P<0.05$.
Subpop = subpopulation; Area = area of subpopulation in km^2; A = number of individuals per subpopulation; H_N = number of haplotypes; H = haplotype diversity; π = nucleotide diversity; D = Tajima's D; Fs = Fu's Fs statistic; r = raggedness index.

Table 5. Simulation showing the probability of survival and retaining rare alleles, with 95% confidence intervals in parentheses, over 50 years in ten subpopulations of *Phelsuma guimbeaui* in the absence of migration.

Subpopulation	Estimated population size	Probability of survival	Probability of retaining rare alleles
L1	1230	0.706 (0.663–0.745)	0.276 (0.238–0.318)
L2	1210	0.740 (0.711–0.767)	0.292 (0.264–0.321)
L3	1450	0.540 (0.509–0.571)	0.191 (0.167–0.217)
L4	720	0.217 (0.192–0.244)	0.046 (0.034–0.061)
L5	1250	0.443 (0.412–0.474)	0.124 (0.105–0.146)
L6	1670	0.633 (0.602–0.663)	0.243 (0.217–0.271)
L7	1060	0.380 (0.350–0.411)	0.116 (0.097–0.138)
L8	760	0.243 (0.217–0.271)	0.084 (0.068–0.103)
L9	1210	0.438 (0.407–0.469)	0.130 (0.110–0.153)
L10	440	0.090 (0.073–0.110)	0.025 (0.017–0.037)

The estimated population size ($N_e \times 10$) with a default rare allele frequency of 0.05 was implemented into the starting parameters of all models.

genotypes is consistent with the extent of landscape fragmentation. Most forest fragments containing subpopulations have been isolated from each other within the last two centuries, suggesting that the population structure occurred as a consequence of recent habitat loss. Alternatively, gene flow may be promoted solely by females, since males are more immobile [78]. However, we consider this is an unlikely explanation of the lack of genetic structure at mtDNA markers because of the low dispersal rates in both sexes of *P. guimbeaui* (S. Buckland, unpublished data) and the observed strong differentiation of nuclear markers.

Implications for conservation

Our genetic data suggest that *P. guimbeaui* was a panmictic population until relatively recently, which has become differentiated through anthropogenic habitat loss and isolation. There is, therefore, no reason to maintain the genetic identity of individual subpopulations and little risk of outbreeding depression should subpopulations be mixed. Conservation management is a priority, since there is a high risk of genetic erosion and extinctions in the next 50 years, even without further habitat loss and other anthropogenic changes. Possible interventions based on our genetic data are: (i) the construction of habitat corridors linking closely-neighbouring subpopulations; (ii) the restoration and expansion of the native habitats in and around existing subpopulations; and (iii) the translocation of individuals to enhance the genetic diversity of viable subpopulations and/or establish one or more new populations.

Continuing habitat loss in Mauritius suggests that some subpopulations are likely to be lost in the next decade: we sampled nine such subpopulations. Given the immediacy of the threat, and that the construction of habitat corridors linking some of them and/or habitat restoration are not realistic options, the only practical solution is to translocate individuals from these nine subpopulations to reinforce viable subpopulations, if there is habitat to support additional geckos, or move them to suitable habitat patches to establish new population(s) [80]. We used AlleleRetain to see which management option would maximise the retention of allelic diversity in any new population that might be established by translocation. This program is designed to examine options when potential new populations are of limited size and cannot be supported by natural immigration [50]. To capture more than 80% of the rare alleles in each subpopulation, initially a minimum of 20 adult geckos would need to be translocated and established from each subpopulation, with assisted migration of 10 geckos each year thereafter. All calculations were based on single subpopulations and it was assumed that each subpopulation would be translocated to a different site. Mixing genetically different subpopulations would increase the proportion of rare alleles and reduce the number of geckos that would need to be translocated.

However, translocation of a selected number of animals is only practical if the host subpopulation is viable in the medium to longer term. Since these nine subpopulations (L1, L3, L4, L5, L7, L8, L9, L10, L13) are highly threatened by further habitat loss, it is unlikely that any will persist for more than a decade. So it may be more logical to translocate as many geckos as possible from each subpopulation in one operation, even if it hastens the ultimate loss of that subpopulation. Nine other subpopulations not sampled during this study (Figure 1) are equally threatened by habitat loss and so the same rationale applies. Data from the subpopulations we sampled strongly suggest that the nine subpopulations we did not sample will also be genetically different as they are small and completely isolated. So their translocation should enhance the overall genetic diversity preserved by conservation action.

Only four of the subpopulations we examined (L2, L6, L11 and L12) were still viable, and they were all within the Black River mountains, the largest area of remnant native forest in Mauritius. There were eight other subpopulations within the Black River mountains, which we did not sample but were probably also viable. Habitat restoration and expansion within this mountain range would support the long-term management of the species and enable *P. guimbeaui* to expand its range in the Black River mountains. The most practical solution would be to create habitat corridors linking those areas of native forest that still harbour relatively large subpopulations of *P. guimbeaui*. While this will increase migration between the 12 viable subpopulations and decrease intra-subpopulation inbreeding, this is a long-term strategy and so assisted migration within the Black River mountains should be considered as an interim option. However, prior to any such intervention, data are needed to confirm the viability and breeding success of translocated geckos following release into a new population. Post-release data are also needed to assess whether greater genetic diversity is retained by releasing geckos "rescued" from non-viable subpopulations into a viable subpopulation or using them to establish a new population.

Conclusions

Although the immediate threat facing *P. guimbeaui* is habitat destruction, our data highlight the importance of genetic studies in guiding conservation management. Maximising the retention of genetic diversity is important and we identified two conservation measures to maximise the retention of genetic diversity in *P. guimbeaui*. We recommend: (i) a short-term rescue action by translocating as many geckos as possible from the 18 subpopulations threatened by imminent habitat loss; and (ii) a long-term action to restore habitats in the Black River mountains to link the 12 potentially viable subpopulations by habitat corridors. Since habitat management is a long-term strategy, especially since *P. guimbeaui* is a habitat specialist relying on high native tree diversity and large tall trees with numerous cavities, assisted migration among the 12 viable subpopulations should reduce the risks of genetic degradation in the short term. However, data are needed on the survival and breeding success of translocated geckos used to reinforce viable subpopulations or to establish new populations prior to the implementing any program of assisted migration.

Supporting Information

Table S1 Simulation results from R package AlleleRetain. The table shows the probability of survival and genetic diversity after 50 years (1000 replications) in different scenarios in each subpopulation (Subpop) of *Phelsuma guimbeaui*: (i) initial number of translocated individuals (StartN) varied from 10–40; (ii) number of assisted migrants after translocation (MigrN) ranged from 0–30; and (iii) frequency at which assisted migrants were translocated (Frequency) confined from one to five years. R and simulations codes are given below the table.

Table S2 Mitochondrial DNA primers in *Phelsuma guimbeaui*.

Table S3 Screening for data quality to select the best loci for analyses of molecular variation in *Phelsuma guimbeaui*.

Table S4 The number of subpopulations of *Phelsuma guimbeaui* not in Hardy-Weinberg equilibrium ($P<0.05$) at the different loci.

Table S5 Null allele frequencies in ten subpopulations of *Phelsuma guimbeaui*.

Table S6 Number of alleles observed for the 20 loci used in the microsatellite analyses of *Phelsuma guimbeaui*.

Table S7 The mean probability of no migration and migration among ten subpopulations of *Phelsuma guimbeaui* obtained in GENECLASS2.

Table S8 Haplotype distribution and frequency in 80 individuals from 13 subpopulations of *Phelsuma guimbeaui*.

Acknowledgments

We thank Sion Henshaw and Stephan Mellier for all their help in the field; Vishnuduth Bachraz, Mannickchand Puttoo and Kevin Ruhomaun from the Mauritian National Parks and Conservation Service for their support; the Mauritian Wildlife Foundation for logistical support; Susana Freitas, Gavin Horsburgh, Andy Krupa, Cheryl Mills, Mauricio Montano Rendon and Rachel Tucker for their advice on the laboratory work and analyses; Ralph and Sylvia Budzinski, whose interest in the protection of *P. guimbeaui* helped initiate this research; the UK Natural Environment Research Council for supporting the laboratory work; and the National Parks and Conservation Service, Ministry of the Agro-Industry, Mauritius, for permission to conduct this research.

Author Contributions

Conceived and designed the experiments: SB NCC JJG SH. Performed the experiments: SB LEG. Analyzed the data: SB CK TB DAD. Wrote the paper: SB SH NCC JJG.

References

1. Frankham R, Ballou JD, Briscoe DA (2010) Introduction to conservation genetics. Cambridge: Cambridge University Press. 643 p.
2. Janečka JE, Tewes ME, Laack LL, Grassman LI, Haines AM, et al. (2008) Small effective population sizes of two remnant ocelot populations (*Leopardus pardalis albescens*) in the United States. Conservation Genetics 9: 869–878. doi: 10.1007/s10592-007-9412-1
3. Slate J, Kruuk LEB, Marshall TC, Pemberton JM, Clutton-Brock TH (2000) Inbreeding depression influences lifetime breeding success in a wild population of red deer (*Cervus elaphus*). Proceedings of the Royal Society of London B 267: 1657–1662. doi: 10.1098/rspb.2000.1192
4. Újvári B, Madsen T, Kotenko T, Olsson M, Shine R, et al. (2002) Low genetic diversity threatens imminent extinction for the Hungarian meadow viper (*Vipera ursinii rakosiensis*). Biological Conservation 105: 127–130. doi: 10.1016/S0006-3207(01)00176-8
5. Liberg O, Andrén H, Pedersen H-C, Sand H, Sejberg D, et al. (2005) Severe inbreeding depression in a wild wolf (*Canis lupus*) population. Biology Letters 1: 17–20. doi: 10.1098/rsbl.2004.0266
6. Bijlsma R, Bundgaard J, Boerema AC (2000) Does inbreeding affect the extinction risk of small populations?: predictions from *Drosophila*. Journal of Evolutionary Biology 13: 502–514. doi: 10.1046/j.1420-9101.2000.00177.x
7. Dixo M, Metzger JP, Morgante JS, Zamudio KR (2009) Habitat fragmentation reduces genetic diversity and connectivity among toad populations in the Brazilian Atlantic Coastal Forest. Biological Conservation 142: 1560–1569. doi: 10.1016/j.biocon.2008.11.016
8. Schtickzelle N, Mennechez G, Baguette M (2006) Dispersal depression with habitat fragmentation in the bog fritillary butterfly. Ecology 87: 1057–1065. doi: 10.1890/0012-9658
9. Da Silva A, Luikart G, Yoccoz NG, Cohas A, Allainé D (2006) Genetic diversity-fitness correlation revealed by microsatellite analyses in European alpine marmots (*Marmota marmota*). Conservation Genetics 7: 371–382. doi: 10.1007/s10592-005-9048-y
10. Peery MZ, Kirby R, Reid BN, Stoelting R, Doucet-Bëer E, et al. (2012) Reliability of genetic bottleneck tests for detecting recent population declines. Molecular Ecology 21: 3403–3418. doi: 10.1111/j.1365-294X.2012.05635.x
11. Cheke A, Hume J (2008) Lost land of the dodo: an ecological history of Mauritius, Réunion & Rodrigues. London: Poyser. 464 p.
12. Arnold EN (2000) Using fossils and phylogenies to understand evolution of reptile communities on islands. Isolated vertebrate communities in the tropics. Bonner Zoologische Monographien 46: 309–323.
13. Buckland S, Horsburgh GJ, Dawson DA, Cole NC, Krupa AP, et al. (2013) Isolation and characterisation of Mauritius lowland day gecko *Phelsuma guimbeaui* microsatellite loci. Conservation Genetics Resources 5: 1013–1018. doi: 10.1007/s12686-013-9957-x
14. Nicholls JA, Double MC, Rowell DM, Magrath RD (2000) The evolution of cooperative and pair breeding in thornbills *Acanthiza* (Pardalotidae). Journal of Avian Biology 31: 165–176. doi: 10.1034/j.1600-048X.2000.310208.x
15. Kenta T, Gratten J, Haigh NS, Hinten GN, Slate J, et al. (2008) Multiplex SNP-SCALE: a cost-effective medium-throughput single nucleotide polymorphism genotyping method. Molecular Ecology Resources 8: 1230–1238. doi: 10.1111/j.1755-0998.2008.02190.x
16. Hoffman JI, Amos W (2005) Microsatellite genotyping errors: detection approaches, common sources and consequences for paternal exclusion. Molecular Ecology 14: 599–612. doi: 10.1111/j.1365-294X.2004.02419.x
17. Park SDE (2002) Trypanotolerance in West African cattle and the population genetic effects of selection. University of Dublin: PhD thesis. 241 p.

18. Van Oosterhout C, Hutchinson WF, Wills DPM, Shipley P (2004) MICRO-CHECKER: software for identifying and correcting genotyping errors in microsatellite data. Molecular Ecology Notes 4: 535–538. doi: 10.1111/j.1471-8286.2004.00684.x
19. Kalinowski ST, Wagner AP, Taper ML (2006) ML-RELATE: a computer program for maximum likelihood estimation of relatedness and relationship. Molecular Ecology Notes 6: 576–579. doi: 10.1111/j.1471-8286.2006.01256.x
20. Anderson EC, Dunham KK (2008) The influence of family groups on inferences made with the program Structure. Molecular Ecology Resources 8: 1219–1229. doi: 10.1111/j.1755-0998.2008.02355.x
21. Antao T, Lopes A, Lopes RJ, Beja-Pereira A, Luikart G (2008) LOSITAN: a workbench to detect molecular adaptation based on a F_{st}-outlier method. BMC Bioinformatics 9 (323). doi: 10.1186/1471-2105-9-323
22. Raymond M, Rousset F (1995) GENEPOP (version 1.2): population genetics software for exact tests and ecumenicism. Journal of Heredity 86: 248–249. doi: cgi/content/short/86/3/248
23. Verhoeven KJF, Simonsen KL, McIntyre LM (2005) Implementing false discovery rate control: increasing your power. Oikos 108: 643–647. doi: 10.1111/j.0030-1299.2005.13727.x
24. Kalinowski ST, Taper ML, Marshall TC (2007) Revising how the computer program CERVUS accommodates genotyping error increases success in paternity assignment. Molecular Ecology 16: 1099–1106. doi: 10.1111/j.1365-294X.2007.03089.x
25. Kalinowski ST (2005) HP-RARE 1.0: a computer program for performing rarefaction on measures of allelic richness. Molecular Ecology Notes 5: 187–189. doi: 10.1111/j.1471-8286.2004.00845.x
26. Pritchard JK, Stephens M, Donnelly P (2000) Inference of population structure using multilocus genotype data. Genetics 155: 945–959.
27. Falush D, Stephens M, Pritchard JK (2003) Inference of population structure using multilocus genotype data: linked loci and correlated allele frequencies. Genetics 164: 1567–1587.
28. Evanno G, Regnaut S, Goudet J (2005) Detecting the number of clusters of individuals using the software STRUCTURE: a simulation study. Molecular Ecology 14: 2611–2620. doi: 10.1111/j.1365-294X.2005.02553.x
29. Earl DA, von Holdt BM (2012) STRUCTURE HARVESTER: a website and program for visualizing STRUCTURE output and implementing the Evanno method. Conservation Genetics Resources 4: 359–361. doi: 10.1007/s12686-011-9548-7
30. Jakobsson M, Rosenberg NA (2007) CLUMPP: a cluster matching and permutation program for dealing with label switching and multimodality in analysis of population structure. Bioinformatics 23: 1801–1806. doi: 10.1093/bioinformatics/btm233
31. Rosenberg NA (2004) DISTRUCT: a program for the graphical display of population structure. Molecular Ecology Notes 4: 137–138. doi: 10.1046/j.1471-8286.2003.00566.x
32. Chen C, Durand E, Forbes F, Francois O (2007) Bayesian clustering algorithms ascertaining spatial population structure: a new computer program and a comparison study. Molecular Ecology Notes 7: 747–756. doi: 10.1111/j.1471-8286.2007.01769.x
33. Piry S, Alapetite A, Cornuet J-M, Paetkau D, Baudouin L, et al. (2004) GENECLASS2: a software for genetic assignment and first-generation migrant detection. Journal of Heredity 95: 536–539. doi: 10.1093/jhered/esh074
34. Rannala B, Mountain JL (1997) Detecting immigration by using multilocus genotypes. Proceedings of the National Academy of Sciences of the USA 94: 9197–9201.
35. Paetkau D, Slade R, Burden M, Estoup A (2004) Genetic assignment methods for the direct, real-time estimation of migration rate: a simulation-based exploration of accuracy and power. Molecular Ecology 13: 55–65. doi: 10.1046/j.1365-294X.2003.02008.x
36. Weir BS, Cockerham CC (1984) Estimating *F*-statistics for the analysis of population structure. Evolution 38: 1358–1370. doi: 10.2307/2408641
37. Peakall R, Smouse PE (2012) GenAlEx 6.5: genetic analysis in Excel. Population genetic software for teaching and research - an update. Bioinformatics 28: 2537–2539. doi: 10.1093/bioinformatics/bts460
38. Excoffier L, Smouse PE, Quattro JM (1992) Analysis of molecular variance inferred from metric distances among DNA haplotypes: application to human mitochondrial DNA restriction data. Genetics 131: 479–491.
39. Hardy OJ, Vekemans X (2002) SPAGeDi: a versatile computer program to analyse spatial genetic structure at the individual or population levels. Molecular Ecology Notes 2: 618–620. doi: 10.1046/j.1471-8278.2002.00305.x
40. Wright S (1943) Isolation by distance. Genetics 28: 114–138.
41. Vaughan RE, Wiehe PO (1937) Studies on the vegetation of Mauritius: I. A preliminary survey of the plant communities. Journal of Ecology 25: 289–343.
42. Page W, D'Argent G (1997) A vegetation survey of Mauritius. Unpublished report commissioned by IUCN, BASEL. Port Louis: Mauritian Wildlife Foundation. 111 p.
43. Nyhagen DF, Kragelund C, Olesen JM, Jones CG (2001) Insular interactions between lizards and flowers: flower visitation by an endemic Mauritian gecko. Journal of Tropical Ecology 17: 755–761.
44. Ovenden JR, Peel D, Street R, Courtney AJ, Hoyle SD, et al. (2007) The genetic effective and adult census size of an Australian population of tiger prawns (*Penaeus esculentus*). Molecular Ecology 16: 127–138. doi: 10.1111/j.1365-294X.2006.03132.x
45. Beerli P (2006) Comparison of Bayesian and maximum-likelihood inference of population genetic parameters. Bioinformatics 22: 341–345. doi: 10.1093/bioinformatics/bti803
46. Cornuet JM, Luikart G (1996) Description and power analysis of two tests for detecting recent population bottlenecks from allele frequency data. Genetics 144: 2001–2014.
47. Gardner MG, Bull CM, Cooper SJB, Duffield GA (2000) Microsatellite mutations in litters of the Australian lizard *Egernia stokesii*. Journal of Evolutionary Biology 13: 551–560. doi: 10.1046/j.1420-9101.2000.00189.x
48. Wilson GA, Rannala B (2003) Bayesian inference of recent migration rates using multilocus genotypes. Genetics 163: 1177–1191.
49. Pritchard JK, Wen X, Falush D (2009) Documentation for *structure* software: Version 2.3. http://kinglab.eeb.lsa.umich.edu/EEID/eeid/evolution/Popgen_EEID_2012/Manuals/STRUCTURE_Manual.pdf.
50. Weiser EL, Grueber CE, Jamieson IG (2012) AlleleRetain: a program to assess management options for conserving allelic diversity in small, isolated populations. Molecular Ecology Resources 12: 1161–1167. doi: 10.1111/j.1755-0998.2012.03176.x
51. Palstra FP, Fraser DJ (2012) Effective/census population size ratio estimation: a compendium and appraisal. Ecology and Evolution 2: 2357–2365. doi:10.1002/ece3.329
52. Austin JJ, Arnold EN, Jones CG (2004) Reconstructing an island radiation using ancient and recent DNA: the extinct and living day geckos (*Phelsuma*) of the Mascarene islands. Molecular Phylogenetics and Evolution 31: 109–122. doi: 10.1016/j.ympev.2003.07.011
53. Rozen S, Skaletsky H (2000) Primer3 on the WWW for general users and for biologist programmers. In: Misener S, Krawetz SA, eds. Methods in molecular biology, vol. 132: bioinformatics methods and protocols. Totowa, NJ: Humana Press. pp 365–386.
54. Kumar S, Nei M, Dudley J, Tamura K (2008) MEGA: a biologist-centric software for evolutionary analysis of DNA and protein sequences. Briefings in Bioinformatics 9: 299–306. doi: 10.1093/bib/bbn017
55. Nei M (1987) Molecular evolutionary genetics. New York: Columbia University Press. 526 p.
56. Rozas J, Sánchez-DelBarrio JC, Messeguer X, Rozas R (2003) DnaSP, DNA polymorphism analyses by the coalescent and other methods. Bioinformatics 19: 2496–2497. doi: 10.1093/bioinformatics/btg359
57. Huelsenbeck JP, Ronquist F (2001) MRBAYES: Bayesian inference of phylogenetic trees. Bioinformatics 17: 754–755. doi: 10.1093/bioinformatics/17.8.754
58. Posada D, Crandall KA (1998) MODELTEST: testing the model of DNA substitution. Bioinformatics 14: 817–818. doi: 10.1093/bioinformatics/14.9.817
59. Clement M, Posada D, Crandall KA (2000) TCS: a computer program to estimate gene genealogies. Molecular Ecology 9: 1657–1659.
60. Tajima F (1989) Statistical method for testing the neutral mutation hypothesis by DNA polymorphism. Genetics 123: 585–595.
61. Fu Y-X (1997) Statistical tests of neutrality of mutations against population growth, hitchhiking and background selection. Genetics 147: 915–925.
62. Slatkin M, Hudson RR (1991) Pairwise comparisons of mitochondrial DNA sequences in stable and exponentially growing populations. Genetics 129: 555–562.
63. Rogers AR, Harpending H (1992) Population growth makes waves in the distribution of pairwise genetic differences. Molecular Biology and Evolution 9: 552–569.
64. Rocha S, Posada D, Harris DJ (2013) Phylogeography and diversification history of the day-gecko genus *Phelsuma* in the Seychelles islands. BMC Evolutionary Biology 13(3). doi: 10.1186/1471-2148-13-3
65. Keller LF, Waller DM (2002) Inbreeding effects in wild populations. Trends in Ecology & Evolution 17: 230–241. doi: 10.1016/S0169-5347(02)02489-8
66. Hartl DL, Clark AG (1997) Principles of population genetics. Sunderland: Sinauer. 542 p.
67. Manel S, Gaggiotti OE, Waples RS (2005) Assignment methods: matching biological questions with appropriate techniques. Trends in Ecology & Evolution 20: 136–142. doi: 10.1016/j.tree.2004.12.004
68. Levy E, Kennington WJ, Tomkins JL, LeBas NR (2012) Phylogeography and population genetic structure of the ornate dragon lizard, *Ctenophorus ornatus*. PLoS One 7 (10): e46351. doi: 10.1371/journal.pone.0046351
69. Balmer O, Ciofi C, Galbraith DA, Swingland IR, Zug GR, et al. (2011) Population genetic structure of Aldabra giant tortoises. Journal of Heredity 102: 29–37. doi: 10.1093/jhered/esq096
70. Wang J-P, Lin H-D, Huang S, Pan C-H, Chen X-L, et al. (2004) Phylogeography of *Varicorhinus barbatulus* (Cyprinidae) in Taiwan based on nucleotide variation of mtDNA and allozymes. Molecular Phylogenetics and Evolution 31: 1143–1156. doi: 10.1016/j.ympev.2003.10.001
71. Landguth EL, Cushman SA, Schwartz MK, McKelvey KS, Murphy M, et al. (2010) Quantifying the lag time to detect barriers in landscape genetics. Molecular Ecology 19: 4179–4191. doi: 10.1111/j.1365-294X.2010.04808.x
72. Wright S (1969) Evolution and the genetics of populations, Volume 2: The theory of gene frequencies. Chicago: University Chicago Press. 520 p.
73. Meirmans PG (2012) The trouble with isolation by distance. Molecular Ecology 21: 2839–2846. doi: 10.1111/j.1365-294X.2012.05578.x
74. Safford RJ (1997) A survey of the occurrence of native vegetation remnants on Mauritius in 1993. Biological Conservation 80: 181–188. doi: 10.1016/S0006-3207(96)00048-1

75. Palsbøll PJ, Peery MZ, Olsen MT, Beissinger SR, Bérubé M (2013) Inferring recent historic abundance from current genetic diversity. Molecular Ecology 22: 22–40. doi: 10.1111/mec.12094
76. Franklin IR, Frankham R (1998) How large must populations be to retain evolutionary potential? Animal Conservation 1: 69–70. 10.1111/j.1469-1795.1998.tb00228.x
77. Busch JD, Waser PM, DeWoody JA (2007) Recent demographic bottlenecks are not accompanied by a genetic signature in banner-tailed kangaroo rats (*Dipodomys spectabilis*). Molecular Ecology 16: 2450–2462. doi: 10.1111/j.1365-294X.2007.03283.x
78. Küpper C, Edwards SV, Kosztolányi A, Alrashidi M, Burke T, et al. (2012) High gene flow on a continental scale in the polyandrous Kentish plover *Charadrius alexandrinus*. Molecular Ecology 21: 5864–5879. doi: 10.1111/mec.12064
79. Brunner PC, Douglas MR, Bernatchez L (1998) Microsatellite and mitochondrial DNA assessment of population structure and stocking effects in Arctic charr *Salvelinus alpinus* (Teleostei: Salmonidae) from central Alpine lakes. Molecular Ecology 7: 209–223. doi: 10.1046/j.1365-294x.1998.00341.x
80. Seddon PJ (2010) From reintroduction to assisted colonization: moving along the conservation translocation spectrum. Restoration Ecology 18: 796–802. doi: 10.1111/j.1526-100X.2010.00724.x

2

Population Differentiation and Hybridisation of Australian Snubfin (*Orcaella heinsohni*) and Indo-Pacific Humpback (*Sousa chinensis*) Dolphins in North-Western Australia

Alexander M. Brown[1*¶], Anna M. Kopps[2,3¶], Simon J. Allen[1], Lars Bejder[1], Bethan Littleford-Colquhoun[2], Guido J. Parra[4,5], Daniele Cagnazzi[6], Deborah Thiele[7], Carol Palmer[8,9], Celine H. Frère[10]

1 Murdoch University Cetacean Research Unit, School of Veterinary and Life Sciences, Murdoch University, Perth, Western Australia, Australia, **2** Centre for Ecology and Conservation, University of Exeter, Penryn, Cornwall, United Kingdom, **3** Marine Evolution and Conservation, Centre for Ecological and Evolutionary Studies, University of Groningen, Groningen, The Netherlands, **4** Cetacean Ecology, Behaviour and Evolution Lab, School of Biological Sciences, Flinders University, Adelaide, South Australia, Australia, **5** South Australian Research and Development Institute, Adelaide, South Australia, Australia, **6** Marine Ecology Research Centre, School of Environment, Science and Engineering, Southern Cross University, Lismore, New South Wales, Australia, **7** Fenner School of Environment & Society, Australian National University, Canberra, Australian Capital Territory, Australia, **8** Marine Ecosystems, Flora and Fauna Division, Department of Land Resource Management, Palmerston, Northern Territory, Australia, **9** Research Institute for the Environment and Livelihoods, Charles Darwin University, Darwin, Northern Territory, Australia, **10** GeneCology Research Centre, University of the Sunshine Coast, Maroochydore DC, Queensland, Australia

Abstract

Little is known about the Australian snubfin (*Orcaella heinsohni*) and Indo-Pacific humpback (*Sousa chinensis*) dolphins ('snubfin' and 'humpback dolphins', hereafter) of north-western Australia. While both species are listed as 'near threatened' by the IUCN, data deficiencies are impeding rigorous assessment of their conservation status across Australia. Understanding the genetic structure of populations, including levels of gene flow among populations, is important for the assessment of conservation status and the effective management of a species. Using nuclear and mitochondrial DNA markers, we assessed population genetic diversity and differentiation between snubfin dolphins from Cygnet (n = 32) and Roebuck Bays (n = 25), and humpback dolphins from the Dampier Archipelago (n = 19) and the North West Cape (n = 18). All sampling locations were separated by geographic distances >200 km. For each species, we found significant genetic differentiation between sampling locations based on 12 (for snubfin dolphins) and 13 (for humpback dolphins) microsatellite loci (F_{ST} = 0.05–0.09; P<0.001) and a 422 bp sequence of the mitochondrial control region (F_{ST} = 0.50–0.70; P< 0.001). The estimated proportion of migrants in a population ranged from 0.01 (95% CI 0.00–0.06) to 0.13 (0.03–0.24). These are the first estimates of genetic diversity and differentiation for snubfin and humpback dolphins in Western Australia, providing valuable information towards the assessment of their conservation status in this rapidly developing region. Our results suggest that north-western Australian snubfin and humpback dolphins may exist as metapopulations of small, largely isolated population fragments, and should be managed accordingly. Management plans should seek to maintain effective population size and gene flow. Additionally, while interactions of a socio-sexual nature between these two species have been observed previously, here we provide strong evidence for the first documented case of hybridisation between a female snubfin dolphin and a male humpback dolphin.

Editor: Tom Gilbert, Natural History Museum of Denmark, Denmark

Funding: The Australian Marine Mammal Centre funded this research (2008-09/19, 2009/19 and 2011/23). Alex Brown was the recipient of a Murdoch University International Postgraduate Scholarship. The funders had no role in study design, data collection and analysis, decision to publish, or preparation of the manuscript.

Competing Interests: The authors have declared that no competing interests exist.

* Email: alex.brown@murdoch.edu.au

¶ These authors share co-first authorship.

Introduction

Maintaining genetic diversity is a key objective of biodiversity conservation [1]. Species of conservation concern are often characterised by small, fragmented populations with restricted gene flow and low genetic diversity [2,3]. Small and fragmented populations with severely restricted gene flow are more vulnerable to the accumulation of deleterious mutations, the loss of genetic diversity through random genetic drift, and inbreeding depression than single populations of the same effective population size [4–7]. Additionally, further isolation and decline of fragmented populations within species may, through mate limitation, increase the probability of hybridisation with related, sympatric species (e.g. [8,9]). These processes may reduce the fitness of populations and impede their ability to adapt to environmental change, resulting in a reduced evolutionary potential and greater risk of extinction [10–12]. Understanding the genetic structure of populations, including levels of gene flow among populations and genetic diversity, is therefore important for the assessment of a species' conservation status as well as the effective management of a species, particularly where anthropogenic activities may contribute to population fragmentation [13].

Inshore dolphins occupying coastal and estuarine areas frequently overlap with areas of high human activity, exposing them to a variety of threats, including habitat loss and degradation, acoustic disturbance, vessel strikes, pollution and incidental capture in fisheries [14]. These threats, combined with the late maturation, slow reproduction, often low abundance and restricted ranges of inshore dolphins, have resulted in priority conservation status being afforded to a number of geographically isolated populations [15–17].

The Australian snubfin dolphin (*Orcaella heinsohni*, 'snubfin dolphin' hereafter) occurs throughout tropical coastal waters of northern Australia and, potentially, Papua New Guinea [18]. The Indo-Pacific humpback dolphin (*Sousa chinensis*, 'humpback dolphin' hereafter) occurs in tropical and temperate inshore waters throughout the Indian and western Pacific Oceans [19], although genetic and morphological data strongly suggest that those in Australian waters are distinct from those in Southeast Asia [20,21]. Throughout their ranges, the conservation status of both species was assessed as 'near threatened' by the IUCN, with caveats noting that additional data would likely result in an elevated status [22,23].

Despite their 'near threatened' conservation listing, the distribution, abundance and population structure of snubfin and humpback dolphins are poorly understood throughout the majority of their ranges in Australian waters. This lack of information is impeding rigorous assessment of their conservation status [24–26]. Studies to date have been largely restricted to the east coast of Australia, primarily in waters adjacent to population centres in Queensland, where snubfin and humpback dolphins exhibit a discontinuous contemporary distribution of small populations of 50–100 animals [24,26–30]. These populations have relatively small ranges of approximately 200–350 km^2 and a preference for inshore habitats of waters <15 m deep and within 5 km of the coast [26,28–32].

Snubfin and humpback dolphins are sympatric throughout most of their distribution in Australia [28,33], which also overlaps that of Indo-Pacific bottlenose dolphins (*Tursiops aduncus*, 'bottlenose dolphin' hereafter). Where species are sympatric, inter-species associations and inter-species mating may facilitate hybridisation. This phenomenon has been reported between several cetacean species (review in [34,35]), particularly among small cetaceans [9,36–38]. To date, no hybrids have been confirmed between snubfin, humpback, or bottlenose dolphins. However, associations between snubfin and humpback dolphins have been reported at several locations along the Queensland coast [30,39], as have associations between humpback and bottlenose dolphins, and snubfin and bottlenose dolphins in north-western Australia [40,41]. In Cleveland Bay, Queensland, the majority (58%, n = 11) of snubfin-humpback dolphin associations were of an aggressive-sexual nature where, in all cases, humpback dolphins were identified as the aggressors [39]. Although the benefits and costs of these interactions are not fully understood, they suggest that inter-specific mating is possible.

Wild hybridisation is typically a conservation concern; when mediated by anthropogenic translocation of species and habitat modification, it has led to the extinction of many animal species and is particularly problematic for species of low abundance [42,43]. Several studies have reported hybridisation events among mammalian species within modified habitats and/or where populations have undergone a decline (e.g. [8,44,45]). However, there is evidence that natural hybridisation may play an important role in the evolution of animals (e.g. [46,47]), as has long been recognised for plants [48].

Examining the structure of populations in the marine environment presents a particular challenge due to the absence of obvious barriers to gene flow, and the highly mobile nature of many marine species. Robust demographic and movement data are often costly and logistically difficult to acquire, while similar challenges exist for the identification of hybridisation through morphological data and observations of species interactions. To this end, molecular tools have been employed to address a variety of questions in mobile marine taxa of conservation and management importance, such as teleost fish (e.g. [49,50]), elasmobranchs (e.g. [51]), marine reptiles (e.g. [52,53]) and marine mammals (e.g. [54–56]). In marine mammals, analyses of molecular markers have often contributed towards the identification of appropriate management units to inform decision-makers (e.g. [57–59]), including the identification of cryptic taxa and genetically-isolated populations of conservation concern (e.g. [18,60–62]). Furthermore, molecular tools have permitted the investigation of hybridisation in the absence of other conclusive evidence (e.g. [9,38]).

Molecular studies of snubfin and humpback dolphins in Australia are largely restricted to investigations of taxonomy [18,20,21,63,64]. The exception is Cagnazzi [30], who examined genetic population structure based on microsatellites of both species sampled at several locations along the Queensland coast. For snubfin dolphins, no structure was found between three populations within a 200 km stretch of coast, but significant differentiation was found between this region and a population approximately 600 km distant. The latter population, which numbers fewer than 100 individuals and is threatened by loss of habitat from port development, has been suggested as qualifying for 'endangered' status under IUCN Red List criteria for regional populations [26]. For humpback dolphins, significant genetic differentiation was detected between almost all putative populations, even when separated by only a few kilometres, such as in the Great Sandy Strait [30]. In contrast, a recent study in Chinese waters found no evidence of genetic population structure in humpback dolphins among three resident populations along a *ca.* 1,000 km stretch of coastline [65].

The lack of information on the genetic population structure of snubfin and humpback dolphins is of particular concern in the north-west of Australia, where data deficiencies are coupled with a resources extraction boom, resulting in widespread and large-scale habitat modification of the inshore environment associated with port development [25,40]. The development of the coastal zone may introduce anthropogenic barriers to dispersal and cause fragmentation of inshore dolphin populations. However, in the absence of any understanding of the genetic diversity or connectivity between populations, the likelihood or significance of these potential effects on inshore dolphins remains unknown. Information on the genetic population structure of these species in this region is essential to determining an appropriate management scale at which to assess potential anthropogenic effects and inform conservation strategies.

In this study, we used mitochondrial DNA (mtDNA) sequence data and nuclear microsatellite markers to examine the genetic diversity and structure of snubfin and humpback dolphins among a limited number of study sites in north-western Australia. In addition to population structure, we also investigated the possible existence of hybrid dolphins across the study area.

Materials and Methods

Ethics statement

Field data collection took place under permits from the WA Department of Local Government Research and Development (U6/2010–2011), the Department of Agriculture and Food (U6/2012–2014), Department of Environment and Conservation (now Department of Parks and Wildlife) WA (SF007596, SF008480, SF008825, SF009119), WA Police (9990071), and with approval from Murdoch, Flinders and the Australian National University Animal Ethics Committees (W2342/10, E297 and A2011/50).

Study sites and sample collection

A total of 110 skin tissue samples were obtained from free-ranging dolphins across north-western Australia between 2008 and 2013 using a biopsy darting system from small research vessels [66]. Snubfin dolphin samples were obtained from Cygnet Bay and Roebuck Bay, and humpback dolphin samples were obtained from Cygnet Bay, the Dampier Archipelago and the North West Cape (Figure 1). To assist in identifying the parental species of a suspected hybrid, we also collected biopsy samples from Indo-Pacific bottlenose dolphins from Cygnet Bay, so as to include all three dolphin species regularly encountered in Cygnet Bay into our analyses. Tissue samples were stored in either 100% ethanol or saturated NaCl/20% dimethyl sulfoxide [67]. Sampled sites represent those accessible by small research vessel and where snubfin or humpback dolphins were sufficiently approachable to distances suitable for successfully obtaining biopsy samples. Samples were primarily collected on an opportunistic basis during research on bottlenose dolphin (*Tursiops* spp.) population structure across north-western Australia [40], and also in parallel to demographic studies of snubfin and humpback dolphins at these locations (Brown *et al.*, unpublished data; Thiele *et al.*, unpublished data).

Genetic analyses

Genomic DNA was extracted using the DNeasy Blood & Tissue Kit (Qiagen) following the manufacturer's instructions. Sex was determined genetically using sex chromosome-specific primers. Loci ZFX and SRY [68] were coamplified in a single PCR reaction. PCR products were run on a 1.5% agarose gel and sex determined based on the number of different fragments amplified.

Mitochondrial DNA (mtDNA) haplotypes were assigned based on a 422 base pair (bp) sequence. The fragment was amplified by the primers dlp1.5 and dlp5 [69]. We followed the PCR conditions described in Bacher *et al.* [70]. Haplotypes were assigned with the software Geneious R6.1 (Biomatters).

We amplified 14 microsatellite loci in four 10 µl volume multiplex PCRs using Qiagen Multiplex Kit^{TM} (Qiagen). The microsatellite markers used here were: DIrFCB1, DIrFCB4 [71], LobsDi_7.1, LobsDi_9, LobsDi_19, LobsDi_21, LobsDi_24, LobsDi_39 [72], SCA9, SCA22, SCA27 SCA39 [73], TexVet5, TexVet7 [74]. We followed the PCR conditions as described in Frère *et al.* [75]. The single stranded PCR products were run on an ABI 3730 DNA Sequencer (Applied Biosystems). Alleles were scored with Genemapper Software 3.7 (Applied Biosystems). We identified duplicate samples, i.e. samples that were genotyped for at least 10 microsatellite loci and matched 95%, using the Microsatellite Toolkit [76] and, from these, we retained the sample with the most complete genotype. Microsatellites were checked for Hardy-Weinberg equilibrium and linkage disequilibrium in GenePop [77].

Several measures of population differentiation were calculated for the sampled study sites. The suspected hybrid and the bottlenose dolphins were excluded from all analyses comparing population structure and diversity within snubfin and humpback dolphins. We calculated F_{ST} values (for microsatellites and mtDNA) and Φ_{ST} values (for mtDNA) in Arlequin [78].

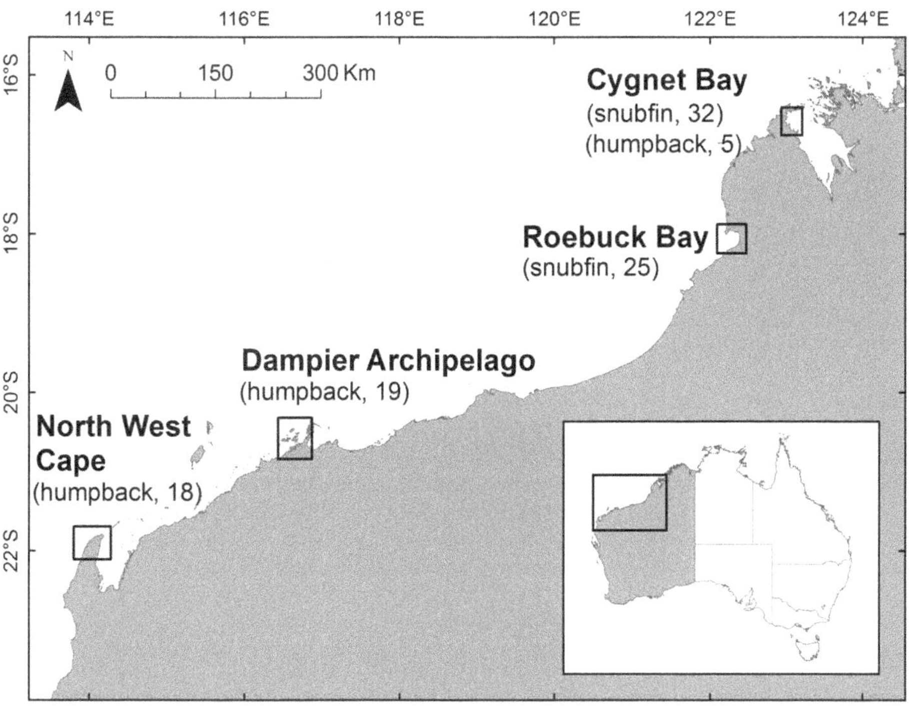

Figure 1. Biopsy sampling locations and sample sizes of Australian snubfin and Indo-Pacific humpback dolphins in north-western Australia.

Contemporary migration rates were calculated in BayesAss 1.3 [79] using 10^7 iterations, a burn-in length of 10^6 and a sampling interval of 1,000 steps. We performed three runs per species with different seeds to confirm that similar mean posterior migration rates and 95% confidence intervals were obtained. An admixture model without information on sampling location was run in STRUCTURE (version 2.2.3 [80,81]) to examine differentiation patterns between populations, with a burn-in length of 10^5 and 10^6 Markov chain Monte Carlo (MCMC) steps. The most likely number of genetically homogeneous clusters (if greater than two) was determined based on 10 iterations for each population (k) = 1–4 by calculating Δk, an ad hoc statistic proposed by Evanno *et al.* [82]. Δk was calculated in STRUCTURE HARVESTER (Web v0.6.93 [83]). We also compared STRUCTURE results to those of the recently published software FLOCK (FLOCK_MSAT 3.0 [84]) using default parameters. Compared to the MCMC-based STRUCTURE, FLOCK uses an iterative method which makes it faster and computationally more efficient.

We calculated the effective population (N_e) sizes for snubfin dolphins based on the linkage disequilibrium method using LNDe v1.31 [85]. For small effective population sizes of <500, the linkage disequilibrium (LD) method has shown to be reliable with the use of 10–20 microsatellite loci and samples of 25–50 individuals [86]. We did not calculate N_e for humpback dolphins because the number of samples per population was less than 25.

An underlying assumption of the linkage-disequilibrium method of estimating N_e is non-overlapping generations. This assumption is violated within the long-lived, polygamous populations examined here, and may lead to a downward bias in estimates of N_e [86–88]. Despite this, Robinson & Moyer [87] showed that for populations with small N_e, the linkage-disequilibrium method performed relatively well for species with overlapping generations under a variety of life history scenarios and sampling strategies. Random sampling of mature individuals, as was the case in the current study, has been shown to produce the best estimates of N_e by LD [87]. The lowest allele frequency considered in the analyses was set to 0.03 to ensure that single copy alleles were filtered out; N_e estimates were correspondingly corrected for downward bias by multiplying the estimate by 1.25 [86,89]. Due to the paucity of information of snubfin dolphin life history traits, we used a correction factor suggested for bottlenose dolphins [89].

We also tested whether any population has recently undergone a bottleneck using a graphical method to detect allele frequency distortion [90] and the software BOTTLENECK (v1.2.02 [91]). We specified 1,000 iterations and used Wilcoxon sign rank tests to assess significance. BOTTLENECK v1.2.02 provides results for three models of the generation of new alleles; the stepwise mutation model (SMM), the infinite allele model (IAM) and the two-phased model of mutation (TPM). In the software manual, the authors recommend the use of TPM for microsatellite datasets; in their paper [92], by contrast, IAM is recommended for microsatellites with fewer than 3 bp repeats. However, TPM is not discussed in the paper.

Hybrid investigation

In Cygnet Bay, we encountered a dolphin that, phenotypically, could not be identified as a humpback, snubfin or bottlenose dolphin. All three of these species are regularly encountered within Cygnet Bay. To confirm hybrid status and to identify the suspected hybrid's parental species, we compared the suspected hybrid's mtDNA haplotype to those of humpback, snubfin and bottlenose dolphins. We also compared the microsatellite genotype of the suspected hybrid to alleles found in the three dolphin species at Cygnet Bay. By doing so, we could assign the parental species of the suspected hybrid based on species-specific alleles. Furthermore, we ran STRUCTURE to obtain a measure of likelihood to which species the suspected hybrid belongs. All samples collected at Cygnet Bay were included in the STRUCTURE analysis using the same parameters as above.

Microsatellite genotypes used in this study are available in the supplementary material and mtDNA haplotype sequences have been archived on GenBank (Accession numbers KJ530719–KJ530740).

Results

Population differentiation

After having removed ten duplicate samples from across the dataset, we conducted the analyses with the following populations and sample sizes: snubfin dolphins from Cygnet Bay (n = 32) and Roebuck Bay (n = 25), and humpback dolphins from Cygnet Bay (n = 5), the Dampier Archipelago (n = 19) and the North West Cape (n = 18). We do not present F_{ST}, Φ_{ST} or contemporary migration rate values for humpback dolphins from Cygnet Bay due to the low sample size. Additionally, we collected one sample of a suspected hybrid and six samples from bottlenose dolphins from Cygnet Bay.

Twelve of the 14 genotyped microsatellite loci were polymorphic in snubfin dolphins and 13 microsatellite loci were polymorphic in humpback dolphins (Table 1, Table S4). On average we genotyped 95% of loci per individual. For both species, none of the microsatellite loci appeared out of Hardy-Weinberg Equilibrium after sequential Bonferroni correction [93], nor linked after sequential Bonferroni correction. We found six mtDNA haplotypes each in snubfin and humpback dolphins (Figure 2). Within species, all population pairs were significantly differentiated based on microsatellites (F_{ST} = 0.05–0.09) and mtDNA loci (F_{ST} = 0.50–0.70, Φ_{ST} = 0.17–0.45) (Table 2). STRUCTURE assigned most individuals sampled at the same location to the same cluster (Figure 3A–3C). For snubfin dolphins, Δk analysis and FLOCK showed that the most likely k was ≤2 (Figure S1). For humpback dolphins, the most likely number of k was four based on STRUCTURE (Figure S1) and three based on FLOCK. Three equals the number of sampled populations.

Contemporary migration rates (i.e. within the last few generations) revealed an estimated proportion of 0.04 (95% CI 0.01–0.10) of snubfin dolphins in Cygnet Bay derived from Roebuck Bay and 0.03 (0.00–0.08) of Roebuck Bay individuals derived from Cygnet Bay. For humpback dolphins, we estimated a proportion of 0.01 (0.00–0.06) individuals from the Dampier Archipelago derived from the North West Cape and 0.13 (0.03–0.24) of North West Cape individuals derived from the Dampier Archipelago.

Effective population size and evidence of bottlenecks

For snubfin dolphins, N_e (95% CI) was estimated to be 49.1 (28.6–112.1) for Cygnet Bay and 56.0 (24.3–77180.6) for Roebuck Bay. The wide confidence intervals are revisited in the discussion. We obtained conflicting results on recent bottlenecks depending on the method used (see Table S3 for P values and Figure S3 for visualisations of potential mode shifts).

Suspected hybrid

The Cygnet Bay individual that could not be visually assigned to species level exhibited a length, girth and light grey colouration typical of adult humpback dolphins in the region. The low, triangular dorsal fin was also indicative of a humpback dolphin, although the position of the dorsal fin was posterior to the mid-point of the body, as in a snubfin dolphin. The surfacing

Table 1. Microsatellite characteristics for snubfin and humpback dolphins.

	N_A	N_E	F_{IS}	H_E	H_O
Snubfin dolphins					
Cygnet Bay	4.25	2.65	0.00	0.57	0.58
Roebuck Bay	4.25	2.88	−0.01	0.58	0.60
Humpback dolphins					
Dampier Archipelago	3.73	2.09	−0.07	0.44	0.46
North West Cape	3.58	2.16	0.06	0.40	0.35

N_A = Number of Alleles, N_E = Number of effective Alleles, F_{IS} = Inbreeding Coefficient, H_E = expected heterozygosity, H_O = observed heterozygosity. Numbers are averages over polymorphic loci. See Tables A1 and A2 for locus specific microsatellite characteristics.

movement was comparable to that of a snubfin dolphin, tilting back the head to breathe, with faint neck creases visible (although without the prominent sunken post-cranial region of a snubfin dolphin). A short rostrum was visible, being noticeably shorter than that of a bottlenose dolphin and far shorter than that of a humpback dolphin (Figure 4).

Over four×one month seasons of photo-identification and biopsy sampling surveys at Cygnet Bay from 2012–2013, the suspected hybrid was observed 22 times on 17 different days (Brown *et al.*, unpublished data). Over these observations, a total of eight hours were spent in the presence of the suspected hybrid; 23% of the time the animal was alone (defined as>100 m from any other individual), 77% in close (<10 m) association with one

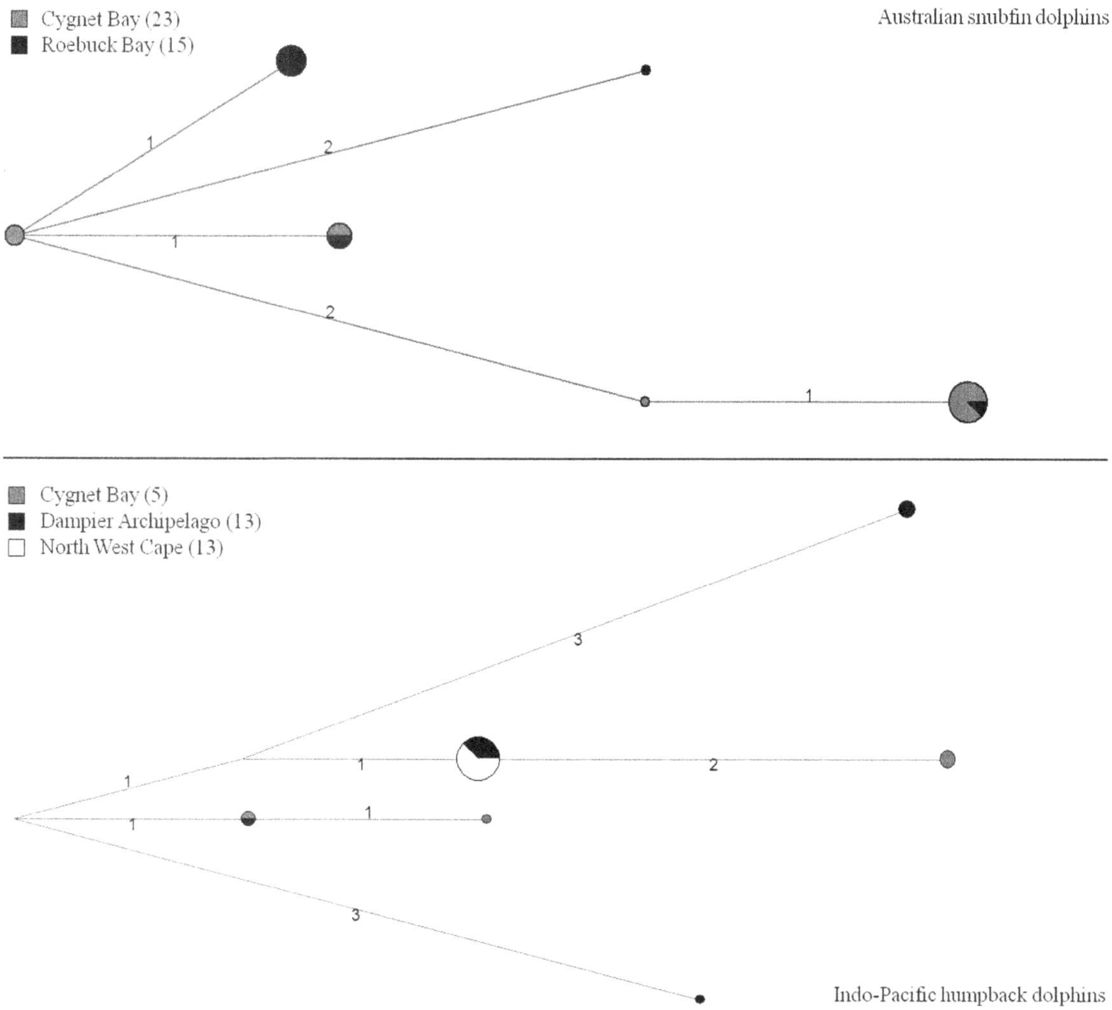

Figure 2. mtDNA networks for snubfin and humpback dolphins. Sample sizes are shown in parentheses. Branch numbers indicate the number of nucleotide differences between mtDNA haplotypes.

Table 2. Genetic differentiation of mtDNA and microsatellite loci.

	Measure of differentiation	mtDNA	microsatellites
Snubfin dolphins	F_{ST}	0.500**	0.091**
(CY-RB)	Φ_{ST}	0.446**	na
Humpback dolphins	F_{ST}	0.699**	0.046**
(DA-NWC)	Φ_{ST}	0.167*	na

Asterisks indicate P values (*$P<0.05$, **$P<0.001$). CY = Cygnet Bay, RB = Roebuck Bay, DA = Dampier Archipelago, NWC = North West Cape. For the mtDNA based estimates a lower sample size was used for both species; 15 samples from RB, 23 samples from CY, and 13 samples each from DA and NWC.

or more snubfin dolphins. Only two brief close associations with humpback and bottlenose dolphins were recorded; in both encounters the suspected hybrid was also in close association with one or more snubfin dolphins. The majority of associations with snubfin dolphins were small groups (<5 individuals) with female individuals (confirmed through genetics or presence of dependent calf). In September 2013, the suspected hybrid was observed on four occasions in larger snubfin dolphin groups (>10 individuals) of mixed sex.

Genetic analyses revealed that the individual was a female and supported its status as a hybrid. The comparison of the hybrid's genotype to alleles found in the three resident dolphin species within Cygnet Bay indicated the majority of alleles (84.4%) found were species-specific. The hybrid shares at least one allele of each microsatellite locus with snubfin dolphins and at least one allele of each microsatellite locus for 11 out of the 14 loci with humpback dolphins (Table 3). At one locus, the hybrid is homozygote and this allele is only shared with snubfin dolphins. At five loci, the hybrid shares an allele with bottlenose dolphins, however, only one of them has not been found in either snubfin or humpback dolphins (Table 3).

STRUCTURE analyses including snubfin, humpback and bottlenose dolphins from Cygnet Bay estimated that the sample originated to 53.4±0.05% (mean of 10 iterations ± SD, k = 3) from a snubfin dolphin, to 46.2±0.05% from a humpback dolphin and to 0.4±0.00% from a bottlenose dolphin (as indicated by the

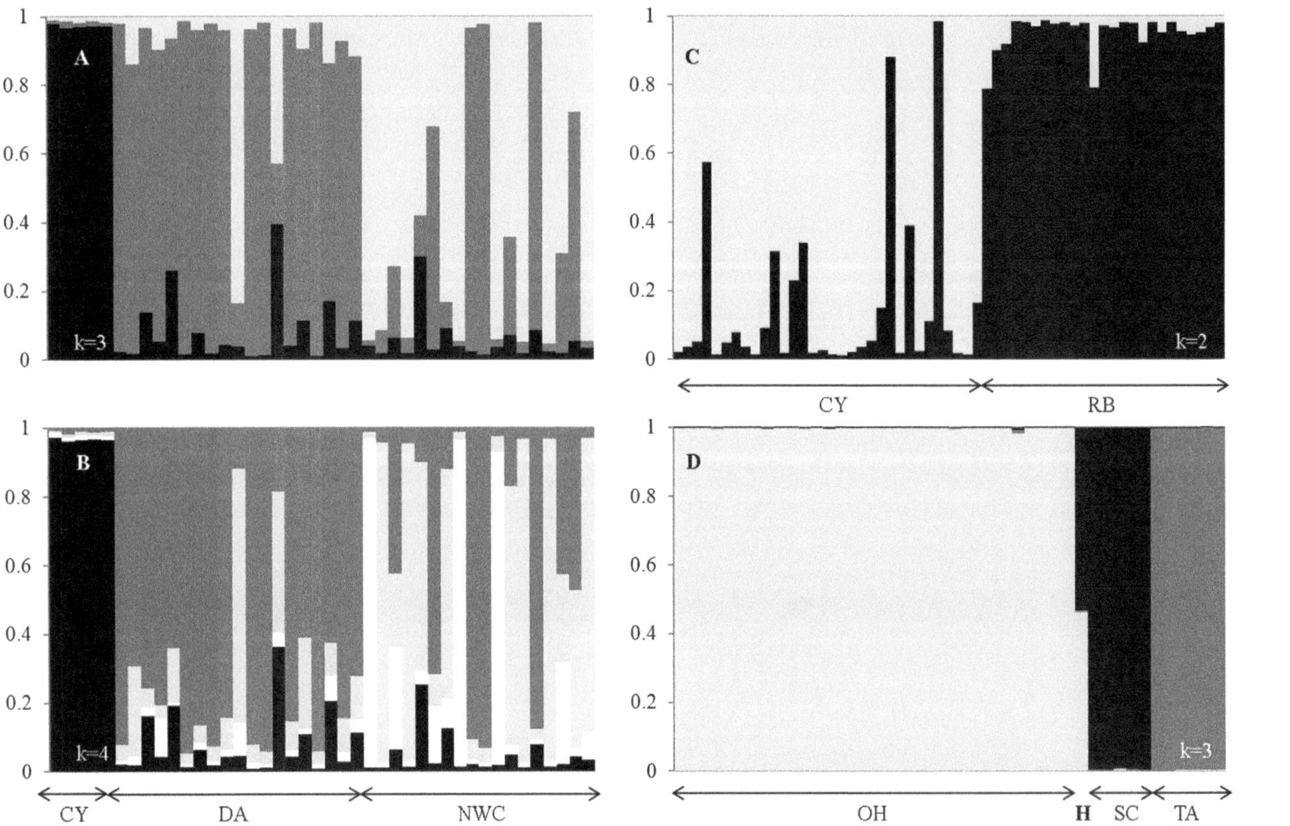

Figure 3. Structure plots for humpback dolphins where k = 3 (A) and k = 4 (B), for snubfin dolphins (C), and the three regularly encountered dolphin species at Cygnet Bay (D). k = number of clusters. Each bar on the x-axis corresponds to an individual. The y-axis indicates the proportion of population/species membership. OH = snubfin dolphins, SC = humpback dolphins, CY = Cygnet Bay, DA = Dampier Archipelago, NWC = North West Cape, RB = Roebuck Bay, H = suspected hybrid, TA = bottlenose dolphin.

Figure 4. Images of hybrid (A1–2), adult snubfin (B1–2), humpback (D1–2) and bottlenose (D1–2) dolphins encountered at Cygnet Bay. Left images show relative dorsal proportions; right images compare head/rostrum characteristics.

proportion of shading in individual bars in Figure 3D). The mtDNA haplotype of the suspected hybrid matched a haplotype found in snubfin dolphins (Figure S4), suggesting that she most likely had a snubfin dolphin mother. The STRUCTURE results and allele comparisons suggest a humpback dolphin father.

From all samples included in this study, other than the hybrid, only one other, a male snubfin dolphin from Roebuck Bay, showed some signs of mixed species ancestry. Images of this individual suggest a normal snubfin dolphin phenotype. STRUCTURE assigned this individual by 16.7% (10 iterations, SD = 0.00) to humpback dolphin (Figure S2) and 83.1% (0.00) to snubfin dolphin. This is suggestive of post-F1 hybrid status, although the small number of microsatellite markers used in this study restricts our interpretation of such results.

Discussion

Population differentiation

We found that snubfin and humpback dolphins showed significant levels of population structure at both the mitochondrial and microsatellite DNA level between the sampling locations. Significant F_{ST} and Φ_{ST} values for snubfin dolphins between Roebuck Bay and Cygnet Bay provide genetic evidence for the presence of discrete populations with limited gene flow. The two populations shared two out of six mtDNA haplotypes and 15 private microsatellite alleles were detected (Table S1). Within each of these sampling locations, STRUCTURE assigned most snubfin individuals to the same cluster. However, three individuals (9%) at Cygnet Bay were predominately assigned to the Roebuck Bay cluster, suggesting that they were Roebuck Bay migrants or of migrant ancestry (Figure 3).

Table 3. Alleles shared by the suspected hybrid and the three resident dolphin species at Cygnet Bay.

	Allele 1			Allele 2			hybrid homozygote	Support of hybrid hypothesis? If not, what are potential explanations?
Locus	**OH**	**SC**	**TA**	**OH**	**SC**	**TA**		
DIrFCB4	y				y			y
DIrFCB5	y				y			y
Lobs7.1	y				y			y
LobsDi9	y			y			y	n; null allele or allele might not have been sampled in SC due to small sample size or rare allele
LobsDi19	y				y			y
LobsDi21	y		y		y			y
LobsDi24	y			y				n; allele 1 or allele 2 might not have been sampled in SC due to small sample size or rare allele
LobsDi39	y				y	y		y
SCA9	y				y			y
SCA22	y	y	y			y		n; allele 2 might not have been sampled in SC and/or OH due to small sample size, or rare allele
SCA27	y		y		y			y
SCA39	y		y		y	y		y
Tex5	y	y		y	y		y	y
Tex7	y				y			y

OH = snubfin dolphin, SC = humpback dolphin, TA = bottlenose dolphin, y = yes, n = no.

Humpback dolphins from the Dampier Archipelago and the North West Cape also exhibit significant population structure with limited gene flow. Significant F_{ST} and Φ_{ST} values were obtained between two sampling locations, and the results of STRUCTURE and FLOCK assigned the majority of animals at these two locations to separate clusters. However, there was some evidence of movement of individuals between sites, particularly from the Dampier Archipelago to the North West Cape, the latter of which included five individuals (26%) predominately assigned to the dominant cluster at the Dampier Archipelago (Figure 3). Humpback dolphins occur along a further 400+km of coastline south of the North West Cape [40]. The results of STRUCTURE at k = 4 further illustrate admixture within North West Cape humpback dolphins and suggest the existence of a potential fourth, not yet sampled, humpback population, possibly to the south of the North West Cape. The sample size for humpback dolphins from Cygnet Bay (n = 5) was too small to calculate meaningful F_{ST} and Φ_{ST} values with samples from the other two locations. However, Cygnet Bay humpback dolphins seem to be genetically differentiated from the other two sampling locations, based on the strong partitioning in the STRUCTURE results. Based on all three sampling sites, two out of six mtDNA haplotypes were shared among dolphins from two out of three different sampling locations, and there were 16 private microsatellite alleles detected (Table S2).

For both species, most contemporary migration rates were low, with estimated proportions of migrants ≤0.04 between sites. Confidence intervals around these estimates were wide, owing to the relatively small sample sizes. However, for most sites, the upper confidence interval of the proportion of migrants was ≤0.1. The exception was migration rates of humpback dolphins from the Dampier Archipelago to the North West Cape, which were slightly higher at 0.13 (95% CI 0.03–0.24) – a result supported by the greater admixture of humpback dolphins at the North West Cape revealed by STRUCTURE. Our confidence in this apparent directionality of gene flow for humpback dolphins between the Dampier Archipelago and the North West Cape is limited by largely overlapping confidence intervals between the two estimates of migration rates. A greater number of samples is required to further investigate the potential source-sink pattern of population structure (e.g. [94]).

Although limited, the photo-identification data available support the findings of population differentiation for snubfin dolphins in this study. Research at Cygnet Bay suggests a high degree of site fidelity for snubfin dolphins, with >80% of individuals resighted across ≥ three of a total of four×one-month field seasons from 2012–2013 (Brown *et al.*, unpublished data). Photo-identification records for snubfin dolphins in Roebuck Bay also suggest a high degree of site fidelity with >40 individuals resighted multiple times over a range of seasons between 2007 and 2012 (Thiele *et al.*, unpublished data). Additionally, these data have not revealed any movement of snubfin dolphin individuals between Cygnet Bay and Roebuck Bay to date (Brown *et al.*, unpublished data; Thiele *et al.*, unpublished data).

Studies on snubfin and humpback dolphins from the east coast of Australia have revealed either a majority of individuals regularly using the same discrete area from year to year [24], or strong site fidelity within resident populations [26,29]. These patterns of site fidelity support our finding of genetic structuring of snubfin and humpback dolphins of north-western Australia. We acknowledge that distances between sampling locations were large (>200 km) and, therefore, cannot rule out that a pattern of isolation-by-distance could explain the significant genetic structuring. However, we cannot test for isolation-by distance based on only two sampling locations for each species.

While acknowledging differences in distances between studies, our results support the conclusions of Cagnazzi [30] for humpback dolphins along the east coast of Queensland, where significant genetic differentiation was found between populations separated by *ca.* 200 km, but also between populations separated by only a few kilometres [29,30].

In contrast to our current results and those of Cagnazzi [30], a study of humpback dolphins in Chinese waters found no evidence of population structure among three resident populations, each separated by approximately 500 km of coastline [65]. Potentially suitable habitat (river mouths) is distributed along much of the coastline [95], and a maximum dispersal distance of 300 km has been recorded for an individual in this region [96]. This suggests that a stepping-stone pattern of gene flow may be occurring, to a level sufficient to prevent differentiation. It was also suggested that gene flow might be of a recently interrupted form, where insufficient time has passed for detectable differentiation to develop [65].

Humpback dolphins have been observed in areas between the sampling locations of this current study [40], although their distribution along the north-western Australian coast remains poorly understood. Individual movements of up to 130 km have been recorded off the east coast of Australia [30]. No obvious natural geographic barriers to dispersal exist along the 350 km of coastline between the Dampier Archipelago and North West Cape, so the significant genetic differentiation found between animals at these two locations may be a result of their geographic separation exceeding individual dispersal distances.

The identification of genetic population structure in snubfin dolphins on the Queensland coast by Cagnazzi [30] was somewhat restricted by the distribution of sampling locations. No structure was found between three relatively close populations (within a 200 km stretch of coast), although significant differentiation was found at a much greater separation of approximately 600 km. Cygnet Bay and Roebuck Bay are separated by approximately 250 km of coastline. Based on our current understanding of the habitat requirements of snubfin dolphins [31], no obvious barriers to dispersal exist between the two sites: the coastline is currently undeveloped and shallow inshore waters are present throughout. Sightings between the two sites are largely restricted to anecdotal reports of small groups immediately north of Roebuck Bay [40]. Two months of boat survey effort along a 30 km stretch of coast between the two sites revealed a low encounter rate of just two sightings of the same pair of snubfin individuals (Brown *et al.*, unpublished data). The maximum reported distance travelled by an individual snubfin dolphin is 70 km [30], suggesting that the geographic distance between Cygnet Bay and Roebuck Bay is likely a key driver of the restricted gene flow documented here.

While barriers to dispersal are rarely obvious in marine habitats, significant genetic structure over relatively small spatial scales has been observed in numerous species of coastal dolphins (e.g. *Tursiops* spp. [97–100]; *Cephalorhynchus hectori* spp. [56]; *Sotalia guianensis* [101]). For bottlenose dolphins (*Tursiops* spp.), a range of environmental, habitat and resource specialisation and social factors have been suggested as drivers of fine-scale population structure (e.g. [99,100,102–105]).

Effective population size and evidence of bottlenecks

For successful conservation strategies, it is important to have an understanding of the effective population size (N_e), which provides an indicator of the number of individuals contributing genes to the next generation [106]. The effective population size is usually lower than the census size, and by definition describes the rate of inbreeding accumulation and loss of genetic diversity [87]. A rule of thumb suggests that N_e should not fall below 50 in the short-term and should be above 500 in the long-term [107]. Mace & Lande [13] suggest that, subject to additional criteria (e.g. population decline), a population of N_e<50 should be considered in a critical state (i.e. 50% probability of extinction within five years or two generations). We found that N_e estimates are close to this theoretical lower limit for snubfin dolphins at Cygnet Bay (N_e = 49.1, 95% CI 28.6–112.1) and Roebuck Bay (56.0, 95% CI 24.3–77180.6). While this may raise conservation concerns, the wide confidence intervals indicate considerable uncertainty in these estimates, particularly for Roebuck Bay. This suggests that sample sizes are too small to accurately estimate N_e and limits our interpretation of these results.

The results on recent bottlenecks are ambiguous for the four sampling sites we investigated. Depending on the mutation model, we obtained significant and non-significant results for each site. The graphical allele frequency distortion method indicated a mode shift of humpback dolphins at the North West Cape. The presence of a recent bottleneck is supported by a low mtDNA diversity (one haplotype) identified at this sampling location. However, under the two-phased model of mutation there was no indication for a recent bottleneck at the North West Cape. The results of our assessments of recent bottlenecks and N_e should be interpreted with caution due to ambiguity and large confidence intervals, respectively.

Hybridisation

We found strong genetic evidence that the suspected hybrid found at Cygnet Bay is the offspring of a snubfin dolphin mother and a humpback dolphin father. While we found that alleles at three microsatellite loci (Table 3) were not shared between the hybrid and humpback dolphins, it is most likely that these alleles also exist in humpback dolphins, but have not been sampled as yet (because only five samples were collected from this species at Cygnet Bay). The absence of these alleles in our samples could also be due to the presence of null alleles, in particular, for the locus LobsDi9 (Table 3).

This is the first documented case of hybridisation between snubfin and humpback dolphins. The hybrid is a female, seemingly fully grown and in good body condition, which associates primarily with snubfin dolphins – her maternal species. Despite a predominance of male sterility among mammalian hybrids (e.g. [108]), there are several examples of fertility among female cetacean hybrids (e.g. within the Genus *Phocoena* [9]; *Balaenoptera* [109,110]; and *Psuedorca* × *Tursiops* [34]) and one record of fertility of a male hybrid of the *Globicephala* genus [38]. In the absence of any evidence of the reproductive history of the snubfin-humpback hybrid identified here, no assessment of her fertility can be made at this stage.

Snubfin and humpback dolphins are sympatric across much of their range, occasionally form mixed groups, and aggressive-sexual inter-specific interactions have been documented [39]. Snubfin-humpback dolphin associations within Cygnet Bay appear to be uncommon and typically affiliative, although one observation of repeated mating attempts by a male humpback dolphin with a female snubfin has been recorded (Brown *et al.*, unpublished data). Frequent hybridisation has been documented between Dall's (*Phocoena dalli*) and harbour (*Phocoena phocoena*) porpoises in a localised area of the northeast Pacific [9]. In all hybrids examined, Willis *et al.* [9] revealed Dall's porpoise to be the maternal species, and suggested that the highly promiscuous male harbour porpoise's indiscriminate pursuit of females of either species could be a driving factor of this hybridisation. In this region, the harbour porpoise is the rarer species, having apparently declined in recent decades [111]. Humpback dolphins, identified as the paternal species of the hybrid in the current study, are the least numerous of the three dolphin species within Cygnet Bay (Brown *et al.*, unpublished data). We hypothesise that the observed propensity of humpback dolphins to initiate aggressive-sexual interactions with snubfin dolphins [39], along with a low availability of conspecific potential mates at the Cygnet Bay study site, are potential drivers of the hybrid dolphin reported here.

Our discovery of a snubfin-humpback dolphin hybrid shows that these two sympatric species are capable of inter-generic hybridisation. There are no indications that snubfin and humpback dolphins interbreed regularly from our data, and molecular studies of these animals on the east coast of Australia have not revealed any evidence of hybridisation, to date [30]. However, total sample sizes are small for both species, with limited survey effort throughout the majority of their range in Australia. This phenomenon likely represents a low-frequency, natural hybridisation, facilitated by a fragmented distribution and potentially low abundance [8,9,45]. Further isolation of already fragmented populations may facilitate further hybridisation and, hence, raise conservation concerns [43].

Conservation and management implications

The definition of populations, stocks or management units (MUs) is typically based on ecological or evolutionary criteria, or a combination of the two [112]. Many different definitions of a population are in use and the criteria used vary according to the purpose for which a population is being defined [112]. Genetic data have been widely used to examine the structure of cetacean populations and to make recommendations on the identification of MUs (e.g. [98,113,114]). Indeed, the level of differentiation we have identified, in terms of significant F_{ST} values, supports the criteria for separate MUs as proposed by Moritz [115]. However, many authors argue that identifying MUs from genetic data alone is unwise (e.g. [112,116,117]), particularly via the use of F_{ST} alone to infer gene flow as it relies on several simplifying assumptions, which typically are not met for natural populations [118,119]. Furthermore, an absence of historical gene flow may not correspond to current demographic isolation, yet it is the contemporary movement of animals which may be more pertinent in conservation and management actions [118]. While a combination of demographic, ecological and genetic data will provide the most robust assessments of MUs (e.g. [117,120,121]), such inter-disciplinary approaches require considerable resources and lengthy time-frames [120].

Palsbøll *et al.* [118] advocate an approach to defining MUs based on a predefined threshold level of genetic divergence, rather than the rejection of panmixia. They encourage a demographic interpretation, with the dispersal rate (i.e. migration rates) of individuals of greater relevance to conservation and management than historical gene flow. A commonly cited threshold for demographic dependence is at least 10% exchange [122]. Among our results, the estimated upper confidence intervals for migrant proportions were ≤0.1 for snubfin dolphins, which supports, with reasonable confidence, the notion of separate MUs based on dispersal rates. The large confidence intervals around our estimated migration rates for humpback dolphins include the value of 0.1, making it difficult to determine if the two sampled locations represent independent MUs based upon proposed dispersal thresholds [118]. A larger number of samples is required to more accurately estimate contemporary migration rates of humpback dolphins.

While based on limited sample sizes, our results suggest that north-western Australian snubfin and humpback dolphins may exist as metapopulations of small, genetically largely isolated population fragments. As such, they are vulnerable to genetic characteristics associated with small, fragmented populations; these include the accumulation of deleterious mutations, the loss of genetic diversity through random genetic drift, inbreeding depression, and a reduced ability to adapt to environmental change [12]. Our data, when combined with our (albeit limited) understanding of their movements, ecology and population structure from elsewhere in their range, suggest that the sampled populations are somewhat isolated and should be managed accordingly. For both species, further data are required to gain a better understanding of their genetic population structure, movements and demographics. However, it would seem appropriate to manage the two sampled populations of snubfin dolphins at Cygnet and Roebuck Bays as independent MUs. Despite the uncertainty around contemporary migration rates between humpback dolphins at the Dampier Archipelago and North West Cape, there is significant population structure and limited gene flow between these sampled populations; in light of the threat of coastal development in this region (described below), we recommend a precautionary approach of managing the sampled populations as independent MUs until further data become available.

Concerns have been raised with regard to the rate of industrial development along the coast of north-western Australia given the lack of appropriate baseline data on inshore dolphins in this region [25,40,41]. A resources boom, focussing on offshore hydrocarbon reserves and terrestrial mineral deposits, has been driving the rapid development of port and coastal processing facilities. The scale of these developments and, in particular, the volume of dredging, is large by global standards. Individual projects are responsible for tens of millions of cubic metres of seafloor dredging; combined dredging volumes for the region are in the hundreds of millions of cubic metres [25,123]. Several such developments (either constructed, under-construction and in-planning) lie within 100 km of the Dampier Archipelago sampling site, while a plan for the world's largest liquefied natural gas processing facility was approved (but subsequently abandoned by the proponents) at a site 50 km north of Roebuck Bay [124].

For tropical inshore dolphins, which are reliant upon the near-shore environment, the habitat modification associated with such coastal development presents multiple pathways for potential effects [14]. For snubfin and humpback dolphins, in particular, data deficiencies are precluding assessment of their conservation status and, therefore, their effective management in this rapidly developing region [25,40]. Given the results presented here, we recommend that conservation actions should include efforts to reduce extinction risk by maintaining effective population size and gene flow. Further restrictions on gene flow or a reduction in effective population size may compromise their evolutionary potential and, therefore, the longevity, of these populations.

Recommendations

We recommend the following conservation actions:

1. *Broad-scale baseline data collection.* Our results are based on a limited sample size, representing a small proportion of the several thousand kilometres of coastline of north-western Australia. The collection of baseline data on the distribution and abundance of inshore dolphins is required to identify and characterise local populations. Similarly, a greater number of biopsy samples across a broader geographic range are required to gain a more detailed understanding of their population genetic structure and connectivity.
2. *Better understanding and protection of identified local populations.* Each local population identified in this study is likely to serve a critical role as a stepping stone for gene flow among a fragmented metapopulation. For each local population, baseline data should be collected on abundance, effective population size, habitat use and potential or realised threatening processes. Data should inform management plans, which identify potential threats to the population, assess the vulnerability of the population against IUCN Red List Criteria, and make recommendations on actions required. Management plans should seek to minimise anthropogenic threats to local populations.
3. *Protection of movement corridors between local populations.* The occasional dispersal of breeding individuals between local populations results in the gene flow required to maintain the evolutionary potential of these small populations of dolphins. As such, proponents of development along the coast should consider their environmental footprint in relation to local populations of snubfin and humpback dolphins and the influence their activities (e.g. prolonged acoustic disturbance) may have on the movement of animals between populations, regardless of the density of animals observed in the vicinity. We strongly urge management agencies and decision-makers (e.g. the Government of Western Australia's Environmental Protection Authority and the Department of Parks and Wildlife) to consider the potential cumulative impacts of multiple developments and other threatening processes.

Supporting Information

Figure S1 Δk plot for snubfin dolphins (A) and humpback dolphins (B). In B, Δk peaks at k = 4 indicating that the most likely number of clusters equals 4.

Figure S2 Structure plot including all samples used for this study. OH = snubfin dolphin, *suspected hybrid, SC = humpback dolphin, TA = bottlenose dolphin, CY = Cygnet Bay, RB = Roebuck Bay, DA = Dampier Archipelago, NWC = North West Cape.

Figure S3 Allele frequency distribution visualising potential mode-shift distortion. The figures are based on 12 microsatellite loci for snubfin dolphins and 13 microsatellite loci for humpback dolphins.

Figure S4 Neighbour-Joining tree of all haplotypes (based on 416 bp) identified in the three resident dolphin populations at Cygnet Bay. TA = bottlenose dolphin, SC = humpback dolphin, OH = snubfin dolphin. The percentage of replicate trees in which the associated taxa clustered together in the bootstrap test (1000 replicates) is shown next to the branches.

Table S1 Locus-specific microsatellite characteristics for snubfin dolphins. N_A = Number of Alleles, N_E = Number of effective Alleles, N_P = Private Alleles, H_E = expected heterozygosity, H_O = observed heterozygosity, * = excluding monomorphic loci.

Table S2 Locus-specific microsatellite characteristics for humpback dolphins. N_A = Number of Alleles, N_E = Number of effective Alleles, N_P = Private Alleles, H_E = expected heterozygosity, H_O = observed heterozygosity, * = excluding monomorphic loci.

Table S3 *P* values (from Wilcoxon sign-rank test) and presence of mode shifts indicating whether dolphins have recently undergone a bottleneck at our sampling locations. Visualisations of potential mode shifts are shown in Figure S3. *H*: heterozygosity; IAM: infinite allele model; SMM: stepwise mutation model; *statistically significant result (*P*<0.05): ¶assessed by BOTTLENECK.

Table S4 Genotype data for snubfin, humpback and hybrid dolphins. OH = snubfin dolphin, SC = humpback dolphin, CY = Cygnet Bay, DA = Dampier Archipelago, EX = - North West Cape, RB = Roebuck Bay, H = hybrid dolphin.

Acknowledgments

The authors wish to thank the Kimberley Marine Research Station, Broome Bird Observatory, Aspen Parks, Arrow Pearl, John Boulton, Jackie Gardiner, Wayne and Ren Young, Mark and Meg Jones and Deb Vincent for providing subsidised or free accommodation for the research teams across north-western Australia. We are grateful to numerous field assistants who volunteered their time to assist with data collection, and also to Michael Krützen for providing his time and an additional research vessel for sample collection around the North West Cape. Kate Bryant and Morten Olsen provided valuable comments on an earlier draft of this manuscript.

Author Contributions

Conceived and designed the experiments: SJA LB GJP CHF. Performed the experiments: AMB AMK SJA LB BLC DC DT CP CHF. Analyzed the data: AMB AMK BLC CHF. Contributed reagents/materials/analysis tools: SJA LB CHF. Wrote the paper: AMB AMK SJA LB GJP DC CHF.

References

1. McNeely JA, Miller KR, Reid WV, Mittermeier RA, Werner TB (1990) Conserving the world's biological diversity. Gland, Switzerland: IUCN. 174 p.
2. Spielman D, Brook BW, Frankham R (2004) Most species are not driven to extinction before genetic factors impact them. Proc. Natl. Acad. Sci. U.S.A. 101: 15261–15264.
3. Frankham R (1995) Conservation genetics. Annu. Rev. Genet. 29: 305–327.
4. Frankham R (1995) Inbreeding and extinction: A threshold effect. Conserv. Biol. 9: 792–799.
5. Lynch M, Conery J, Burger R (1995) Mutation accumulation and the extinction of small populations. Am. Nat. 146: 489–518.
6. Keller LF, Waller DM (2002) Inbreeding effects in wild populations. Trends Ecol. Evol. 17: 230–241.
7. Reed DH (2004) Extinction risk in fragmented habitats. Anim. Conserv. 7: 181–191.
8. Lehman N (1991) Introgression of coyote mitochondrial DNA into sympatric North American gray wolf populations. Evolution 45: 104–119.
9. Willis PM, Crespi BJ, Dill LM, Baird RW, Hanson MB (2004) Natural hybridization between Dall's porpoises (*Phocoenoides dalli*) and harbour porpoises (*Phocoena phocoena*). Can. J. Zoo. 82: 828–834.
10. Reed DH, Frankham R (2003) Correlation between fitness and genetic diversity. Conserv. Biol. 17: 230–237.
11. O'Grady JJ, Brook BW, Reed DH, Ballou JD, Tonkyn DW, et al. (2006) Realistic levels of inbreeding depression strongly affect extinction risk in wild populations. Biol. Conserv. 133: 42–51.
12. Frankham R, Ballou JD, Briscoe DA (2010) Introduction to Conservation Genetics (Second Edition). New York: Cambridge University Press. 618 p.
13. Mace GM, Lande R (1991) Assessing extinction threats: toward a reevaluation of IUCN threatened species categories. Conserv. Biol. 5: 148–157.
14. Jefferson TA, Hung SK, Würsig B (2009) Protecting small cetaceans from coastal development: Impact assessment and mitigation experience in Hong Kong. Mar. Policy 33: 305–311.
15. Reeves RR, Jefferson TA, Karczmarski L, Laidre K, O'Corry-Crowe G, et al. (2008) *Orcaella brevirostris*. IUCN Red List of Threatened Species Version 2013-1: IUCN.
16. Reeves RR, Dawson SM, Jefferson TA, Karczmarski L, Laidre K, et al. (2013) *Cephalorhynchus hectori*. IUCN Red List of Threatened Species Version 2013-1: IUCN.
17. Rojas-Bracho L, Reeves RR, Jaramillo-legorreta A, Taylor BL (2008) *Phocoena sinus*. IUCN Red List of Threatened Species Version 2013-1: IUCN.
18. Beasley I, Robertson KM, Arnold P (2005) Description of a new dolphin, the Australian snubfin dolphin *Orcaella heinsohni* sp. N. (Cetacea, Delphinidae). Mar. Mamm. Sci. 21: 365–400.
19. Parra GJ, Ross GJB (2009) Humpback dolphins: S. chinensis and S. teuszii. In: Perrin WF, Bernd W, Thewissen JGM, editors. Encyclopaedia of Marine Mammals (Second Edition). London: Academic Press. 576–582.
20. Frère CH, Seddon J, Palmer C, Porter L, Parra GJ (2011) Multiple lines of evidence for an Australasian geographic boundary in the Indo-Pacific humpback dolphin (*Sousa chinensis*): Population or species divergence? Conserv. Genet. 12: 1633–1638.
21. Mendez M, Jefferson TA, Kolokotronis S-O, Krützen M, Parra GJ, et al. (2013) Integrating multiple lines of evidence to better understand the evolutionary divergence of humpback dolphins along their entire distribution range: a new dolphin species in Australian waters? Mol. Ecol. 22: 5936–5948.
22. Reeves RR, Dalebout ML, Jefferson TA, Karczmarski L, Laidre K, et al. (2008) *Sousa chinensis*. IUCN Red List of Threatened Species Version 2013-1: IUCN.
23. Reeves RR, Jefferson TA, Karczmarski L, Laidre K, O'Corry-Crowe G, et al. (2008) *Orcaella heinsohni*. IUCN Red List of Threatened Species Version 2013-1: IUCN.
24. Parra GJ, Corkeron PJ, Marsh H (2006) Population sizes, site fidelity and residence patterns of Australian snubfin and Indo-Pacific humpback dolphins: Implications for conservation. Biol. Conserv. 129: 167–180.
25. Bejder L, Hodgson A, Loneragan N, Allen S (2012) Coastal dolphins in north-western Australia: The need for re-evaluation of species listings and short-comings in the Environmental Impact Assessment process. Pac. Conserv. Biol. 18: 22–25.
26. Cagnazzi D, Parra GJ, Westley S, Harrison PL (2013) At the Heart of the Industrial Boom: Australian Snubfin Dolphins in the Capricorn Coast, Queensland, Need Urgent Conservation Action. PloS one 8: e56729.
27. Corkeron P, Morissette N, Porter L, Marsh H (1997) Distribution and status of hump-backed dolphins, *Sousa chinensis*, in Australian waters. Hong Kong: Hong Kong University Press.
28. Parra GJ, Corkeron PJ, Marsh H (2004) The Indo-Pacific Humpback Dolphin, *Sousa chinensis* (Osbeck, 1765), in Australian Waters: A Summary of Current Knowledge. Aquat. Mamm. 30: 197–206.
29. Cagnazzi DDB, Harrison PL, Ross GJB, Lynch P (2011) Abundance and site fidelity of Indo-Pacific Humpback dolphins in the Great Sandy Strait, Queensland, Australia. Mar. Mamm. Sci. 27: 255–281.
30. Cagnazzi D (2011) Conservation Status of Australian snubfin dolphin, *Orcaella heinsohni*, and Indo-Pacific humpback dolphin, *Sousa chinensis*, in the Capricorn Coast, Central Queensland, Australia [PhD]. Lismore, Australia: Southern Cross University. 199 p.
31. Parra GJ (2006) Resource partitioning in sympatric delphinids: Space use and habitat preferences of Australian snubfin and Indo-Pacific humpback dolphins. J. Anim. Ecol. 75: 862–874.
32. Parra GJ, Schick R, Corkeron PJ (2006) Spatial distribution and environmental correlates of Australian snubfin and Indo-Pacific humpback dolphins. Ecography 29: 396–406.
33. Parra GJ, Azuma C, Preen AR, Corkeron PJ, Marsh H (2002) Distribution of Irrawaddy dolphins, *Orcaella brevirostris*, in Australian waters. Raffles Bull. Zool: 141–154.
34. Schaurich M, Vieira Lopes F, de Oliveira L (2012) Hybridization phenomenon in cetacean and pinniped species. Neotropical Biology and Conservation 7: 199–209.
35. Bérubé M (2009) Hybridism. In: Perrin WF, Würsig B, Thewissen JGM, editors. Encyclopedia of Marine Mammals. Second ed. San Diego: Academic Press. 588–592.
36. Herzing DL, Moewe K, Brunnick BJ (2003) Interspecies interactions between Atlantic spotted dolphins, *Stenella frontalis*, and bottlenose dolphins, *Tursiops truncatus*, on Great Bahama Bank, Bahamas. Aquat. Mamm. 29: 335–341.
37. Acevedo-Gutiérrez A, DiBerardinis A, Larkin S, Larkin K, Forestell P (2005) Social interactions between tucuxis and bottlenose dolphins in Gandoca-Manzanillo, Costa Rica. Latin Am. J. Aquat. Mammals 4: 49–54.
38. Miralles L, Lens S, Rodríguez-Folgar A, Carrillo M, Martín V, et al. (2013) Interspecific Introgression in Cetaceans: DNA Markers Reveal Post-F1 Status of a Pilot Whale. PloS one 8: e69511.
39. Parra GJ (2005) Behavioural ecology of Irrawaddy, *Orcaella brevirostris* (Owen in Gray, 1866), and Indo-Pacific humpback dolphins, *Sousa chinensis* (Osbeck, 1765), in northeast Queensland, Australia: a comparative study [PhD]. Townsville, Australia: James Cook University. 375 p.
40. Allen SJ, Cagnazzi DD, Hodgson AJ, Loneragan NR, Bejder L (2012) Tropical inshore dolphins of north-western Australia: Unknown populations in a rapidly changing region. Pac. Conserv. Biol. 18: 56–63.

41. Brown A, Bejder L, Cagnazzi D, Parra GJ, Allen SJ (2012) The North West Cape, Western Australia: A potential hotspot for Indo-Pacific humpback dolphins *Sousa chinensis*? Pac. Conserv. Biol. 18: 240–246.
42. Rhymer JM, Simberloff D (1996) Extinction by hybridization and introgression. Annu. Rev. Ecol. Syst. 27: 83–109.
43. Allendorf FW, Leary RF, Spruell P, Wenburg JK (2001) The problems with hybrids: Setting conservation guidelines. Trends Ecol. Evol. 16: 613–622.
44. Carr SM, Ballinger SW, Derr JN, Blankenship LH, Bickham JW (1986) Mitochondrial DNA analysis of hybridization between sympatric white-tailed deer and mule and deer in west Texas. Proc. Natl. Acad. Sci. U.S.A. 83: 9576–9580.
45. Bérubé M, Aguilar A (1998) A new hybrid between a blue whale, *Balaenoptera musculus*, and a fin whale, *B. physalus*: Frequency and implications of hybridization. Mar. Mamm. Sci. 14: 82–98.
46. Amaral AR, Lovewell G, Coelho MM, Amato G, Rosenbaum HC (2014) Hybrid Speciation in a Marine Mammal: The Clymene Dolphin (*Stenella clymene*). PloS one 9: e83645.
47. Dowling TE, Secor CL (1997) The role of hybridization and introgression in the diversification of animals. Annu. Rev. Ecol. Syst. 28: 593–619.
48. Stebbins GL (1950) Variation and Evolution in Plants. New York: Columbia University Press.
49. Knutsen H, Jorde PE, André C, Stenseth NC (2003) Fine-scaled geographical population structuring in a highly mobile marine species: The Atlantic cod. Mol. Ecol. 12: 385–394.
50. Mariani S, Hutchinson WF, Hatfield EMC, Ruzzante DE, Simmonds EJ, et al. (2005) North Sea herring population structure revealed by microsatellite analysis. Mar. Ecol. Prog. Ser. 303: 245–257.
51. Dudgeon CL, Blower DC, Broderick D, Giles JL, Holmes BJ, et al. (2012) A review of the application of molecular genetics for fisheries management and conservation of sharks and rays. J. Fish Biol. 80: 1789–1843.
52. Carreras C, Godley BJ, León YM, Hawkes LA, Revuelta O, et al. (2013) Contextualising the Last Survivors: Population Structure of Marine Turtles in the Dominican Republic. PloS one 8: e66037.
53. Dutton PH, Roden SE, Stewart KR, LaCasella E, Tiwari M, et al. (2013) Population stock structure of leatherback turtles (*Dermochelys coriacea*) in the Atlantic revealed using mtDNA and microsatellite markers. Conserv. Genet. 14: 625–636.
54. Garcia-Rodriguez AI, Bowen BW, Domning D, Mignucci-Giannoni AA, Marmontel M, et al. (1998) Phylogeography of the West Indian manatee (*Trichechus manatus*): How many populations and how many taxa? Mol. Ecol. 7: 1137–1149.
55. Rosenbaum HC, Brownell Jr RL, Brown MW, Schaeff C, Portway V, et al. (2000) World-wide genetic differentiation of Eubalaena: Questioning the number of right whale species. Mol. Ecol. 9: 1793–1802.
56. Hamner RM, Pichler FB, Heimeier D, Constantine R, Baker CS (2012) Genetic differentiation and limited gene flow among fragmented populations of New Zealand endemic Hector's and Maui's dolphins. Conserv. Genet. 13: 987–1002.
57. LeDuc RG, Weller DW, Hyde J, Burdin AM, Rosel PE, et al. (2002) Genetic differences between western and eastern North Pacific gray whales (*Eschrichtius robustus*). J. Cetacean Res. Manage. 4: 1–5.
58. Segura I, Rocha-Olivares A, Flores-Ramirez S, Rojas-Bracho L (2006) Conservation implications of the genetic and ecological distinction of *Tursiops truncatus* ecotypes in the Gulf of California. Biol. Conserv. 133: 336–346.
59. Wiemann A, Andersen LW, Berggren P, Siebert U, Benke H, et al. (2010) Mitochondrial Control Region and microsatellite analyses on harbour porpoise (*Phocoena phocoena*) unravel population differentiation in the Baltic Sea and adjacent waters. Conserv. Genet. 11: 195–211.
60. Pichler FB, Dawson SM, Slooten E, Baker CS (1998) Geographic isolation of Hector's dolphin populations described by mitochondrial DNA sequences. Conserv. Biol. 12: 676–682.
61. Natoli A, Birkun A, Aguilar A, Lopez A, Hoelzel AR (2005) Habitat structure and the dispersal of male and female bottlenose dolphins (*Tursiops truncatus*). Proc. R. Soc. B 272: 1217–1226.
62. Mirimin L, Miller R, Dillane E, Berrow SD, Ingram S, et al. (2011) Fine-scale population genetic structuring of bottlenose dolphins in Irish coastal waters. Anim. Conserv. 14: 342–353.
63. Frère CH, Hale PT, Porter L, Cockcroft VG, Dalebout ML (2008) Phylogenetic analysis of mtDNA sequences suggests revision of humpback dolphin (*Sousa* spp.) taxonomy is needed. Mar. Freshw. Res. 59: 259–268.
64. Palmer C, Murphy SA, Thiele D, Parra GJ, Robertson KM, et al. (2011) Analysis of mitochondrial DNA clarifies the taxonomy and distribution of the Australian snubfin dolphin (*Orcaella heinsohni*) in northern Australian waters. Mar. Freshw. Res. 62: 1303–1307.
65. Lin W, Chang L, Frère CH, Zhou R, Chen J, et al. (2012) Differentiated or not? An assessment of current knowledge of genetic structure of *Sousa chinensis* in China. J. Exp. Mar. Biol. Ecol. 416–417: 17–20.
66. Krützen M, Barre LM, Möller LM, Heithaus MR, Simms C, et al. (2002) A biopsy system for small cetaceans: Darting success and wound healing in *Tursiops* spp. Mar. Mamm. Sci. 18: 863–878.
67. Amos W, Hoelzel AR (1991) Long term preservation of whale skin for DNA analysis. Rep. Int. Whal. Comm. 13: 99–104.
68. Gilson A, Syvanen M, Levine K, Banks J (1998) Deer gender determination by polymerase chain reaction: Validation study and application to tissues, bloodstains, and hair forensic samples from California. Calif. Fish Game 84: 159–169.
69. Baker CS, Perry A, Bannister JL, Weinrich MT, Abernethy RB, et al. (1993) Abundant Mitochondrial-DNA Variation and Worldwide Population-Structure in Humpback Whales. Proc. Natl. Acad. Sci. U.S.A. 90: 8239–8243.
70. Bacher K, Allen S, Lindholm A, Bejder L, Krützen M (2010) Genes or Culture: Are Mitochondrial Genes Associated with Tool Use in Bottlenose Dolphins (*Tursiops* sp.)? Behav. Genet. 40: 706–714.
71. Buchanan FC, Friesen MK, Littlejohn RP, Clayton JW (1996) Microsatellites from the beluga whale *Delphinapterus leucas*. Mol. Ecol. 5: 571–575.
72. Cassens I, Van Waerebeek K, Best PB, Tzika A, Van Helden AL, et al. (2005) Evidence for male dispersal along the coasts but no migration in pelagic waters in dusky dolphins (*Lagenorhynchus obscurus*). Mol. Ecol. 14: 107–121.
73. Chen L, Yang G (2009) A set of polymorphic dinucleotide and tetranucleotide microsatellite markers for the Indo-Pacific humpback dolphin (*Sousa chinensis*) and cross-amplification in other cetacean species. Conserv. Genet. 10: 697–700.
74. Rooney AP, Merritt DB, Derr JN (1999) Microsatellite diversity in captive bottlenose dolphins (*Tursiops truncatus*). J. Hered. 90: 228–231.
75. Frère CH, Krzyszczyk E, Patterson EM, Hunter S, Ginsburg A, et al. (2010) Thar she blows! A novel method for DNA collection from cetacean blow. PloS one 5: e12299.
76. Park SDE (2001) Trypanotolerance in West African Cattle and the Population Genetic Effects of Selection [PhD]. Dublin, Republic of Ireland: University of Dublin.
77. Rousset F (2008) GENEPOP'007: a complete re-implementation of the GENEPOP software for Windows and Linux. Mol. Ecol. Resour. 8: 103–106.
78. Excoffier L, Laval G, Schneider S (2005) Arlequin (version 3.0): An integrated software package for population genetics data analysis. Evolutionary Bioinformatics 1: 47–50.
79. Wilson GA, Rannala B (2003) Bayesian Inference of Recent Migration Rates Using Multilocus Genotypes. Genetics 163: 1177–1191.
80. Falush D, Stephens M, Pritchard JK (2003) Inference of Population Structure Using Multilocus Genotype Data: Linked Loci and Correlated Allele Frequencies. Genetics 164: 1567–1587.
81. Pritchard JK, Stephens M, Donnelly P (2000) Inference of Population Structure Using Multilocus Genotype Data. Genetics 155: 945–959.
82. Evanno G, Regnaut S, Goudet J (2005) Detecting the number of clusters of individuals using the software STRUCTURE: a simulation study. Mol. Ecol. 14: 2611–2620.
83. Earl DA, von Holdt BM (2012) STRUCTURE HARVESTER: a website and program for visualizing STRUCTURE output and implementing the Evanno method. Conservation Genet Resour 4: 359–361.
84. Duchesne P, Turgeon J (2012) FLOCK Provides Reliable Solutions to the "Number of Populations" Problem. J. Hered. 103: 734–743.
85. Waples RS, Do C (2008) LDNe: a program for estimating effective population size from data on linkage disequilibrium. Mol. Ecol. Resour. 8: 753–756.
86. Waples RS, Do C (2010) Linkage disequilibrium estimates of contemporary Ne using highly variable genetic markers: a largely untapped resource for applied conservation and evolution. Evol. Appl. 3: 244–262.
87. Robinson JD, Moyer GR (2013) Linkage disequilibrium and effective population size when generations overlap. Evol. Appl. 6: 290–302.
88. Palstra FP, Ruzzante DE (2008) Genetic estimates of contemporary effective population size: what can they tell us about the importance of genetic stochasticity for wild population persistence? Mol. Ecol. 17: 3428–3447.
89. Ansmann IC, Lanyon JM, Seddon JM, Parra GJ (2013) Monitoring Dolphins in an Urban Marine System: Total and Effective Population Size Estimates of Indo-Pacific Bottlenose Dolphins in Moreton Bay, Australia. PloS one 8: e65239.
90. Luikart G, Allendorf FW, Cornuet JM, Sherwin WB (1998) Distortion of allele frequency distributions provides a test for recent population bottlenecks. J. Hered. 89: 238–247.
91. Piry S, Luikart G, Cornuet J (1996) BOTTLENECK (A program for detecting recent effective population size reductions from allele data frequencies). Laboratoire de Modélisation et de Biologie Evolutive. INRA-URLB. 488 rue de la Croix-Lavit, 34090 Montpellier Cedex, France.
92. Cornuet JM, Luikart G (1996) Description and power analysis of two tests for detecting recent population bottlenecks from allele frequency data. Genetics 144: 2001–2014.
93. Rice WR (1989) Analyzing Tables of Statistical Tests. Evolution 43: 223–225.
94. Andreasen AM, Stewart KM, Longland WS, Beckmann JP, Forister ML (2012) Identification of source-sink dynamics in mountain lions of the Great Basin. Mol. Ecol. 21: 5689–5701.
95. Chen H, Zhai K, Chen J, Chen Y, Wen H, et al. (2008) A preliminary investigation on genetic diversity of *Sousa chinensis* in the Pearl River Estuary and Xiamen of Chinese waters. Journal of Genetics and Genomics 35: 491–497.
96. Jefferson TA, Hung SK (2004) A Review of the Status of the Indo-Pacific Humpback Dolphin (*Sousa chinensis*) in Chinese Waters. Aquat. Mamm. 30: 149–158.
97. Krützen M, Sherwin WB, Berggren P, Gales N (2004) Population structure in an inshore cetacean revealed by microsatellite and mtDNA analysis: Bottlenose dolphins (*Tursiops* sp.) in Shark Bay, Western Australia. Mar. Mamm. Sci. 20: 28–47.

98. Sellas AB, Wells RS, Rosel PE (2005) Mitochondrial and nuclear DNA analyses reveal fine scale geographic structure in bottlenose dolphins (*Tursiops truncatus*) in the Gulf of Mexico. Conserv. Genet. 6: 715–728.
99. Rosel PE, Hansen L, Hohn AA (2009) Restricted dispersal in a continuously distributed marine species: common bottlenose dolphins *Tursiops truncatus* in coastal waters of the western North Atlantic. Mol. Ecol. 18: 5030–5045.
100. Ansmann IC, Parra GJ, Lanyon JM, Seddon JM (2012) Fine-scale genetic population structure in a mobile marine mammal: Inshore bottlenose dolphins in Moreton Bay, Australia. Mol. Ecol. 21: 4472–4485.
101. Hollatz C, Vilaça ST, Redondo RAF, Marmontel M, Baker CS, et al. (2011) The Amazon River system as an ecological barrier driving genetic differentiation of the pink dolphin (*Inia geoffrensis*). Biol. J. Linn. Soc. 102: 812–827.
102. Bilgmann K, Möller LM, Harcourt RG, Gibbs SE, Beheregaray LB (2007) Genetic differentiation in bottlenose dolphins from South Australia: Association with local oceanography and coastal geography. Mar. Ecol. Prog. Ser. 341: 265–276.
103. Möller LM, Wiszniewski J, Allen SJ, Beheregaray LB (2007) Habitat type promotes rapid and extremely localised genetic differentiation in dolphins. Mar. Freshw. Res. 58: 640–648.
104. Wiszniewski J, Beheregaray LB, Allen SJ, Möller LM (2010) Environmental and social influences on the genetic structure of bottlenose dolphins (*Tursiops aduncus*) in Southeastern Australia. Conserv. Genet. 11: 1405–1419.
105. Kopps AM, Ackermann C, Sherwin WB, Allen SJ, Bejder L, et al. (2014) Cultural transmission of tool use combined with habitat specialisations leads to fine-scale genetic structure in bottlenose dolphins Proc. R. Soc. B 281, 20133245. (doi:10.1098/rspb.2013.3245).
106. Wright S (1969) Evolution and the Genetics of Populations, Volume 2: Theory of gene frequencies. Chicago: University of Chicago Press. 520 p.
107. Franklin I (1980) Evolutionary change in small populations. In: Soulé M, Wilcox B, editors. Conservation biology: an evolutionary-ecological perspective: Sunderland: Sinauer. 135–150.
108. Wu CI, Johnson NA, Palopoli MF (1996) Haldane's rule and its legacy: Why are there so many sterile males? Trends in Ecology and Evolution 11: 281–284.
109. Spilliaert R, Vikingsson G, Arnason U, Palsdottir A, Sigurjonsson J, et al. (1991) Species hybridization between a female blue whale (*Balaenoptera musculus*) and a male fin whale (*B. physalus*): Molecular and morphological documentation. J. Hered. 82: 269–274.
110. Glover KA, Kanda N, Haug T, Pastene LA, Øien N, et al. (2013) Hybrids between common and Antarctic minke whales are fertile and can back-cross. BMC Genet. 14: 25.
111. Baird RW (2003) Update COSEWIC status report on the harbour porpoise *Phocoena phocoena* (Pacific Ocean population) in Canada, *in* COSEWIC assessment and update status report on the harbour porpoise *Phocoena phocoena* (Pacific Ocean population) in Canada. Ottawa. 1–22 p.
112. Waples RS, Gaggiotti O (2006) What is a population? An empirical evaluation of some genetic methods for identifying the number of gene pools and their degree of connectivity. Mol. Ecol. 15: 1419–1439.
113. Mendez M, Rosenbaum HC, Subramaniam A, Yackulic C, Bordino P (2010) Isolation by environmental distance in mobile marine species: molecular ecology of franciscana dolphins at their southern range. Mol. Ecol. 19: 2212–2228.
114. Bilgmann K, Möller LM, Harcourt RG, Gales R, Beheregaray LB (2008) Common dolphins subject to fisheries impacts in Southern Australia are genetically differentiated: implications for conservation. Anim. Conserv. 11: 518–528.
115. Moritz C (1994) Defining 'evolutionarily significant units' for conservation. Trends Ecol. Evol. 9: 373–375.
116. Crandall KA, Bininda-Emonds ORR, Mace GM, Wayne RK (2000) Considering evolutionary processes in conservation biology. Trends Ecol. Evol. 15: 290–295.
117. Paetkau D (1999) Using genetics to identify intraspecific conservation units: A critique of current methods. Conserv. Biol. 13: 1507–1509.
118. Palsbøll PJ, Berube M, Allendorf FW (2007) Identification of management units using population genetic data. Trends Ecol. Evol. 22: 11–16.
119. Pearse DE, Crandall KA (2004) Beyond FST: Analysis of population genetic data for conservation. Conserv. Genet. 5: 585–602.
120. Olsen MT, Andersen LW, Dietz R, Teilmann J, Härkönen T, et al. (2014) Integrating genetic data and population viability analyses for the identification of harbour seal (*Phoca vitulina*) populations and management units. Mol. Ecol. 23: 815–831.
121. Taylor BL, Dizon AE (1999) First policy then science: Why a management unit based solely on genetic criteria cannot work. Mol. Ecol. 8: S11–S16.
122. Hastings A (1993) Complex interactions between dispersal and dynamics: lessons from coupled logistic equations. Ecology 74: 1362–1372.
123. EPA (2013) Environmental Protection Authority 2012–13 Annual Report. Perth, Western Australia: EPA. 85 p.
124. DMP (2014) Major Resource Projects - March 2013. Department of Mines and Petroleum, Western Australia. Available from: http://www.dmp.wa.gov.au/5590.aspx. Accessed: 07 March 2014.

3

Genetic Population Structure of Local Populations of the Endangered Saltmarsh Sesarmid Crab *Clistocoeloma sinense* in Japan

Takeshi Yuhara[1]*, Masako Kawane[2], Toshio Furota[3]

1 Department of Environmental Science, Graduate School of Science, Toho University, Funabashi, Chiba, Japan, **2** *KYOUSEI* Science Center for Life and Nature, Nara Women's University, Nara, Japan, **3** Faculty of Sciences, Tokyo Bay Ecosystem Research Center, Toho University, Funabashi, Chiba, Japan

Abstract

During recent decades, over 40% of Japanese estuarine tidal flats have been lost due to coastal developments. Local populations of the saltmarsh sesarmid crab *Clistocoeloma sinense*, designated as an endangered species due to the limited suitable saltmarsh habitat available, have decreased accordingly, being now represented as small remnant populations. Several such populations in Tokyo Bay, have been recognised as representing distributional limits of the species. To clarify the genetic diversity and connectivity among local coastal populations of Japanese *Clistocoeloma sinense*, including those in Tokyo Bay, mitochondrial DNA analyses were conducted in the hope of providing fundamental information for future conservation studies and an understanding of metapopulation dynamics through larval dispersal among local populations. All of the populations sampled indicated low levels of genetic diversity, which may have resulted from recent population bottlenecks or founder events. However, the results also revealed clear genetic differentiation between two enclosed-water populations in Tokyo Bay and Ise-Mikawa Bay, suggesting the existence of a barrier to larval transport between these two water bodies. Since the maintenance of genetic connectivity is a requirement of local population stability, the preservation of extant habitats and restoration of saltmarshes along the coast of Japan may be the most effective measures for conservation of this endangered species.

Editor: Erik Sotka, College of Charleston, United States of America

Funding: The present research was supported by a grant from the Mikimoto Fund for Marine Ecology. The funders had no role in study design, data collection and analysis, decision to publish, or preparation of the manuscript. (http://www.jwrc.or.jp/shintaku/2.htm).

Competing Interests: The authors have declared that no competing interests exist.

* E-mail: takeshi.yuhara@gmail.com

Introduction

Many estuarine ecosystems worldwide have been severely disrupted by human activities, such as overexploitation, pollution and land reclamation [1]. In Japan, more than 40% of tidal flats, including estuarine in saltmarshes, have been lost, mainly during the latter half of the 20th century [2]. The destruction of estuarine mudflats and saltmarshes by land reclamation have resulted in the reduction of local populations of many mudflat/saltmarsh-specific species [3]. For example, large-scale reclamation of tidal flats and saltmarshes along the entire coastline of Tokyo Bay, conducted during the late 1960s and early 1970s [4], resulted in the loss of most of the saltmarshes in the high intertidal zone by the end of the 1970's. Accordingly, most mudflat/saltmarsh-specific benthic species in Tokyo Bay have since become extinct or endangered [5].

The saltmarsh sesarmid crab *Clistocoeloma sinense* Shen, 1933 is distributed throughout estuaries in subtropical and temperate eastern Asia, Japan [6], Korea [7], China [8], and Taiwan [9]. Although distributed along the Pacific coasts of central and western Honshu, Shikoku, Kyushu and Okinawa, the species is designated as an endangered species in Japan by the Japanese Association of Benthology [2] because of the very significant decrease in suitable habitat. Furthermore, Tokyo Bay, representing the northern and easternmost limits of the species geographical distribution, is separated by a distance of ca. 250 km from the nearest coastal population (in Mikawa Bay). Because the genetic structures of populations at the limits of geographical distribution are generally characterized by increased genetic isolation and differentiation [10], the conservation of such populations requires an understanding of gene flow among edge and central populations, and their connectivity through larval transport.

This investigation of local population genetic structures of the endangered saltmarsh crab *Clistocoeloma sinense* along the Japanese coast utilized mitochondrial DNA (mtDNA) analyses in order to clarify the genetic diversity and relationships/connectivity among local populations of the species. Based on the investigation, we discuss a focus on conservation strategies for the geographically isolated Tokyo Bay populations.

Materials and Methods

Ethics Statement

In 2011 and 2012, sampling of *Clistocoeloma sinense* was conducted by hand-digging on saltmarshes at 8 localities in inner Tokyo Bay and 13 localities on saltmarshes along the western Japanese main Islands from Shizuoka Prefecture to Saga Prefecture, and on Okinawa Island, thereby covering the entire geographical range of *Clistocoeloma sinense* in Japan. In total, 418 samples of *Clistocoeloma sinense* were collected from the 8 Tokyo Bay localities and 9 other localities (Fig. 1, Table 1). No crabs were collected from Izu Peninsula, Okayama, Fukuoka or Okinawa Island. Ethical approval for this research was not required under any local, national or international laws, because all of the animals used in the study were invertebrates. Field collections in two localities included within wildlife protection areas (localities f and g) were authorized by Chiba Prefecture and Tokyo Metropolitan Area local governments. All samples were collected in strict accordance with good animal practice as defined by the relevant national and/or local animal welfare bodies. The third ambulatory leg only was dissected, using the autotomy methods so as to reduce any harmful effects on the sampled individuals, which were then all returned to their places of capture. All pereopod samples were preserved in 99% ethanol immediately after collection.

DNA Sequencing

Total genomic DNA was extracted from the ambulatory leg musculature using a CTAB protocol [11]. The target DNA segment of a portion of the mtDNA cytochrome c oxidase subunit I (COI) was amplified by polymerase chain reaction (PCR) with the primers mtd10 5′-TTGATTTTTTGGTCATCCAGAAGT-3′ [12] and C/N2769 5′-TTAAGTCCTA-GAAAATGTTGGGGA-3′ [13]. PCR amplification was conducted in a total volume of 20 μL containing 0.16 μL of TaKaRa ExTaq (5 units/μL), 2.0 μL of 10× Ex Taq buffer, 1.6 μL of dNTP mixture (2.5 mM each); 1.0 μL of each primer (10 mM), and 2.0 μL of template. PCR conditions comprised 35 cycles of denaturation (94°C, 30 s), annealing (45°C, 30 s), and extension (72°C, 60 s) on thermal cyclers in a GeneAmp 9700 PCR System (Applied Biosystems). Amplification products were checked for size by loading 3 μL on a 1.5% agarose gel (GenePure LE) with 0.5 μg/mL ethidium bromide. Subsequent product sequencing was performed at the Dragon Genomics Center [Takara Bio, Otsu, Japan]. Sequencing reactions followed the manufacturer's suggested protocol. All final sequences were obtained from both strands of gene segments for verification. DNA segments of 528 base pairs of the partial mtDNA COI gene were sequenced and used for DNA analyses. All of the newly obtained sequences were deposited in the DNA Data Bank of Japan (accession numbers AB813717– AB813725).

Data Analysis

The number of mtDNA COI haplotypes, and the haplotype and nucleotide diversities were calculated for each local population of *Clistocoeloma sinense* by using the computer software program Arlequin Ver. 3.1 [14]. The structure of each population was examined by analysis of molecular variance (AMOVA, [15]), using Arlequin with a model of one group of populations. Pairwise F_{ST} (fixation index) values based on uncorrected sequence differences [15,16] computed using Arlequin were used as indices of genetic differentiation between populations. Significance was calculated for 100,000 permutations using Arlequin and corrected by the false discovery rate method [17]. Population growth evaluations were based on Tajima's D test [18,19] and Fu's Fs statistic [20], using Arlequin. In addition, the fit of the mismatch distribution was compared to that expected under the spatial expansion model with Arlequin. The sum of squared deviations (SSD) and raggedness index (rg) between observed and expected distributions were used as test statistics, the significance being assessed after 1,000 bootstraps. Index of the age of population expansion (τ) was also calculated by Arlequin. To test for evidence of Isolation by Distance, a Mantel test was performed on genetic distance ($F_{ST}/1-F_{ST}$) against geographic distance (minimum coastline distance) between all pairs of sampling localities, using the Isolation by Distance Web Service (Version 3.23) [21]. The significance of Mantel's Z-test statistic was based on 10,000 permutations.

Results

Nine different mtDNA COI haplotypes, with a total of eight variable sites, were identified from the sequenced samples ($n = 418$, Fig. 1, Table 2). No insertions or deletions were found. The nine haplotypes included two that were dominant (Cs1 and Cs2), being found at all of the localities represented. The most frequent haplotype (Cs2) was found at 12 localities (Tokyo Bay, localities a to h; southern Shikoku and Kyushu, localities 5 and 7 to 9),

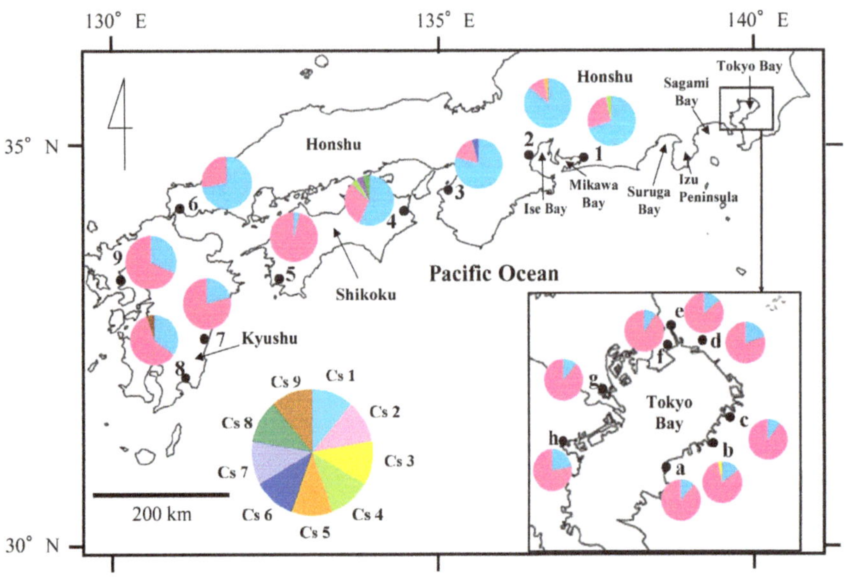

Figure 1. Sampling localities and pie chart representation of mtDNA COI haplotype frequencies for *Clistocoeloma sinense*. See also Tables 1 and 2 for further details.

Table 1. Sampling localities and sample size of *Clistocoeloma sinense*.

Locality	Prefecture	Locality number	Latitude	Longitude	n
Obitsu R.	Chiba	a	35° 24′ N	139° 54′ E	28
Shiizu R.	Chiba	b	35° 28′ N	140° 01′ E	28
Tamasaki	Chiba	c	35° 31′ N	140° 03′ E	30
Yatsu tidlflat	Chiba	d	35° 40′ N	140° 00′ E	20
Edogawa R.	Chiba	e	35° 42′ N	139° 55′ E	29
Shinhama lagoon	Chiba	f	35° 40′ N	139° 55′ E	30
Tokyo Port Wild Bird Park	Tokyo	g	35° 34′ N	139° 46′ E	27
Yokohama Port	Kanagawa	h	35° 27′ N	139° 37′ E	18
Kamita R.	Aichi	1	34° 41′ N	137° 19′ E	27
Tanaka R.	Mie	2	34° 47′ N	136° 33′ E	28
Kinokawa R.	Wakayama	3	34° 14′ N	135° 33′ E	24
Katsuura R.	Tokushima	4	34° 02′ N	134° 34′ E	23
Souzu R.	Ehime	5	32° 57′ N	132° 33′ E	28
Koya R.	Yamaguchi	6	34° 03′ N	131° 03′ E	18
Hitotsuse R.	Miyazaki	7	32° 03′ N	131° 29′ E	24
Yoshida R.	Miyazaki	8	31° 27′ N	131° 12′ E	20
Tagori R.	Saga	9	32° 57′ N	130° 12′ E	16

ranging in frequency from 0.60 to 0.96. Haplotype Cs1 was dominant at five localities (numbers 1 to 4 and 6 in central Honshu), ranging in frequency from 0.57 to 0.86. The remaining six haplotypes each occurred only once, at six sampling localities. Haplotype (h) and nucleotide (π) diversity values tended to be very low at most localities, with values ranging from 0.0714 to 0.6087 and from 0.000135 to 0.001392, respectively. The Katsuura River population (locality 4) exhibited the greatest genetic diversity ($h = 0.6087$ and $\pi = 0.001392$), followed by the population from Yoshida River (locality 8) ($h = 0.5421$ and $\pi = 0.001146$). In contrast, almost all of the Tokyo Bay populations were characterized by low genetic diversity (h ranging from 0.1862 to 0.3368; π ranging from 0.000353 to 0.000693). The parsimony network of the mtDNA COI haplotypes of *Clistocoeloma sinense* consisted of two major haplotypes (separated by a single mutational step) and seven minor haplotypes, separated from the major haplotypes by a single mutational step (Fig. 2). Table 2 summarizes the distribution of haplotypes per population.

The AMOVA analysis revealed significant overall population differentiation (among populations: % variation = 33.82; within populations: % variation = 66.18, $\Phi_{ST} = 0.338$, $P<0.001$). Pairwise F_{ST} values (Table 3) showed a tendency toward genetic differentiation between populations ($P<0.001$ in comparisons of 60 out of all 136 pairs) and had the following pattern: (i) differentiation between Tokyo Bay populations (localities a to h) and central Honshu populations (localities 1 to 4 and 6); (ii) differentiation between central Honshu populations (localities 1 to 4 and 6) and southern Shikoku and Kyushu populations (localities 5 and 7 to 9), excluding the Yoshida River population (locality 8); and (iii) no differentiation between the Tokyo Bay populations (localities a to h) and southern Shikoku and Kyushu populations (localities 5 and 7 to 9), excluding the Yoshida River population (locality 8).

In the demographic analyses (Table 4), Tajima's D and Fu's Fs statistics were negative for Tokyo Bay to southern Shikoku population (locality 5), and result of SSD p-value in those populations indicated that each population had undergone a sudden population expansion. Under the assumption of the spatial expansion hypothesis, the value of τ for Tokyo Bay population was lower compared to that of other populations (except for southern Shikoku population). These results show that Tokyo Bay population was critically affected by bottleneck/founder effect in recent years. With respect to Isolation by Distance, no significant relationship between genetic distance and linear geographic distance was apparent within the longitudinal gradient under study (Fig. 3; Mantel Z-test = 30905.65, 10,000 randomizations, $r = 0.0078$, $P = 0.592$).

Discussion

Low Genetic Diversity

All populations of the endangered sesarmid crab *Clistocoeloma sinense* in estuarine saltmarshes located from central Honshu to Kyushu showed low levels of genetic diversity, as measured by haplotype and nucleotide diversities, compared with those of other intertidal brachyuran crabs (Table 5). Grant and Bowen [22] classified marine fishes into four categories on the basis of different combinations of high and low values for haplotype diversity and nucleotide diversity of mtDNA sequences. *Clistocoeloma sinense*, with its low values for both haplotype and nucleotide diversity, is characteristic of the first category, which may have occurred through a recent population bottleneck or founder event involving a single or at most, a limited number of mtDNA linkages.

The Tokyo Bay populations, located at the northern (and eastern) geographical limits of the species, were characterized by lower genetic diversity than the other local populations. Similar genetic patterns have been previously reported for other marine invertebrate populations representing distributional limits in Japanese waters, viz. the tideland snail *Batillaria zonalis* [23], swimming crab *Portunus trituberculatus* [24], and fiddler crab *Uca arcuata* [25]. Because of the urbanization of coastal areas about Tokyo Bay, about 90% of former tidal flats, including saltmarshes, have been reclaimed, and the remaining limited area habitats

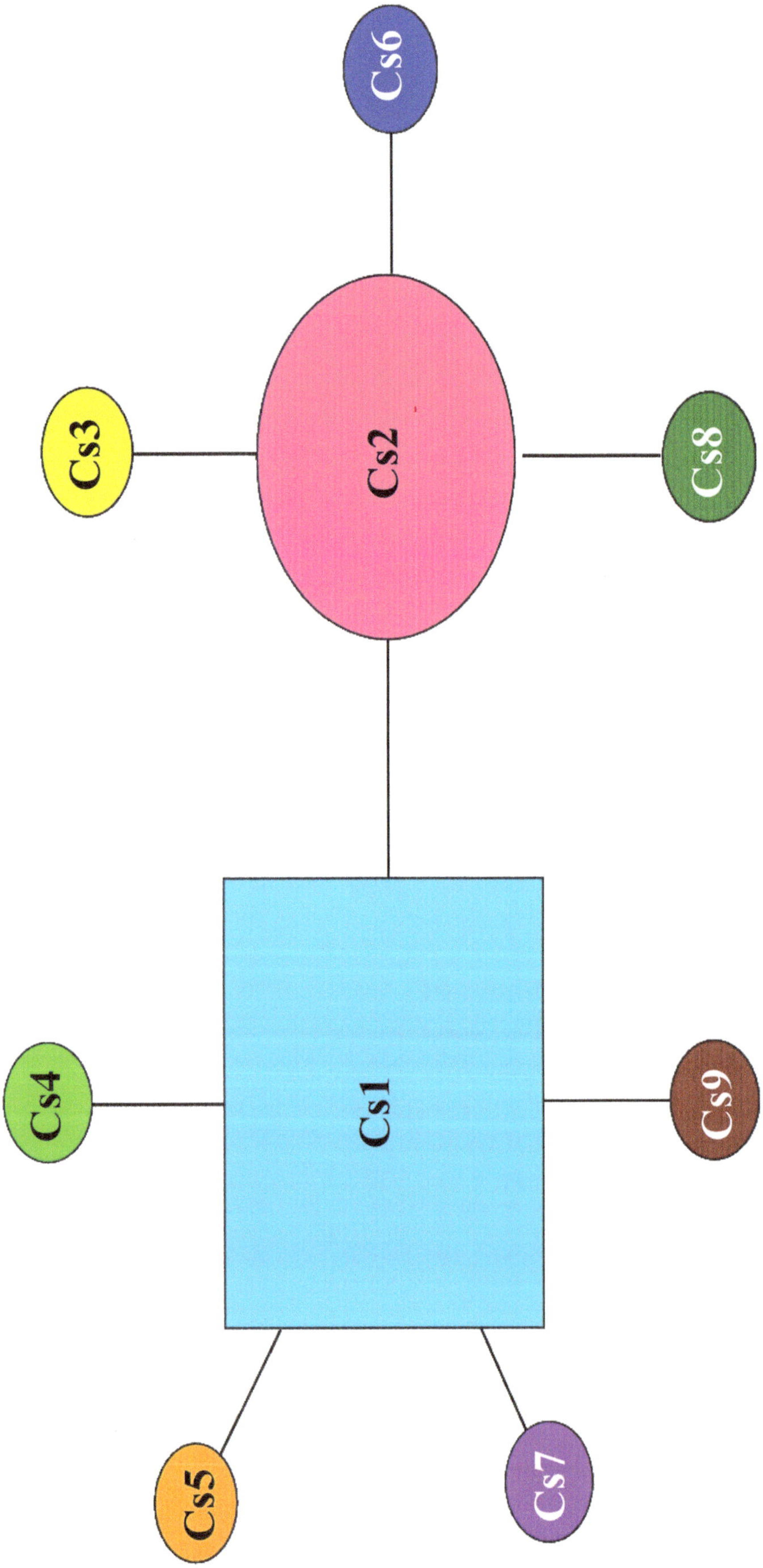
Cs6
Cs3
Cs2
Cs8
Cs4
Cs1
Cs9
Cs5
Cs7

Figure 2. Statistical parsimony network of mtDNA COI haplotypes from *Clistocoeloma sinense* specimens. Each circle and a square represent a unique haplotype. Haplotypes (Cs1–Cs9) correspond to Table 2. Circles and a square size proportional to number of haplotypes.

isolated [4]. As a result, the population range of saltmarsh-specific species has been reduced, the majority of such populations in Tokyo Bay being now endangered or extinct [4].

Low Gene Flow

The AMOVA analysis revealed significant overall population differentiation. Differences were apparent by the F_{ST} analysis between the (i) Tokyo Bay populations and central Honshu populations and (ii) central Honshu populations and southern Shikoku and Kyushu populations, but not between the Tokyo Bay populations and southern Shikoku and Kyushu populations. According to the Isolation by Distance analysis (Fig. 3), however, there was no significant relationship between genetic distance and linear geographic distance within the longitudinal gradient examined, probably due to the lack of genetic differentiation between the Tokyo Bay and Kyushu populations, despite their broad geographic separation. Removal of Tokyo Bay data from the results indicated that the remaining local populations of *Clistocoeloma sinense* may comprise two genetically deviated groups occurring in the eastern and western parts of the species overall distribution range (Fig. 1). To verify this possibility, IBD was analyzed without Tokyo Bay data. However, no significant relationship between genetic distance and linear geographic distance was apparent within the longitudinal gradient (Figure S1; Mantel Z-test = 7859.62, 10,000 randomizations, $r = 0.167$, $P = 0.154$).

Marine species are generally characterized by metapopulation structures that have a large population size, high dispersal capacity during the pelagic larval stages and extensive distribution [26]. The apparent lack of barriers to dispersal in the marine environment often effectively reduces genetic heterogeneity among populations which makes the differentiation of discreet regional populations difficult [27]. However, some studies have reported that although some intertidal brachyuran crabs have a planktonic larval stage, gene flow between local populations is limited [28,29]. Similarly, the genetic analysis in the present study revealed genetic differentiation among local populations of Japanese *Clistocoeloma sinense*, indicating limited gene flow between distant local populations, despite their planktonic larval development [30,31].

The genetic differentiation between geographically distant local populations in Tokyo Bay and Mikawa Bay, demonstrated herein, parallels similar examples of genetic differentiation, also shown by mtDNA analysis, in other marine animal species occurring in Tokyo Bay (including Sagami Bay) and along the coast of western Honshu, viz. the intertidal goby *Chaenogobius annularis* [32], surfperch *Ditrema jordani* [33], and brackish water crab *Deiratonotus cristatus* [29]. Dispersal of these species may be impeded by the extensive land mass of the Izu Peninsula, interrupting gene flow. On the other hand, genetic differentiation between local populations separated by the peninsula has not been found in the intertidal mudskipper *Periophthalmus modestus* [34], Japanese mitten crab *Eriocheir japonica* [35], intertidal rocky shore crab *Gaetice depressus* [28], intertidal mudflat crab *Chiromantes dehaani* [28], or intertidal ocypodid crab *Ilyoplax pusilla* [36]. As indicated by these examples, genetic differentiation patterns can vary between marine species with planktonic larval stages.

Each of the aforementioned intertidal brachyuran crabs have four or more zoea and one megalopa stages [31,37], whereas

Table 2. Haplotype composition, haplotype diversity (Mean ± SD) and nucleotide diversity (Mean ± SD) of 17 *Clistocoeloma sinense* populations.

Locality number	Haplotype									n	Haplotype diversity	Nucleotide diversity
	Cs1	Cs2	Cs3	Cs4	Cs5	Cs6	Cs7	Cs8	Cs9			
a	3	25								28	0.1984±0.0924	0.000376±0.000537
b	4	23	1							28	0.3148±0.1024	0.000616±0.000713
c	3	27								30	0.1862±0.0881	0.000353±0.000517
d	4	16								20	0.3368±0.1098	0.000638±0.000739
e	4	25								29	0.2463±0.0935	0.000466±0.000606
f	3	27								30	0.1862±0.0881	0.000353±0.000517
g	3	24								27	0.2051±0.0947	0.000389±0.000548
h	4	14								18	0.3660±0.1124	0.000693±0.000781
1	19	7		1						27	0.4530±0.0869	0.000896±0.000895
2	24	3			1					28	0.2619±0.1022	0.000511±0.000640
3	19	4				1				24	0.3587±0.1096	0.000810±0.000845
4	13	7		1			1	1		23	0.6087±0.0818	0.001392±0.001196
5	1	27								28	0.0714±0.0652	0.000135±0.000310
6	13	5								18	0.4248±0.0993	0.000805±0.000854
7	5	19								24	0.3442±0.0987	0.000652±0.000742
8	7	12							1	20	0.5421±0.0763	0.001146±0.001061
9	5	11								16	0.4583±0.0954	0.000868±0.000902

Locality number as given in Fig. 1 and Table 1.

Table 3. Pairwise F_{ST} values between *Clistocoeloma sinense* populations. Significant probabilities following collection with the FDR method for multiple tests were in bold type.

Locality Number	a	b	c	d	e	f	g	h	1	2	3	4	5	6	7	8
a																
b	−0.027															
c	−0.035	−0.023														
d	−0.009	−0.030	−0.001													
e	−0.032	−0.032	−0.028	−0.029												
f	−0.035	−0.023	−0.034	−0.001	−0.028											
g	−0.038	−0.029	−0.036	−0.013	−0.034	−0.036										
h	0.005	−0.024	0.014	−0.054	−0.021	0.014	0.000									
1	**0.540**	**0.465**	**0.555**	**0.399**	**0.498**	**0.555**	**0.532**	**0.368**								
2	**0.723**	**0.651**	**0.733**	**0.612**	**0.686**	**0.733**	**0.717**	**0.587**	0.032							
3	**0.601**	**0.524**	**0.616**	**0.464**	**0.559**	**0.616**	**0.594**	**0.433**	−0.027	0.000						
4	**0.394**	**0.326**	**0.410**	**0.251**	**0.353**	**0.413**	**0.385**	**0.220**	−0.021	0.088	0.004					
5	0.002	0.028	−0.003	0.096	0.029	−0.003	0.005	0.124	**0.645**	**0.811**	**0.706**	**0.501**				
6	**0.561**	**0.468**	**0.577**	**0.400**	**0.511**	**0.577**	**0.552**	**0.366**	−0.044	0.044	−0.034	−0.033	**0.689**			
7	0.000	−0.023	0.008	−0.048	−0.021	0.008	−0.004	−0.050	**0.394**	**0.601**	**0.458**	**0.252**	0.101	**0.394**		
8	0.165	0.104	**0.181**	0.039	0.123	**0.181**	0.157	0.016	0.155	**0.359**	**0.209**	0.056	**0.296**	0.132	0.038	
9	0.087	0.031	0.102	−0.024	0.045	0.102	0.080	−0.041	**0.257**	**0.488**	**0.322**	0.124	**0.240**	**0.244**	−0.025	−0.039

Clistocoeloma sinense has only three zoea and one megalopa stage [30,31], suggesting that the latter may have a shorter laraval duration and, therefore, reduced dispersal ability, compared with the other species. However, the larvae of some intertidal brachyuran crabs, even those with four or more zoea stages, remain near their spawning areas, subsequently recruiting to their original or nearby population [38,39]. Since the larvae of *Clistocoeloma sinense* may have a relatively short planktonic life, it is likely that they remain near their parental population, with little chance of extensive dispersal. This may well have resulted in the genetic differentiation apparent among the regional populations. In the enclosed water body of Tokyo Bay, however, little genetic differentiation was apparent among the discrete local populations, suggesting that strong gene flow occurred between them via larval dispersal. Short distance larval dispersal in an enclosed water body

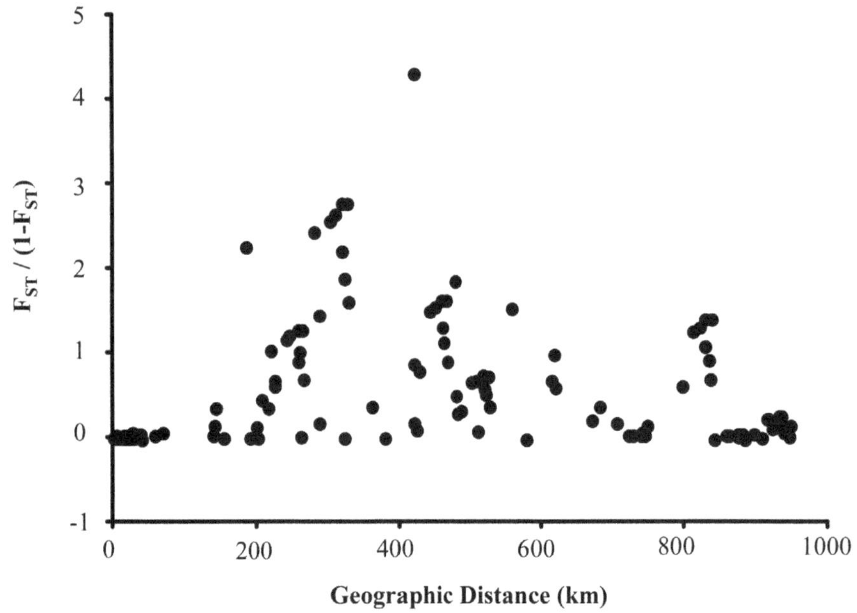

Figure 3. Isolation by distance of *Clistocoeloma sinense* samples. Genetic distances ($F_{ST}/1{-}F_{ST}$) plotted against geographical distances (minimal coastline distance) among between all localities.

Table 4. Neutrality tests (Tajima's D) and demographic estimates for mismatch distributions under the spatial expansion model (SSD, rg and τ).

Population	Neutrarity		Demographic		
Locality Number	Tajima's D	Fu's Fs	SSD	rg	τ
Tokyo Bay	−0.385	−0.433	**0.002**	0.327	0.287
1	−0.187	−0.114	0.017	0.184	0.606
2	−0.994	−1.128	0.001	0.295	0.315
3	−0.444	−0.382	0.000	0.177	0.460
4	−0.891	−1.742	0.020	0.164	0.880
5	−1.151	−1.155	0.000	0.740	0.172
6	0.870	1.039	0.010	0.203	0.608
7	0.480	0.847	0.003	0.216	0.459
8	0.173	0.153	0.021	0.177	0.765
9	1.034	1.096	0.014	0.217	0.678

SSD: sum of squared deviations between observed and expected distributions; rg: raggedness index; τ: expansion parameter. Significant values after sequential Bonferroni Method are indicated in bold.

may be adaptive for the maintaining of small local populations as a regional metapopulation.

No local populations of *Clistocoeloma sinense* were found along the coast between Tokyo Bay and Mikawa Bay (a distance of ca. 250 km), despite the existence of small saltmarshes habitat in some estuaries on Shizuoka Prefecture [40]. Some intertidal brachyuran crabs, including *Chiromantes dehaani* and *Eriocheir japonica* are, in contrast, distributed along this stretch of coastline, in Sagami Bay or Suruga Bay [28,35], which are both characterized by extensive beaches. These crab species are generalists in terms of habitat preference, utilizing various substrate types over wide intertidal range including fresh water zone. By comparison, *Clistocoeloma sinense* is estuarine habitat specialist, utilizing only the saltmarsh habitat in the highest intertidal zone. The discontinuity of local populations of some other habitat specialists, apparently also a consequence of the steep rocky shore lined Izu Peninsula [29], has been observed in other coastal marine animal species, including *Chaenogobius annularis*, *Ditrema jordani*, and *Deiratonotus cristatus*. Thus, suitable habitat continuity along the coast is important to *Clistocoeloma sinense* for the maintenance of local population networks of marine and brackish water habitat specialists.

The similar genetic structures of the widely separated Tokyo Bay and Kyushu populations of *Clistocoeloma sinense* is suggestive of recent gene flow between the two populations. Unintentional human mediated introduction of the crab to Tokyo Bay from Kyushu is another possibility, as suggested by the genetic analysis of Japanese populations of the venerid bivalve *Phacosoma japonicum* [41]. Sato [41] suggested that juvenile of *Phacosoma japonicum* appear to have been accidentally transported to Tokyo Bay from Ariake Bay (Western Kyushu) together with juvenile short neck clams intended for aquaculture on the tidal flats. Unintended introductions of animal species from foreign countries, including *Phacosoma japonicum* together with short neck clams, have been reported elsewhere, including the tidal flat purse crab *Philya pisum* and other tidal flat animals [42]. These introductions may have occurred when the short neck clams were caught. In the case of *Clistocoeloma sinense*, however, introduction mediated by short neck clam aquaculture is unlikely as these two species live in quite different habitats; the short neck clam lives in sandy sediment on middle to lower tidal flats whereas the crab lives in muddy sediment under higher salt marsh vegetation. Furthermore, aquaculture of transplanted short neck clam began post 1980s in Tokyo Bay [43], whereas Tokyo Bay populations of *Clistocoeloma sinense* have been found in 1952 [44].

Ballast water and ship hull fouling are thought to be major vectors for the introduction of non-indigenous marine animals [45]. In Tokyo Bay, introduced populations of the crabs, *Pyromaia tuberculata*, *Carcisus estuarii*, *Phithropanogeus harrisii* and *Acantholoblus pacificus*, following their transportation in ballast water and/or ship hull fouling, have been found since 1980 [46]. These above crabs

Table 5. Comparison of genetic diversity of mtDNA COI of intertidal brachyuran crabs. Every species was analyzed using the same primers in this study.

Species	Sample size	Number				Diversity		Source
		Locality	Haplotypes	Shared haplotypes	Unique haplotypes	Haplotype (h)	Nucleotide (π)	
Scylla serrata	77	4	24	5	19	0.370–0.850	0.1700–0.4600	[47]
Chiromantes dehaani	85	9	25	11	14	0.700–1.000	0.0016–0.0081	[28]
Neohelice granulata	69	11	34	3	31	0.454–0.955	0.0013–0.0096	[48]
Gaetice depressus	80	8	27	4	23	0.511–0.933	0.0012–0.0042	[28]
Ptychognathus ishii	76	6	20	4	16	0.455–0.889	0.0012–0.0024	Kawane et al. unpub.
Deiratonotus cristatus	113	13	46	15	31	0.000–0.889	0.0000–0.0230	[29]
Deiratonotus japonicus	71	10	34	5	29	0.400–1.000	0.0007–0.0064	[28]
Deiratonotus kaoriae	42	2	8	4	4	0.503–0.594	0.0010–0.0056	[49]
Ilyoplax pusilla	124	6	40	3	37	–	–	[36]
Macropthalmus banzai	124	6	83	30	53	0.673–1.000	0.0019–0.0116	[50]
Ocypode ceratophthalma	85	8	26	7	16	0.700–1.000	0.0085–0.0203	[28]
Clistocoeloma sinense	**418**	**17**	**9**	**3**	**6**	**0.071–0.609**	**0.00014–0.00139**	**Present study**

Every species was used by the same primers in this study.

are subtidal species that can live underwater on ship hulls, a habitat that is not suitable for intertidal crabs, such as *Clistocoeloma sinense*. The possibility of introduction of *Clistocoeloma sinense* in ballast water is also low, since most domestic Japanese ships before the 1950s did not have ballast water tanks. Accordingly, populations of *Clistocoeloma sinense* may have been established naturally before at least 1950, but reasons for the genetic similarity between the Tokyo Bay and Kyushu populations remain unknown. Furthermore, the low genetic diversity and demographic analysis results for these populations suggest that these largely separated local populations may have been established coincidently by the introduction of larvae from unknown another habitat. The haplotype similarity between the Tokyo Bay and Kyushu populations may have resulted from naturally occurring genetic drift. The present data, however, is insufficient for inferring population connectivity. Because microsatellite DNA analysis is likely to be a useful tool for estimating contemporary gene flow among populations, such should be considered in future studies aiming to elucidate the metapopulation structures of *Clistocoeloma sinense*.

Implications for Conservation of the Tokyo Bay Regional Populations

These findings provide the first substantial genetic information on the endangered sesarmid crab *Clistocoeloma sinense*, such being basic for future conservation and management of local populations of this species in Japanese waters. The Tokyo Bay regional populations of *Clistocoeloma sinense* may warrant special conservation status, as they are already exhibiting relatively low levels of diversity. By way of comparison, the tidal flat snail *Cerithidea cingulata*, formerly with a distribution pattern similar to that of *Clistocoeloma sinense*, occurring in both Tokyo Bay and Mikawa Bay until the 1990s, was not found in Tokyo Bay or along the coastline between the two bays [4]. Planktonic larvae of *Cerithidea cingulata* populations still extant in Mikawa Bay probably cannot disperse to Tokyo Bay due to the lack of suitable interjacent habitat, as in the case of *Clistocoeloma sinense*.

There would appear to be few barriers to the continuation of gene flow between the Tokyo Bay populations of *Clistocoeloma sinense* through larval dispersal. However, the preservation of extant habitats, restoration of saltmarshes and creation of new suitable habitats around the coastline of Tokyo Bay would likely be the most effective measures for conserving this endangered saltmash species.

Supporting Information

Figure S1 Isolation by distance of *Clistocoeloma sinense* samples. Genetic distances ($F_{ST}/1- F_{ST}$) plotted against geographical distances (minimal coastline distance) without data of Tokyo Bay (Locality Number a~h).

Acknowledgments

We gratefully thank the following persons who kindly provided information on the distribution of *Clistocoeloma sinense* during this study: A. Nomoto (IDEA Consultants, Inc.), T. Koga (Wakayama University), T. Kimura (Mie University) and S. Kawaida (Mie University) for providing the species distribution information during this study. We would like to thank M. Hamaguchi (National Research Institute of Fisheries and Environment of Inland Sea) and Y. Tatara (Toho University) for technical support with DNA analysis. We also thank A. Abe, K. Kato, Y. Koike and M. Tanaka (Toho University) for support during field investigations, and express our gratitude to K. Okoshi, M. Kagami and M. Taru (Toho University) for their valuable advice and suggestions. We also thank Dr. Graham S. Hardy (Ngunguru, New Zealand) revised an early draft of the manuscript. Special thanks are given to the Gyotoku Wildlife Sanctuary (Shinhama lagoon) and Tokyo Port Wild Bird Park for granting permits to conduct this study, which was also supported by the Mikimoto Fund for Marine Ecology.

Author Contributions

Conceived and designed the experiments: TY MK. Performed the experiments: TY MK. Analyzed the data: MK. Contributed reagents/materials/analysis tools: TY MK. Wrote the paper: TY MK. Performed major editorial support: TF.

References

1. Lotze HK, Lenihan HS, Bourque BJ, Bradbury RH, Cooke RG, et al. (2006) Depletion, degradation, and recovery potential of estuaries and coastal seas. Science 312: 1806–1809.
2. Japanese Association of Benthology (2012) Threatened Animal of Japanese Tidal Flats: Red Data Book of Seashore Benthos. Tokai University Press, Kanagawa. 285p (In Japanese).
3. Sato M (2010) Anthropogenic decline of the peculiar fauna of estuarine mudflats in Japan. Plank. Benth. Res 5 (Supplement): 202–213.
4. Furota T (2000) Extinction and conservation of mollusks in enclosed waters, a consideration from population decline in the batillariid snails in Tokyo Bay. Kaiyo (supplement) 20: 74–82 (In Japanese).
5. Furota T (2007) Tokyo Bay. In: Iijima A, editer. The 7th National Survey on the Natural Environment: Shallow Sea Survey (Tidal Flats). Biodiversity Center, Nature Conservation Bureau, Ministry of the Environment, Yamanashi Prefecture. 39–47 (In Japanese).
6. Sakai T (1976) Crabs of Japan and the Adjacent Seas, Kodansha, Tokyo. 773p (In Japanese).
7. Kamita T (1941) Studies of the Decapod Crustaceans of Chosen. The Fisheries Society of Chosen, Keiju. 289p (In Japanese).
8. Dai AY, Yang SL (1991) Crabs of the China Seas, China. Ocean Press, Beijing. 682p.
9. Hsueh PW, Huang JF (1996) A new record of *Clistocoeloma sinense* Shen, 1933, from Taiwan (Decapoda, Brachyura, Grapsidae, Sesarminae), with notes on its distribution and ecology. Crustaceana 69: 63–70.
10. Sexton JP, McIntyre PJ, Angert AL, Rice KJ (2009) Evolution and ecology of species range limits. Annual Review of Ecology, Evolution, and Systematics 40: 415–436.
11. Bucklin A (1992) Use of formalin-preserved samples for molecular analysis. Newsl Crustacean Mol Tech 2: 3.
12. Roehrdanz RL (1993) An improved primer for PCR amplification of mitochondrial DNA in a variety of insect species. Insect Mol Biol 2: 89–91.
13. Gopurenko D, Hughes JM, Keenan CP (1999) Mitochondrial DNA evidence for rapid colonization of the Indo-West Pacific by the mud crab *Scylla serrata*. Mar Biol 134: 227–233.
14. Excoffier L, Laval G, Schneider S (2006) Arlequin ver 3.1. An integrated software package for population genetics data analysis. Computational and Molecular Population Genetics Lab, University of Berne, Switzerland.
15. Excoffier L, Smouse P, Quattro J (1992) Analysis of molecular variance inferred from metric distances among DNA haplotypes: Application to human mitochondrial DNA restriction data. Genetics 131: 479–491.
16. Wright S (1951) The genetical structure of populations. Ann Eugen 15: 323–354.
17. Benjamini Y, Hochberg Y (1995) Controlling the false discovery rate: A practical and powerful approach to multiple testing. J R Statist Soc B 57: 289–300.
18. Tajima F (1989) Statistical method for testing the neutral mutation hypothesis by DNA polymorphism. Genetics 123: 585–595.
19. Tajima F (1989) The effect of change in population size on DNA polymorphism. Genetics 123: 597–601.
20. Fu YX (1997) Statistical tests of neutrality of mutations against population growth, hitchhiking and background selection. Genetics 147: 915–925.
21. Jensen JL, Bohonak AJ, Kelley ST (2005) Isolation by Distance Web Service.
22. Grant WS, Bowen BW (1998) Shallow population histories in deep evolutionary lineages of marine fishes: insights from sardines and anchovies and lessons for conservation. Heredity 89: 415–426.
23. Kojima S, Kamimura S, Iijima A, Kimura T, Mori K, et al. (2005) Phylogeography of the endangered tideland snail *Batillaria zonalis* in the Japanese and Ryukyu Islands. Ecol Res 20: 686–694.
24. Imai H, Numachi K (2002) Genetic variability of swimming crab based on PCR-RFLP analysis of mitochondrial DNA d-loop region. Suisanzoshoku 50: 1–7 (In Japanese with English abstract.)
25. Aoki M, Naruse T, Cheng HJ, Suzuki Y, Imai H (2008) Low genetic variability in an endangered population of fiddler crab *Uca arcuata* on Okinawajima Island: analysis of mitochondrial DNA. Fish Sci 74: 330–340.

26. Kumar G, Kunal SP, Menezes MR, Meena RM (2012) Single genetic stock of kawakawa *Euthynnus affinis* (Cantor, 1849) along the Indian coast inferred from sequence analyses of mitochondrial DNA D-loop region. Conserv Genet 13: 1119–1131.
27. Palumbi SR (1992) Marine speciation on a small planet. Trends in Ecology and Evolution 7: 114–118.
28. Kawane M, Wada K, Watanabe K (2008) Comparisons of genetic population structures in four intertidal brachyuran species of contrasting habitat characteristics. Mar Biol 156: 193–203.
29. Kawamoto M, Wada K, Kawane M, Kamada M (2012) Population subdivision of the brackish-water crab *Deiratonotus cristatus* on the Japanese coast. Zool Sci 29: 21–29.
30. Saba M (1972) "Umore-benkeigani" (*Clistocoeloma merguiense* de Man) no koukihassei. Mie-seibutsushi 22: 25–29 (In Japanese).
31. Cuesta JA, Guerao G, Liu HC, Schubart CD (2006) Morphology of the first zoeal stages of eleven Sesarmidae (Crustacea, Brachyura, Thoracotremata) from the Indo-West Pacific, with a summary of familial larval characters. Invertebr Reprod Dev 49: 151–173.
32. Hirase S, Ikeda M, Kanno M, Kijima A (2012) Phylogeography of the intertidal goby *Chaenogobius annularis* associated with paleoenvironmental changes around the Japanese Archipelago. Mar Ecol Prog Ser 450: 167–179.
33. Katafuchi H, Kai Y, Nakabo T (2011) Genetic divergence in *Ditrema jordani* (Perciformes: Embiotocidae) from the Pacific coast of southern Japan inferred from mitochondrial DNA sequences. Ichthyol Res 58: 90–94.
34. Mukai T, Sugimoto M (2006) Genetic population structure of the mudskipper, *Periophthalmus modestus*, in Japan inferred from mitochondrial DNA sequence variations. Japan J Ichthy 53: 151–158 (In Japanese with English abstract).
35. Yamasaki I, Yoshizaki G, Yokota M, Strüssmann CA, Watanabe S (2006) Mitochondrial DNA variation and population structure of the Japanese mitten crab *Eriocheir japonica*. Fish Sci 72: 299–309.
36. Yamada A, Furukawa F, Wada K (2009) Geographical variations in waving display and barricade-building behaviour, and genetic population structure in the intertidal brachyuran crab *Ilyoplax pusilla* (de Haan, 1835). J Nat Hist 43: 17–34.
37. Yatsuzuka K (1957) Study of brachyuran zoea, artificial rearing and development. Suisangaku Shusei: 571–591 (In Japanese).
38. Jones MB, Epifanio CE (2005) Patches of crab megalopae in the mouth of Delaware Bay: An analysis of spatial scales. J Shellfish Res 24: 261–267.
39. Díaz-Cabrera E, Hernández-Miranda E, Hernández CE, Quiñones RA (2012) Mesoscale β diversity and spatial nestedness of crustacean larvae in the coastal zone off central southern Chile: population and community implications. ICES J Mar Sci 69: 429–438.
40. Murase A, Yuhara T (2012) Record of an endangered gastropod, *Cerithidea rhizophorarum*, from estuarine habitat of Ogamo River, southern Izu Peninsula, central Japan, Bull Biogeogr Soc Jpn 67: 261–264 (In Japanese with English abstract).
41. Sato S (1996) Genetic variability and population structure of *Phacosoma japonicum* (Bivalvia: Veneridae). Venus 55: 51–63.
42. Okoshi K (2004) Alien species introduced with imported clams: the clam-eating moon snail *Euspira fortunei* and other unintentionally introduced species. Jpn J Benth 59: 74–82 (In Japanese with English abstract).
43. Toba M (2002) Present status of the fishery for the clam *Ruditapes philippinarum* in Chiba Prefecture, Tokyo Bay. Jpn J Benth 57: 145–150 (In Japanese with English abstract).
44. Gamo S (1965) An Ecological Study of the Crabs inhabiting the Estuaries in southern Kanto District. Res Crust 2: 91–101 (In Japanese with English abstract).
45. Lockwood JL, Hoopes MF, Marchetti MP (2007) Invasion Ecology. Oxford, UK: Blackwell Publishing. 304 p.
46. Komai T, Furota T (2013) A new introduced crab in the western North Pacific: *Acanthobulus pacificus* (Crustacea: Decapoda: Brachuruya: Panopeidae), collected from Tokyo Bay, Japan. Mar Biodivers Rec 6: e96.
47. Fratini S, Vannin M (2002) Genetic differentiation in the mud crab *Scylla serrata* (Decapoda: Portunidae) within the Indian Ocean. J Exp Mar Biol and Ecol 272: 103–116.
48. Ituarte RB, D'Anatro A, Luppi TA, Ribeiro PD, Spivak ED, et al. (2012) Population structure of the SW Atlantic estuarine crab *Neohelice granulata* throughout its range: a genetic and morphometric study. Estuar Coast 35: 1249–1260.
49. Kawane M, Wada K, Umemoto A, Miura T (2012) Genetic population structure and life history characteristics of the rare brackish-water crab *Deiratonotus kaoriae* Miura, Kawane and Wada, 2007 (Brachyura: Camptandriidae) in western Japan. J Crust Biol 32: 119–125.
50. Aoki M, Nakagawa Y, Kawamoto M, Wada K (2012) Population divergence of the sentinel crab *Macrophthalmus banzai* is consistent with regional breeding season differences. Zool Sci 29: 821–826.

4

Characterizing the Phylogenetic Tree Community Structure of a Protected Tropical Rain Forest Area in Cameroon

Stéphanie Manel[1,2*], Thomas L. P. Couvreur[3,4], François Munoz[5], Pierre Couteron[6], Olivier J. Hardy[7], Bonaventure Sonké[4]

1 Aix Marseille Université, IRD, LPED UMR 151, Marseille, France, **2** Centre de coopération internationale en recherche agronomique pour le développement, UMR AMAP, Montpellier, France, **3** Institut de Recherche pour le Développement, UMR DIADE, Montpellier, France, **4** Département des Sciences Biologiques, Laboratoire de Botanique Systématique et d'Ecologie, Université de Yaoundé I, Ecole Normale Supérieure, Yaoundé, Cameroon, **5** Université Montpellier 2, UMR AMAP, Montpellier, France, **6** Institut de Recherche Pour le Développement, UMR AMAP, Montpellier, France, **7** Université Libre de Bruxelles, Faculté des Sciences, Evolutionary Biology and Ecology Brussels, Belgium

Abstract

Tropical rain forests, the richest terrestrial ecosystems in biodiversity on Earth are highly threatened by global changes. This paper aims to infer the mechanisms governing species tree assemblages by characterizing the phylogenetic structure of a tropical rain forest in a protected area of the Congo Basin, the Dja Faunal Reserve (Cameroon). We re-analyzed a dataset of 11538 individuals belonging to 372 taxa found along nine transects spanning five habitat types. We generated a dated phylogenetic tree including all sampled taxa to partition the phylogenetic diversity of the nine transects into alpha and beta components at the level of the transects and of the habitat types. The variation in phylogenetic composition among transects did not deviate from a random pattern at the scale of the Dja Faunal Reserve, probably due to a common history and weak environmental variation across the park. This lack of phylogenetic structure combined with an isolation-by-distance pattern of taxonomic diversity suggests that neutral dispersal limitation is a major driver of community assembly in the Dja. To assess any lack of sensitivity to the variation in habitat types, we restricted the analyses of transects to the terra firme primary forest and found results consistent with those of the whole dataset at the level of the transects. Additionally to previous analyses, we detected a weak but significant phylogenetic turnover among habitat types, suggesting that species sort in varying environments, even though it is not predominating on the overall phylogenetic structure. Finer analyses of clades indicated a signal of clustering for species from the Annonaceae family, while species from the Apocynaceae family indicated overdispersion. These results can contribute to the conservation of the park by improving our understanding of the processes dictating community assembly in these hyperdiverse but threatened regions of the world.

Editor: Andrew Hector, University of Oxford, United Kingdom

Funding: PC and BS are partly funded by "PEERS ISMOBIAC" through support provided by the Agence inter-établissements de recherche pour le développement - Programme Doctorant du Sud (AIRD-DPF). The funders had no role in study design, data collection and analysis, decision to publish or preparation manuscript.

Competing Interests: The authors have declared that no competing interest exist.

* Email: stephanie.manel@univ-amu.fr

These authors contributed equally to this work.

Introduction

Tropical rain forests are the most biodiverse terrestrial ecosystems on Earth, containing over 50% of known terrestrial biodiversity packed in just 7-10% of the Earth's surface [1,2]. The level of biodiversity is remarkable both locally (alpha diversity) and in terms of variation in space (beta diversity; e.g. [3,4]). Despite the importance of rain forests for terrestrial biodiversity, the drivers of diversity gradients within and between the world's main rain forest areas remain poorly understood [5,6]. In addition, large areas of rain forests, among which the central African block, are poorly explored by scientists, giving a fragmentary view of spatial diversity patterns, even for well-studied organisms such as plants [7].

In the last century, rain forests have been overexploited in many parts of the world leading to their alteration, fragmentation, and in some areas, complete destruction [7]. The consequences of these changes include biodiversity loss and increased atmospheric carbon dioxide concentration resulting in climate change, due to the conversion of high-carbon storage forest to low-carbon storage agriculture [8]. As a consequence, there is an urgent need to better understand the processes that sustain the biological diversity in tropical rain forests [9].

Biodiversity is classically assessed at species level (e.g. [10]), from the observation of the presence/absence of species (i.e. species occurrence) or species abundance in transects or ecological plot surveys (e.g. [11]). However, biodiversity assessments based on species counts and their relative abundance statistics provide little information regarding the functional diversity of the ecosystem under study, since they do not acknowledge the variation in their ecological niches [12,13,14,15]. Estimation of phylogenetic or functional diversity in addition to species diversity has been recognized as improving our understanding of the niche-based

Table 1. Permutation used to test taxonomic and phylogenetic structure in relation with our hypotheses.

Hypothesis	Permutations	Tests
(i) If community assembly is dominated by limited dispersal, no phylogenetic structure should be detected among transects, while isolation-by-distance is expected in taxonomic beta diversity	Model 1-3x Whole dataset - transect. Permutation of individuals between transects within habitat types.	I_{ST}
	Model 1 s Whole dataset-transect Permutation of species in phylogeny	B_{ST} - Π_{ST}
(i) Isolation-by-distance is expected in taxonomic beta diversity.	Model 2–3x Whole dataset -habitat. Permutations of individuals among habitats	I_{ST}
(ii) If environmental filtering differed among habitat types a pattern of phylogenetic clustering between habitats should be detected	Model 1s - Whole dataset-habitat. Permutations of species in phylogeny	B_{ST} - Π_{ST}
(iv) Environmental filtering and competitive exclusion may simultaneously occur and cancel out to yield apparently "neutral" patterns		
(iii) If competitive exclusion/niche differentiation prevents the co-occurrence of related species locally, generating a patchwork distribution of functionally equivalent species, phylogenetic overdispersion might be detected within transects, at least within a habitat type	Model 1 s -TPF. Permutations of species in phylogeny within TPF only	B_{ST} - Π_{ST}
(iv) Environmental filtering and competitive exclusion may simultaneously occur and cancel out to yield apparently "neutral" patterns		

processes leading to the observed patterns of present day biodiversity [15,16]. Those estimations also help to better conserve phylogenetic diversity (e.g. [14,15]). In this context, phylogenetic relatedness is classically considered as a proxy of functional relatedness, because the closer the species are in the phylogeny, the more likely they have inherited similar traits from a common ancestor. As a consequence the consideration of phylogenetic diversity informs on ecosystem functioning and adaptability. This 'ecophylogenetic' approach is therefore a relevant basis for conservation purposes [17].

Recently, progress in phylogeny reconstruction either from DNA sequences or from existing resources has accelerated theoretical and methodological advances in ecophylogenetics [18]. The number of studies on phylogenetic alpha and beta diversity in tropical tree communities has increased in the last two or three years [9]. Those studies have focused on partitioning diversity into spatial and environmental components (e.g. [19,20,21]) or by comparing the alpha and beta components of phylogenetic diversity (e.g. [18,20]).

Recent studies carried out on tropical rain forest trees reported that (i) phylogenetic turnover (i.e. a spatial turnover of the dominance of clades) is associated with habitat or environmental differentiation [19,20,22,23]; (ii) functional traits [22] and climate niche proxies [23] usually display a significant, although sometimes weak, phylogenetic structure; (iii) phylogenetic turnover parallels functional turnover, although with a weaker strength [22]. It is noteworthy that most of these studies were carried out in regions with substantial climatic and/or edaphic gradients, so that environmental filtering effects favoring functional and phylogenetic clustering may predominate over competitive exclusion effects that might lead to functional and phylogentic overdispersion (but see [24]).

In this paper we will focus on a region located in the margin of the Congo Basin which is home to the second largest tropical rain forest after the Amazon basin, with a high level of species diversity and endemism [25]. In recent years, several countries have created national parks in an effort to conserve rain forest biodiversity (e.g. Gabon). The Dja Faunal Reserve (DFR) is a UNESCO world heritage reserve of 526,000 ha located 250 kilometers south east from the Cameroon capital Yaoundé. The reserve was established in 1950 and is the largest protected rain forest in Cameroon [26]. There is no steep climatic or edaphic macrogeographic gradient across the DFR. However, its topography characterized by half-orange shaped hills generates a heterogeneity of soil hydromorphy, so that distinct edaphic habitats can be recognized. Besides, natural or past human disturbances (gap dynamics; ancient agricultural fields) are recognized locally by affecting the structure of the vegetation. In a previous study, Hardy and Sonké [27] assessed the role of dispersal and niche differentiation in shaping tree species turnover along nine transects covering the DFR. To this end, they quantified the impact of spatial distance and habitat differentiation on the probability that pairs of individuals are conspecific. With the exception of pioneer species, they found a pattern of isolation by distance, i.e. spatial species clustering due to the combined effect of limited dispersal and local ecological drift [10]. Habitat differentiation was also found to be a major determinant of the spatial pattern but had a lower impact than spatial distance per se. These results suggest that in the DRF, the degree of species aggregation might be better determined by dispersal-assembly rules rather than by niche-assembly rules, at least for the common species. Because this study, which was based on taxonomic diversity, did not take into account phylogenetic diversity between species, a reanalysis of the dataset used in Hardy and Sonke [27], but accounting for phylogenetic variation will provide further insights in how niche-based processes constrains the composition of communities [23]. Because macrogeographic environmental gradients are weak across the DFR compared to most previous works carried out at a similar scale in tropical rain forests, patterns of phylogenetic structure might differ.

The main objective of the paper was to investigate the relative importance of niche-based and dispersal-based processes governing tree species assemblage within the Dja Faunal Reserve. We used an ecophylogenetic approach [14,15] to provide a phylogenetic quantification of biodiversity in this area for better conservation strategies. We wanted to address the question: can we detect phylogenetic or species turnover across the reserve? Specifically, we elaborated our approach to test the following four hypotheses:

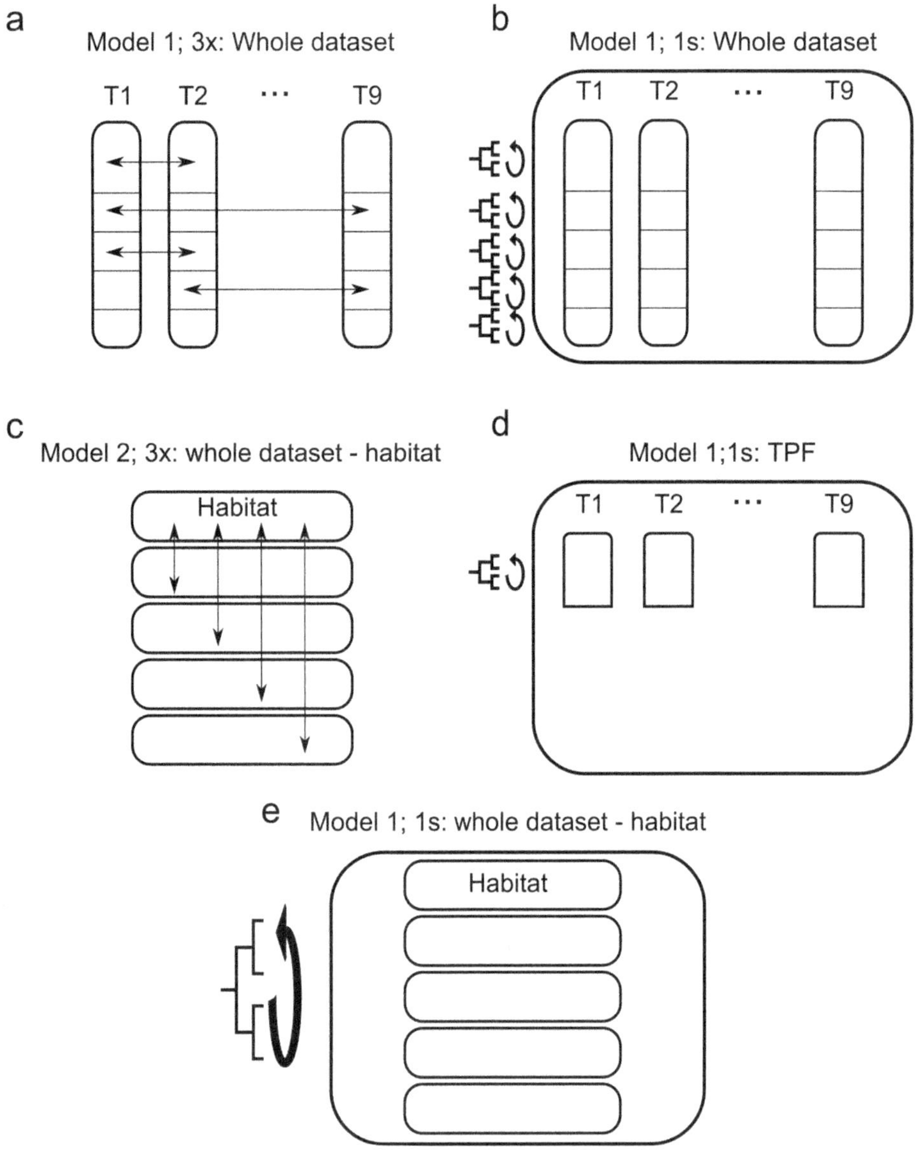

Figure 1. Schema of the 3 types of randomization used to test taxonomic and phylogenetic structure. In model 1-3x, individuals were randomized among transects or species within each habitat type (a). In model 2–3x, individuals or species were randomized among habitats (c). These models of permutation aimed to test for taxonomic turnover using I_{ST}. Phylogenetic structure (B_{ST} and Π_{ST}) was tested using a model 1 s which randomized the observed species across the tips of the phylogenetic tree (b, d, e). Randomization were respectively done for the whole dataset (a,b), for the habitat data set (c, e) and for TPF only (d).

(i) If community assembly is dominated by limited dispersal, no phylogenetic structure should be detected between transects, while isolation-by-distance is expected in taxonomic beta diversity.

(ii) If environmental filtering differed among habitat types a pattern of phylogenetic clustering in habitat should be detected. This interpretation assumes phylogenetic niche conservatism of relevant traits [22,23].

(iii) If competitive exclusion prevents the co-occurrence of related species locally, generating a patchwork distribution of functionally equivalent species, phylogenetic overdispersion might be detected within transects, at least within a habitat type.

(iv) Environmental filtering and competitive exclusion may simultaneously occur and cancel each other out to yield apparent "neutral" patterns.

A critical issue for testing the relative imprint of these processes is to define sampling units that are relevant according to the scale of the processes. Here we considered the nine transects of Hardy and Sonké [27] that provide information on forest tree composition at two levels: among transects and among habitat types.

To address the four hypotheses above, we therefore partitioned taxonomic and phylogenetic diversity within and between transects, as well as within and between habitat types. For this we applied the statistical framework developed by Hardy and Senterre [20] for characterizing and testing the phylogenetic structure of transects and habitats types using appropriate randomization procedures [28] (Table 1, Figure 1).

Materials and Methods

1. Study site and tree communities

The Dja Faunal Reserve (DFR) is situated between latitudes 2°50'–3°30' N and longitudes 12°20'–13°40 E in southeastern Cameroon. About two-thirds of the reserve's perimeter is demarcated by the Dja River, forming a natural boundary. In this reserve, tree species have been inventoried along nine 5 km long and 5 m wide transects (Figure 2), in which all species with a diameter at breast height bigger than 10 cm were identified and mapped. This dataset (sampling, preliminary taxonomic analysis) is described in Sonké [29] and Sonké and Couvreur [30]. In total 11546 individuals were inventoried belonging to 312 identified species and 60 taxa identified to generic level only (and considered as a morphospecies, 372 total taxa included in the analyses). All nomenclatural criteria regarding species names and families followed Sonké and Couvreur [30]. The vegetation in the reserve has a 30–40 m tall canopy with emergent trees rising up to 60 m [26]. Detailed descriptions of the structure and species composition of the mixed forests can be found in Sonké [31] and Sonké and Couvreur [30].

The nine transects were established across five broad types of forest (i.e. habitat) depending on soil hydromorphy and vegetation structure [32]. Terra firme forests were established on non-hydromorphic soils and subdivided into three successional types, namely (i) terra firme primary forest (74% of total individuals); (ii) secondary forest (8% of total individuals) ; (iii) gaps (4% of total individuals). Conversely, two hydromorphic types were defined as (iv) swamps (11%) and (v) flooded forest (3%).

We considered two sampling levels for subsequent phylogenetic and taxonomic composition analyses. First, we analyzed the variation within and between the 9 transects, such as the comparison of transects represented the largest spatial scale of beta variation. Second, the variation across the five habitats within transects conveyed the imprint of local niche-based processes. All analyses were performed on three datasets: 1) all individuals within each transect (= whole data - transect); 2) only individuals found in habitat (i) terra firme primary forest = ("TPF") within each transect; 3) the five different habitats were treated independently of transects (i.e. transects were merged = "habitat") (= whole data – habitat). This allowed us to test the different assumptions stated in the introduction (Table 1). Our analysis is different than the one of Hardy and Sonké [27] who estimated the probability that two randomly chosen individuals belong to the same species according to the distance separating them (1) on the combined samples of all transects using individual coordinates and then (2) within the three main habitat types.

2. Phylogenetic relationships of the DFR tree community

A phylogenetic tree of 372 species was generated in three steps. First, all species were grafted onto a comprehensive phylogenetic tree using the program PHYLOMATIC v3 [33] (http://

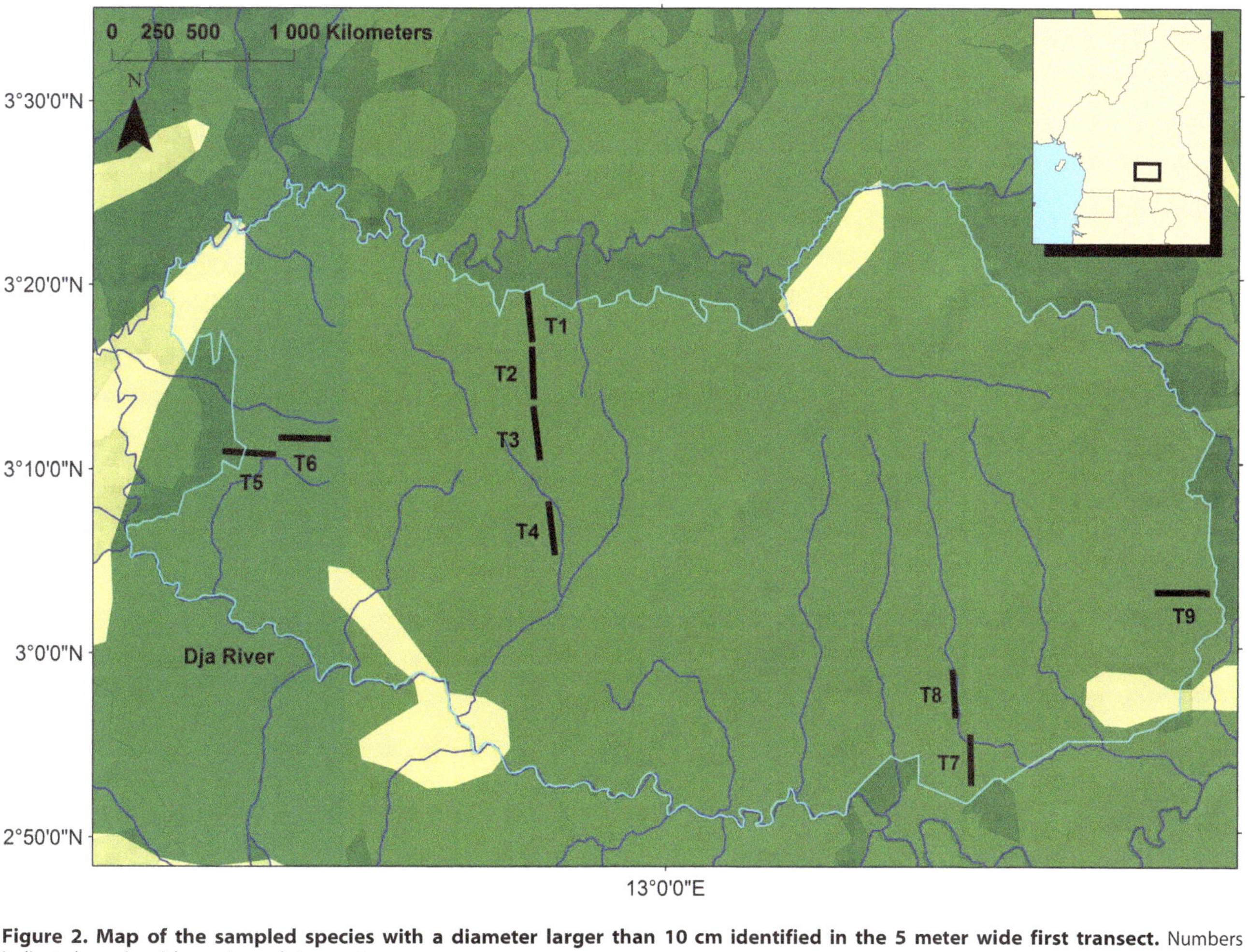

Figure 2. Map of the sampled species with a diameter larger than 10 cm identified in the 5 meter wide first transect. Numbers indicated species richness in each transect.

phylodiversity.net/phylomatic/). The program generated a tree in which the family relationships of the sampled species followed the angiosperm phylogeny APG III [34] version R20120829. We then manually resolved the generic relationships within most of the families based on specific molecular phylogenies (Table 2), using the software Mesquite [35]. Only relationships that were supported with bootstrap values of more than 70% were taken into account. For families where no phylogenetic information was available or for which the published phylogeny did not provide enough insights into the relationships between genera, generic relationships where left unresolved (polytomies). Finally, we used the branch adjustment algorithm BLADJ implemented in Phylocom [36] to scale the branch lengths based on a set of node age estimates from several publications (Table 2). For this part, we first used the dated phylogeny of Wikstrom et al [37] for major nodes. We also used family level dated trees to further constrain certain nodes (Table 2). Intra- and interspecific branch lengths were assumed to be 0 (i.e. relationships between species and within species are unknown and unresolved).

3. Species and phylogenetic structure analyses

We used the measures of phylogenetic distinctness and differentiation within and between transects /habitats introduced by Hardy & Senterre [20]. These statistics are based on the additive partitioning of Rao entropy [38], which lead to differentiation coefficients between transects/habitats that are analogous to the coefficients expressing genetic differentiation among populations in population genetics.

Tests of phylogenetic structure can be biased when there is a non-random phylogenetic distribution of species abundance at regional scale (i.e., in the overall dataset) [28]. Therefore, to test if abundant species were randomly distributed across the phylogeny, we first calculated the Abundance Phylogenetic Deviation (APD) statistic [28]. When APD <0, species abundances are over-dispersed, whereas when APD >0, species abundances are clustered (abundant species mainly belong to one or a few clades).

We re-estimated taxonomic diversity for phylogenetic analyses since we used a different strategy from the one used in Hardy and Sonké [27]. We calculated the probabilities that two individuals belonged to different species (Simpson-Gini diversity index) within a transect/habitat (D_{IS}) and between transects/habitats (D_{IT}), as well as the mean phylogenetic distances (based on the divergence time) between individuals (an index of phylogenetic diversity) within transect/habitat (D_{PS}), and between transects/habitats (D_{PT}). $I_{ST} = (D_{IT}\text{-}D_{IS})/D_{IT}$ then expresses the species turnover between transects/habitats. Taxonomic clustering in transects/habitat is expected to be reflected by $I_{ST}>0$, while taxonomic overdispersion should result in $I_{ST}<0$. $P_{ST} = (D_{PT}\text{-}D_{PS})/D_{PT}$ expresses the combined effect of species and phylogenetic turnover. However, as it is difficult to interpret, we do not consider this quantity in our interpretations. In addition, we estimated the mean phylogenetic distances between two non-conspecific individuals sampled at local (i.e. within transect/habitat) or regional scale (i.e. between transects/habitats), respectively denoted as D_{BS} and D_{BT}, so that $B_{ST} = (D_{BT}\text{-}D_{BS})/D_{BT}$ expressed phylogenetic turnover between transects/habitats independently of species turnover [38]. $B_{ST} >0$ under local phylogenetic clustering while $B_{ST} <0$ under local phylogenetic overdispersion.

These estimators require abundance data, and rare species are underemphasized, while the distribution of rare species can also bring useful information on species assembly rules. Thus, we used measures of phylogenetic distinctness based on species incidence [20]. Δ_{PS} is defined as the mean phylogenetic distance between distinct species within transects/habitats and Δ_{PT} between transects/habitats (i.e. mean phylogenetic distance between

Table 2. References to phylogenetic trees and chronograms used to manually resolve relationships and identify calibration points in families with three or more species sampled in this study.

Family	Phylogenetic relationships	Calibration points
angiosperms		[37]
Annonaceae	[52]	[48]
Apocynaceae	[58]	
Bignoniaceae	[59]	
Chrysobalanceae	[60]	
Clusiaceae	[61]	
Combretaceae	[62]	
Euphorbiaceae	[63]	
Leguminosea: Caesalpinoideae	[64];	
Leguminosea: Mimosoideae	[65]	[65]
Malvaceae	[66] and http://www.malvaceae.info	
Meliaceae	[67,68]	
Myristicaceae	[69]	
Mytaceae	[70]	
Olecaceae	[71]	
Phyllantaceae	[72]	
Rubiaceae	[73]	[73]
Rutaceae	[74]	
Sapindaceae	[75]	[75]

distinct species sampled from two transects/habitats, averaged over all pairs of transects/habitats). Hence a coefficient analogue to B_{ST} is defined, $\Pi_{ST} = (\Delta_{PT}-\Delta_{PS})/\Delta_{PT}$ which expresses phylogenetic turnover by the gain of phylogenic distance between species occurring in different sites compared with species occurring in the same site. Π_{ST} is equivalent to B_{ST} but neglects species abundances. This coefficient excludes comparisons of a species with itself. All the estimations were performed using the software SPACoDi [39].

4. Testing species and phylogenetic structure

To test for species turnover and phylogenetic structure, we used 3 models of randomization (Table 1, Figure 1). In model 1-3x, we randomized individuals among transects or species within each habitat type (Figure 1a). In model 2–3x, we randomized individuals or species among habitats (Figure 1c). These models of permutation aimed to test for taxonomic turnover using I_{ST}. The models 3x in which community composition was randomized but not the position of taxa in the phylogeny has been shown to be biased to test the phylogenetic structure [28]. Thus, to test for phylogenetic structure, we used a third permutation model (model 1s, Figure 1 b, d, e). The model 1s randomizes the observed species across the tips of the phylogenetic tree and allowed testing for phylogenetic structure using B_{ST} and Π_{ST} (Table 1).

We undertook 999 permutations for each model, providing 999 estimations of the above differentiation coefficients under those null models. Deviations of observed coefficient from random coefficients were used to test whether $I_{ST}=0$, $B_{ST}=0$ or $\Pi_{ST}=0$. A significant test for I_{ST} is expected at least under hypothesis (i) between transects or habitats with the model 1–3x. Under hypothesis (ii) (habitat filtering dominates), we expect $B_{ST}>0$ and $\Pi_{ST}>0$ with the whole dataset- habitat (model 1s); under hypothesis (iii) (competitive exclusion between related species dominates), we expect $B_{ST}<0$ and $\Pi_{ST}<0$ between transects at least within TPF habitat; while no phylogenetic significant tests should be obtained under hypotheses (i, neutral assembly with limited dispersal) and (iv, compensation between ii and iii) (Table 1). Mantel tests were used to test the relations between pairwise taxonomic (I_{ST}) or phylogenetic distances (B_{ST} and Π_{ST}) and geographic distances among the 9 transects using the R package vegan [40]. A significant test with I_{ST} but not with B_{ST} or Π_{ST} is expected under hypothesis (i).

Finally, to assess the robustness of the results with respect to the taxonomic scale investigated, and possibly assess whether hypothesis (iv, compensatory effects between habitat filtering and competitive exclusion) might hold, partial randomization of the data between transects was performed on certain clades defined as species rich which were arbitrarily defined as families containing 10 or more sampled species. We also looked at Eudicot and Magnoliales clades. For each clade the coefficients described above were calculated under the 1s model (999 randomization of tree tips). This was done by using the spacodi.per.nodes function in the SpacodiR [41].

Results

1. Phylogenetic tree of the DFR

For 17 families, phylogenetic studies allowed the resolution of most relationships between genera (Table 2). A total of 23 calibration points (Table 2) were used to generate the dated phylogenetic tree of the DFR. The tree was produced using the iTOL web application [42,43] (Figure 3).

2. Species and phylogenetic structure analyses

The Abundance Phylogenetic Deviation (APD) estimations were not significant for the 3 datasets (Table 3) indicating that abundant species were randomly distributed across the phylogeny at the scale of the study area.

The probability that two randomly selected individuals belonged to different species (D_{IS}) was high for all our 3 datasets (0.9805, 0.9789 and 0.9798 for the whole dataset-transect, TPF and the whole dataset- habitat respectively) (Table 2). The mean divergence time between individuals was $D_{PS}=129.49$ million years (Myr), 129.83 Myr and 126.84 Myr respectively for the three datasets. The mean divergence times between species (Δ_{PS}) was 129.75 Myr, 130.01 Myr and 128.97 Myr (Table 2). According to these coefficients, most diversity occurred within transect or habitat, the between contribution being always less than or equal to 0.6% for the whole dataset-transect ($I_{ST}=0.0067$, $B_{ST}=0.0002$, $\Pi_{ST}=0.0002$), 0.7% for TPF ($I_{ST}=0.0079$, $B_{ST}=0.0002$, $\Pi_{ST}=0.0004$) and 1.1% for the whole dataset-habitat ($I_{ST}=0.0079$, $B_{ST}=0.0032$, $\Pi_{ST}=0.0005$). When the coefficients were calculated using taxonomic ranks to produce a surrogate of phyletic distances, estimates of phylogenetic distinctness (Δ_{PS} and Δ_{PT}) were only slightly different, and estimates of phylogenetic differentiation between transects (Π_{ST}) were only slightly affected (Table 3).

The distribution of divergence times between individuals within a transect and within habitat showed that more than half of the pairs of individuals diverged between 160 and 179 Myr ago (Figure 4).

3. Testing species and phylogenetic structure

Whatever the randomization model used (models 1 or 2–3x), all the tests done on taxonomic differentiation were significant ($I_{ST}>0$), indicating species turnover (Table 4). In the case of model 1–3 x, on the whole dataset at transect level, the test indicated limited dispersal between transects within habitat. The model 2–3x on the whole dataset at habitat level suggested that the turnover is also due to a habitat effect because in the habitat dataset geographic distances are meaningless.

Concerning tests on phylogenetic structure, we found no significant deviations from random of the observed phylogenetic statistics under model 1 s at the transect level, so that $B_{ST}=0$ and $\Pi_{ST}=0$ for the whole dataset-transect and for the TPF dataset. When analyzing the whole dataset at habitat level, the model 1 s still indicated $\Pi_{ST}=0$, but (marginally) significant $B_{ST}>0$ indicating low phylogenetic turnover for abundant species among habitats (Table 4).

Mantel tests indicated that pairwise taxonomic distances (I_{ST}, species turnover across space) were significantly correlated to pairwise geographic distances both for the whole dataset-transect as well as for TPF only ($r=0.74$, p-val $=9\times10^{-4}$ and $r=0.77$, p-val $=0.0013$). The relations between pairwise phylogenetic and geographic distances were never significant.

4. Variation in phylogenetic structure across the phylogeny

Ten families (sensu AGP III) contained ten or more species [30]. Out of those, only two families exhibited a significant phylogenetic turnover between transects (I_{ST} or P_{ST}) under the null model 1 s: Annonaceae ($B_{ST}=0.0009$, p-value $=0.04$; $I_{ST}=0.0256$) and Apocynaceae ($B_{ST}=-0.0004$, p-value $=0.019$; $I_{ST}=0.0241$). Hence, small but significant phylogenetic clustering was identified in Annonaceae (as B_{ST} is positive); whereas overdispersion was detected in Apocynaceae (as B_{ST} is negative). This indicated that

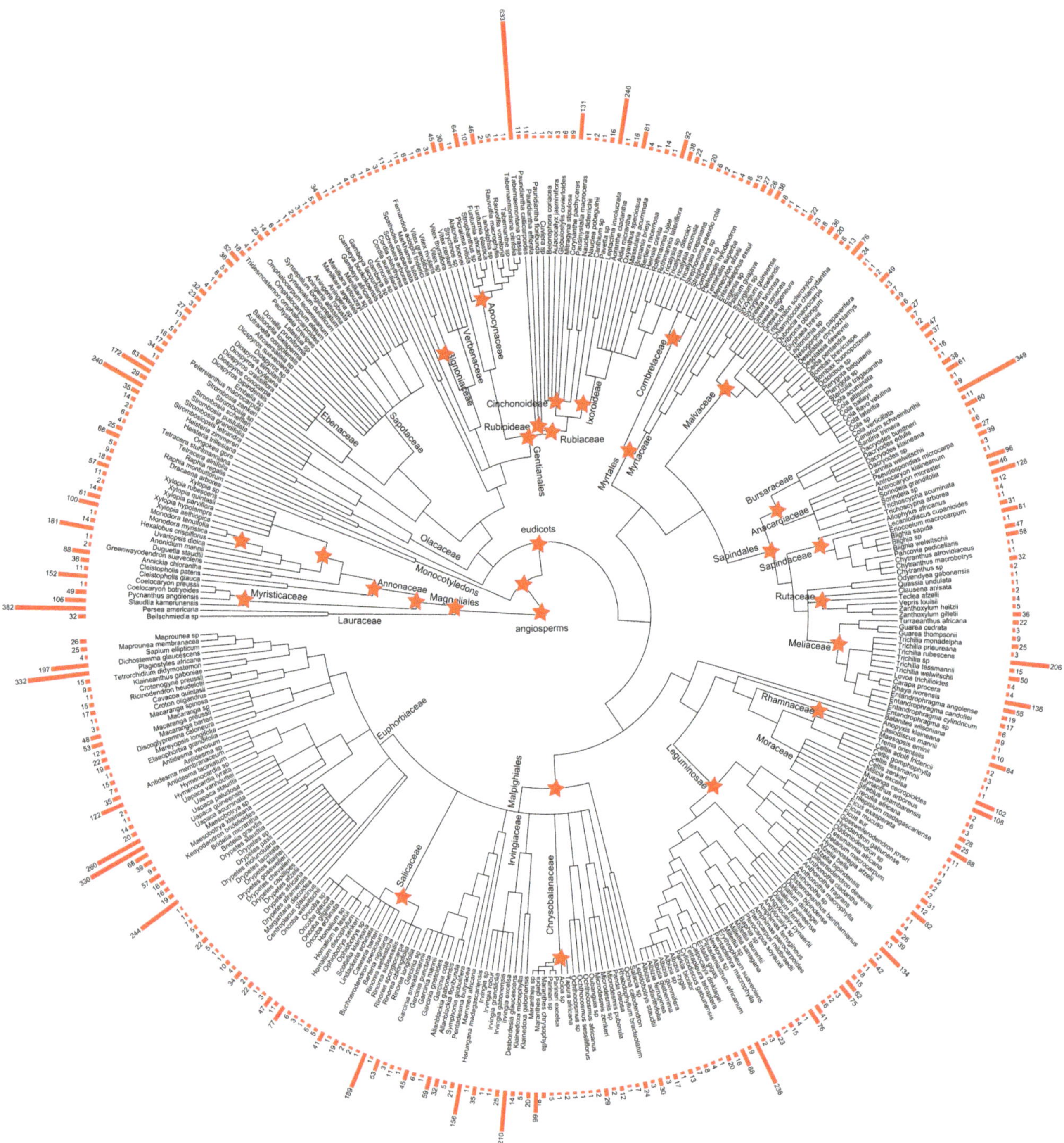

Figure 3. Chronogram of the tree flora of the Dja Faunal Reserve. Branches are proportional to time. Red stars indicate calibration points used in the BALDJ analysis. Species names are indicated in the tips, with their respective abundance information for the whole dataset. Graphic created with using the iTOL web application.

Annonaceae species were more related within transects than species taken from different transects, while Apocynaceae species were more related among than within transects. Finally, our results show no significant phylogenetic turnover for most of the species rich families (8 out of 10) indicating a neutral pattern at the transect level, and probably contributing mostly to the global neutral observed pattern.

Discussion

The present study aimed to better understand the processes underlying taxonomic and phylogenetic diversity and turnover in the Dja Faunal Reserve. We investigated both the alpha and beta diversity components in order to infer mechanisms of local community assembly, as well as the nature of the turnover of species across space, based from the analysis on 9 transects

Table 3. Partition of taxonomic and phylogenetic diversity within and between the 9 transects from the Dja Faunal Reserve (11538 individuals belonging to 372 species) for the whole dataset and for TPF only.

Whole dataset	**APD = −0.034562**	**(pval = *0.0750*)**	
Coefficients based on	Average within site diversity α	Total diversity (γ = α+β)	Differentiation (β/γ)
Species identity and abundance	D_{IS} = 0.9805	D_{IT} = 0.9871	I_{ST} = 0.0067
Species phylogeny and abundance	D_{PS} = 129.4888	D_{PT} = 130.3916	P_{ST} = 0.0069
	D_{BS} = 132.097	D_{BT} = 132.097	B_{ST} = 0.0002
Species phylogeny and incidence	Δ_{PS} = 129.7548	Δ_{PT} = 129.7781	Π_{ST} = 0.0002
TPF dataset	APD = −0.030372	(pval = *0.0880*)	
Coefficients based on	Local diversity α	Total diversity (γ = α+β)	Differentiation (β/γ)
Species identity and abundance	D_{PS} = 129.8353	D_{PT} = 130.8939	P_{ST} = 0.0081
Species phylogeny and abundance	D_{PS} = 129.8353	D_{PT} = 130.8939	P_{ST} = 0.0081
	D_{BS} = 132.64	D_{BT} = 132.672	B_{ST} = 0.0002
Species phylogeny and incidence	Δ_{PS} = 130.0129	Δ_{PT} = 130.0629	Π_{ST} = 0.0004
Whole dataset-habitat	APD = −0.016572	(pvalue = *0.758*)	
Coefficients based on	Local diversity α	Total diversity (γ = α+β)	Differentiation (β/γ)
Species identity and abundance	D_{IS} = 0.9798	D_{IT} = 0.9876	I_{ST} = 0.0079
Species phylogeny and abundance	D_{PS} = 126.8430	D_{PT} = 128.2658	P_{ST} = 0.0111
	D_{BS} = 129.4628	D_{BT} = 129.8765	B_{ST} = 0.0032
Species phylogeny and incidence	Δ_{PS} = 128.9733	Δ_{PT} = 129.0418	Π_{ST} = 0.0005

TPF: terra firme primary forest; APD: mean abundance phylogenetic deviation index.

sampled throughout the reserve at both level, transect and habitat type.

1. Phylogenetic structure among transects in the Dja Faunal Reserve

The main result of our study is the absence of phylogenetic structure between transects across the DFR when considering the whole dataset (Table 4). This random phylogenetic pattern suggests that competitive exclusion does not prevent the co-occurrence of related species within transects, and there is no strong environment constraint preventing the co-occurrence of distant species [14]. However it does not mean that these processes do not occur, since their entangled effects can lead to not different from random patterns. Our result keeps true when looking just at species found on terra firme primary rain forest (TPF) (Table 4). Finally, the absence of phylogenetic structure within a habitat type is consistent with the hypothesis that processes structuring habitats are dominated by dispersal assembly rules, independently from the species niche attributes or that filtering and competition apparently cancel out.

The existence of a dispersal limitation is further confirmed by the species turnover increasing with spatial distance (isolation by distance) for both datasets at transect levels, while the phylogenetic turnover is insensitive to distance. Using only taxonomic data for the same set of species and with different statistics, Hardy and Sonké [27] also found isolation by distance to be a major driver of community variation in the DFR. This pattern is consistent with a neutral model of community dynamics assuming species equivalence in terms of patterns but not automatically in terms of processes [44]. Previous analyses of beta diversity in tropical forests confirmed that the taxonomic turnover found between forest plots separated by 0.2 to 50 km are consistent with the expectation of the neutral model [10], indicating that dispersal is an important factor in community assemblage rules in these ecosystems. The sampled transects covered most of the delimited DFR area with a maximum distance between sites of 105 km (Figure 2) which justifies interpreting the results at the scale of the whole reserve. As phylogenetic turnover in tropical tree communities is expected to be related to species sorting along an environmental gradient (e.g. [22,23]), the absence of phylogenetic turnover among transects indicates that, at coarse grain, no environmental variation influenced the variation in forest composition across space. This would agree with the overall homogeneity of environmental conditions found in the DFR [27]. For example, elevation, which has been shown to be an important factor in species composition in tropical rain forests [20], varies between 600 and 700 meters, and does not generate significant environmental variation. Thus even though there is a significant species turnover across the reserve, phylogenetic relatedness remains equivalent at different places across the Dja rejecting our hypothesis (iii) of the presence of competitive exclusion, and confirming our hypothesis (i), that our communities are marked by dispersal assembly rules. However, we still need to be cautious with our conclusions since an alternative explanations of the fact that no phylogenetic sorting was found among transects might actually not have anything to do with community assembly mechanisms, but rather with the resolution of the phylogeny. Because relationships are only resolved to family/genus level, if species belonging to the same genus are functionally distinct and are sorted among sites we would not be able to tell it with this phylogeny.

Even though our dataset contained only 9 transects, which is fairly limiting, we nevertheless have an extensive coverage of the reserve (Figure 2), and have sampled over 11 000 individuals for 372 species or morphospecies. In addition, significant values of I_{ST} and significant Mantel tests between species and geographic distances among transects indicated that spatial variation among species has been captured in the study. I_{ST} values pointed however towards lower values than those observed for more fragmented wet forests such as in the Western Ghats of India, where I_{ST} values

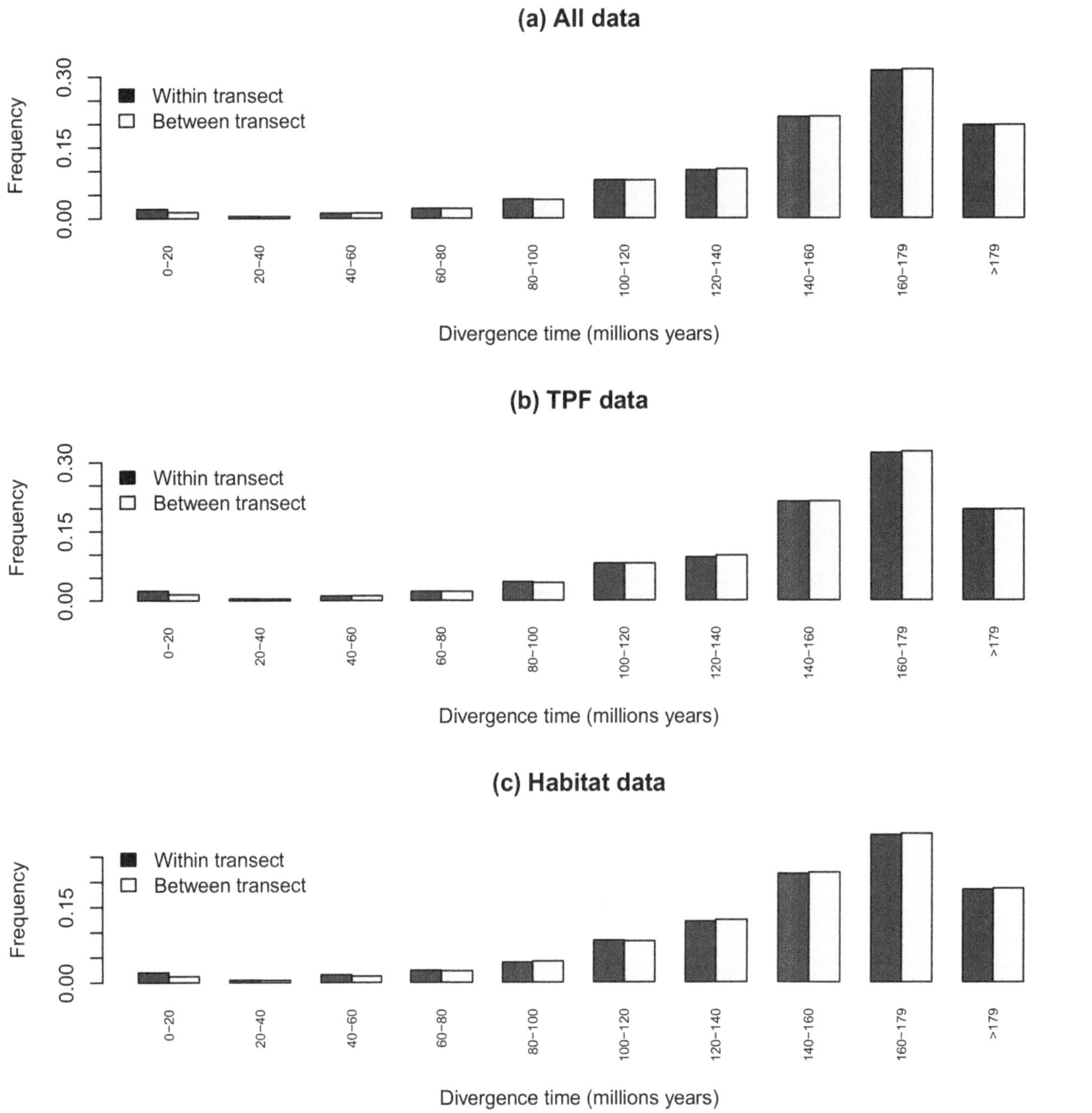

Figure 4. Decomposition within and among plots for Simpson's diversity indices according to the divergence time i.e. frequency distribution of divergence times between individuals from different species for pairs of individuals sampled (a) within transect or between transects for whole data, (b) within transect or between transects for TPF: terra firma primary forest, (c) within habitat or between transects for whole data.

were found above 0.027 (computations from the results of [45], against 0.0067 to 0.0079 here (Table 3); I_{ST} is an increasing function of relative community differentiation). The same pattern is observed when considering Π_{ST} (incidence data integrating phylogeny) which is 2–4 10^{-4} in the Dja Reserve against 34 10^{-4} for the dataset in India and 13 10^{-4} in the Panama Canal watershed [23]. This may indicate that differentiation between localities is not very pronounced in the continuous Dja forest.

2. Phylogenetic structure among habitat types in the Dja Faunal Reserve

At habitat level, a signal of phylogenetic turnover was barely significant (P-value $=0.05$) with abundance data but not with incidence data ($B_{ST}>0$ and $\Pi_{ST} \sim 0$). Since the most constraining habitat types are of limited extent, this may explain why there are some significant patterns with abundance data and not with incidence data. This result suggests that species are sorting not just because of limited colonization but also because of environmental variation due to habitat heterogeneity. Indeed, phylogenetic

Table 4. Testing species and phylogenetic structure within and between transects/habitats.

Dataset	Randomisation	Hypothesis / Results	Interpretation
Wholedata-transect	Model 1-3x	$H_0{:}I_{ST}=0$ pval = 0.000 ***	Limited dispersal between transects within habitat
		$H_1{:}I_{ST}>0$	
	Model 1s	$H_0{:}B_{ST}=0$ pval = 0.348 NS	No phylogenetic turnover
		$H_0{:}\ \Pi_{ST}=0$ pval = 0.090 NS	
Whole data-habitat	Model 2-3x	$H_0{:}I_{ST}=0$ pval = 0.000 ***	Taxonomic differentiation between habitats: reflect a filtering habitat effect on species
		$H_1{:}I_{ST}>0$	
	Model 1s	$H_0{:}B_{ST}=0$ pval = 0.05 * (H1: $B_{ST}>0$)	Low phylogenetic turnover for abundant species among habitats
		$H_0{:}\Pi_{ST}=0$ pval = 0.225 NS	
TPF	Model 1s	$H_0{:}\ B_{ST}=0$ pval = 0.364 NS	No phylogenetic turnover
		$H_0{:}\ \Pi_{ST}=0$ pval = 0.064 NS	No phylogenetic turnover

Details of model permutation are given in Table 1. P-values are given after 999 permutations of individuals or species in model of permutation. NS: non significant. Stars indicate the level of the significance.

turnover was expected among habitats because some of the environmental gradients distinguishing between habitat types in the reserve, such as water availability and anoxic stresses (flooded or swamp vegetation versus terra firme vegetation), are known to be strong filtering factors influencing tropical forest community structures [19]. This confirms our second hypothesis (ii): environmental filtering differs among habitat types and relevant selected traits could be phylogenetically conserved although this would have to be further tested using traits dataset.

3. Variation in phylogenetic structure across the phylogeny

To date few studies have investigated community phylogenetic structure in African rain forests. In a study of 28 one-hectare plots in mature rain forest in Monte Alén National Park (Equatorial Guinea), Hardy and Senterre [20] found a phylogenetic clustering structure which was attributed to adaptations of local species to elevation gradients. They also found that most of the signal is related to ancient clade subdivisions, with most of the individuals pairs (between and among plots) occurring between 100 and 120 Myr. In our analysis, we find a comparable situation where most subdivisions between individual pairs occur between 140–160 Myr. The difference could be related to the different calibrations used to date of the phylogenetic trees. Hardy and Senterre [20] used ages from Davis et al. [46] while we used the more conservative value of Wikstrom *et al* [37] to constrain the origin of the angiosperms (150 Myr versus 179 Myr) in addition to several other calibrations points based on detailed family-specific dating analyses (see methods). Approaches whereby DNA sequence data is generated for the whole sampling would possibly provide better resolution at shallower nodes and hence better address more recent signals [47]. However the resolution of our tree is good for ancient lineages but poor for recent ones. As a consequence this pattern should be interpreted with care.

Our results indicate that Annonaceae species are more related within transects than species taken from different transects. This result could be real and not just an artifact of phylogenetic resolution as the phylogenetic tree for Annonaceae is well resolved [48,49]. Clustering of Annonaceae was also found in another study of African phylogenetic structure in Equatorial Guinea [20] in which phylogenetic differentiation was shown to be correlated with elevation. They also indicated that the number of magnoliid (which includes Annonaceae) species per plot was correlated with altitude. Interestingly, the DFR has very little elevation variability [27] and thus the significant phylogenetic differentiation detected in Annonaceae of the Dja would have to result from a different process than altitudinal gradients. Moreover, in contrast to South America, lowland Annonaceae are more or as diverse at mid latitudinal as indicated in a survey of an elevation gradient in Mont Cameroon [50] . On the contrary, we observed an opposite pattern for Apocynaceae since species among transects appear more related than within indicating an overdispersion of phylogenetic pattern. This result might be linked to the fact that the species *Tabernaemontana crassa* (Apocynaceae) is the most abundant species inventoried in the reserve and strongly present in transects [51].The rest the Apocynaceae species are not well represented (1–64 individuals/species) and less well represented across all transects. In both cases, Annonaceae and Apocynaceae have a large number of liana species [52,53] that have not been inventoried in this study and thus more detailed sampling and tests should be carried out before a link to any evolutionary pattern can be done. However, the more in depth analyses looking at the species rich clade as the Annonaceae indicates that some phylogenetic sorting is occurring among transects and can be detected when more data is available.

4. Phylogenetic diversity and conservation

A classical measure of phylogenetic diversity (signal) is Faith's PD [12] which measures the total phylogenetic branch length (i.e. amount) of evolutionary history in the studied community. This measure is equivalent in our work to Δ_{PS} which also does not account for species abundance. However, Δ_{PS} is a measure of phylogenetic distinctness, but has the advantage not to be influenced by species richness [20].

Recent literature has debated the interest of adding phylogenetic diversity evaluation in conservation planning [17,54,55]. One general agreement in favor of taking phylogenetic diversity into account is to conserve all components of biodiversity including evolutionary information, and that the explicit consideration of biodiversity as comprising evolving and related lineages would add power and robustness to measures of biodiversity for conservation [17]. Specifically, adding phylogenetic estimation in conservation strategy would result in maximization of the set of species to be conserved [55]. According to our results, most diversity occurred within transect or habitat, the between contribution being always less than or equal to 1.1% (whole dataset-habitat).

Conclusions

The preservation of tropical rainforest is an ethical, political and practical concern and biodiversity assessment should be a major focus in nature preservation programs [56]. Indeed, faced with high anthropic pressure in tropical forest, the number and extent of protected areas have increased across the tropics [57]. The objective of such protected areas is to conserve a sufficient sample of the world's biodiversity.

Few conservation policies consider phylogenetic diversity as an important component probably because the added value of phylogenetic diversity for nature conservation remains unclear [55] due to a lack of consensus between various measures and a difficulty to interpret the results in terms of conservation perspectives [54,55]. Here, we detected a random phylogenetic pattern between transects at the scale of the Dja Faunal Reserve, possibly because of a common history and weak environmental variation. We also showed that geographic distance encompassed species turnover. In addition, our phylogenetic based analysis added new results to the previous study of Hardy and Sonké [27] using the same dataset, by detecting a weak but significant phylogenetic turnover signal among habitats reflecting a filtering effect of the habitat. Our results can contribute to the conservation of the park by providing insights into the processes driving community assembly. Notably, the prevalence of patterns compatible with dispersal assembly highlights the need to conservation schemes that allow for sufficiently large conservation areas. Future studies should investigate more plots to be based on a hierarchical sampling plan considering spatial variation within transects in order to better interpret the phylogenetic structure.

Acknowledgments

We thank the editor and the anonymous reviewers for their comments. OJH is senior research associate of the Belgian Fund for Scientific Research F.R.S.-FNRS. SM is a junior member of the Institut Universitaire de France.

Author Contributions

Conceived and designed the experiments: BS. Performed the experiments: BS. Analyzed the data: TC SM OH. Contributed reagents/materials/analysis tools: FM. Wrote the paper: SM TC PC FM OH.

References

1. Pimm SL, Raven P (2000) Biodiversity - Extinction by numbers. Nature 403: 843–845.
2. Gibson L, Lee TM, Koh LP, Brook BW, Gardner TA, et al. (2011) Primary forests are irreplaceable for sustaining tropical biodiversity. Nature 478: 378–381.
3. Koleff P, Lennon JJ, Gaston KJ (2003) Are there latitudinal gradients in species turnover? Global Ecology and Biogeography 12: 483–498.
4. Kraft NJB, Comita LS, Chase JM, Sanders NJ, Swenson NG, et al. (2011) Disentangling the drivers of beta diversity along latitudinal and elevational gradients. Science 333: 1755–1758.
5. Givnish TJ (1999) On the causes of gradients in tropical tree diversity. Journal of Ecology 87: 193–210.
6. Parmentier I, Malhi Y, Senterre B, Whittaker RJ, Alonso A, et al. (2007) The odd man out? Might climate explain the lower tree alpha-diversity of African rain forests relative to Amazonian rain forests? Journal of Ecology 95: 1058–1071.
7. Burgess N, Kuper W, Mutke J, Brown J, Westaway S, et al. (2005) Major gaps in the distribution of protected areas for threatened and narrow range Afrotropical plants. Biodiversity and Conservation 14: 1877–1894.
8. Lewis SL (2006) Tropical forests and the changing earth system. Philosophical Transactions of the Royal Society B-Biological Sciences 361: 195–210.
9. Swenson NG (2013) The assembly of tropical tree communities the advances and shortcomings of phylogenetic and functional trait analyses. Ecography 36: 264–276.
10. Condit R, Pitman N, Leigh EG, Chave J, Terborgh J, et al. (2002) Beta-diversity in tropical forest trees. Science 295: 666–669.
11. Gentry AH (1988) Tree species richness of upper amazonian forests. Proceedings of the National Academy of Sciences of the United States of America 85: 156–159.
12. Faith DP (1992) Conservation evaluation and hylogenetic diversity Biological Conservation 61: 1–10.
13. Swenson NG, Stegen JC, Davies SJ, Erickson DL, Forero-Montana J, et al. (2012) Temporal turnover in the composition of tropical tree communities: functional determinism and phylogenetic stochasticity. Ecology 93: 490–499.
14. Webb CO, Ackerly DD, McPeek MA, Donoghue MJ (2002) Phylogenies and community ecology. Annual Review of Ecology and Systematics 33: 475–505.
15. Mouquet N, Devictor V, Meynard CN, Munoz F, Bersier LF, et al. (2012) Ecophylogenetics: advances and perspectives. Biological Reviews 87: 769–785.
16. McGill BJ, Enquist B, Weiher E, Westoby M (2006) Rebuilding community ecology from functional traits. Trends in Ecology and Evolution 14: 178–185.
17. Diniz-Filho JAF, Loyola RD, Raia P, Mooers AO, Bini LM (2013) Darwinian shortfalls in biodiversity conservation. Trends in ecology & evolution (Personal edition).
18. Swenson NG, Erickson DL, Mi XC, Bourg NA, Forero-Montana J, et al. (2012) Phylogenetic and functional alpha and beta diversity in temperate and tropical tree communities. Ecology 93: S112–S125.
19. Fine PVA, Kembel SW (2011) Phylogenetic community structure and phylogenetic turnover across space and edaphic gradients in western Amazonian tree communities. Ecography 34: 552–565.
20. Hardy OJ, Senterre B (2007) Characterizing the phylogenetic structure of communities by an additive partitioning of phylogenetic diversity. Journal of Ecology 95: 493–506.
21. Swenson NG, Anglada-Cordero P, Barone JA (2011) Deterministic tropical tree community turnover: evidence from patterns of functional beta diversity along an elevational gradient. Proceedings of the Royal Society B-Biological Sciences 278: 877–884.
22. Baraloto C, Hardy OJ, Paine CET, Dexter KG, Cruaud C, et al. (2012) Using functional traits and phylogenetic trees to examine the assembly of tropical tree communities. Journal of Ecology 100: 690–701.
23. Hardy OJ, Couteron P, Munoz F, Ramesh BR, Pelissier R (2012) Phylogenetic turnover in tropical tree communities: impact of environmental filtering, biogeography and mesoclimatic niche conservatism. Global Ecology and Biogeography 21: 1007–1016.
24. Mayfield MM, Levine JM (2010) Opposing effects of competitive exclusion on the phylogenetic structure of communities. Ecology Letters 13: 1085–1093.
25. White F (1983) The vegetation of Africa. A descriptive memoir to accompagn the UNESCO/AEFTFAT/UNSO vegatation map of Africa. Paris, Copedith.
26. McGinley M (2008) Dja Faunal Reserve. In Cleveland, C.J. (ed) Encyclopedia of Earth. United Nations Environment Programme–World Conservation Monitoring Centre. http://www/eoearth.org/article/Dja_Faunal_Reserve,_Cameroon.
27. Hardy OJ, Sonke B (2004) Spatial pattern analysis of tree species distribution in a tropical rain forest of Cameroon: assessing the role of limited dispersal and niche differentiation. Forest Ecology and Management 197: 191–202.
28. Hardy OJ (2008) Testing the spatial phylogenetic structure of local communities: statistical performances of different null models and test statistics on a locally neutral community. Journal of Ecology 96: 914–926.
29. Sonké B (1998) Etudes floristiques et structurales des forêts de la Réserve de Faune du Dja (Cameroun). Thèse de Ph. D. Recherche, Université Libre deBruxelles: 129–130.
30. Sonké B, Couvreur TLP (2014) Tree diversity of the Dja Faunal Reserve, South Caleroon. Biodiversity Data Journal. 2: e1049. DOI: 10.3897/BDJ.2.e1049
31. Sonké B (2004) Forêts de la réserve du Dja (Cameroun). Etudes floristiques et structurales. Scripta Botanica Belgica 32: 1–144.
32. Lebrun J, Gilbert GS (1954) Une classification écologique des forêts du Congo. Publication de l'Institut National pour l'étude AGronomique du Congo Belge 63: 9–89.
33. Webb CO, Donoghue MJ (2005) Phylomatic: tree assembly for applied phylogenetics. Molecular Ecology Notes 5: 181–183.
34. TAP G (2009) An update of the Angiosperm Phylogeny Group classification for the orders and families of flowering plants: APG III. Botanical Journal of the Linnean Society 161: 105–121.
35. Maddison W, Maddison D (2009) Mesquite: A modular system for evolutionary analysis. version 2.7. http://mesquiteproject.org.
36. Webb CO, Ackerly DD, Kembel SW (2008) Phylocom: software for the analysis of phylogenetic community structure and trait evolution. Bioinformatics 24: 2098–2100.
37. Wikstrom N, Savolainen V, Chase MW (2001) Evolution of the angiosperms: calibrating the family tree. Proceedings of the Royal Society B-Biological Sciences 268: 2211–2220.

38. Hardy OJ, Jost L (2008) Interpreting and estimating measures of community phylogenetic structuring. Journal of Ecology 96: 849–852.
39. Hardy OJ (2010) SPACoDi 0.10: a program dor spatial & phylogenetic analysis of community diversity. AVailable at http://ebe.ulb.ac.be/ebe/Software.html.
40. Team RDC (2012) R: A Language and Environment for Statistical Computing. R Foundation for Statistical Computing.
41. Eastman JM, Paine CET, Hardy OJ (2011) spacodiR: structuring of phylogenetic diversity in ecological communities. Bioinformatics 27: 2437–2438.
42. Letunic I, Bork P (2007) Interactive Tree Of Life (iTOL): an online tool for phylogenetic tree display and annotation. Bioinformatics 23: 127–128.
43. Letunic I, Bork P (2011) Interactive Tree of Life v2: online annotation and display of phylogenetic trees made easy. Nucleic Acids Research 39: W475–W478.
44. Hubbell S (2001) A unified neutral theory of biodiversity and Biogeography. Princeton University Press, Princeton.
45. Munoz F, Couteron P, Ramesh BR (2008) Beta diversity in spatially implicit neutral models: A new way to assess species migration. American Naturalist 172: 116–127.
46. Davies TJ, Barraclough TG, Chase MW, Soltis PS, Soltis DE, et al. (2004) Darwin's abominable mystery: Insights from a supertree of the angiosperms. Proceedings of the National Academy of Sciences of the United States of America 101: 1904–1909.
47. Kress WJ, Erickson DL, Swenson NG, Thompson J, Uriarte M, et al. (2010) Advances in the Use of DNA Barcodes to Build a Community Phylogeny for Tropical Trees in a Puerto Rican Forest Dynamics Plot. PLoS ONE 5.
48. Couvreur TLP, Pirie MD, Chatrou LW, Saunders RMK, Su YCF, et al. (2011) Early evolutionary history of the flowering plant family Annonaceae: steady diversification and boreotropical geodispersal. Journal of Biogeography 38: 664–680.
49. Chatrou LW, Pirie MD, Erkens RHJ, Couvreur TLP, Neubig KM, et al. (2012) A new subfamilial and tribal classification of the pantropical flowering plant family Annonaceae informed by molecular phylogenetics. Botanical Journal of the Linnean Society 169: 5–40.
50. Bele MY, Focho DA (2011) Inventory and distribution of the Annonaceae along elevation gradient on Mount Cameroon. Journal of Horticulture and Forestry 10: 307–319.
51. Sonké B, Couvreur TLP (2014) Tree diversity of the Dja Faunal Reserve, South Cameroon. Biodiversity Data Journal 2: e1049.
52. Chatrou L, Pirie M, Erkens R, Couvreur T, Neubig KM, et al. (2012) A new subfamilial and tribal classification of the pantropical flowering plant family Annonaceae informed by molecular phylogenetics. Botanical Journal of the Linnean Society 169: 5–40.
53. Lahaye R, Civeyrel L, Speck T, Rowe N (2005) Evolution of shrub-like growth forms in the lianoid subfamily Secamonoideae (Apocynaceae s.l.) of Madagascar: phylogeny, biomechanics, and development. American Journal of Botany 92: 1381–1396.
54. Rolland J, Cadotte MW, Davies J, Devictor V, Lavergne S, et al. (2012) Using phylogenies in conservation: new perspectives. Biology Letters 8: 692–694.
55. Winter M, devictor V, Schweiger O (2013) Phylogenetic diversity and nature conservation : where are we? Trends in Ecology & Evolution 28: 199–204.
56. Chave J, Wiegand K, Levin S (2002) Spatial and biological aspects of reserve design. Environmental Modeling & Assessment 7: 115–122.
57. Jenkins CN, Joppa L (2009) Expansion of the global terrestrial protected area system. Biological Conservation 142: 2166–2174.
58. Potgieter K, Albert VA (2001) Phylogenetic Relationships within Apocynaceae s.l. Based on trnL Intron and trnL-F Spacer Sequences and Propagule Characters. Annals of the Missouri Botanical Garden 88: 523–549.
59. Olmstead RG, Zjhra ML, Lohmann LG, Grose SO, Eckert AJ (2009) A molecular phylogeny and classification of Bignoniaceae. American Journal of Botany 96: 1731–1743.
60. Yakandawala D, Morton CM, Prance GT (2010) Phylogenetic Relationships of the Chrysobalanaceae Inferred from Chloroplast, Nuclear, and Morphological Data1. Annals of the Missouri Botanical Garden 97: 259–281.
61. Gustafsson MHG, Bittrich V, Stevens PF (2002) Phylogeny of Clusiaceae Based on rbcL sequences. International Journal of Plant Sciences 163: 1045–1054.
62. Tan F, Shi S, Zhong Y, Gong X, Wang Y (2002) Phylogenetic relationships of Combretoideae (Combretaceae) inferred from plastid, nuclear gene and spacer sequences. Journal of Plant Research 115: 475–481.
63. Wurdack KJ, Hoffmann P, Chase MW (2005) Molecular phylogenetic analysis of uniovulate Euphorbiaceae (Euphorbiaceae sensu stricto) using plastid RBCL and TRNL-F DNA sequences. American Journal of Botany 92: 1397–1420.
64. Bruneau A, Forest F, Herendeen PS, Klitgaard BB, Lewis GP (2001) Phylogenetic Relationships in the Caesalpinioideae (Leguminosae) as inferred from chloroplast trnL intron sequences. Systematic Botany 26: 487–514.
65. Lavin M, Herendeen PS, Wojciechowski MF (2005) Evolutionary rates analysis of leguminosae implicates a rapid diversification of lineages during the tertiary. Systematic Biology 54: 575–594.
66. Alverson WS, Whitlock BA, Nyffeler R, Bayer C, Baum DA (1999) Phylogeny of the core Malvales: evidence from ndhF sequence data. American Journal of Botany 86: 1474–1486.
67. Muellner AN, Pennington TD, Koecke AV, Renner SS (2010) Biogeography of Cedrela (Meliaceae, Sapindales) in central and south America. American Journal of Botany 97: 511–518.
68. Muellner AN, Samuel R, Johnson SA, Cheek M, Pennington TD, et al. (2003) Molecular phylogenetics of Meliaceae (Sapindales) based on nuclear and plastid DNA sequences. American Journal of Botany 90: 471–480.
69. Sauquet H, Doyle JA, Scharaschkin T, Borsch T, Hilu KW, et al. (2003) Phylogenetic analysis of Magnoliales and Myristicaceae based on multiple data sets: implications for character evolution. Botanical Journal of the Linnean Society 142: 125–186.
70. Wilson PG, O'Brien MM, Heslewood MM, Quinn CJ (2005) Relationships within Myrtaceae sensu lato based on a *mat* K phylogeny. Plant Systematics and Evolution 251: 3–19.
71. Malécot V, Nickrent DL (2008) Molecular Phylogenetic Relationships of Olacaceae and Related Santalales. Systematic Botany 33: 97–106.
72. Wurdack KJ, Hoffmann P, Samuel R, de Bruijn A, van der Bank M, et al. (2004) Molecular phylogenetic analysis of Phyllanthaceae (Phyllanthoideae pro parte, Euphorbiaceae sensu lato) using plastid RBCL DNA sequences. American Journal of Botany 91: 1882–1900.
73. Bremer B, Eriksson T (2009) Time tree of Rubiaceae: phylogeny and dating the family, subfamilies, and tribes. International Journal of Plant Sciences 170: 766–793.
74. Groppo M, Pirani JR, Salatino MLF, Blanco SR, Kallunki JA (2008) Phylogeny of Rutaceae based on twononcoding regions from cpDNA. American Journal of Botany 95: 985–1005.
75. Buerki S, Forest F, Acevedo-Rodríguez P, Callmander MW, Nylander JAA, et al. (2009) Plastid and nuclear DNA markers reveal intricate relationships at subfamilial and tribal levels in the soapberry family (Sapindaceae). Molecular Phylogenetics and Evolution 51: 238–258.

5

Prioritizing Populations for Conservation Using Phylogenetic Networks

Logan Volkmann[1], Iain Martyn[2], Vincent Moulton[3], Andreas Spillner[4], Arne O. Mooers[1,2]*

1 Department of Biological Sciences, Simon Fraser University, Burnaby, British Columbia, Canada, 2 Interdisciplinary Research in the Mathematical and Computational Sciences (IRMACS) Centre, Simon Fraser University, Burnaby, British Columbia, Canada, 3 School of Computing Sciences, University of East Anglia, Norwich, United Kingdom, 4 Department of Mathematics and Computer Science, University of Greifswald, Greifswald, Germany

Abstract

In the face of inevitable future losses to biodiversity, ranking species by conservation priority seems more than prudent. Setting conservation priorities within species (i.e., at the population level) may be critical as species ranges become fragmented and connectivity declines. However, existing approaches to prioritization (e.g., scoring organisms by their expected genetic contribution) are based on phylogenetic trees, which may be poor representations of differentiation below the species level. In this paper we extend evolutionary isolation indices used in conservation planning from phylogenetic trees to phylogenetic networks. Such networks better represent population differentiation, and our extension allows populations to be ranked in order of their expected contribution to the set. We illustrate the approach using data from two imperiled species: the spotted owl *Strix occidentalis* in North America and the mountain pygmy-possum *Burramys parvus* in Australia. Using previously published mitochondrial and microsatellite data, we construct phylogenetic networks and score each population by its relative genetic distinctiveness. In both cases, our phylogenetic networks capture the geographic structure of each species: geographically peripheral populations harbor less-redundant genetic information, increasing their conservation rankings. We note that our approach can be used with all conservation-relevant distances (e.g., those based on whole-genome, ecological, or adaptive variation) and suggest it be added to the assortment of tools available to wildlife managers for allocating effort among threatened populations.

Editor: Tom Gilbert, Natural History Museum of Denmark, Denmark

Funding: Funded by the Natural Sciences and Engineering Research Council of Canada (Discovery grant to AOM), http://www.nserc-crsng.gc.ca/, Simon Fraser University (fellowship to LAV and IM), and Yale University (YIBS fellowship, AOM). The funders had no role in study design, data collection and analysis, decision to publish or preparation of the manuscript.

Competing Interests: The authors have declared that no competing interests exist.

* E-mail: amooers@sfu.ca

Introduction

Extinctions due to human impacts are now unavoidable: even optimistic scenarios predict significant changes in biodiversity by the year 2100 [1,2], with most extinction starting with the loss of isolated populations [3,4].

One prime conservation goal is to preserve genetic variation [5,6], both as a representation of past evolution and raw material for future evolution [7] and, potentially, as a surrogate for improved ecosystem function [8]. However, not all genetic lineages are equally important, with more isolated lineages warranting additional interest because of their expected contribution to total variation [5,8,9]. Indices of evolutionary isolation have been developed to rank species on a phylogenetic tree based on unique and shared evolutionary history (*e.g.*, [10–13]). These metrics use rooted phylogenetic trees with edge lengths as input (Figure 1), and rank tips with less shared history as requiring more urgent conservation attention. For example, the Zoological Society of London has made this approach operational in their "Edge of Existence" programme (www.edgeofexistence.org). In the United States, taxonomic distinctiveness is one of several explicit criteria for prioritizing conservation attention [14]. The extension to populations within species would seem to be straightforward.

Below the species level, Ryder [15] advocated the use of *evolutionarily significant units* (ESUs) to identify populations with genetic variation in need of long-term conservation; this was expanded by Moritz [16] and Waples [17] to the concepts of *management units* (MUs) and *distinct population segments* (DPS), respectively, for species that had undergone more recent range fragmentation. All these population-based approaches have enjoyed wide usage in population genetic studies (*e.g.*, [18–21]), and are the basis for identifying populations worthy of protection in law. Importantly, ESUs assume that the relationships among populations can be represented by a bifurcating tree. However, bifurcating trees often fail to capture the relationships among populations [22]. DPSs and MUs can deal with populations that have more complex interrelationships (Figure 1), but neither of these designations is designed to prioritize among populations. This would seem a major shortcoming if populations do need to be prioritized for conservation attention.

Previous authors have shown that the logic of measuring and maximizing phylogenetic diversity [6], which forms the basis for tree-based prioritization schemes, can be generalized to phylogenetic networks [23–25]. Here we show that the prioritization approaches for trees can also be adapted for populations within species by extending evolutionary isolation indices from trees to networks. We develop efficient algorithms to compute these

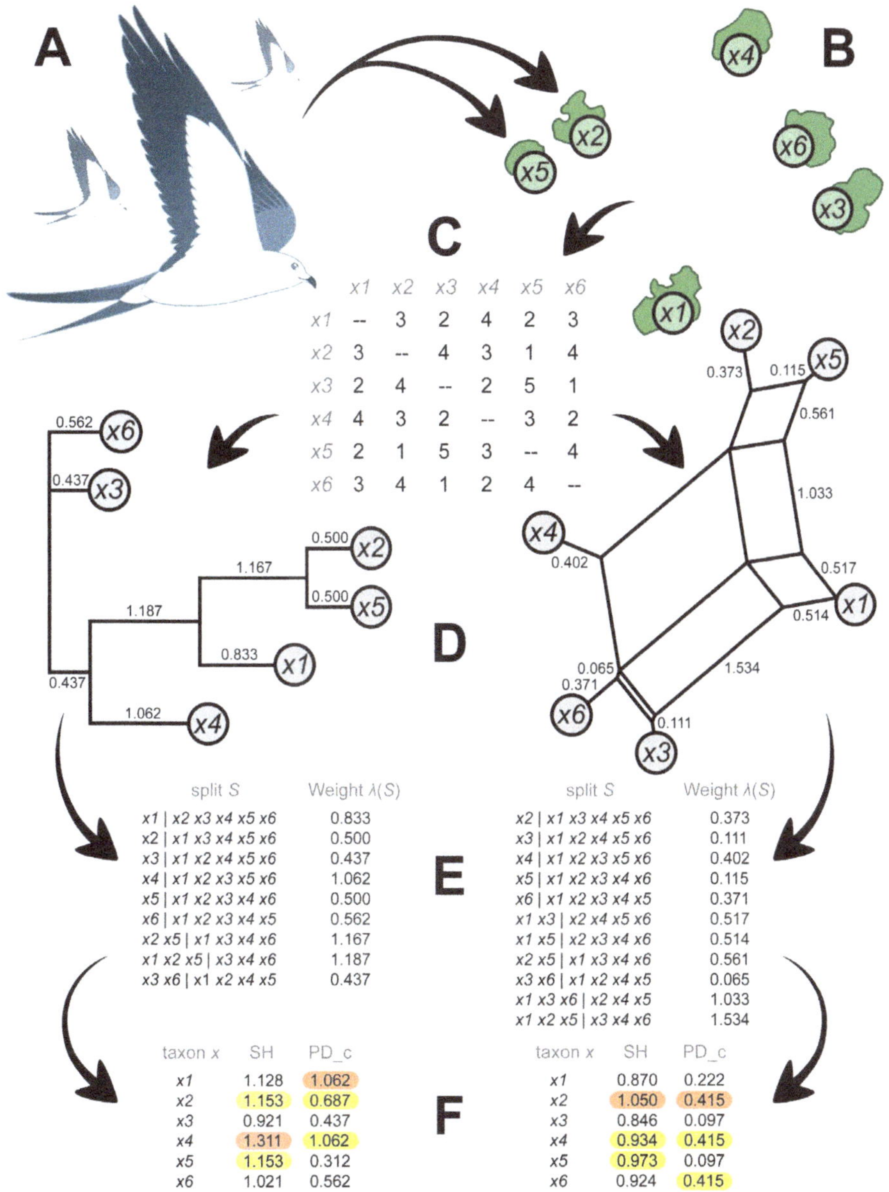

Figure 1. Using pairwise distances to rank species or populations. Consider a hypothetical group of taxa (A)—a set of closely-related species or populations of a single species—that is distributed across several islands in an archipelago (B). Differences among the taxa, labeled *x1* through *x6*, can be organized into a pairwise distance matrix (C). We can represent this matrix either as a phylogenetic tree or as a phylogenetic network (D), where a set of weighted *splits* describes the relationships among the taxa (E). Altogether, these splits represent the group's phylogenetic diversity (PD). By selecting subsets of splits that exclude a given taxon, we can calculate each taxon's contribution to the total PD of the tree or network (F). The *Shapley metric* (SH) and *expected PD complementarity* (PD_c) are different approaches for ranking taxa based on split data. Note that the highest-scoring taxa (highlighted values) can differ considerably depending on the type of metric used and whether the splits come from a tree or network. We discuss the reasons for these differences and methods for ranking taxa in Section (*ii*) of the main text.

indices for NeighborNet networks [26,27], and illustrate their use with heuristic data from two imperiled species, the spotted owl (*Strix occidentalis* Xantus de Vesey 1860) and the mountain pygmy-possum (*Burramys parvus* Broom 1896). The new approach to assessing population differentiation might be of immediate practical use to those tasked with managing discrete populations of a threatened species, and may allow for new policy associated with conservation triage [28].

Methods

We present our approach for prioritizing populations in three steps. First, we briefly review the various approaches for measuring diversity and evolutionary isolation on bifurcating trees of taxa. We then review the properties of NeighborNet networks as a representation of pairwise evolutionary distances and describe how to prioritize taxa by their expected contribution to biodiversity. In File S1, we outline efficient algorithms for estimating evolutionary isolation on NeighborNet networks. Finally, we illustrate the new method of population prioritization using two small published datasets.

(i) Diversity measures on trees and networks

The concept of evolutionary isolation can be understood in terms of a species' biological distinctiveness, which we might measure by comparing its adaptive or non-adaptive traits to those of related species. More generally, our goal is to measure a taxon's contribution to the current and/or future "diversity" in a set of taxa. Several different approaches for quantifying such diversity have been proposed. One of the earliest, described by Weitzman [23], is *expected diversity*. Rather than score taxa individually, this approach seeks to identify the set of taxa that will retain the most diversity on a future tree, given some measure of diversity and a probability of persistence for each potential combination of taxa. Although Weitzman's original diversity metric was rather general, he did consider an example of biological character-state differences that could be represented on a phylogenetic tree.

On such a tree, every taxon contributes an amount of unique evolutionary information denoted by the length of the branch (or *edge*) linking it to all other taxa (Figure 1) [6,23]. This length may be calibrated in units of time (*e.g.*, millions of years) or in raw or inferred genetic distances. Looking specifically at biological systems, Witting and Loeschcke [29] and Faith and Walker [30] combined Weitzman's [23] expected diversity framework with Faith's [6] concept of *phylogenetic diversity* (PD), the latter which specifically calculates the sum of all branch lengths on a tree (see next section). Like Weitzman [23], this *expected PD* approach can be used to identify a set of taxa that maximizes the amount of total tree length retained, given a set of extinction probabilities for the tips.

The related *k of n problem* [6] seeks to identify the most diverse subset of k taxa (*i.e.*, the one that maximizes PD) on a tree of size n. Faith [31] and Weitzman [32] explored the special case where $k = 1$, which Faith [33] refers to as the *PD complementarity* of a given taxon.

An independently-derived approach based on Game Theory ([10], first published 2005) explicitly considers the individual contribution of each taxon to future diversity. Like Weitzman's [23] expected diversity framework, all possible subsets of taxa on a tree may persist. By calculating the amount of unique information each taxon contributes to future subsets (*i.e.*, the average length of the edge linking the taxon to all possible future trees), one can rank taxa in order of their relative impact on future diversity. This *Shapley metric* (SH) is almost identical to the ad-hoc *evolutionary distinctness* (ED) metric used by the Zoological Society of London in their Edge of Existence programme (www.edgeofexistence.org). The major difference between the two is that the ED metric is explicitly measured on a rooted tree, as opposed to the more general undirected graph that SH takes as input [34].

The Shapley metric was further refined by Steel *et al.* [35] and named HED (for *heightened evolutionary distinctiveness*). HED is the expected contribution of a given taxon to future subsets of taxa where the subsets are weighted by their probability of persistence. In this case, the focal taxon is assumed to persist (*i.e.*, its probability of extinction does not affect its HED score). On trees, HED is formally equivalent to a form of PD complementarity where the contribution of a taxon is measured with respect to all possible subsets, each weighted by their probability of persistence [33]. Weitzman [32] also arrived at this formulation ten years earlier, which he termed the "distinctiveness" of a taxon, in the context of his "Noah's Ark Problem" of biodiversity preservation. Using Faith's [33] terminology, HED, which combines the concepts of expected PD with PD complementarity, might be considered *expected PD complementarity*.

As a final antecedent, Minh *et al.* [24,25,36] extended PD to phylogenetic networks and presented algorithms for solving the k of n problem to maximize diversity for a given subset size. They referred to this metric as *split diversity* (SD).

In this context it should be possible to measure the PD contribution of individual taxa on a phylogenetic network. Critically for our purposes, the two metrics we use here (SH and HED) do not require a rooted phylogenetic tree, and so can be adapted to networks in the same way that PD indices can [24,25,27,36]. SH and HED are formally defined in File S1 and discussed further below. In short, if we do not have probabilities of extinction for taxa, we assume all future subsets of taxa are equally likely, and calculate SH. If we can estimate (even broadly) the probabilities of persistence of all taxa, we can weight future subsets by their probability, and use HED.

(ii) Interpreting phylogenetic networks, Shapley values, and HED

NeighborNet [26] is a method that permits the representation of pairwise distances between taxa in the form of a network. An important property of NeighborNet networks is that they permit the representation of relationships among the underlying taxa that cannot be depicted on any phylogenetic tree. For example, to the extent that populations exchange migrants, the between-population genetic distance data (F_{ST}) may yield many alternative trees, none of which accurately reflect the actual relationships among these populations (*e.g.*, [37]). The NeighborNet framework, by contrast, accommodates for such phylogenetic uncertainty and will always yield a single network with positive edge lengths, permitting calculation of SH and HED. If a pairwise distance matrix is tree-like (*i.e.*, yields only one possible phylogeny) the resulting NeighborNet output will resemble a phylogenetic tree. Where there is no tree-like history, a network representation should be more informative. Indeed, for many distance matrices (including Example A below, results not shown), the assumptions necessary to produce a tree are not met, and a neighbor-joining tree, for example, produces negative edge lengths. Here, a network representation would definitely be preferred [26].

An example of a very simple matrix of pairwise distances and the resulting network is depicted in Figure 1. Each edge or set of parallel edges in the network corresponds to a partition of the underlying set of taxa into two non-overlapping subsets, called a *split* (S). The *edge length* reflects the *weight* of the split ($\lambda(S)$)—in other words, a component of the pairwise distance (F_{ST}, for example) separating any two taxa. Thus, just as a phylogenetic tree represents a collection of weighted splits (Σ) [38], where each branch of the tree denotes a split, a NeighborNet network represents a weighted collection of splits of the underlying set of taxa. As Figure 1 illustrates, the distance between two tips on a network (*i.e.*, the shortest path between two taxa) represents the observed distance in the distance matrix.

Whether represented on a tree or a network, every split system contains information on the overall diversity of its constituent taxa [5,23]. The conservation planning metric *phylogenetic diversity* (PD) [6] can be calculated for split systems as

$$PD_{(\Sigma,\lambda)}(Y) = \sum_{\substack{A|B\in\Sigma \\ A\cap Y\neq 0, B\cap Y\neq 0}} \lambda(A|B)$$

where Y is a subset of taxa on the tree or network and $\lambda(A|B)$ is the weight of the split between two non-overlapping groups A and

B of taxa. Note that the overall PD for both trees [6,39] and networks [24,25,36] is simply the sum of all split weights (Figure 1).

A very simple approach for measuring an individual taxon's PD contribution, illustrated in Figure 1, is to consider the change in PD when this taxon is removed from the tree or network [40]. This *PD complementarity* (PD_c) metric can be expressed as

$$PD\_c(x) = PD_{(\Sigma,\lambda)}(X) - PD_{(\Sigma,\lambda)}(\bar{X})$$

where X is the set of all taxa in the tree or network and $\bar{X}$ is the subset where a given taxon x has been removed from the underlying distance matrix.

We can also extend the metrics SH and HED from trees to NeighborNet networks using similar ideas for extending PD calculations from trees (*e.g.*, [6,27,29]) to networks (*e.g.*, [24,25,36,41]). On a tree, the Shapley value (ψ_x^{sh}) for taxon x can be defined as the mean split weight of the set of splits defining $x|\bar{X}$, where $\bar{X}$ represents all unique possible subsets of the taxon set X that do not contain x. Importantly, Haake *et al.* [10] present a formal proof that the Shapley value for x can also be calculated as a weighted sum of all the edge lengths on a tree, with the weights determined by the sizes of the sets containing x. This can be presented compactly using split notation as

$$\psi_x^{sh}(\Sigma,\lambda) = \sum_{S\in\Sigma} \frac{|\bar{S}(x)|}{|X||S(x)|}\lambda(S)$$

where (Σ,λ) is the set of splits defined by the network and their weights, $|X|$ is the total number of taxa, $|S(x)|$ is the size of a split set containing the taxon x, $|\bar{S}(x)|$ is the size of the complementary set that does not contain x, and $\lambda(S)$ (following the notation from Minh *et al.* [24,25]) is the split weight, equal to the edge length separating $S(x)$ from $\bar{S}(x)$. To calculate the Shapley value for taxon $x1$ in the network in Figure 1, we take the first split $x2|x1,x3,x4,x5,x6$ to be composed of $|S(x1)| = 5$ and $|\bar{S}(x1)| = 1$ and $\lambda(S) = 0.373$, the second split $x3|x1,x2,x4,x5,x6$ to be composed of $|S(x1)| = 5$ and $|\bar{S}(x1)| = 1$ and $\lambda(S) = 0.111$ and so on. With a taxon set containing six elements, $|X| = 6$ and the Shapley value for taxon $x1$ is 0.870 (Figure 1).

As with a phylogenetic tree, the sum of Shapley values will always equal the sum of all parallel split weights in the network. Because the shape of a network reflects the relative distances among its taxa, we should expect outlying taxa (*i.e.*, those connected to the rest of the network by long edges, like taxon $x4$) to show higher values for ψ_x^{sh}. Thus, the Shapley values calculated for a network can reflect the relative degree of isolation of each taxon based on molecular, morphological, or any other relevant distance measure.

Though conceptually similar, the calculation of HED (ψ_x^{hed}) is somewhat more complex, as it accounts for differences in the probability of extinction $p(y)$ for each taxon:

$$\psi_x^{head}(\Sigma,\lambda) =$$

$$\sum_{S\in\Sigma}\left(\prod_{y\in(S(s)-\{x\})} p(y)\right)\cdot\left(1-\left(\prod_{y\in(\bar{S}(x))} p(y)\right)\right)\cdot\lambda(S)$$

Here, the first product operator considers $p(y)$ for every taxon in $S(x)$ but excludes $p(y)$ for taxon x itself [27,35]. The second product operator considers $p(y)$ for every taxon in $\bar{S}(x)$. Unlike SH, the sum of HED scores will not equal the sum of split weights in the split system. We also note that $\lambda(S)$ will influence HED more strongly than $p(y)$ for outlying taxa. Thus, the ranking order for highly isolated populations should be similar for SH and HED, regardless of which populations have a higher extinction probability.

A more detailed mathematical treatment of the SH and HED metrics and efficient algorithms for their computation are given in File S1. For the datasets in this paper, we used the implementation of NeighborNet in the SplitsTree software package [41] to compute networks. For a given matrix of pairwise distances, this yields the network together with the corresponding collection of weighted splits. We also developed custom R scripts (available in File S1) [42] to compute SH and HED on the outputs from SplitsTree.

(iii) Application

We present SH and HED ranking for two datasets based on putatively neutral genetic markers. In the first example (*A*), the size of each population (and hence the probability of extinction for each population) is not known. In the second example (*B*), population sizes are known, allowing us to estimate separate probabilities of extinction for each population.

We selected our two examples based on the following criteria: (1) The species as a whole is of conservation interest (*i.e.*, vulnerable, endangered, or critically endangered), (2) its distribution is fragmented (*i.e.*, we can define multiple populations), (3) sampling efforts have covered its entire range, and (4) genetic analyses have been published or the raw sequence data made publicly available.

Readers should note that the primary goals of this article are to introduce and illustrate our network ranking approach, not to advocate new management decisions for the taxa described below.

Example A.. Spotted owls (*Strix occidentalis*) are distributed throughout late-succession conifer forests in western North America [43]. Four subspecies are currently recognized (Figure 2a): *S. o. caurina* from southern British Columbia to northwest California, *S. o. occidentalis* in California and Nevada, *S. o. lucida* in Utah, Colorado, Arizona, New Mexico, and northern Mexico, and *S. o. juanaphillipsae* in central Mexico [44,45]. Populations in the United States continue to decline due largely to poor timber harvesting practices, but also as a result of climate change and the westward expansion of barred owls (*S. varia* Barton 1799) [46]. *S. o. caurina* (the northern spotted owl) and *S. o. lucida* (the Mexican spotted owl) are threatened subspecies under the United States' Endangered Species Act, and *S. o. occidentalis* (the California spotted owl) is a subspecies of special concern in the state of California [47]. Spotted owls in the American Southwest "sky islands" (mostly *S. o. lucida*) are particularly fragmented and perhaps most suitable for population-level conservation [48]. Although genetic data for the Mexican subspecies remain poor, we can construct a reasonably complete representative phylogenetic network for subspecies in the United States.

Spotted owl mitochondrial sequences were obtained from Genbank (accession numbers AY833608–AY833644, AY836774–AY836776, DQ230843–DQ230888) and aligned in Mega v. 5 [49] using MUSCLE [50]. These sequences comprise about 1105 bp of the control (D-loop) region and represent 86 haplotypes from 32 populations in the United States and Mexico (Figure 2b; Table 1) [48,51]. We ran a standard analysis of molecular variance (AMOVA) [52] on all 298 aligned sequences in Arlequin v. 3.5 [53] using the Kimura 2-Parameter model [54] to compute distances among haplotypes (Φ_{ST}). This procedure

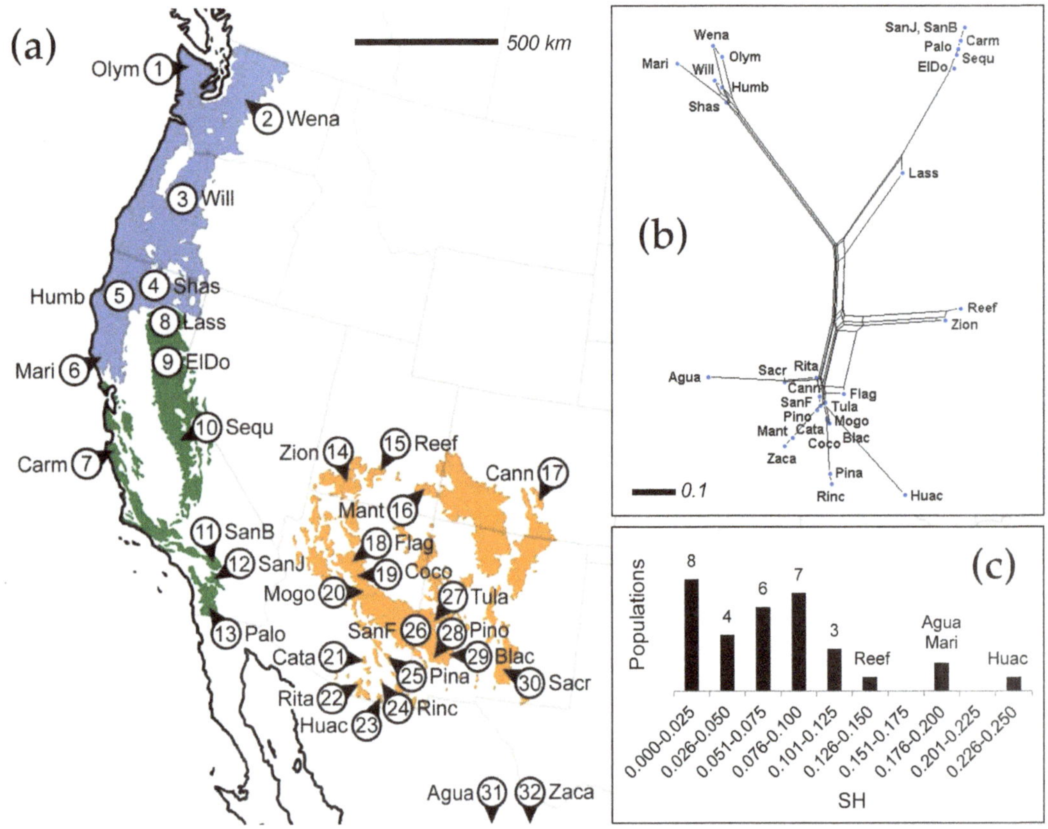

Figure 2. Conservation prioritization of spotted owl (*Strix occidentalis*) populations. (a) Distribution of spotted owls in the United States and the populations sampled by Barrowclough *et al.* [48,51]. Shaded areas denote suitable habitat based on forest cover data [73]. Colors denote the subspecies *S. o. caurina* (blue), *S. o. occidentalis* (green), and *S. o. lucida* (orange). Populations 31 and 32 represent the *S. o. juanaphillipsae* subspecies in Mexico (range not shown). (b) NeighborNet of sampled populations based on mtDNA differentiation (pairwise Φ_{ST} values). (c) Histogram of SH values, highlighting the populations with the highest scores. See Table 1 for an explanation of abbreviations used.

generated a pairwise differentiation matrix for the 32 populations (Table S1). A NeighborNet based on this matrix (Figure 2b) [26] was then constructed in SplitsTree v. 4.11 [41] under default assumptions. Negative Φ_{ST} values were treated as being equal to zero. Because the size of each population is not known, for the purposes of illustration, we gave each population an extinction probability $p_i = 0.5$ when calculating HED—an approach similar to the "PD50" metric used by FISHBASE (www.fishbase.org) [55].

Example B. Mountain pygmy-possums (*Burramys parvus*) are alpine specialists restricted to three small regions of the Australian Alps (Figure 3a). The species depends on block streams and block fields found above 1,400 meters—habitats less than 10 km^2 in total extent [56]. The areas where mountain pygmy-possums still occur are particularly sensitive to destruction and fragmentation. Surveys conducted in the 1990s estimated the adult population size to be 2,600 [57]. A decade later this number had decreased to below 2,000 [56], with signs of continued decline [58]. At present, the IUCN lists mountain pygmy-possums as critically endangered [59].

Because of its restricted distribution and high extinction risk, the species has been subject to extensive population genetic research [58,60–62]. Unlike our example with spotted owls, direct estimates of population sizes are available, within-population sample sizes are uniformly large, and genetic data are available across the mountain pygmy-possums' entire range. This provides us with an opportunity to compare SH to HED and assess the effect of variable population sizes on conservation ranking.

We used a published matrix of genetic differentiation (F_{ST}) based on data from 8 microsatellite loci [58] to construct a phylogenetic network for 13 mountain pygmy-possum populations (Figure 3b). Our methods for generating NeighborNet outputs, and for computing SH and HED, were the same as above.

We modeled the probabilities of extinction for individual populations (p_i) of a given size (n_i) as a negative exponential

$$p_i \propto e^{-cn_i}$$

where the constant of proportionality c is $-\ln(P)/N$, with P being the probability that the entire species goes extinct and N being the total census size of the species (the sum of n_i). We used a conservative 100-year extinction probability for the entire species, $P = 0.4$, to derive HED (see [63]).

Results

Example A

As expected for a set of lineages with a recent history of gene flow, the network for spotted owls is quite non tree-like (Figure 2b). However, populations with the greatest degrees of genetic differentiation, relative to all other populations, occupy nodes subtending the longest edges. Populations at relatively isolated

Table 1. Spotted owl populations sampled by Barrowclough *et al.* [48,51] and ranked by Shapley value (SH) and heightened evolutionary distinctiveness (HED).

Pop.	Code	Subspecies	State	Sampling Locality	*n* ind.	*n* hap.	SH	HED
1	Huac	*lucida*	AZ	Huachuca Mountains	5	2	0.242	7.431E-03
2	Agua	*juanaphillipsae*	—	Aguascalientes, Sierra Fria, Mexico	1	1	0.191	3.983E-03
3	Mari	*caurina*	CA	Marin County	8	3	0.177	4.067E-03
4	Reef	*lucida*	UT	Capitol Reef National Park	9	4	0.133	3.683E-03
5	Wena	*caurina*	WA	Cascade Range	10	8	0.111	1.777E-03
6	Olym	*caurina*	WA	Olympic Peninsula	10	4	0.106	1.729E-03
7	Zion	*lucida*	UT	Zion National Park	7	4	0.105	2.851E-03
8	SanB	*occidentalis*	CA	San Bernardino Mountains	15	1	0.091	9.418E-04
9	SanJ	*occidentalis*	CA	Mount San Jacinto	15	1	0.091	9.418E-04
10	Rinc	*lucida*	AZ	Rincon Mountains	8	4	0.089	2.542E-03
11	Will	*caurina*	OR	Willamette National Forest	15	8	0.081	7.621E-04
12	Carm	*occidentalis*	CA	Carmel Valley	10	1	0.079	5.187E-04
13	ElDo	*occidentalis*	CA	El Dorado National Forest	15	4	0.077	6.214E-04
14	Zaca	*juanaphillipsae*	—	Zacatecas, Sierra de Urica, Mexico	1	1	0.076	2.079E-03
15	Palo	*occidentalis*	CA	Mount Palomar	8	1	0.074	3.994E-04
17	Humb	*caurina*	CA	Humboldt and Siskiyou Counties	30	11	0.072	4.528E-04
16	Sequ	*occidentalis*	CA	Sierra National Forest	15	6	0.071	3.422E-04
18	Pina	*lucida*	AZ	Pinaleno Mountains Graham County	4	2	0.071	1.605E-03
19	Shas	*caurina*	CA	Klamath and Shasta National Forests	16	8	0.067	3.638E-04
20	Mant	*lucida*	UT	Manti-La Sal National Forest	2	2	0.054	1.276E-03
21	Sacr	*lucida*	NM	Sacramento Mountains	8	6	0.047	1.284E-03
22	Lass	*occidentalis*	CA	Lassen National Forest	11	6	0.041	1.006E-04
23	Flag	*lucida*	AZ	San Fransisco Peaks	4	4	0.027	2.144E-04
24	Blac	*lucida*	NM	Black Range	8	6	0.026	3.461E-04
25	Coco	*lucida*	AZ	Coconino Plateau	15	9	0.022	1.039E-04
26	Cata	*lucida*	AZ	Santa Catalina Mountains	5	3	0.021	6.359E-05
27	Pino	*lucida*	NM	Pinos Altos Mountains	5	4	0.020	2.832E-05
28	SanF	*lucida*	NM	San Fransisco Mountains	7	4	0.017	1.020E-06
29	Cann	*lucida*	CO	Near Canon City	4	4	0.017	8.505E-07
30	Rita	*lucida*	AZ	Santa Rita Mountains	4	4	0.017	1.374E-05
31	Mogo	*lucida*	AZ	Mogollon Mesa	8	6	0.017	8.578E-07
32	Tula	*lucida*	NM	Tularosa Mountains	15	12	0.017	2.034E-06

Number of individuals (*n* ind.), number of haplotypes (*n* hap.), SH, and HED scores from the present study are reported.

nodes, such as those from Mount San Jacinto and the Huachuca Mountains, share few mutations with neighboring populations and subsequently exhibit higher pairwise Φ_{ST} values (Table 2; Table S1). Conversely, the (uncorrected) pairwise Φ_{ST} values for closely-related populations are either negative (as great as -1) or close to zero, indicating higher levels of genetic differentiation within these populations than among them [52].

We observe strong geographic structure across the United States consistent with current subspecific designations (Figure 2b). Populations of *S. o. lucida* exhibit a more star-like phylogenetic network that may reflect historical isolation in the "sky islands" of the American Southwest [48]. The intermediate position of the Lassen National Forest population, in contrast, may be due to its location near the point of contact between southern *S. o. caurina* and northern *S. o. occidentalis* [51].

The results of our SH and HED ranking are shown in Table 1. As expected, populations at relatively isolated tips score higher than those closer to the interior of the network (Figure 2b, c). The rankings are highly consistent between the two metrics (Spearman rank correlation = 0.91), and the same populations receive top ranking for both SH and HED.

Example B

As with spotted owls, the most genetically differentiated populations of mountain pygmy-possums, namely those in the northern and southern areas of their range, occupy nodes that are separated from most other populations by long edges (Figure 3b). Overall the structure of our network is in good agreement with the species' present distribution (Figure 3a). Given the habitat requirements and limited dispersal ability of mountain pygmy-possums, it is not likely that Mount Buller and Kosciusko National Park still exchange migrants with the Bogong High Plains [58]. In contrast, the close grouping of central populations in our

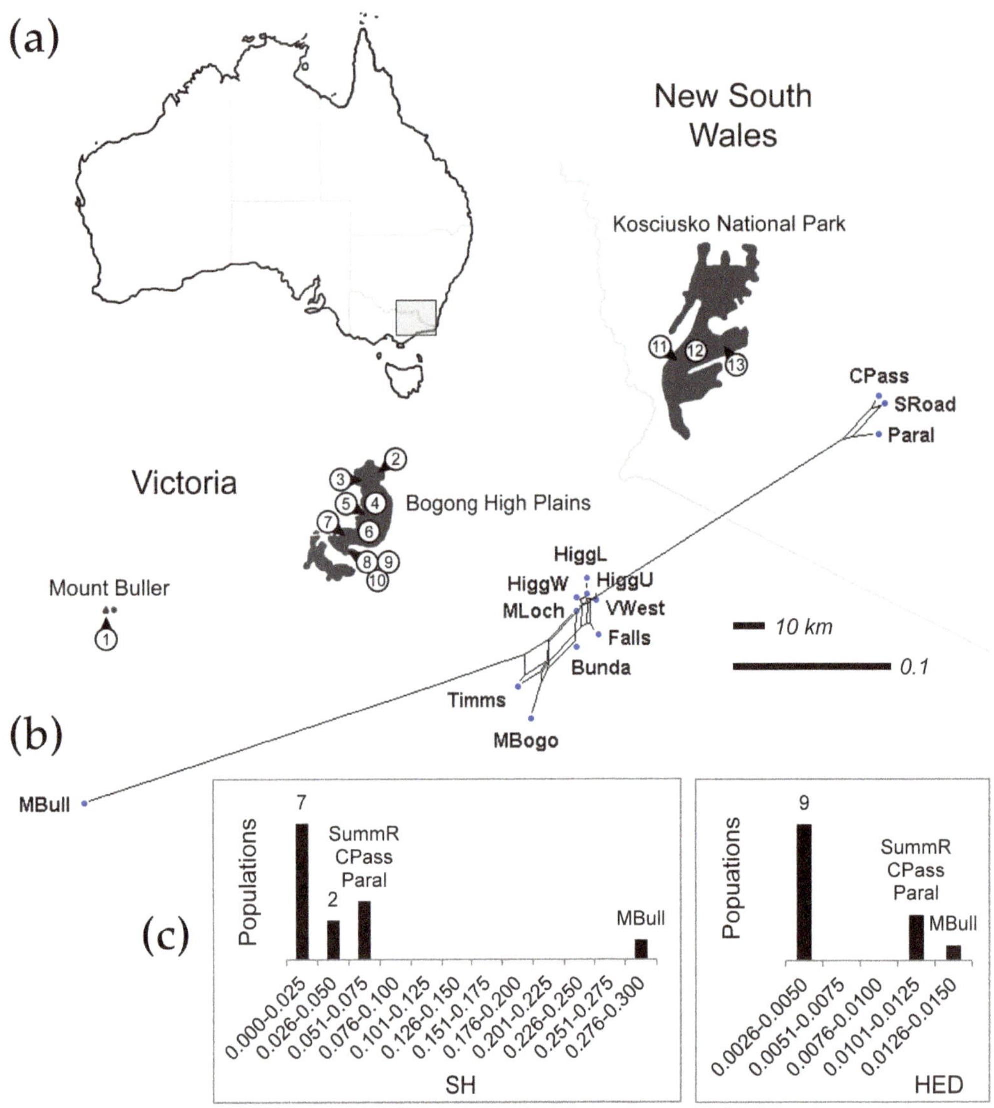

Figure 3. Conservation prioritization of mountain pygmy-possum (*Burramys parvus*) populations. (a) Distribution of mountain pygmy-possums in Australia (gray inset), showing populations sampled by Mitrovski *et al.* [58]. Shaded areas denote suitable habitat above 1,400 m. (b) NeighborNet of sampled populations based on microsatellite differentiation (pairwise F_{ST} values). (c) Histograms of SH and HED values, highlighting the populations with the highest scores. See Table 2 for an explanation of abbreviations used.

phylogenetic network, and subsequently their low SH and HED, is consistent with a shared history and/or recent gene flow.

The ranking results are shown in Table 2. Again, the phylogenetic network for mountain pygmy-possums reflects geographic distribution. Although we did not make *a priori* group assignments based on sampling location, the 13 populations still partition into northern, central, and southern regions. Again, outlying populations on the network tend to receive higher SH and HED scores. Unsurprisingly, the small and isolated Mount Buller population consistently ranks highest. For HED, no bias towards small or large populations is apparent; populations with high extinction probabilities do not necessarily receive high scores [35]. Again, although ranking order changes slightly between SH and HED, the two methods provide roughly equivalent rankings (Spearman rank correlation $=0.97$, Figure 3c). High-ranking populations are similar in both cases.

We note that SH and HED calculations on a network consider a taxon's distance from *all other* taxa. Thus, although the three northern populations are closely related to each other, they still receive high SH and HED scores because of the long branches separating them from the central and southern populations (Table 2; Figure 3c).

Discussion

The premise of conservation below the species level is not novel. Faith's original [6] discussion of prioritizing taxa also considered populations on a tree. Several economically-important taxa have received population-level management since the late 1980s, *e.g.*, Atlantic salmon (*Salmo salar* Linnaeus 1758) [64], brown trout (*Salmo trutta* Linnaeus 1758) [65] and yellowfin tuna (*Thunnus albacares* Bonnaterre 1788) [66]. Managing species at the population level implies at least an informal ranking scheme, one which would rely, for example, on estimates of habitat patch size or effective population size [67]. Habitat degradation, climate change, and the demands of a growing human population have

Table 2. Mountain pygmy-possum populations sampled by Mitrovski *et al.* [58] and ranked by Shapley value (SH) and heightened evolutionary distinctiveness (HED).

Pop.	Code	Region	Sampling Locality	*n* ind.	*n* all.	*r*	*N*	*pi*	SH	HED
1	MBull	Southern	Mount Buller	66	3.38	2.29	150	0.9072	0.292	1.497E-02
2	CPass	Northern	Charlottes Pass	44	6.00	5.21	45	0.9712	0.072	1.153E-02
3	Paral	Northern	Paralyser	40	6.63	5.77	22	0.9858	0.071	1.089E-02
4	SummR	Northern	Summit Road	43	6.13	5.16	25	0.9839	0.068	1.138E-02
5	MBogo	Central	Mount Bogong	42	6.50	5.77	100	0.9372	0.045	4.497E-03
6	Timms	Central	Timm Spur	120	6.88	5.32	120	0.9251	0.028	3.859E-03
7	Falls	Central	Falls Creek	35	6.63	5.73	30	0.9807	0.020	3.051E-03
8	HiggL	Central	Mount Higginbotham L	17	5.25	5.25	50	0.9681	0.019	2.850E-03
9	Bunda	Central	Bundara	78	7.25	5.45	120	0.9251	0.018	2.858E-03
10	HiggW	Central	Mount Higginbotham W	59	7.63	6.01	250	0.8502	0.013	2.659E-03
11	VWest	Central	Pretty Valley West	69	7.00	5.95	50	0.9681	0.010	2.541E-03
12	HiggU	Central	Mount Higginbotham U	56	7.25	5.76	50	0.9681	0.010	2.420E-03
13	Mloch	Central	Mount Loch	93	7.00	5.80	400	0.7714	0.010	2.646E-03

Number of individuals (*n* ind.), number of alleles (*n* all.), allelic richness (*r*), and adult population sizes (*N*) are reported from previously-published data. Probabilities of extinction (*pi*, with $P = 0.4$), SH, and HED scores from the present study are also shown.

ensured the continued fragmentation of species' ranges over the next century (see, *e.g.*, pikas (*Ochotona princeps* Richardson 1828) [68]). In the midst of such rapid change, managing an imperiled species over its entire range may no longer be feasible, such that population rankings may be necessary.

Phylogenetic diversity measures have previously been adapted for non-treelike population genetic data (*e.g.*, [24,25,36]). However, the PD complementarity scores that can be obtained from these methods are contingent, *i.e.*, subject to change if extinction alters the shape of the network. Ours is the first ranking scheme to consider a taxon's contribution to all possible future networks (*sensu* Weitzman [23]), a potentially relevant framework for preserving future biodiversity. Given the stochastic nature of extinction, the general ranking systems offered by SH and HED may be more useful to wildlife managers than those that only consider the present structure of a phylogenetic network. Unlike previous approaches based on PD (*e.g.*, [24,25,36]), SH and HED rankings allow one to lengthen or shorten the list of taxa to conserve in the event that resources become more or less available.

Molecular techniques are now inexpensive and robust enough to make population genetic sampling a standard component of conservation planning, and we argue that a phylogenetic network approach offers insight into a species' population structure complementary to the current statistical assessments of differentiation employed by MUs and DPSs [16,17]. We encourage researchers to employ such networks in future population genetic studies to provide conservation agencies with more informative analysis of datasets. Genotyping at multiple loci will provide more accurate estimates of population differentiation and allow for more sophisticated analyses of conservation-relevant processes such as recent demographic history and gene flow [69].

We acknowledge that the mathematical shortcomings of Φ_{ST} and F_{ST} estimators [70] may influence the magnitude and ranking of SH and HED scores, depending on the number of loci measured and the distribution of genetic diversity in a set of taxa. Our intention here is not to solve these theoretical problems but to demonstrate our network-based prioritization method with existing data. Newly-developed metrics such as Jost's D can be used to calculate SH and HED just as readily as traditional Φ_{ST} and F_{ST} distances, and we encourage the use of such unbiased estimators whenever such data are available. Indeed, any conservation metric of difference (*e.g.*, ecological, genomic, adaptability) can be used.

Several properties of the networks described here invite further investigation. In both our heuristic datasets, geographically peripheral populations are more genetically isolated, meaning they would rank highly on SH and HED. However, this was based on only very few putatively neutral markers. Two related questions concern how processes such as demographic history and current patterns of gene flow map onto genetic isolation as we measure it here, and also how phylogenetic networks map onto networks produced from ecological data (*e.g.*, niche use differences among populations).

We do not advocate relying solely on genetic isolation when deciding where and how resources should be allocated at the population level. Total population genetic diversity (*i.e.*, number of haplotypes) might also be considered. We note that in our examples, low-ranking populations tend to be geographically close to one another, meaning that their haplotypes are often shared. Important differences in ecology and adaptability [71] and current and future connectivity [72], must also be considered. However, our network approach and ranking system based on genetic differentiation can supplement existing systems of MUs and DPSs to improve the conservation of evolutionarily distinct populations in a world of increasing pressures and limited resources.

Supporting Information

Table S1 Pairwise genetic distances (Φ_{ST}) for spotted owl (*Strix occidentalis*) populations based on data from Barrowclough *et al.* [48,51], with negative values set to zero.

Acknowledgments

We thank the Isaac Newton Center (Cambridge University) for facilitating our collaboration, and M Hart, JB Joy, M Steel, three anonymous reviewers and the FAB Lab at Simon Fraser University for discussion and comments on the manuscript. All R scripts used in this paper are available in File S1.

Author Contributions

Conceived and designed the experiments: AOM VM AS. Performed the experiments: LV IM AS. Analyzed the data: LV AOM. Contributed reagents/materials/analysis tools: IM AS. Wrote the paper: LV AOM.

References

1. Sala OE, Chapin FS, Armesto JJ, Berlow E, Bloomfield J, et al. (2000) Global biodiversity scenarios for the year 2100. Science 287: 1770–1774.
2. Jenkins M (2003) Prospects for biodiversity. Science 302: 1175–1177.
3. Hughes JB, Daily GC, Ehrlich PR (1997) Population diversity: its extent and extinction. Science 278: 689–692.
4. Ceballos G, Ehrlich PR (2002) Mammal population losses and the extinction crisis. Science 296: 904–907.
5. Vane-Wright RI, Humphries CJ, Williams PH (1991) What to protect?—Systematics and the agony of choice. Biol Conserv 55: 235–254.
6. Faith DP (1992) Conservation evaluation and phylogenetic diversity. Biol Conserv 61: 1–10.
7. Bowen BW, Roman J (2005) Gaia's handmaidens: the Orlog model for conservation biology. Conserv Biol 19: 1037–1043.
8. Redding DW, Mooers AØ (2010) Can systematists help decide the relative worth of bits of biodiversity? Systematist 32: 4–8.
9. May RM (1990) Taxonomy as destiny. Nature 347: 129–130.
10. Haake CJ, Kashiwada A, Su FE (2008) The Shapley value of phylogenetic trees. J Math Biol 56: 479–497.
11. Pavoine S, Ollier S, Dufour AB (2005) Is the originality of a species measurable? Ecol Lett 8: 579–586.
12. Redding DR, Mooers AØ (2006) Incorporating evolutionary measures into conservation prioritization. Conserv Biol 20: 1670–1678.
13. Isaac NBJ, Turvey ST, Collen B, Waterman C, Baillie JEM (2007) Mammals on the EDGE: conservation priorities based on threat and phylogeny. PLoS One 2: e296.
14. Fay JJ, Thomas WL (1983) Endangered species listing and recovery priority guidelines. Fed Regist 48: 43098–43105.
15. Ryder OA (1986) Species conservation and systematics: the dilemma of subspecies. Trends Ecol Evol 1: 9–10.
16. Moritz C (1994) Defining 'evolutionarily significant units' for conservation. Trends Ecol Evol 9: 373–375.
17. Waples RS (1991) Pacific salmon, *Oncorhynchus* spp., and the definition of 'species' under the endangered species act. Mar Fish Rev 53: 11–22.
18. Firestone KB, Elphinstone MS, Sherwin WB, Houlden BA (1999) Phylogeographical population structure of tiger quolls *Dasyurus maculatus* (Dasyuridae: Marsupialia), an endangered carnivorous marsupial. Mol Ecol 8: 1613–1625.
19. Gorbics CS, Bodkin JL (2001) Stock structure of sea otters (*Enhydra lutris kenyoni*) in Alaska. Mar Mammal Sci 17: 632–647.
20. Haig SM, Wagner RS, Forsman ED, Mullins TD (2001) Geographic variation and genetic structure in spotted owls. Conserv Genet 2: 25–40.
21. Jensen EL, Govindarajula P, Russello MA (2013) When the shoe doesn't fit: applying conservation unit concepts to western painted turtles at their northern periphery. Conserv Genet. doi: 10.1007/s10592-013-0535-2.
22. Fitch WM (1997) Networks and viral evolution. J Mol Evol 44: S65–S75.
23. Weitzman ML (1992) On diversity. Q J Econ 107: 363–405.
24. Minh BQ, Pardi F, Klaere S, von Haeseler A (2009) Budgeted phylogenetic diversity on circular split systems. IEEE/ACM Trans Comput Biol Bioinform 6: 22–29.
25. Minh BQ, Klaere S, von Haeseler A (2009) Taxon selection under split diversity. Syst Biol 57: 586–594.
26. Bryant D, Moulton V (2004) Neighbor-Net: an agglomerative method for the construction of phylogenetic networks. Mol Biol Evol 21: 255–265.
27. Martyn I, Kuhn TS, Mooers AØ, Moulton V, Spillner A (2012) Computing evolutionary distinctiveness indices in large scale analysis. Algorithms Mol Biol 7: 6.
28. Bottrill MC, Joseph LN, Carwardine J, Bode M, Cook C, et al. (2008) Is conservation triage just smart decision making? Trends Ecol Evol 23: 649–654.
29. Witting L, Loeschcke V (1995) The optimization of biodiversity conservation. Biol Conserv 71: 205–207.
30. Faith DP, Walker PA (1996) DIVERSITY – TD. In: Faith DP, Nicholls AO, editors. BioRap: rapid assessment of biodiversity. Volume 3: tools for assessing biodiversity priority areas. Canberra: The Australian BioRap Consortium. pp. 63–74.
31. Faith DP (1996) Conservation priorities and phylogenetic pattern. Conserv Biol 10: 1286–1289.
32. Weitzman ML (1998) The Noah's Ark Problem. Econometrica 66: 1279–1298.
33. Faith DP (2008) Threatened species and the potential loss of phylogenetic diversity: conservation scenarios based on estimated extinction probabilities and phylogenetic risk analysis. Conserv Biol 22: 1461–1470.
34. Hartmann K (2013) The equivalence of two phylogenetic biodiversity measures: the Shapley value and the Fair Proportion index. J Math Biol 67: 1163–1170.
35. Steel M, Mimoto A, Mooers AØ (2007) Hedging our bets: the expected contribution of species to future phylogenetic diversity. Evol Bioinform Online 3: 237–244.
36. Minh BQ, Klaere S, von Haeseler A (2007) Phylogenetic diversity on split networks. Technical Report NI07090-PLG. Cambridge: Isaac Newton Institute. 19 p.
37. Poczai P, Mátyás KK, Szabó I, Varga I, Hyvönen J, et al. (2011) Genetic variability of thermal *Nymphaea* (Nymphaeaceae) populations based on ISSR markers: implications on relationships, hybridization, and conservation. Plant Mol Biol Rep 29: 906–918.
38. Buneman P (1971) The recovery of trees from measures of dissimilarity. In: Hodson FR, Kendall DG, Tautu P, editors. Mathematics in the archaeological and historical sciences. Edinburgh: Edinburgh University Press. pp. 387–395.
39. Steel M (2005) Phylogenetic diversity and the greedy algorithm. Syst Biol 54: 527–529.
40. Faith DP (1994) Phylogenetic pattern and the quantification of organismal biodiversity. Phil Trans R Soc Lond B 345: 45–58.
41. Huson DH, Bryant D (2006) Application of phylogenetic networks in evolutionary studies. Molec Biol Evol 23: 254–267.
42. R-Development Core Team (2009) R: a language and environment for statistical computing. Vienna: R Foundation for Statistical Computing. 1706 p.
43. Gutiérrez RJ, Franklin AB, Lahaye WS (1995) Spotted owl (*Strix occidentalis*). In: Poole A, editor. The birds of North America online. Ithaca: Cornell Lab of Ornithology. Available: http://bna.birds.cornell.edu/bna/species/179. Accessed 28 November 2011.
44. Dickerman RW (1997) Geographic variation in southwestern US and Mexican spotted owls, with the description of a new subspecies. In: Dickerman RW, editor. The era of Allen R. Phillips: a festschrift. Albuquerque: Horizon Communications. pp. 45–48.
45. Barrowclough GF, Gutiérrez RJ, Groth JG (1999) Phylogeography of spotted owl (*Strix occidentalis*) populations based on mitochondrial DNA sequences: gene flow, genetic structure, and a novel biogeographic pattern. Evolution 53: 919–931.
46. US Fish and Wildlife Service (2011) Revised recovery plan for the northern spotted owl (*Strix occidentalis caurina*). Portland, OR: US Fish and Wildlife Service. 277 p.
47. Gould GI (1985) Management of spotted owls by the California Department of Fish and Game. In: Gutiérrez RJ, Carey AB, editors. Ecology and management of the spotted owl in the Pacific Northwest. Portland, OR: US Department of Agriculture, Forest Service. pp. 21–26.
48. Barrowclough GF, Groth JG, Mertz LA, Gutiérrez RJ (2006) Genetic structure of Mexican spotted owl (*Strix occidentalis lucida*) populations in a fragmented landscape. Auk 123: 1090–1102.
49. Tamura K, Peterson D, Peterson N, Stecher G, Nei M, et al. (2011) MEGA5: molecular evolutionary genetics analysis using maximum likelihood, evolutionary distance, and maximum parsimony methods. Mol Biol Evol 28: 2731–2739.
50. Edgar RC (2004) MUSCLE: multiple sequence alignment with high accuracy and high throughput. Nucl Acids Res 32: 1792–1797.
51. Barrowclough GF, Groth JG, Mertz LA, Gutiérrez RJ (2005) Genetic structure, introgression, and a narrow hybrid zone between northern and California spotted owls (*Strix occidentalis*). Mol Ecol 14: 1109–1120.
52. Excoffier L, Smouse PE, Quattro JM (1992) Analysis of molecular variance inferred from metric distances among DNA haplotypes: application to human mitochondrial DNA restriction data. Genetics 131: 479–491.
53. Excoffier L, Laval G, Schneider S (2005) Arlequin (version 3.0): An integrated software package for population genetics data analysis. Evol Bioinform Online 1: 47–50.
54. Kimura M (1980) A simple method for estimating evolutionary rates of base substitutions through comparative studies of nucleotide sequences. J Mol Evol 16: 111–120.
55. Faith DP, Reid CAM, Hunter J (2004) Integrating phylogenetic diversity, complementarity, and endemism for conservation assessment. Conserv Biol 18: 255–261.
56. Heinze D, Broome LS, Mansergh I (2004) A review of the ecology and conservation of the mountain pygmy-possum *Burramys parvus* (Broom 1896). In: Hillis DM, Moritz C, Mable BK, editors. The biology of the Australian possums and gliding possums. Sunderland, MA: Sinauer Associates. pp. 254–267.
57. Mansergh I, Broome L (1994) The mountain pygmy-possum of the Australian Alps, Australia. Sydney: New South Wales University Press. 114 p.
58. Mitrovski P, Heinze DA, Broome L, Hoffmann AA, Weeks AR (2007) High levels of variation despite genetic fragmentation in populations of the endangered mountain pygmy-possum, *Burramys parvus*, in alpine Australia. Mol Ecol 16: 75–87.

59. Menkhorst P, Broome L, Driessen M (2008) Burramys parvus. IUCN Red List of Threatened Species. Available: http://www.iucnredlist.org/details/full/3339/0. Accessed 28 November 2011.
60. Osborne MJ, Norman JA, Christidis L, Murray ND (2000) Genetic distinctness of isolated populations of an endangered marsupial, the mountain pygmy-possum, *Burramys parvus*. Mol Ecol 9: 609–613.
61. Mitrovski P, Heinze DA, Guthridge K, Weeks AR (2005) Isolation and characterization of microsatellite loci from the Australian endemic mountain pygmy-possum, *Burramys parvus* Broom. Mol Ecol Notes 5: 395–397.
62. Mitrovski P, Hoffmann AA, Heinze DA, Weeks AR (2008) Rapid loss of genetic variation in an endangered possum. Biol Lett 4: 134–138.
63. Mooers AØ, Faith DP, Maddison WP (2008) Converting endangered species categories to probabilities of extinction for phylogenetic conservation prioritization. PLoS One 3: e3700.
64. Fontaine PM, Dodson JJ, Bernatchez L, Slettan A (1997) A genetic test of metapopulation structure in Atlantic salmon (*Salmo salar*) using microsatellites. Can J Fish Aquat Sci 54: 2434–2442.
65. Ferguson A (1989) Genetic differences among brown trout, *Salmo trutta*, stocks and their importance for the conservation and management of the species. Freshw Biol 21: 35–46.
66. Ward RD (1995) Population genetics of tunas. J Fish Biol 47: 259–280.
67. Lande R, Barrowclough GF (1987) Effective population size, genetic variation, and their use in population management. In: Soulé ME, editor. Viable populations for conservation. Cambridge: Cambridge University Press. pp. 87–123.
68. Galbreath KE, Hafner DJ, Zamudio KR (2009) When cold is better: climate-driven elevation shifts yield complex patterns of diversification and demography in an alpine specialist (American pika, *Ochotona princeps*). Evolution 63: 2848–2863.
69. Marko PB, Hart MW (2011) The complex analytical landscape of gene flow inference. Trends Ecol and Evol 26: 448–456.
70. Jost L (2008) GST and its relatives do not measure genetic differentiation. Mol Ecol 17: 4015–4026.
71. Bonin A, Nicole F, Pompanon F, Miaud C, Taberlet P (2007) Population adaptive index: a new method to help measure intraspecific genetic diversity and prioritize populations for conservation. Conserv Biol 21: 697–708.
72. Geffen E, Waidyaratne S, Dalén L, Angerbjörn A, Vila C, et al. (2007) Sea ice occurrence predicts genetic isolation in the artic fox. Mol Ecol 16: 4241–4255.
73. US Geological Service (2013) The national atlas of the United States of America. Washington, DC: US Department of Interior, Geological Service. Available: http://www.nationalatlas.gov/biology.html. Accessed 28 November 2011.

6

Genetic Structure and Hierarchical Population Divergence History of *Acer mono* var. *mono* in South and Northeast China

Chunping Liu[1,2]❡, Yoshiaki Tsuda[3]❡, Hailong Shen[1*¤], Lijiang Hu[1], Yoko Saito[2], Yuji Ide[2]

1 State Key Laboratory of Forest Genetics and Breeding, Northeast Forestry University, Harbin, Heilongjiang Province, China, **2** Laboratory of Forest Ecosystem Studies, Department of Ecosystem Studies, Graduate School of Agriculture and Life Science, The University of Tokyo, Tokyo, Japan, **3** Program in Plant Ecology and Evolution, Department of Ecology and Genetics, Evolutionary Biology Centre, Uppsala University, Uppsala, Sweden

Abstract

Knowledge of the genetic structure and evolutionary history of tree species across their ranges is essential for the development of effective conservation and forest management strategies. *Acer mono* var. *mono*, an economically and ecologically important maple species, is extensively distributed in Northeast China (NE), whereas it has a scattered and patchy distribution in South China (SC). In this study, the genetic structure and demographic history of 56 natural populations of *A. mono* var. *mono* were evaluated using seven nuclear microsatellite markers. Neighbor-joining tree and STRUCTURE analysis clearly separated populations into NE and SC groups with two admixed-like populations. Allelic richness significantly decreased with increasing latitude within the NE group while both allelic richness and expected heterozygosity showed significant positive correlation with latitude within the SC group. Especially in the NE region, previous studies in *Quercus mongolica* and *Fraxinus mandshurica* have also detected reductions in genetic diversity with increases in latitude, suggesting this pattern may be common for tree species in this region, probably due to expansion from single refugium following the last glacial maximum (LGM). Approximate Bayesian Computation-based analysis revealed two major features of hierarchical population divergence in the species' evolutionary history. Recent divergence between the NE group and the admixed-like group corresponded to the LGM period and ancient divergence of SC groups took place during mid-late Pleistocene period. The level of genetic differentiation was moderate (F_{ST}= 0.073; G'_{ST} = 0.278) among all populations, but significantly higher in the SC group than the NE group, mirroring the species' more scattered distribution in SC. Conservation measures for this species are proposed, taking into account the genetic structure and past demographic history identified in this study.

Editor: Fei-Hai Yu, Beijing Forestry University, China

Funding: This study was partially supported by Heilongjiang Provincial Key Technologies R & D Program (GA09B202-06) and by the Special Funding from the Support Plan for Distinguished Researchers in Northeast Forestry University, China. The funders had no role in study design, data collection and analysis, decision to publish, or preparation of the manuscript.

Competing Interests: The authors have declared that no competing interests exist.

* E-mail: shenhl-cf@nefu.edu.cn

¤ Current address: School of Forestry, Northeast Forestry University, Harbin, China

❡ These authors contributed equally to this work.

Introduction

Northeast China contains the country's most important forest areas, but since the late 19th century forests in this region have been severely disturbed by wars, fires, deforestation, overexploitation and changes in land use to agriculture and grazing. Consequently, many tree species are now threatened [1]. Some conservation measures, such as designating areas as non-commercial forest zones and forbidding clear-cutting in degraded sites, have been applied in this region, but degradation of forests is still a serious threat [2–3]. The genetic diversity of historical lineages cannot be recovered if it is lost [4]. Thus, it is essential to understand the biological and genetic features of the tree species in a region's forests [5–6] in order to formulate effective conservation and restoration programs that will maintain the diversity required to adapt to environmental changes [4,9–10]. This requires robust assessments of the genetic diversity, structure and demographic history of species [7–8]. Further, the focus must be on preserving the overall genetic diversity and natural evolutionary processes in meta-populations rather than the diversity within separate subpopulations [11]. Thus, large-scale genetic structure analysis covering species' ranges has been applied to numerous forest tree species, especially species that are economically, ecologically, and/or genetically important for a specific region [5–6,8,11–16].

The genetic diversity and population structure of various tree species in Northeast China have been evaluated using several types of molecular markers, for example *Phellodendron amurense* (AFLP) [17], *Quercus mongolica* (RAPD, AFLP and ISSR) [18–20], *Tilia amurensis* (ISSR) [21], and *Fraxinus mandshurica* (SSR) [14]. The genetic diversity of *Q. mongolica* [19] and *F. mandshurica* [14] has been shown to decrease with increasing latitude. Hu *et al.* [14] postulated that the detected cline in *F. mandshurica* populations arose from northward expansion of the forest biomes in the mid-Holocene from a single refugium located in eastern China during the last glacial maximum (LGM) [22]. This putative post-LGM population expansion to the north from southern refugium is believed to have shaped the modern genetic structure of many species in temperate zones in the northern hemisphere [23–24]. However, it is still not clear whether the decrease in genetic diversity with increasing latitude in the northern populations is a common pattern for tree species in Northeast China. Moreover, no coalescence-based genetic analysis has been previously applied to plant species in Northeast China, thus it is uncertain whether the reductions in genetic diversity are related to changes in species distribution driven by climate changes during the LGM. Recent forest genetics studies have suggested that an ancient demographic event predating the LGM has continuing effects on modern genetic diversity and structure [25–28]. Therefore, it is essential to evaluate the past population demography over sufficient time to elucidate the evolutionary history of a tree species and apply the acquired knowledge when formulating forest management and conservation strategies. Recently, a very powerful and flexible approach, Approximate Bayesian Computation (ABC), has been developed to estimate demographic and historical parameters (e.g. effective population size, divergence time) and quantitatively compare alternative scenarios [29]. This approach can be used to address many important questions in ecology and evolutionary biology [29].

Acer mono (mono maple) is a common deciduous tree species that is widely distributed in east Asia, where there are many varieties [30–32], and an important companion tree species in the temperate mixed Korean pine and broadleaved forests of Northeast China [33]. It is an economically, ecologically, and culturally important species because of the unique properties of its wood, potential for production of sugar, oil, and medicinal materials, excellence for landscape planting and forestation of barren hills, and especially its attractiveness as an ornamental due to its vibrant red autumn foliage and fog-tolerance [34]. In China, *A. mono* is mainly distributed in Northeast China [32], but there are also scattered populations in forests of North China and the middle and lower reaches of the Yangtze River (Fig. 1). The natural resources of *A. mono* have decreased, and almost disappeared, at some sites due to: (1) harvesting of its wood for timber or fuel; (2) inappropriate forest management activities such as removal of all undergrowth species during thinning; and (3) of most importance recently, transplantation from natural forests to urban areas for use as ornamental, shade, or landscape trees. The conservation and sustainable management of natural populations of *A. mono*, as well as breeding and artificial planting of the species, are therefore now receiving more attention. However, few molecular ecology studies have addressed the genetic diversity and structure of natural populations of *A. mono* in China, although several have examined the reproduction and mating patterns of natural populations of the species in Japan [35–37].

We therefore conducted large-scale genetic analysis of *A. mono* using nuclear SSR markers to: 1) evaluate the wide-range genetic structure of 56 populations selected to cover its distribution in northeastern and southern China, 2) infer its population demographics using an ABC approach and 3) formulate conservation and management strategies for this natural genetic resource in Northeast China, taking into account its current genetic structure and inferred evolutionary history.

Materials and Methods

Ethics Statement

No specific permission was required for sampling activities in most populations because we sampled in the wild and only a few leaves were collected from each tree, except for some populations located in National Nature Reserves (listed after the corresponding populations in the "*Sampling and study sites*" section), in which the sampling activities were permitted by the relevant authorities. The field studies did not involve disturbance of any endangered or protected species.

Sampling and Study Sites

We collected leaves of *Acer mono* var. *mono* from 48 natural populations, covering almost its entire distributions in Northeast China (hereafter NE populations), and eight in the rest of China (hereafter southern China, SC), where there are scattered, low-density populations (Fig. 1).

The NE populations were distributed in two major mountain ranges (numbers in brackets indicate assigned designations): (i) the Xiaoxing'anling Mountains (1–12), and (ii) the Changbai Mountains (13–48) including the main range (33–42, 44), branches such as the Zhangguangcailing (13–24) and Laoyeling Mountains (27, 31,32), and ranges extending outwards such as the Wandashan (25, 26, 28–30), Longgangshan (43, 45–47), and Qianshan Mountains (48) (Fig. 1). The individuals present at a location were treated as "a population", a common approach in traditional genetic structure studies, thus several locations were examined within these ranges in NE. The SC populations are separated by much greater geographical distances than the NE populations, and their distributions are discontinuous because they are all located in separate mountain ranges: the Yanshan (49; Wulingshan National Nature Reserve, Hebei Province, China), Baihuashan (50), Zhongtiaoshan (51), Qinling (52), Jigongshan (53; Jigongshan National Nature Reserve, Henan Province, China), Langyashan (54; Langyashan National Nature Reserve, Anhui Province, China), Tianmushan (55; Tianmushan National Nature Reserve, Zhejiang Province, China), and Xuefengshan (56; Kanglong National Nature Reserve, Hunan Province, China) Mountains (Fig. 1). Individuals within each population were separated from each other by at least 30 m. In total, leaves from 1702 individuals were collected and stored in silica gel at room temperature until required for DNA extraction. The height and diameter at breast height of each individual were recorded when sampling.

DNA Extraction and Genotyping

DNA was extracted using a modified CTAB method [38]. Nuclear SSR markers were selected from primers previously developed for Japanese *A. mono* [35] and other *Acer* species [39–41]. Seven of these markers (Am116, Am118, Am258, Am340, Am607 and Am742 [35], and Aca24 [39]) were found to be sufficiently stable and polymorphic and therefore used in this study.

Primers were mixed to a final concentration of 2 μM, except for primer Am258 (4 μM). The target sequences were then amplified in 5 μL reaction mixtures containing 2.5 μL of 2× Qiagen Multiplex PCR Master Mix, 0.5 μL of the primer mixture, 1.5 μL H_2O, and 0.5 μL template DNA by PCR (95°C for 15 min, followed by 30 cycles of 94°C for 30 s, 58°C for 90 s, and 72°C for 60 s, with a final extension at 60°C for 30 min) using a TaKaRa TP600 Thermal Cycler. All PCR products were genotyped using an ABI 3130 Genetic Analyzer and the results were analyzed using GeneScan analysis software and Genotyper software version 3.5 (Applied Biosystems).

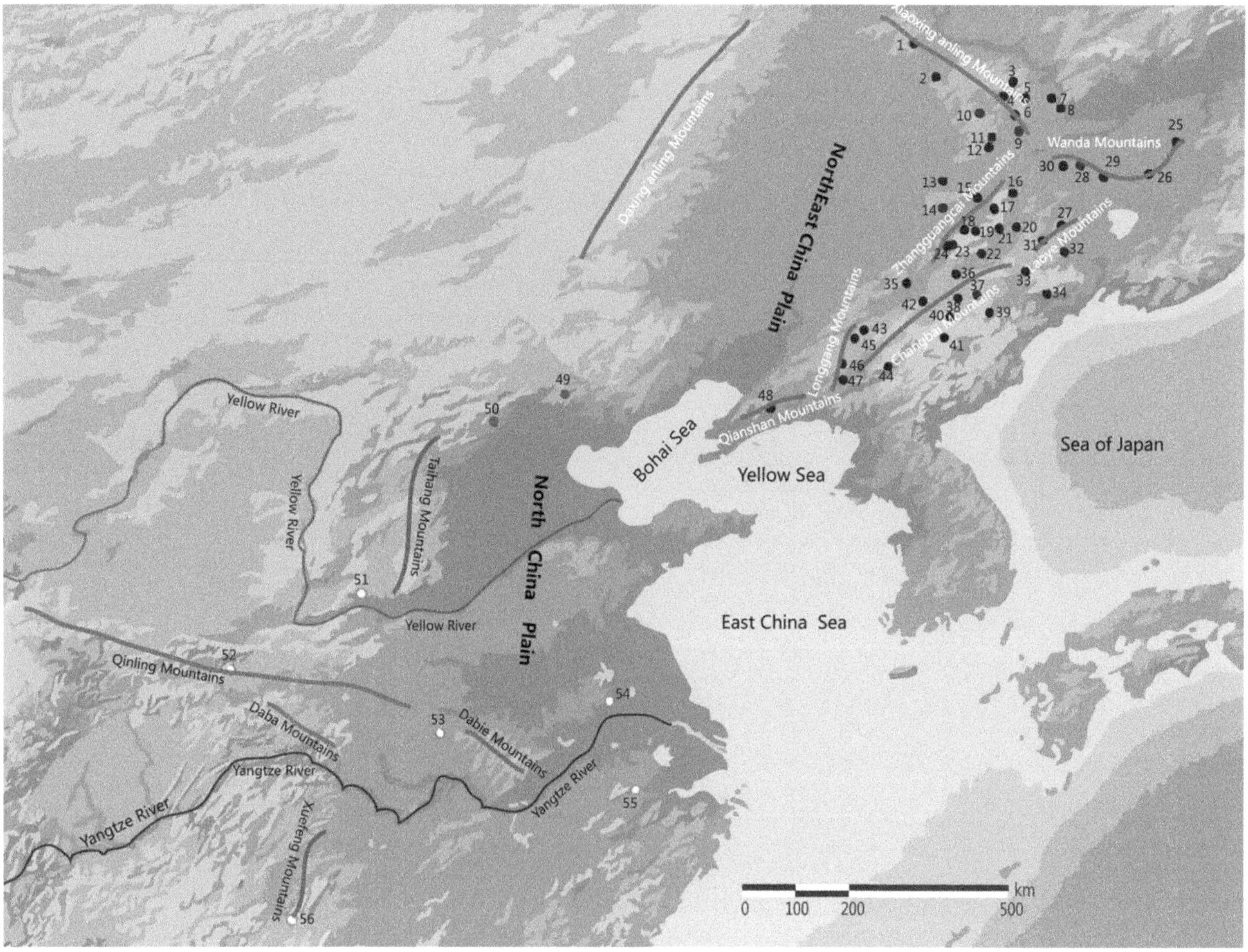

Figure 1. Locations of the 56 populations of *Acer mono* and geography of the study regions in China.

Genetic Diversity within Populations

The FSTAT 2.9.3.2 software package [42] was used to evaluate the genetic diversity of each population by calculating the total number of alleles detected (N_A), range of allele size (R_{AS}), allelic richness (A_R), expected heterozygosity (H_E), and fixation index (F_{IS}). In addition, significant deviations from Hardy-Weinberg equilibrium, as indicated by deviations of F_{IS} from zero, and genotypic disequilibrium for all locus pairs in each population, were tested by randomization using FSTAT. Rare allelic richness (RA_R; [8]) and private allelic richness (PA_R; [8]) were calculated. Pearson Correlation Analysis, implemented in SPSS 13.0 for Windows [43], was applied to determine possible correlations between genetic diversity parameters (A_R, RA_R, PA_R and H_E) and geographical gradients (of altitude, latitude and longitude). Since clear genetic differentiation was detected between the NE group (Populations 1–48) and the SC group (Populations 51–56) in the NJ tree and by STRUCTURE analysis (see *Results*), average values of A_R, H_E, F_{IS} and F_{ST} for each group and their differences between the two groups were tested using permutation tests in FSTAT. RA_R and PA_R of each group and the differences between the two groups were tested by One-way ANOVA using SPSS.

Genetic Differentiation and Structure

The genetic differentiation among populations was evaluated by calculating F_{ST} values [44], and their significance was tested by comparison to 95% and 99% confidence intervals acquired from 1000 bootstraps. Pairwise F_{ST} values were also calculated, and the significance of pairwise population differentiation was tested by randomizing multilocus genotypes between the pairs using FSTAT. The source of genetic variation was determined by analysis of molecular variance analysis (AMOVA) [45], using GENALEx6.1 [46]. Standardized genetic differentiation (G'_{ST}) was then calculated following Hedrick [47].

Isolation by distance (IBD) [48] was evaluated with GenAlEx 6.4 [46] using the method of Rousset (1997) [49], which tests for significant association between pairwise population differentiation ($F_{ST}/(1-F_{ST})$) and the natural logarithms of direct minimum geographic distance between populations. The genetic relationships among populations were evaluated by generating a Neighbor-Joining (NJ) tree based on the D_A genetic distances [50], using Populations 1.2.30 beta software [51]. Statistical confidence in the topology of the tree was evaluated using 1000 bootstraps derived using the same software. The NJ tree was reconstructed on a topographic map using Mapmaker and GenGIS2 software packages [52]. Individual-based genetic structure was evaluated by Bayesian clustering as described by Pritchard*et al.* [53] using STRUCTURE 2.3.3 (hereafter, STRUCTURE analysis) [54]. We assumed a pre-assigned number of genetic clusters (K) ranging from 1 to 20. All runs involved 100,000 Markov Chain Monte Carlo (MCMC) generations, after a burn-in period of 100,000 iterations according to Hubisz *et al.* [55], based on the LOCPRIOR model described by Hubisz *et al.* [55], an admixture model and the correlated allele frequencies model (hereafter, the *F*-model) described by Falush *et al.* [56]. Ten runs were performed for each value of *K*. The optimum number of *K* was evaluated by calculating ΔK values [57] and applying the methods suggested in the original paper describing STRUCTURE analysis [53], which involves evaluating genetic structure by comparing mean log likelihoods of postulated models. Bar charts for the proportions of the membership coefficient of each individual in STRUCTURE analysis over 10 runs for each *K* were summarized using CLUMPP software [58] and visualized using DISTRUCT software [59].

Population Demographic Analysis

DIYABC v1.0.4.39 and v2.0 software [60–61] were used to infer the species' demographic history by a two-step ABC approach. Populations covering the whole species' distribution were first analyzed to infer the history of the topmost hierarchical genetic structure indicated by the STRUCTURE analysis (see *Results*, Fig. 2). Further ABC analysis then focused solely on SC populations as clear genetic structure was still detected among them in not only the STRUCTURE analysis but also the NJ tree (see *Results*, Fig. 2).

A) Demographic inferences across the whole species range. To simplify scenarios for ABC analysis, we defined three populations based on the results of STRUCTURE analysis: Pop1 (Northeastern populations 1–48), Pop2 (admixed populations 49 and 50) and Pop3 (Southern populations 51–56). All sampled individuals of Pop2 (42 individuals) and Pop3 (177 individuals) were used for analysis. However, for Pop1, 200 individuals were randomly selected from the pool of 48 NE populations, as genetic structure among them was quite weak (F_{ST}= 0.029, Fig. 2C) and use of all individuals (1483 in total) would have been computationally expensive without substantially increasing the amount of information acquired. We examined four simple population demography scenarios (Fig. 3A), taking into account the results of the NJ tree and the STRUCTURE analysis. In these scenarios, t# and N# refer to timescale (expressed as generation time) and effective size of the corresponding populations (Pops1, 2, 3 or the ancestral population) during each time period (e.g. 0– t1, t1– t2). The scenarios are as follows:

Scenario 1 *(isolation with admixture model):* Pop2 was generated by admixture of Pops 1 and 3 at t1, then Pop1 merged with Pop3 at t2. Based on the results of STRUCTURE analysis, this scenario was expected to be the most likely one.

Scenario 2 *(hierarchical split model 1):* Northern Pop1 merged with Pop2 at t1 and subsequently with southern Pop3 at t2.

Scenario 3 *(hierarchical split model 2)*: Southern Pop3 merged with Pop2 at t1 and subsequently with Northern Pop1 at t2.

Scenario 4 *(simple split model):* all three populations diverged at the same time, t2.

In all scenarios, Pop2 was specified as the population to be traced back to an ancestral population (as required for DIYABC analysis). In addition, we assumed a bottleneck in the past at t3, which we positioned before the divergence point in all scenarios. We assumed past population growth and set the ancestral population size (Na in scenario 1 and Nb in scenarios 2–4) to be smaller than that of other populations. In DIYABC, we employed the higher mutation rate in the generalized stepwise mutation model (GSM) [62] with the lower mutation rate of single nucleotide indels (SNI) for modeling the mutation of SSRs. Default prior values were used for all these parameters (Table S1). The mean values of expected heterozygosity (H_E) and number of alleles (N_A) were used as the summary statistics for each of the three populations, while for summary statistics for each pair of populations from the three groups we used H_E, N_A, genotype likelihood and F_{ST}. A million simulations were performed for each scenario. After the total of three million simulations, the most-likely scenario was evaluated by comparing posterior probabilities using logistic regression. The goodness of fit of the scenario was assessed by the option "model checking" with principal component analysis (PCA) in DIYABC, which measures the discrepancy between model and real data.

B) Demographic inferences in South China. Because the STRUCTURE analysis indicated that ΔK was highest at K= 4 and detected clear genetic structure (see *Results*, Fig. 2D), we defined four populations: SC1 (Pops 51 and 52), SC2 (Pops53 and 54), SC3 (Pop 55) and SC4 (Pop 56) to simplify demographic analysis of the SC populations. We examined the hierarchical split models (scenarios 1–3) and simple split model (scenario 4) using DIYABC v2.0 as shown in Fig. 3B. In the hierarchical split model, the most ancient split was set between SC3 and the others because SC3 was the most highly genetically differentiated of the SC populations according to the NJ tree (Fig. 2B). After the split from SC3, all permutations of pairs of SC1, 2 and 4 populations were set in the hierarchical split models in scenarios 1–3. The effective population sizes of SC1, 2, 3 and 4 were designated N1, N2, N3 and N4, respectively (setting N1> N2> N3> N4, according to their genetic diversity). Default prior values were used for all these parameters (Table S2), and the same approach was applied as for the whole-range analysis described above.

Results

Genetic Diversity of Populations

In total, 223 alleles were scored over the seven loci among the 1702 sampled *A. mono* individuals. The allelic diversity was highly variable across loci, with N_A ranging from 14 to 45 (32±12 on average), A_R from 4.824 to 16.778 (10.867±3.981 on average), and H_E from 0.587 to 0.861 (0.737±0.110 on average) (Table 1). The fixation index (F_{IS}) for individual loci ranged from –0.116 to 0.223 and the average over all loci was 0.053. No significant genotypic linkage disequilibrium for pairwise-loci was detected within any population (P>0.05).

The genetic diversity of each population is presented in Table 2. Values of N_A, R_A, P_A, A_R, RA_R, PA_R and H_E ranged from 39 to 84, 8 to 55, 0 to 7, 5.181 to 11.210, 1.002 to 6.031, 0 to 1.293 and 0.535 to 0.840, respectively (Table 2). Although significant deviation of the F_{IS} value from zero was detected in 11 of 56 populations, there was no geographic pattern in the distributions of F_{IS} values (Table 2). When the correlations between genetic variation parameters (A_R and H_E) and geographical gradients (latitude, longitude and altitude) were evaluated, significant positive correlation was detected only between H_E and latitude (P<0.05, Table 3). However, when geographical pattern of genetic diversity was analyzed within each of the NE and SC regions, different trends were detected: A_R significantly decreased with increasing latitude within NE populations (P<0.01), while both A_R and H_E increased together with latitude within SC populations (P<0.05, Table 3).

Population Genetic Structure

The F_{ST} among all populations was 0.073, with 95% and 99% confidence intervals (CI) ranging from 0.056 to 0.093 and 0.051 to 0.1, respectively, and 66.35% of pairwise-F_{ST}values were significant (P<0.05). The overall value of F_{ST} corresponded to a standardized genetic differentiation (G'_{ST}) of 0.278, suggesting

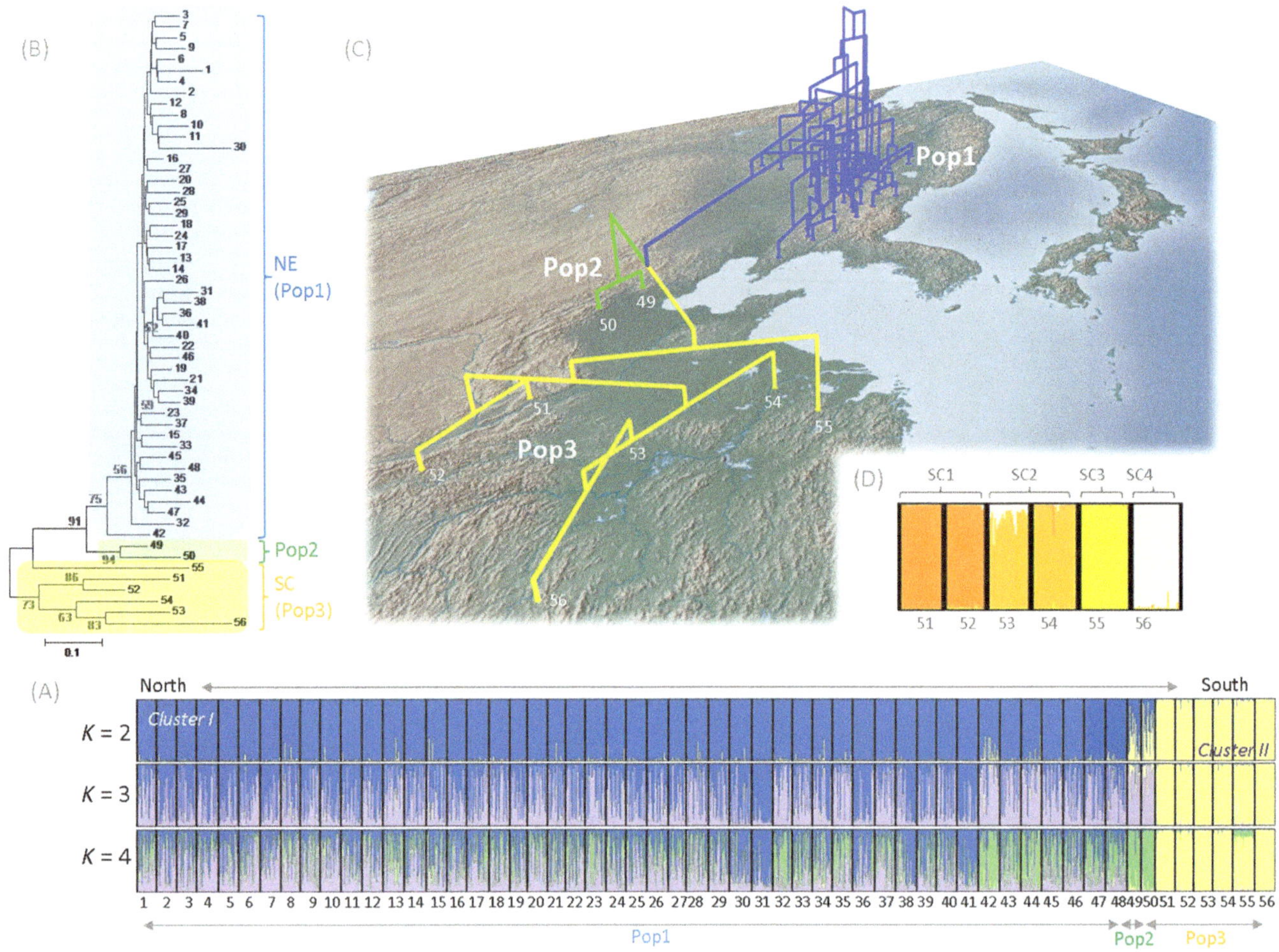

Figure 2. Results of STRUCTURE analysis and Neighbor-joining (NJ) tree of the 56 populations of *Acer mono* and STRUCTURE analysis of the South China populations. A, the proportion of the membership coefficient for each individual from 56 *A. mono* populations for the inferred clusters when $K = 2$ to 4 according to the STRUCTURE analysis; B and C, Neighbor-joining (NJ) tree of 56 populations based on D_A distance (Nei et al. 1983), showing bootstrap values exceeding 50%, definition of the three populations (Pop1, 2, and 3) used in DIYABC (A, B, C); D, the proportion of the membership coefficient for each individual from 6 SC populations for the inferred clusters when $K = 4$ according to the STRUCTURE analysis.

moderate population structure. Separate analysis of the genetic variation in NE and SC populations, based on results of the STRUCTURE analysis and the NJ tree (see below), showed that H_E and RA_R values were significantly higher for NE than SC population ($H_E = 0.743$ and 0.676, $RA_R = 3.825$ and 2.799, respectively; Table 4). In contrast, PA_R and F_{ST} values were significantly higher for SC than NE populations ($PA_R = 0.309$ and 0.006, $F_{ST} = 0.207$ and 0.029, respectively), but no significant differences in A_R was detected (Table 4). The F_{IS} values of both groups were almost zero (0.052 in NE and 0.042 in SC) and there was no significant differentiation between them (Table 4). The G'_{ST} values for the NE and SC groups were 0.114 and 0.722, respectively, suggesting strong genetic structure in the SC group. Thus, 96.4% of the pairwise-F_{ST} values were significant ($P<0.05$) in the SC group while the proportion was much lower (56.2%) in the NE group. Moreover, separate analysis of the pattern of genetic diversity within the groups showed that genetic variation decreased with increasing latitude in NE populations, but increased in SC populations (Table 3). Analysis of molecular variance based on these two groups showed that within-population variation, variation among populations in each region and between-region variation accounted for 76%, 16% and 8% of the total detected genetic variation, respectively ($P = 0.01$). Mantel tests indicated that there was significant isolation by distance in all populations ($R^2 = 0.4241, P = 0.000$), but not within either the NE or SC populations ($R^2 = 0.009, P = 0.090$ for NE populations; $R^2 = 0.081, P = 0.060$ for SC populations).

According to the whole-range STRUCTURE analysis, the probability of the LnP(D) data increased progressively with increases in K up to $K = 8$, where it started to plateau (Fig. S1). However, the results for $K>4$ showed complex admixture patterns and none of the clusters were specific to any populations or regions. Since ΔK indicated that the optimal value for K was 2 (Fig. S1), in the following text discussion of population genetic structure is mainly based on results acquired with $K = 2$. In this case, all NE populations clustered into one group, and populations 51–56 clustered into the other group. Populations 49 and 50, located in Northern China, adjacent to the regions occupied by both groups, appeared to possess admixed-like features of the two groups (Fig. 2A, $K = 2$). However, with values of $K = 3$ and 4, although

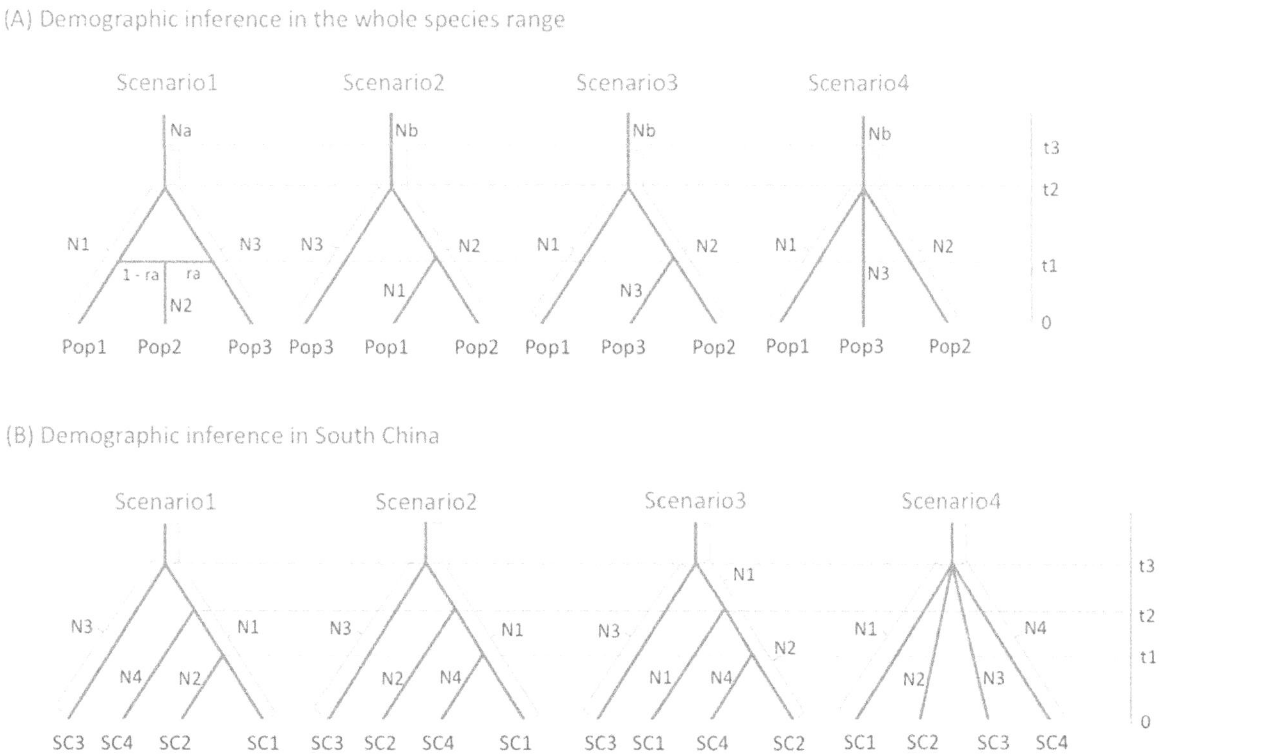

Figure 3. The four demographic scenarios for all populations (A) and South China populations (B) examined in DIYABC. t# is time scale measured in generations and N# is effective population size of the corresponding populations (A, Pop 1, 2, 3, a and b; B, SC1, 2, 3 and 4) during each time period (e.g. 0–t1, t1–t2, t2–t3).

genetic structure was not clear over all NE populations, these two populations were assigned to a specific cluster. The NJ tree showed similar genetic structure to that detected by the STRUCTURE analysis, with two main groups corresponding to the NE and SC populations (Fig. 2B, C). Populations 49 and 50, for which STRUCTURE analysis suggested admixture when $K = 2$, fell between the two groups (Fig. 2B, C). In addition, most of the nodes in the NE populations were poorly supported by the bootstraps due to weak population structure, but most of the nodes for Populations 49 and 50, and SC populations, were supported by higher bootstrap values (Fig. 2B), indicating strong genetic structure. Because of higher differentiation, STRUCTURE analysis on the 6 populations in SC was conducted. These six populations divided into four clusters, according to the highest ΔK (Fig. 2D) [57], and when $K = 6$ they were clearly divided into six population-specific clusters (Fig. S2A, $K = 6$). Moreover, separate STRUCTURE analysis of NE populations detected weak genetic structure among populations associated with the mountain ranges, notably the 12 populations located in the Xiaoxing'anling Mountains (1–12) were almost entirely dominated by a single cluster (Fig. S2B, $K = 2$).

Table 1. Genetic characteristics of the seven nSSR loci examined in all sampled individuals from 56 natural populations of *A. mono*.

Loci	R_{AS} (bp)	N_A	A_R	H_E	F_{ST}
Aca24	90–113	14	4.824	0.587	0.088
Am116	236–298	45	16.778	0.833	0.077
Am118	153–189	18	7.301	0.610	0.121
Am258	164–259	39	9.628	0.740	0.071
Am340	164–242	36	12.120	0.702	0.072
Am607	130–180	27	11.997	0.861	0.035
Am742	150–203	44	13.424	0.829	0.060

Abbreviations: R_{AS}, range of allele sizes; N_A, number of alleles detected; A_R, allelic richness; H_E, expected heterozygosity.

ABC-based Inferences of Population Demography

In DIYABC analysis, based on the whole range STRUCTURE analysis, scenario 1, (the isolation with admixture model) was expected to be the most likely. However, the highest posterior probability was found for scenario 2 (the hierarchical split model 1), and the value (0.7341, 95% CI: 0.7079–0.7602) was much higher than those for scenario 1 (0.2557), 3 (0.0032) and 4 (0.0071) (Table 5A). In scenario 2, the median values of effective population size of N1 (Pop1), N2 (Pop2), N3 (Pop 3), and Nb (ancestral population) were 6570, 3700, 8700 and 1470, respectively (Table S3). Estimated median times of recent divergence (t1 for Pop 1 and Pop2) and ancient divergence (t2 for southern Pop3 and the others) were 681 and 2960 generations ago, respectively. Assuming the generation time of *A. mono* to be 30 years, the divergence times of t1 and t2 correspond to 20,430 and 88,800 years ago, respectively. The median time for a pronounced increase in population size (t3) was estimated at 6480 generations ago (194,400 years with a generation time of 30 years). However, the posterior distribution pattern suggested that this estimate was

Table 2. Locations and sizes of individuals in the 56 sampled populations of *Acer mono* and estimates of genetic diversity within populations.

Mountains		Locations	Lat (°N)	Long (°E)	Alt (m)	Height (m)	DBH (cm)	N	N_A	R_A	P_A	A_R	RA_R	PA_R	H_E	F_{IS}
Xiaoxing'anling		1 Heihe	49.590	126.786	460–480	6.2±2.3	9.30±4.32	29	43	8	0	5.992	1.002	0	0.684	−0.001
		2 Zhanhe	48.726	127.386	400–528	4.6±1.4	6.38±2.79	30	52	22	0	6.905	2.627	0	0.704	0.005
		3 Wuyiling	48.605	129.450	429–476	9.1±2.6	12.77±10.76	30	64	28	1	8.215	3.105	0.017	0.748	0.191*
		4 Wuying	48.234	129.201	366–430	11.0±2.8	12.81±6.54	33	65	35	0	7.964	3.691	0	0.693	−0.118
		5 Xinqing	48.193	129.780	455–550	14.0±2.5	19.90±7.16	30	60	24	0	7.791	2.705	0	0.703	−0.037
		6 Meixi	47.752	129.511	360–496	11.1±3.3	19.91±10.86	32	68	36	0	8.462	3.900	0	0.753	0.129*
		7 Hebei	48.187	130.483	234–388	13.3±2.8	15.49±7.48	31	67	37	1	8.325	4.043	0.016	0.735	0.009
		8 Luobei	47.917	130.731	199–226	11.1±3.1	11.80±4.67	29	69	28	0	9.071	3.245	0	0.790	0.246*
		9 Jinshantun	47.344	129.618	254–297	9.4±2.5	–	30	65	23	1	8.518	2.544	0.017	0.762	0.019
		10 Cuiluan	47.796	128.567	332–377	12.2±3.2	15.16±6.35	31	68	37	0	8.456	4.033	0	0.761	0.104
		11 Dailing	47.182	128.895	380–440	10.6±2.3	18.39±6.40	31	69	33	1	8.693	3.561	0.016	0.752	0.051
		12 Langxiang	46.922	128.821	337–418	10.7±2.1	18.36±8.09	31	65	36	0	8.430	4.292	0	0.743	0.075
Changbai	ZH	13 Mulan	46.051	127.585	75–108	10.8±2.8	17.94±8.54	32	76	44	1	9.198	4.638	0.016	0.762	0.075
		14 Maoershan	45.342	127.584	300–434	13.4±2.4	19.30±6.19	33	70	38	0	8.806	4.247	0	0.746	0.077
		15 Fangzheng	45.613	128.512	472–529	7.4±2.8	8.06±4.51	32	71	43	1	9.162	5.158	0.016	0.761	0.176*
		16 Baomashan	45.729	129.460	474–517	17.5±5.4	20.62±10.40	31	63	30	0	8.218	3.513	0	0.733	0
		17 Shuguang	45.339	128.958	107–233	10.3±3.1	20.33±10.70	34	76	38	1	9.331	3.931	0.015	0.765	0.149*
		18 Weihe	44.785	128.160	319–367	14.2±8.0	23.88±16.15	31	66	30	0	8.533	3.396	0	0.792	0.11
		19 Yabuli	44.761	128.465	467–526	13.3±7.8	10.83±7.85	26	62	27	0	8.469	3.477	0	0.745	0.019
		20 Chaihe	44.865	129.556	270–319	7.0±2.3	13.27±8.25	30	66	30	0	8.460	3.336	0	0.749	0.015
		21 Hailin	44.822	129.097	558–615	11.0±7.6	11.01±9.65	28	61	25	0	8.235	3.088	0	0.714	−0.105
		22 Ning'an	44.184	128.626	577–704	7.3±4.8	9.82±5.32	30	70	31	0	8.929	3.405	0	0.718	−0.068
		23 Shanhetun	44.402	127.854	290–451	15.9±4.2	19.33±8.67	32	84	55	1	10.168	6.031	0.016	0.792	0.081
		24 Shahe	44.369	127.705	322–363	10.8±3.2	24.66±11.14	30	72	28	1	9.143	2.892	0.017	0.779	0.096
	LY	27 Bamiantong	44.911	130.760	129–151	6.6±2.9	11.15±6.05	27	67	30	1	8.896	4.476	0.019	0.750	0.09
		31 Muling	44.518	130.262	96–151	8.5±2.8	14.13±6.85	31	52	32	0	8.486	3.497	0	0.535	−0.249
		32 Suiyang	44.225	130.847	402–466	11.0±2.6	17.07±8.69	29	71	35	1	8.685	3.409	0.017	0.792	0.085
	WD	25 Yingchun	47.051	133.800	241–327	13.7±4.5	15.26±7.80	31	70	39	1	9.987	5.426	0.016	0.752	0.063
		26 Dongfanghong	46.244	133.079	189–250	13.2±2.9	13.47±3.32	33	67	32	0	8.377	3.388	0	0.760	0.054
		28 Baoshan	46.471	131.268	108–154	14.2±3.9	18.72±11.26	33	83	51	0	7.687	2.849	0	0.776	0.135*
		29 Shuangyashan	46.161	131.880	333–592	14.1±4.2	15.60±7.78	31	65	30	0	6.411	3.559	0	0.756	0.141*
		30 Huanan	46.447	130.805	410–586	7.2±2.3	7.96±2.80	33	59	25	1	9.069	3.940	0.015	0.684	0.12
	CB	33 Wangqing	43.729	129.797	429–479	12.9±6.9	13.90±9.67	30	68	26	0	8.831	2.875	0	0.766	0.13*
		34 Hunchun	43.153	130.372	296–338	9.3±7.6	11.94±8.02	29	70	36	0	8.884	4.044	0	0.751	−0.109
		35 Yongji	43.417	126.613	554–617	11.6±5.3	17.87±9.69	30	74	34	0	9.384	3.705	0	0.777	0.069
		36 Huangnihe	43.647	127.942	469–534	13.4±5.9	19.64±10.08	33	57	27	0	7.045	2.771	0	0.716	−0.054
		37 Dashitou	43.127	128.503	616–660	10.3±3.0	12.02±4.56	31	69	40	0	8.660	4.524	0	0.768	0.107
		38 Dunhua	43.035	127.993	705–761	12.4±4.6	17.22±8.20	32	58	35	1	7.177	3.894	0.016	0.660	−0.109
		39 Helong	42.667	128.845	556–619	5.7±5.5	6.55±7.16	29	60	27	0	7.782	3.083	0	0.701	−0.012
		40 Lushuihe	42.554	127.779	735–771	12.1±6.2	16.66±10.35	31	75	47	1	9.344	5.348	0.016	0.742	−0.037
		41 Songjianghe	42.018	127.632	910–936	16.1±5.7	17.62±6.63	32	54	29	0	6.799	3.233	0	0.628	−0.081
		42 Hongshi	42.957	127.045	441–464	14.5±4.8	17.09±9.18	32	74	39	0	9.285	4.302	0	0.789	0.044
		44 Wunvfeng	41.268	126.122	473–586	12.5±4.5	11.82±5.02	31	84	44	1	11.007	5.307	0.016	0.840	0.078
	LG	43 Meihekou	42.196	125.480	459–503	12.9±4.4	18.81±9.18	32	73	38	0	9.241	4.252	0	0.807	0.148*
		45 Qingyuan	41.989	125.236	575–714	15.8±4.7	17.24±7.44	32	71	38	1	9.015	4.317	0.016	0.781	0.051
		46 Xinbin	41.331	124.913	673–821	13.7±4.6	16.70±7.05	31	73	46	0	9.211	5.357	0	0.749	−0.064
		47 Kuandian	40.920	124.927	496–634	11.5±3.0	14.25±4.92	32	84	50	0	10.287	5.442	0	0.821	0.190*
	QS	48 Xiuyan	40.183	122.998	320–406	7.5±2.6	12.90±4.97	32	75	42	1	9.432	4.724	0.016	0.809	0.122

Table 2. Cont.

Mountains	Locations	Lat (°N)	Long (°E)	Alt (m)	Height (m)	DBH (cm)	N	N_A	R_A	P_A	A_R	RA_R	PA_R	H_E	F_{IS}
Yanshan	49 Chengde	40.552	117.490	938–1029	6.0±2.8	8.67±6.34	21	58	28	1	8.286	4.000	0.024	0.755	0.062
Baihuashan	50 Beijing	39.836	115.576	1159–1264	7.9±2.7	8.93±3.55	21	58	29	5	8.286	4.143	0.703	0.710	0.176*
Zhongtiaoshan	51 Yuanqu	35.398	111.975	1409–1796	4.4±2.7	7.84±12.03	29	75	35	4	9.655	3.969	0.341	0.771	0.074
Qinling	52 Ningdong	33.443	108.479	1216–1842	5.3±1.8	4.48±2.93	27	86	44	7	11.210	5.216	1.293	0.818	0.039
Jigongshan	53 Lijiazhai	31.813	114.075	318–534	5.2±3.7	7.07±10.45	28	53	16	1	7.178	1.905	0.018	0.704	0.058
Langyashan	54 Chuzhou	32.077	118.806	100–200	10.6±6.2	12.47±10.14	30	50	15	1	6.579	1.592	0.017	0.680	0.016
Tianmushan	55 Lin'an	30.360	119.422	987–1095	4.3±1.8	5.73±1.89	33	55	27	1	6.902	2.918	0.015	0.550	0.079
Xuefengshan	56 Huaihua	27.017	110.104	503–1180	4.0±2.3	7.38±5.67	30	39	11	2	5.181	1.192	0.170	0.564	−0.022

Abbreviations: ZH, Zhangguangcailing Mts.; LY, Laoyeling Mts.; WD, Wandashan Mts.; CB, Changbai Mts.; LG, Longgangshan Mts.; QS, Qianshan Mts.; DBH, diameter at breast height; Lat, latitude; Long, longitude; Alt, altitude; N, number of individuals; N_A, total number of alleles detected; R_A, number of rare alleles; P_A, number of private alleles; A_R, allelic richness based on at least 21 diploid individuals of the smallest population; RA_R, rare allelic richness based on 21 diploid individuals of the smallest population; PA_R, private allelic richness based on 21 diploid individuals of the smallest population;H_E, average expected heterozygosity; F_{IS}, Wright's inbreeding coefficient, where *indicates significant deviation from 0 when $\alpha = 0.05$; –indicates missing data.

not robust (Fig. S3). Estimated median mean mutation rates of SSR and SNI were estimated to be 4.80×10^{-4} and 8.41×10^{-5}, respectively (Table S3). None of the expected heterozygosity (H_E) and number of alleles (N_A) values for any population – or H_E, N_A, genotype likelihood and F_{ST} values for any population-pair – estimated using the acquired posterior distributions significantly differed from observed values, suggesting that scenario 2 was a good fit (Table S4). This was corroborated by the PCA, which yielded a large cloud of data from the prior and observed datasets, centered around a small cluster from the posterior predictive distribution (Fig. S4).

The DIYABC based on SC populations STRUCTURE analysis of the SC populations ($K = 4$) found the highest posterior probability for the hierarchical population split scenario 3 (0.9262, 95% CI: 0.9140-0.9384; Table 5B), in which the median values of effective population size of N1 (SC1), N2 (SC2), N3 (SC3) and N4 (SC4) were 8930, 3890, 2390 and 866, respectively (Table S5). The median estimated values of time of t1, t2 and t3 were 651, 2960 and 6330 generations ago, respectively (Table S5), while the median estimated mean mutation rates of SSR and SNI were 6.29×10^{-4} and 5.80×10^{-7}, respectively (Table S5). The posterior probability (Fig. S5), summary statistics (Table S6) and the PCA results (Fig. S6) all suggested that scenario 3 provides a good fit to the data.

Discussion

Genetic Diversity of *A. mono* Populations

In Northeast China, although the forest zone is threatened, nSSR studies have shown that several tree species maintain moderate to strong genetic diversity, similar to that of *A. mono* ($H_E = 0.738\pm0.058$), e.g. *Fraxinus mandshurica* ($H_E = 0.746\pm0.051$) [14] and *Juglans mandshurica* ($H_E = 0.756$) [63]. Significant positive correlations between genetic diversity (H_E) and latitude was detected over all the 56 populations (Table 3). However, as both A_R and H_E showed significant positive correlation with latitude in SC group (Table 3), the smaller population size and the greater geographical distances between populations in SC than in NE may have strongly contributed to this overall pattern. Thus, different trends were detected within NE populations and A_R values significantly decreased with increasing latitude. Although it was not significant, H_E also showed the similar negative correlation with latitude. As mentioned in the Introduction, the reductions in genetic diversity with increasing latitude observed in NE populations of *Q. mongolica* [19] and *F. mandshurica* [14] were also found in *A. mono*, indicating that it may be a common trend for tree species in this region. According to previous studies [64], the genetic diversity of most species declines from central to marginal populations. We therefore inferred a probable route for the expansion of *A. mono*, from the central section of China toward the north and south. However, the lower diversity, especially the low rare allelic richness in populations 1 and 53–56, may not be mainly due to margin effects. For population 1, extremely low winter temperature may contribute to low genetic diversity by hindering the growth of seedlings and young saplings [14], or even killing them, thereby reducing the population size and rare allelic richness. For populations 53–56, besides the patchy and scattered distribution of the species in their region, another factor contributing to lower rare allelic richness may be landscape fragmentation (one of the major factors causes of dramatic loss of biodiversity globally) by blocking gene flow among populations, altering the biogeographical environments and diminishing available habitats [65].

Although inbreeding will increase homozygosity in the short term and affect diversity in the longer term [66], neither H_E nor A_R in the 11 populations for which F_{IS} values significantly deviated from 0 (Table 2) were significantly lower than others. Thus, reduction of genetic diversity induced by inbreeding was not suggested in this study. Instead of inbreeding, demographic factor such as Wahlund effect would be better explanation to higher values of F_{IS} detected in this study.

Genetic Structure and Demographic History of *A. mono* Populations

The F_{ST} and G'_{ST} values indicated that differentiation was much lower among NE populations ($F_{ST} = 0.029$; $G'_{ST} = 0.114$) than among SC populations ($F_{ST} = 0.201$; $G'_{ST} = 0.722$). For the NE populations, the relatively short geographic distances between populations and the continuous distribution of mountains provide a more uniform landscape that probably facilitated gene exchange among populations [69], explaining the higher diversity and lower differentiation observed. The weak genetic structure and reductions in genetic diversity with latitude among the NE populations suggest that they expanded from a single lineage. Interestingly,

Table 3. Pearson Correlation Coefficients between genetic variation and geographical gradients.

	All populations		NE populations		SC populations	
	A_R	H_E	A_R	H_E	A_R	H_E
Latitude	0.133	0.289*	−0.416**	−0.225	0.814*	0.844*
Longitude	0.07	0.154	−0.218	−0.24	−0.463	−0.491
Altitude	0.05	−0.115	−0.159	−0.145	0.726	0.406

Values indicate correlation indices, * and **indicate significant correlation when $\alpha = 0.05$ and 0.01, respectively.

Table 4. Comparison of genetic diversity parameters: allelic richness (A_R), rare allelic richness (RA_R), private allelic richness (PA_R), fixation index (F_{IS}), averaged expected heterozygosity (H_E), and Weir & Cockerham's F_{ST} (1984) between the Northeastern (NE) and the Southern (SC) group.

Group	A_R	RA_R	PA_R	F_{IS}	H_E	F_{ST}
NE (Population Nos. 1–48)	8.59	3.825	0.006	0.052	0.743	0.029
SC (Population Nos. 51–56)	7.78	2.799	0.309	0.042	0.676	0.207
P-value	0.104	0.024	0.000	0.789	0.009	0.002

Table 5. Posterior probability of each scenario and the corresponding 95% confidence interval based on logistic estimation by DIYABC.

Scenario	Posterior probability	95% CI (lower − upper)
A) Analysis across the whole range		
1	0.2557	0.2297–0.2817
2	0.7341	0.7079–0.7602
3	0.0032	0.0022–0.0041
4	0.0071	0.0055–0.0087
B) Analysis of South China (SC) populations		
1	0.0247	0.0190–0.0305
2	0.0272	0.0211–0.0333
3	0.9262	0.9140–0.9384
4	0.0219	0.0165–0.0272

similar observations (no significant substructure and latitudinal reduction of genetic diversity among mountain ranges in Northeast China) have also been reported for an ash species, *F. mandshurica* [14]. The patterns of both species could be explained by reduction of genetic diversity to the north due to expansion from a single refugium in the last ice age. STRUCTURE analysis focused solely on NE populations detected weak genetic structure among mountain ranges (Fig. S2B, $K = 2$). More specifically, the clustering pattern shown by the barplots indicated that the 12 populations located in the Xiaoxing'anling Mountains (1–12) were almost entirely dominated by a single cluster. Thus, the weak genetic structure in NE populations was probably due to the mountain ranges limiting gene flow, as they often do for tree species [70]. Since *A. mono* is an insect-pollinated species [71], and its large winged seeds are dispersed by wind, the potential for gene flow may be more sensitive to geographic barriers than that of wind-pollinated and wind-dispersed species such as birch (*Betula maximowicziana*), which was examined by Tsuda *et al.* [70]. In addition, the STRUCTURE analysis and NJ tree separated all populations into three groups (NE and SC populations, and those from adjacent regions). The analysis indicated that the two populations (49 and 50) located in northern China were admixed by STRUCTURE analysis when $K = 2$, but they appeared as an additional, independent genetic group in the results with $K = 3$ and 4. Since the isolation by admixture scenario was not supported by the ABC analysis, we propose that these two populations constitute an independent group.

The demographic history of plant species associated with the LGM and species refugia in Northeastern China has been discussed [14,63], but no coalescent-based genetic analysis has been previously applied. Thus, the results of ABC analysis obtained in this study provide valuable foundations for addressing the demographic history of *A. mono* over a long time scale. The estimated timing of t3 (a median value of 6480 generations, corresponding to 194,400 years) was not robust (Fig. S3), but since the effective population size increased to 3700 (N2) from 1470 (Nb) at t3, the results clearly indicate that the population expanded rapidly at or around this time.

The ancient divergence of Pop3 at t2, may have been caused by the Taihang Mountains acting as a barrier. These Mountains have a north-south orientation (36–40°N, 112–115°E) and form a major natural boundary in eastern China. The southern Taihang Mountain marginal fault zone has been a major boundary of neotectonic deformation since the mid-late Pleistocene, dividing northern China latitudinally, and the south region containing the deformation system that includes the south Qinling Mountains [72]. This division might explain not only the distribution of cpDNA haplotypes of *Juglans mandshurica* in the south and north of the Taihang and Qinling Mountains (Fig. 1) [63], but also the divergence of Pop3. In further support of this hypothesis, maternally-inherited mtDNA patterns of *Pinus tabulaeformis* are reportedly divided into three geographical groups, separated latitudinally by the Yellow River and longitudinally by the Taihang Mountains [12]. Pop1 and Pop2 were separated by the Northeast Plain, the largest Plain in China with a total area of 350,000 km^2 (40° 25′ to 48° 40′N, 118°40′ to 128°E), and their estimated divergence time was estimated at 20,430 a BP. The Northeast Plain was covered by coniferous forest/steppe during the early late Pleistocene and broad-leaved forest during the middle late Pleistocene, but at the end of the Pleistocene (42,000 ~ 20,300 a BP), because of the colder and drier climate, most of the Northeast Plain was covered by sand, and the vegetation changed to open forest/steppe [73]. Thus, Pop1 and 2 appear to have diverged during this period, when the species occupied refugia before the LGM.

STRUCTURE analysis divided SC populations (Pop3) into four subgroups. The strong genetic structure in these populations was probably caused by a combination of the greater geographic distances separating them [16,67] and the presence of geographic barriers such as discontinuous mountain ridges [68–69] and large plains and valleys [25] (Fig. 1). These factors presumably isolated the populations and thus promoted population differentiation by limiting the potential for gene exchange. DIYABC analysis of the SC populations provided estimated divergence time (t1, t2 and t3) of each subgroup (SC1, SC2, SC3 and SC4) but the estimate for t3 (189,900 years) was not robust (Fig. S5). SC2 and SC4 apparently diverged from SC1 at 88,800 a BP (t2), possibly due to a major uplift of the Qinling Mountains. These mountains form a major boundary between north and south China, ranging west-east (32–35°N, 103–113°E) and rose strongly during the Quaternary, especially the Late Quaternary (100,000 a BP), when they reportedly rose up to 5.4 mm per year [74–75]. Thus, if we accept t3 as accurate, we can assume that the Qinling Mountains rose several tens or hundreds meters during the t3 to t2 period, which could have been a major contributor to the divergence of SC1 and SC2.

Implications for Genetic Breeding, Conservation, and Silviculture of *A. mono*

A shelter wood system of forest management has been found to have little or no impact on genetic diversity and mating in *Fagus sylvatica* [76]. However, extensive logging can significantly alter the level and distribution of genetic variation in *Acer saccharum* [77]. The cited studies indicate that silvicultural activities may affect genetic diversity, but in ways that vary according to species and management type.

Natural resources of *A. mono* in China are under increasing pressure due to its use as a source of wood and various other materials. Hence, specific natural varieties are being selected for artificial planting and silviculture in Northeast China and elsewhere. We only used neutral genetic markers in the present study, but the information obtained is potentially highly informative for the breeding, conservation and silviculture of *A. mono*, especially for the designation of conservation units. Since clear genetic differentiation was found between NE and SC populations, these two groups should be treated as independent conservation units. Although admixture was found for Populations 49 and 50 in the STRUCTURE analysis when $K=2$, the ABC-based analysis suggested that these two populations were not generated by the admixture of NE and SC populations. Thus, populations in this region of northern China located between NE and SC populations should also be treated as an independent conservation unit. Seeds and nursery stocks of local provenance should be used in reforestation efforts in each area to avoid nonlocal introductions and genetic homogenization [11,78]. Moreover, since strong genetic structure was detected in SC populations, further local-scale studies in this region are needed to develop appropriate strategies for conservation and the designation of natural reserves. In addition, *ex situ* conservation efforts, which could involve collecting seeds from populations in this area, may be useful. Since Populations 51 and 52 showed high levels of diversity and more private alleles than other populations, natural reserve areas should be established in these regions. In contrast, the NE populations maintain relatively high levels of genetic diversity and densities of individuals within populations, but low genetic differentiation among populations, implying that the species should be able to recover without assistance (if further disturbance is avoided) in this region. Selection of elite trees and the development of varieties for specific purposes are likely to be more important than provenance selection for conserving *A. mono* in Northeast China. To obtain the thorough overview required to develop an effective conservation strategy for this species further studies are required, including evaluation of the extent of its adaptive genetic variation.

Supporting Information

Figure S1 Mean values of ln P(D) and standard deviations obtained from 10 runs for each value of $K = 1–20$, and distributions of ΔK (Evanno et al. 2005) for $K = 1–20$.

Figure S2 Results of additional runs of STRUCTURE analysis focused on the six SC and 48 NE populations. A, $K = 6$, in which the clustering corresponded to each of the six SC populations; B, $K = 2$ (according to the highest ΔK) for the 48 NE populations.

Figure S3 Prior and posterior distributions for each parameter obtained by DIYABC analysis of populations across the whole range.

Figure S4 Principal Component Analysis (PCA) score plot obtained from DIYABC analysis of populations across the whole range.

Figure S5 Prior and posterior distributions for each parameter obtained by DIYABC analysis of in South China populations.

Figure S6 Principal component analysis (PCA) score plot obtained from DIYABC of South China populations.

Table S1 Prior distributions of the parameters used in the whole-range DIYABC analysis.

Table S2 Prior distributions of the parameters used in the South China DIYABC analysis.

Table S3 Demographic parameters obtained from the whole-range DIYABC analysis.

Table S4 Comparison of summary statistics for the observed data set and posterior simulated data sets.

Table S5 Demographic parameters obtained from DIYABC analysis of South China populations.

Table S6 Comparison of summary statistics for the observed data set and simulated posterior data sets.

Acknowledgments

We are very grateful to Dr. Jian Cong, Mr. Xiuliang Zhang, Mr. Jianwei Xu, Mr. Lingdong Wu, Miss XiaojieZhai, Miss Xiaoli Shen, Miss Yanlin Li, Miss Yan Liu, Miss Qinqin Yang, Mr. Cunxue Lin, and Miss Lin Shan for their contributions to the extensive and difficult field work. We also thank many anonymous members of local institutes and forestry stations for their help with sampling.

Author Contributions

Conceived and designed the experiments: HS YI CL YS YT LH. Performed the experiments: CL LH HS. Analyzed the data: CL YT. Contributed reagents/materials/analysis tools: HS YI. Wrote the paper: CL YT HS. Revised the manuscript: CL YT HS YS YI.

References

1. Fu LG (1992) China Plant Red Data Book, Rare and Endangered Plants (Vol. 1). Beijing: Chinese Science Press. 385–387p (in Chinese).
2. Li WH (2004) Degradation and restoration of forest ecosystems in China. For Ecol Manage 201: 33–41.
3. Zhu JJ, Mao ZH, Hu LL, Zhang JX (2007) Plant diversity of secondary forests in response to anthropogenic disturbance levels in montane regions of northeastern China. J For Res 12: 403–416.
4. Moritz C (2002) Strategies to protect biological diversity and the evolutionary process that sustain it. Syst Biol 51: 238–254.
5. Escudero A, Iriondo JM, Torres ME (2003) Spatial analysis of genetic diversity as a tool for plant conservation. Biol Conserv 113: 351–365.
6. Newton AC, Allnutt TR, Gillies ACM, Lowe AJ, Ennos RA (1999) Molecular phylogeography, intraspecific variation and the conservation of tree species. Trends Ecol Evol 14: 140–145.
7. Geburek T (1997) Isozymes and DNA markers in gene conservation of forest trees. BiodiversConserv 6: 1639–1654.
8. Tsuda Y, Ide Y (2005) Wide-range analysis of genetic structure of *Betula maximowicziana*, a long-lived pioneer tree species and noble hardwood in the cool temperate zone of Japan. Mol Ecol 14: 3929–3941.
9. Crandall KA, Bininda-Emonds ORP, Mace GM, Wayne RK (2000) Considering evolutionary processes in conservation biology. Trends EcolEvol 15: 290–295.
10. Frankel OH (1974) Genetic conservation: our evolutionary responsibility. Genetics 78: 53–65.
11. Steinitz O, Robledo-Arnuncio JJ, Nathan R (2012) Effects of forest plantations on the genetic composition of conspecific native Aleppo pine populations. Mol Ecol 21: 300–313.
12. Chen K, Abbott RJ, Milne RI, Tian XM, Liu J (2008) Phylogeography of *Pinus tabulaeformis* Carr. (Pinaceae), a dominant species of coniferous forest in northern China. Mol Ecol 17: 4276–4288.
13. Heuertz M, Hausman JF, Hardy OJ, Vendramin GG, Frascaria-Lacoste N, et al. (2004) Nuclear microsatellites reveal contrasting patterns of genetic structure between Western and Southeastern European populations of the common ash (*Fraxinus excelsior* L). Evolution 58: 976–988.
14. Hu LJ, Uchiyama K, Shen HL, Saito Y, Tsuda Y, et al. (2008) Nuclear DNA microsatellites reveal genetic variation but a lack of phylogeographical structure in an endangered species, *Fraxinus mandshurica*, across North-east China. Ann Bot 102: 195–205.
15. Hu LJ, Uchiyama K, Shen HL, Saito Y, Ide Y (2010) Multiple-scaled spatial genetic structures of *Fraxinus mandshurica* over a riparian-mountain landscape in Northeast China. Conserv Genet 11: 77–87.
16. Logossa ZA, Camus-Kulandaivelu L, Allal F, Vaillant A, Sanou H, et al. (2011) Molecular data reveal isolation by distance and past population expansion for the shea tree (*Vitellariaparadoxa* C.F. Gaertn) in West Africa. Mol Ecol 20: 4009–4027.
17. Yan ZF, Zhang BG, Zhang Z, Yu JL (2006) Genetic diversity in wild populations of *Phellodendron amurense*, a rare and endangered medicinal plant, detected by AFLP. Chinese Biodiversity Science14: 488–497 (in Chinese with English abstract).
18. Xia M, Zhou XF, Zhao SD (2001) RAPD analysis on genetic diversity of natural populations of *Quercus mongolica*. Scientia Silvae Sinicae 37: 126–133 (In Chinese with English abstract).
19. Zhang J, Wu D, Wang CL, Qu HJ, Zhou XZ, et al. (2007) Genetic diversity analysis of *Quercus mongolica* populations with Inter-Simple Sequence Repeats (ISSR) technique. Biodiversity Science 15: 292–299 (in Chinese with English abstract).
20. Li YW, Gu WC, Zhou SL (2003) AFLP analysis on genetic diversity of *Quercus mongolica* populations. Scientia Silvae Sinicae 39: 29–36 (in Chinese with English abstract).
21. Mu LQ, Liu YN (2007) Genetic diversity of *Tulia amurensis* populations indifferent geographical distribution regions. J Plant Ecol (Chinese Version) 31: 1190–1198 (in Chinese with English abstract).
22. Yu G, Prentice IC, Harrison SP, Sun X (1998) Pollen-based biome reconstructions for China at 0 ka and 6 ka. J Biogeogr 25: 1055–1069.
23. Hewitt G (2000) The genetic legacy of the Quaternary ice ages. Nature 405: 907–913.
24. Hewitt G (2004) Genetic consequences of climatic oscillations in the Quaternary. Philos Trans R Soc Lond B Biol Sci 359: 183–195.
25. Magri D, Vendramin GG, Comps B, Dupanloup I, Geburek T, et al. (2006) A new scenario for the Quaternary history of European beech populations, palaeobotanical evidence and genetic consequences. New Phytol 171: 199–221.
26. Magri D, Fineschi S, Bellarosa R, Buonamici A, Sebastiani F, et al. (2007) The distribution of *Quercus suber* chloroplast haplotypes matches the palaeogeographical history of the western Mediterranean. Mol Ecol 16: 5259–5266.
27. Lascoux M, Pyhäjärvi T, Källman T, Savolainen O (2008) Past demography in forest trees: what can we learn from nuclear DNA sequences that we do not already know? Plant Ecol Divers 1: 209–215.
28. Ingvarsson PK (2008) Multilocus patterns of nucleotide polymorphism and the demographic history of *Populus tremula*. Genetics 180: 329–340.
29. Bertorelle G, Benazzo A, Mona S (2010) ABC as a flexible framework to estimate demography over space and time: some cons, many pros. Mol Ecol 19: 2609–2625.
30. Hsu T (1992) Variation patterns and systematics of *Acer mono* Maxim.Guihaia 12: 229–234 (in Chinese with English abstract).
31. Satake G, Hara K, Hara H, Watari S, Tominali U (1993) Wild flowers of Japan: woody plants. Tokyo: Heibonsha (in Japanese).
32. Zheng W (2004) Sylva Sinica (vol. 4). Beijing: Chinese publishing House of Forestry. 4258–4260p (in Chinese).
33. Dong SL (1985) Study on Aceraceae of northeastern China. Bulletin of Botanical Research 5: 97–111 (in Chinese).
34. Hsu T (1988) Evaluation on *Acer* plant resources in China. Resource Development & Market 4: 51–54.
35. Kikuchi S, Shibata M (2008) Development of polymorphic microsatellite markers in *Acer mono* Maxim.Mol Ecol Resour 8: 339–341.
36. Kikuchi S, Shibata M, Tanaka H, Yoshimaru H, Niiyama K (2009) Analysis of the disassortative mating pattern in a heterodichogamous plant, *Acer mono* Maxim. using microsatellite markers. Plant Ecol 204: 43–54.
37. Shibata M, Kikuchi S, Tanaka H, Sueyoshi M, Yoshimaru H, et al. (2009) Effects of population density, sex morph, and tree size on reproduction in a heterodichogamous maple, *Acer mono*, in a temperate forest of Japan. Ecol Res 24: 1–9.
38. Lian CL, Oishi R, Miyashita N, Nara K, Nakaya H, et al. (2003) Genetic structure and reproduction dynamics of *Salix reinii*during primary succession on Mount Fuji, as revealed by nuclear and chloroplast microsatellite analysis. Mol Ecol 12: 609–618.
39. Terui H, Lian CL, Saito Y, Ide Y (2006) Development of microsatellite markers in *Acer capillipes*. Mol Ecol Notes 6: 77–79.
40. Segarra-Moragues JG, Gleiser G, González-Candelas F (2008) Isolation and characterization of microsatellite loci in *Acer opalus* (Aceraceae), a sexually-polymorphic tree, through an enriched genomic library. Conserv Genet 9: 1059–1062.
41. Pandey M, Gailing O, Fischer D, Hattemer DH, Finkeldey R (2004) Characterization of microsatellite markers in sycamore (*Acer pseudoplatanus*L.). Mol Ecol Notes 4: 253–255.
42. Goudet J (2001) FSTAT, a program to estimate and test gene diversities and fixation indices (ver. 2.9.3). Available: http://www2.unil.ch/popgen/softwares/fstat.html.
43. SPSS Inc. (2004) SPSS for Windows, Release 13.0, September 1, Chicago, IL, SPSS Inc.
44. Weir BS, Cockerham CC (1984) Estimating *F*-Statistics for the analysis of population structure. Evolution 38: 1358–1370.
45. Excoffier L, Smouse PE, Quattro JM (1992) Analysis of molecular variance inferred from metric distances among DNA haplotypes: application to human mitochondrial DNA restriction data. Genetics 131: 479–491.
46. Peakall R, Smouse PE (2007) GenAlEx V6.1, *Genetic Analysis inExcel. Population Genetic Software for Teaching and Research*. Canberra: Australian National University Press.
47. Hedrick PW (2005) A standardized genetic differentiation measure. Evolution 59: 1633–1638.
48. Wright S (1943) Isolation by distance. Genetics 28: 114–138.
49. Rousset F (1997) Genetic differentiation and estimation of geneflow from *F*-statisticsunder isolation by distance. Genetics 145: 1219–1228.
50. Nei M, Tajima F, Tateno Y (1983) Accuracy of estimated phylogenetic trees from molecular data. II. Gene frequency data. J Mol Evol 19: 153–170.
51. Langella O (2007) Populations 1.2.30: Population genetic software (individuals or populations distances, phylogenetic trees).France. Available: http://bioinformatics.org/~tryphon/populations/.
52. Parks DH, Porter M, Churcher S, Wang S, Blouin C, et al. (2009) GenGIS: A geospatial information system for genomic data. Genome Res 19: 1896–1904.
53. Pritchard JK, Stephens M, Donnelly P (2000) Inference of population structure using multilocus genotype data. Genetics 155: 945–959.
54. Pritchard JK, Wen XQ, Falush D (2010) Documentation for structure software: Version 2.3. Available: http://pritch.bsd.uchicago.edu/software/structure_v.2.3.1.html.
55. Hubisz M, Falush D, Stephens M, Pritchard JK (2009) Inferring weak population structure with the assistance of sample group information. Mol Ecol 9: 1322–1332.
56. Falush D, Stephens M, Pritchard JK (2003) Inference of population structure using multilocus genotype data: Linked loci and correlated allele frequencies. Genetics 164: 1567–1587.
57. Evanno G, Regnaut S, Goudet J (2005) Detecting the number of clusters of individuals using the software STRUCTURE: a simulation study. Mole Ecol 14: 2611–2620.
58. Jakobsson M, Rosenberg NA (2007) CLUMPP: a cluster matching and permutation program for dealing with label switching and multimodality in analysis of population structure. Bioinformatics 23: 1801–1806.
59. Rosenberg NA (2004) Distruct: a program for the graphical display of population structure. Mol Ecol Notes 4: 137–138.

60. Cornuet JM, Santos F, Beaumont MA, Robert CP, Marin JM, et al. (2008) Inferring population history with DIYABC: a user-friendly approach to Approximate Bayesian Computations. Bioinformatics 24: 2713–2719.
61. Cornuet JM, Ravigné V, Estoup A (2010) Inference on population history and model checking using DNA sequence and microsatellite data with the software DIYABC (v1.0). BMC Bioinformatics 11: 401.
62. Estoup A, Jarne P, Cornuet JM (2002) Homoplasy and mutation model at microsatellite loci and their consequences for population genetics analysis. Mol Ecol 11: 1591–1604.
63. Bai WN, Liao WJ, Zhang DY (2010) Nuclear and chloroplast DNA phylogeography reveal two refuge areas with asymmetrical gene flow in a temperate walnut tree from East Asia. New Phytol 188: 892–901.
64. Eckert CG, Samis KE, Lougheed SC (2008) Genetic variation across species' geographical ranges: the central–marginal hypothesis and beyond. Mol Ecol 17: 1170–1188.
65. Liu JF, Xiao WF, Jiang ZP, Feng X, Li XY (2005) A study on the influence of landscape fragmentation on biodiversity. Forest Research 18: 222–226 (in Chinese with English abstract).
66. Charlesworth D (2003) Effects of inbreeding on the genetic diversity of populations. Philosophical Trans actions of the Royal Society Lond. B 358: 1051–1070.
67. Petit RJ, Aguinagalde I, de Beaulieu JL, Bittkau C, Brewer S, et al. (2003) Glacial refugia, Hotspots but not melting pots of genetic diversity. Science 300: 1563–1565.
68. Godbout J, Jaramillo-Correa JP, Beaulieu J, Bousquet J (2005) A mitochondrial DNA minisatellite reveals the postglacial history of jack pine *(Pinus banksiana)*, a broad-range North American conifer. Mol Ecol 14: 3497–3512.
69. Lu ZX, Wang YH, Peng YH, Korpelainen H, Li CY (2006) Genetic diversity of *Populuscathayana*Rehd. populations in southwestern China revealed by ISSR markers. Plant Sci 170: 407–412.
70. Tsuda Y, Sawada H, Ohsawa T, Nakao K, Nishikawa H, et al. (2010) Landscape genetic structure of *Betula maximowicziana* in the Chichibu mountain range, central Japan. Tree Genet Genomes 6: 377–387.
71. Matsui K (2010) Pollination ecology of four *Acer* species in Japan with special reference to bee pollinators. Plant Species Biol 6: 117–120.
72. Zhang Y, Yang N, Ma Y (2003) Neotectonics in the southern part of the Taihang uplift, North China. Journal of Geomechanics 9: 313–329 (in Chinese with English abstract).
73. Lin N, Tang J, Bian J, Yang J (1999) The Quaternary environmental evolution and the problem of desertification in Northeast Plain. Quaternary Sciences 5: 448–454 (in Chinese with English abstract).
74. Xue X, Li H, Li Y, Liu H (2004) The new data of the uplifting of Qinling Mountains since the Middle Pleistocene. Quaternary Sci 24: 82–87 (in Chinese with English abstract).
75. Xue X, Li W, Liu L (2002) The northward shift of Weihe River and the uplift of Qinling Mountains. Journal of Northeast University (Natural Science Edition) 32: 451–454 (in Chinese with English abstract).
76. Buiteveld J, Vendramin GG, Leonardi S, Kamer K, Geburek T (2007) Genetic diversity and differentiation in European beech (*Fagus sylvatica* L.) stands varying in management history. For Ecol Manage 247: 98–106.
77. Baucom RS, Estill JC, Cruzan MB (2005) The effect of deforestation on the genetic diversity and structure in *Acer saccharum* (Marsh): evidence for the loss and restructuring of genetic variation in a natural system. Conserv Genet 6: 39–50.
78. Naydenov KD, Naydenov MK, Tremblay F, Alexandrov A, Aubin-Fournier LD (2011) Patterns of genetic diversity that result from bottlenecks in Scots Pine and the implications for local genetic conservation and management practices in Bulgaria. New For 42: 179–193.

7

Genetic Diversity and Ecological Niche Modelling of Wild Barley: Refugia, Large-Scale Post-LGM Range Expansion and Limited Mid-Future Climate Threats?

Joanne Russell[1], Maarten van Zonneveld[2], Ian K. Dawson[1]*, Allan Booth[1], Robbie Waugh[1], Brian Steffenson[3]

1 Cell and Molecular Sciences, The James Hutton Institute, Invergowrie, Scotland, United Kingdom, 2 Regional Office for the Americas, Bioversity International, Cali, Colombia, 3 Department of Plant Pathology, University of Minnesota, Saint Paul, Minnesota, United States of America

Abstract

Describing genetic diversity in wild barley (*Hordeum vulgare* ssp. *spontaneum*) in geographic and environmental space in the context of current, past and potential future climates is important for conservation and for breeding the domesticated crop (*Hordeum vulgare* ssp. *vulgare*). Spatial genetic diversity in wild barley was revealed by both nuclear- (2,505 SNP, 24 nSSR) and chloroplast-derived (5 cpSSR) markers in 256 widely-sampled geo-referenced accessions. Results were compared with MaxEnt-modelled geographic distributions under current, past (Last Glacial Maximum, LGM) and mid-term future (anthropogenic scenario A2, the 2080s) climates. Comparisons suggest large-scale post-LGM range expansion in Central Asia and relatively small, but statistically significant, reductions in range-wide genetic diversity under future climate. Our analyses support the utility of ecological niche modelling for locating genetic diversity hotspots and determine priority geographic areas for wild barley conservation under anthropogenic climate change. Similar research on other cereal crop progenitors could play an important role in tailoring conservation and crop improvement strategies to support future human food security.

Editor: Mingliang Xu, China Agricultural University, China

Funding: These authors have no support or funding to report.

Competing Interests: The authors have declared that no competing interests exist.

* E-mail: iankdawson@aol.com

Introduction

Ecological niche modelling of the distributions of crop wild relatives in present, past and future climates can provide important insights into the past evolution and future trajectories of crop progenitors and domesticates [1]. Non-overlaps between current and predicted future distributions may reveal populations at particular threat from anthropogenic climate change [2]. At the same time, it has been suggested that overlaps between modelled past and present distributions may indicate refugial areas rich in genetic diversity, although this is a theory that requires wider validation [3–5]. In both instances, distributional differences may indicate wild genetic resources of particular importance for conservation and for breeding of the domesticated crop, in order to respond to new environmental pressures [6].

Wild barley (*Hordeum vulgare* ssp. *spontaneum*), the progenitor of the agriculturally important domesticated *H. vulgare* ssp. *vulgare* [7], provides an excellent opportunity to explore the utility of ecological niche modelling for supporting conservation and use. One reason is that its extensive natural distribution, which covers a range of environments across the Fertile Crescent and Central Asia [8], has been widely sampled for seed. This seed has been made available for genotyping and is an important resource for characterising genetic variation, to assist the cultivated crop to respond to anthropogenic climate change and other production challenges [9,10]. Another reason for wild barley's utility is that a wide range of molecular tools are available to describe genetic diversity. These tools include single nucleotide polymorphisms (SNPs [11]), nuclear simple sequence repeats (nSSRs [12]) and chloroplast simple sequence repeats (cpSSRs [13]) that were developed initially for studying cultivated barley, but can also be used to characterise the wild resource. If centres of diversity in wild barley (as described by these tools) are spatially coincident with habitat common to past and present modelled distributions, then this would support the utility of niche modelling for locating genetic refugia. On the other hand, if centres of maximum variation are outside areas of common past-present habitat, then the utility of niche modelling for locating genetic diversity would be weakened.

In this paper, we explore this issue by combining ecological niche modelling, based on the MaxEnt procedure [14], with a spatial analysis of SNP, nSSR and cpSSR data sets, using various geographic information systems [15,16,17]. Our intention is to build a greater understanding of the impacts of climate change on wild barley, and to provide information to help manage natural stands better in the context of environmental change. In turn, this will support breeding to adapt the domesticated barley crop to future climate. Our analysis is based on a range-wide, fully geo-referenced collection of 256 wild barley accessions sampled from 19 countries, and involves distribution modelling under three conditions: current climate, climate at the Last Glacial Maximum (the LGM) and future climate for the 2080s under anthropogenic scenario A2. We begin to explore the possible utility of linkage disequilibrium analysis for discriminating between alternative hypotheses for describing and explaining spatial genetic structure, and discuss the merits and limitations of the methods we employ. An approach similar to that described in this paper could be used to improve the management of other progenitors of domesticated cereals originating in the Fertile Crescent and Central Asia. This is important for supporting future global human food security in the context of anthropogenic climate change [18].

Materials and Methods

The Wild Barley Collection

The Wild Barley Diversity Collection (WBDC) is the most comprehensive geo-referenced collection of *H. vulgare* ssp. *spontaneum* currently subject to wide characterisation [19]. Our sampling of 256 individuals from the WBDC included 19 countries and was designed to cover as much of the accepted natural distribution of the taxon as possible [8]. Sampling extended from North Africa through the Fertile Crescent into Central Asia (Fig. 1 and Table S1). Sampling did not include Tibet, where wild barley very different from that to the west is found [20,21], as very few geo-referenced samples (as required for ecological niche modelling purposes) are available from there. Most accessions originated from the International Center for Agricultural Research in the Dry Areas (ICARDA), Aleppo, Syria, and were assembled by the former gene bank curator there, Dr Jan Valkoun. The majority of accessions were sampled from the wild in the fifty-year period 1953 to 2002, especially in the ten-year periods of 1983 to 1992 and 1993 to 2002 (79 and 101 accessions, respectively). These collection periods correspond well with when the weather station data that are used to support the interpolation of bioclimatic variables for ecological niche modelling were obtained (see below, [22]). For some accessions with early collection dates, latitudes and longitudes used in the current study are based on the interpretation of passport site-description data rather than actual given GPS coordinates. These accessions are therefore likely to be less precisely located.

Assembling Molecular Marker Data Sets

Three molecular marker data sets were analysed in the current study. First, SNP data derived from two Illumina barley oligonucleotide pool assay platforms were used (see [11,23] for a description of these platforms, referred to as BOPAs 1 and 2 or collectively as BOPA SNPs). Here, a subset of 2,505 mostly chromosome-position-mapped BOPA SNPs from an existing study on the WBDC ([24], to investigate disease resistance traits) was used. 'Ascertainment bias' can confound the interpretation of BOPA SNP results when comparing domesticated and wild barley genetic resources [25]. In the current study, however, which only involved comparing different portions of wild barley's range, no significant confounding effect is expected (see discussion in the study by Russell et al. [26], which compared landrace and wild barleys in the Fertile Crescent using BOPA SNPs). Second, we characterised variation *de novo* at 24 of the barley nSSR loci described by Ramsay et al. [12], using the methods given there. Third, we determined variation *de novo* at five of the cpSSR loci designed for *Hordeum* by Provan et al. [13], using the methods of Comadran et al. [27]. A list of all 2,534 loci used in the current study is given in Table S1.

Analysing Molecular Marker Data

Spatial autocorrelation analysis. Spatial autocorrelation analysis using SPAGeDi [28] was undertaken to assess the relationship between inter-individual genetic identities of the 256 tested wild barley accessions and geographic distances. Separate analyses were carried out for BOPA SNPs, nSSRs and cpSSRs. Ritland's [29] kinship coefficient was employed to quantify average pairwise genetic identity based on 20 geographic distance classes of equal sample size. Whether or not individual kinship values were different from expectations (under a random spatial distribution of genetic variation) was assessed by a randomisation test with 1,000 permutations. Kinship values were regressed against the natural logarithm of distance classes to estimate the overall extent of spatial genetic structure. The significance of the regression slope was determined by 1,000 random permutations of locations.

STRUCTURE analysis. STRUCTURE analysis was not designed for predominantly inbreeding species such as barley, but it has been widely applied to cultivated and wild barley populations to reveal interesting genetic features (see discussion in [26]). Here, BOPA SNP and nSSR data sets were each analysed with STRUCTURE 2.2 [30] to assign accessions to one of K groups for different values of K. Each analysis was based on 25,000 'burn-in' replications and a further 25,000 Markov chain Monte Carlo steps (initial trial runs indicated that these numbers of replications were sufficient to ensure the convergence of key parameters). After trial runs, K was set at five because log $\Pr(X/K)$ values in STRUCTURE had started to plateau at this point [30]. (Note that for our purposes it is more important to capture the major genetic divisions within data sets than to determine an 'absolute' value for K.) The 'no admixture' model in STRUCTURE was used to assign single states to individuals. Other analysis options were kept at default settings. STRUCTURE was run five times for each data set and the most common group assignments used as the basis for the interpretation of results (most accessions placed in the same groups in separate runs).

Circular neighbourhood analysis. To overlay genetic diversity onto geographic maps we employed DIVA-GIS 7.3 [15] (www.diva-gis.org) and ArcGIS 10 [16] (www.esri.com/software/arcgis/). Two approaches were used, the first based on allelic (or haplotype) richness and the second based on K groupings. In the first, allelic (BOPA SNP and nSSR) and haplotype (cpSSR) richness estimates were calculated for groups of accessions. Groups were circumscribed using a circular neighbourhood diameter of four degrees and a grid size of 30 minutes (method described in [6]). This allowed us to capture sufficient collection sites within neighbourhoods to estimate genetic parameters with some confidence. To account for varying sampling intensity in geographic space, which otherwise affects diversity estimates [31], rarefaction to a sample size of 10 individuals in neighbourhoods was undertaken using ADZE [17]. In the second approach, K groupings ($K=5$) revealed by STRUCTURE for BOPA SNPs and nSSRs were used instead of allelic/haplotype richness estimates, in order to reveal genetic differentiation at a local geographic scale. Apart from this, the method of analysis was the same as applied in the first approach.

Circular neighbourhood analysis with rarefaction, as conducted in both the above approaches, has the advantage of allowing unbiased comparisons of genetic diversity across geographic space. However, it necessarily excludes accessions from analysis where sampling intensity is low. In the current study, a total of 38 accessions were thereby excluded (including from North Africa, southwestern Iran, Afghanistan and Azerbaijan).

Ecological Niche Modelling

Although MaxEnt [14] has a number of well-documented limitations [32], it is reported to predict the natural distributions of

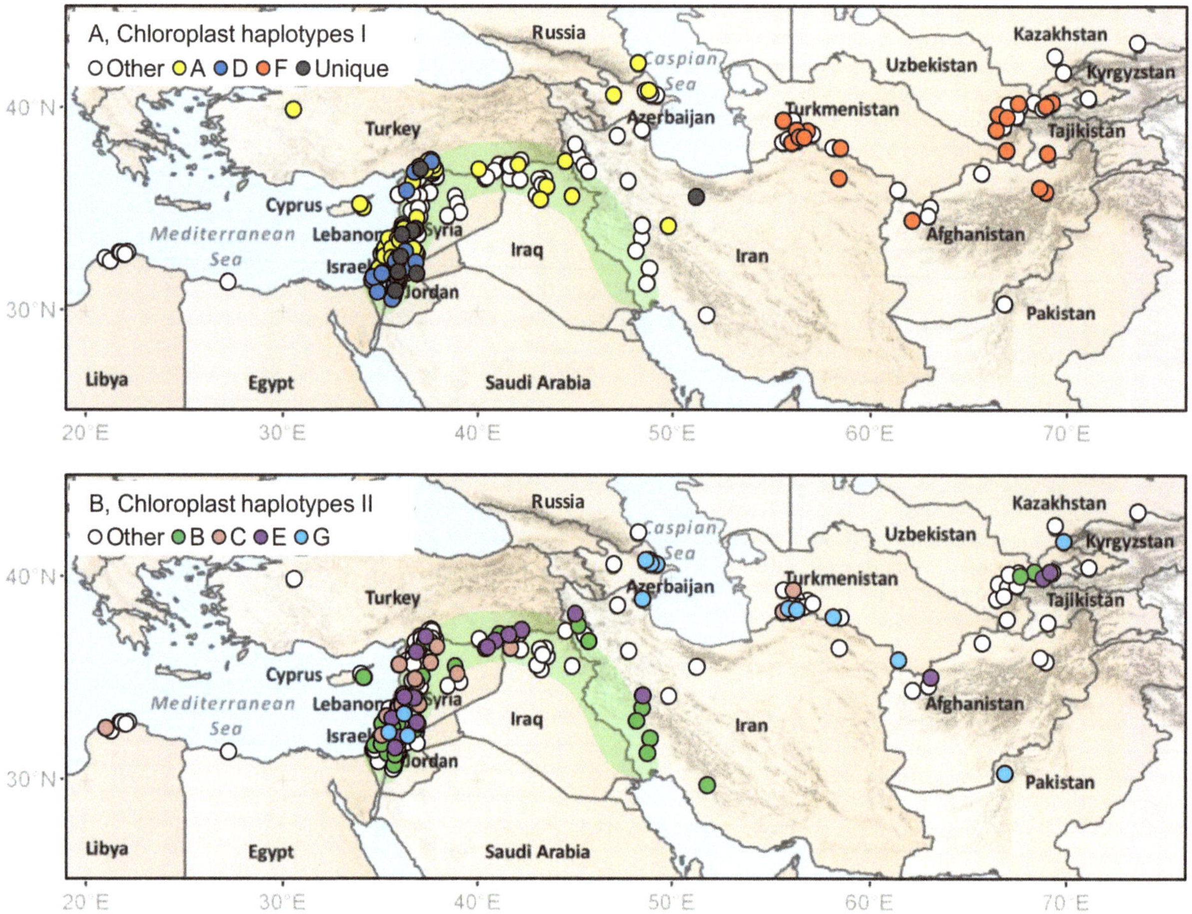

Figure 1. 256 wild barley accessions sampled for genetic analysis and ecological niche modelling. Sampling covered 19 countries and much of the geographic range of wild barley. Superimposed on the positions of accessions are cpSSR haplotype designations for seven common haplotypes (frequency ≥0.05 across all accessions). A, distribution of three common, clearly geographically-differentiated, haplotypes. The distribution of 10 unique haplotypes is also shown. B, distribution of the other four common haplotypes. In both A and B, other sampled accessions are indicated by white circles ('Other'). In total, 31 chloroplast haplotypes were revealed, as described in the Results and Discussion. The approximate dimensions of the Fertile Crescent, a region considered crucial in the development of agriculture and where dense stands of wild barley can occur [8], are indicated by green shading for reference purposes (see [26] for a discussion of barley domestication). The coordinates of sampled accessions and full chloroplast haplotype data are given in Table S1.

plants well when based on 'presence only' location data compared to other ecological niche modelling approaches [33–35]. This is especially so when modelling is based on location data from a limited number of sites (less than 700 [34]). We therefore employed MaxEnt 3.3.1 to model geographic distributions for wild barley under current, past and future climates, in a manner similar to van Zonneveld et al. [36]. This involved extracting data for 19 bioclimatic variables for each of the 256 wild barley accession collection sites from WorldClim [22] (www.worldclim.org/) (extracted values for variables listed in Table S1). These 19 variables are derived from monthly temperature and rainfall values and include seasonality and limiting environmental factors. Our modelling was bounded by longitudes of 18.63 and 80.42 degrees east, latitudes of 21.58 and 47.96 degrees north. The output of MaxEnt is a grid map with each cell assigned a probability of taxon presence [37]. Modelled distribution areas were restricted to the threshold suitability value of maximum training sensitivity plus specificity recommended by Liu et al. [38]. In total, 13 accessions were excluded from modelling because they were identified to occur at 'outlier' sites (see Table S1).

For the LGM (~21,000 years before present), modelling was based on CCSM and MIROC models (available at WorldClim, [2]; the results of models were averaged to provide overall estimates). The LGM is believed to have been an influential period in determining contemporary patterns of genetic variation in many plant species and much modelling of past distributions has therefore been based on it [2,3,5]. For future climate, modelling was based on the 2080s period (2070 to 2099) and the medium- to high-emission trajectory A2 for anthropogenic global warming. The 2080s A2 scenario has been widely used in modelling to provide insights on a timescale and threat level that is useful for planning purposes [39,40]. Nineteen general circulation models (GCM) were used for future climate (again, results were averaged across models). Data on future climate projections were provided by the CGIAR Climate Change, Agriculture and Food Security Research Programme (CCAFS) and downscaled with the Delta method [39].

For current, past and future distribution modelling, 2.5-minute downscaled climate layers were employed, which is the same resolution as used by Waltari et al. [2] for past-climate distribution projections. Based on the overall geographic scale of our sample range, we consider this degree of resolution sufficient for our study.

Results and Discussion

Isolation-by-Distance Alone Does Not Explain the Observed Genetic Structure in Wild Barley

Individual scores for all 256 wild barley accessions for 2,505 BOPA SNPs, 24 nSSRs and five cpSSRs are provided in Table S1. The overall quality of the data was high, with the mean level of missing data (including ambiguous calls) across all markers less than 1%. 2,363 BOPA SNPs were polymorphic and all 24 nSSRs and 5 cpSSRs. A mean of 16.7 alleles per locus (ranging from 3 to 54) was revealed at nSSRs. Length variation at cpSSR products indicated a series of single base differences, with combined data (summing differences across separate products) revealing 31 haplotypes. Seven haplotypes could be defined as common (A to G, occurring at a frequency ≥0.05), while 10 were unique. Compared to haplotype A, haplotypes B, C, D and E differed by a single nucleotide length at one cpSSR product, while haplotypes F and G differed by single nucleotide length differences at two products. Of the common haplotypes, A, D and F showed very clear geographic structuring but the others did not (compare Figs. 1A and B). Haplotype A occurred throughout the Fertile Crescent but not further east, D occurred only in the Eastern Mediterranean region of the Fertile Crescent, while F occurred only in Turkmenistan and further east.

Spatial autocorrelation analysis has been widely applied to assess genetic structure in plant species and to describe deviations from isolation-by-distance expectations (e.g. [78,79]). It can also be a useful method for comparing different molecular marker data sets compiled on the same taxon, as we do here for wild barley (Fig. 2). In our analysis, a degree of geographic-based genetic structure is evident for BOPA SNPs, nSSRs and cpSSR haplotypes ($P<0.01$ in a test for overall structure in each case). The decrease in similarity observed with geographic distance is, however, not a simple trend for any of our three data sets. Differences in profiles are also observed between marker types. For both BOPA SNPs and nSSRs, an increase in similarity at a distance class of around 1,000 km is observed, after which similarity declines again. For cpSSR haplotypes, an obvious increase in similarity is also observed at a distance class of around 500 km. Spatial autocorrelation analysis therefore indicates that a simple isolation-by-distance model does not fully explain genetic structure across the geographic range of wild barley tested (as indicated also, e.g., by the distribution of accessions among BOPA SNP STRUCTURE groups, as shown in Fig. 3). In such situations, climate change-related expansions and contractions in range could have a role in determining patterns of variation [3,5].

The differences we observed for spatial autocorrelation analysis profiles for nuclear BOPA SNPs and nSSRs compared to maternally-inherited cpSSRs may indicate the more restricted role of seed when compared to pollen in gene flow (even though wild barley is predominantly self-pollinated and so pollen-mediated gene flow is expected to be relatively low [41]). The smaller effective population size of the organellar genome compared to the nuclear genome may also be a factor in determining the differences observed [42].

The Spatial Distribution of Genetic Variation Corresponds With Niche Modelling in Locating Diversity Hotspots and is Consistent With Post-LGM Expansion in Central Asia

Geographic information systems are underutilised in genetic diversity studies, but they can be very effective in expressing variation in geographic and environmental space [43–45]. Our analysis is the first on wild barley to use circular neighbourhoods with rarefaction to account for differences in sampling intensity across geographic space. These differences otherwise skew the visualisation and interpretation of genetic diversity, as illustrated by nSSR analyses of cacao (*Theobroma cacao* [5]) and the cherimoya fruit tree (*Annona cherimola* [6]) in South America. The results of our analyses of wild barley are given in Figures 4 and 5. Figure 4, based on allelic/haplotype richness estimates, demonstrates that BOPA SNPs, nSSRs and cpSSRs all provide similar profiles of diversity across geographic space. In each case, higher sample-size-corrected values of richness were observed in the Eastern Mediterranean region than in Central Asia. Circular neighbourhood analysis based on chloroplast haplotypes therefore clearly corresponds with the distribution of unique haplotypes shown in Figure 1 (nine of 10 unique haplotypes occurred in the Eastern Mediterranean region). Figure 5 (A, B), which shows levels of BOPA SNP and nSSR diversity based on *K* group richness, also indicates higher diversity (greater genetic differentiation at a local level) in the Eastern Mediterranean than in Central Asia (as also evident from individual *K* group assignments in Fig. 3).

Our findings are consistent with the limited previous molecular marker research (uncorrected for sampling intensity) comparing wild barley from the Eastern Mediterranean region and environs with Central Asia. For example, Volis et al. [46] measured lower variation in wild barley in Central Asia (samples from Turkmenistan only) than in the Eastern Mediterranean using isozymes, while Fu and Horbach [47] found the same based on nSSRs. Our analysis provides comprehensive evidence to reinforce these observations and confirms the status of the Eastern Mediterranean wild barley stands as important resources for conservation and evaluation [9,10].

Our intention in this study is to compare patterns of genetic variation in wild barley with the modelled distributions of the taxon under all three conditions of current, past and future climates, something which to our knowledge has not been undertaken before for any member of the genus *Hordeum*. The results of our ecological niche modelling are presented in Figure 6, from which several interesting observations can be drawn in relation to the genetic data, as set out in this section for the current-past climate comparison and in the next section for the current-future climate comparison.

Considering first the modelled present-day geographic distribution of wild barley, visual assessment confirms that the accessions included in our study for genetic analysis provide good coverage of most of the taxon's range in the Fertile Crescent and Central Asia (compare Fig. 6A with Fig. 1). The most obvious exception is the 'peak' of the Fertile Crescent in southern Turkey, where it is known that important stands of wild barley occur [8]. Although not sampled in this study, these Turkish stands should be incorporated in future work. A comparison of the modelled present-day geographic distribution of wild barley with that projected for the LGM (Figs. 6B, D) suggests that since the LGM suitable habitat has been lost in areas that include southeastern Iran and northern Saudi Arabia. At the same time, the comparison indicates that at the LGM wild barley was, just as it is now, widely present in the region bordering the Eastern Mediterranean coast. A comparison of current and past modelled distributions also indicates habitat gains since the LGM. Particular areas identified in this regard include the northern Iraq portion of the Fertile Crescent and, especially, a large part of Central Asia.

The apparent relatively recent range expansion of wild barley in Central Asia as revealed by ecological niche modelling is consistent with our findings of lower levels of genetic diversity in the region

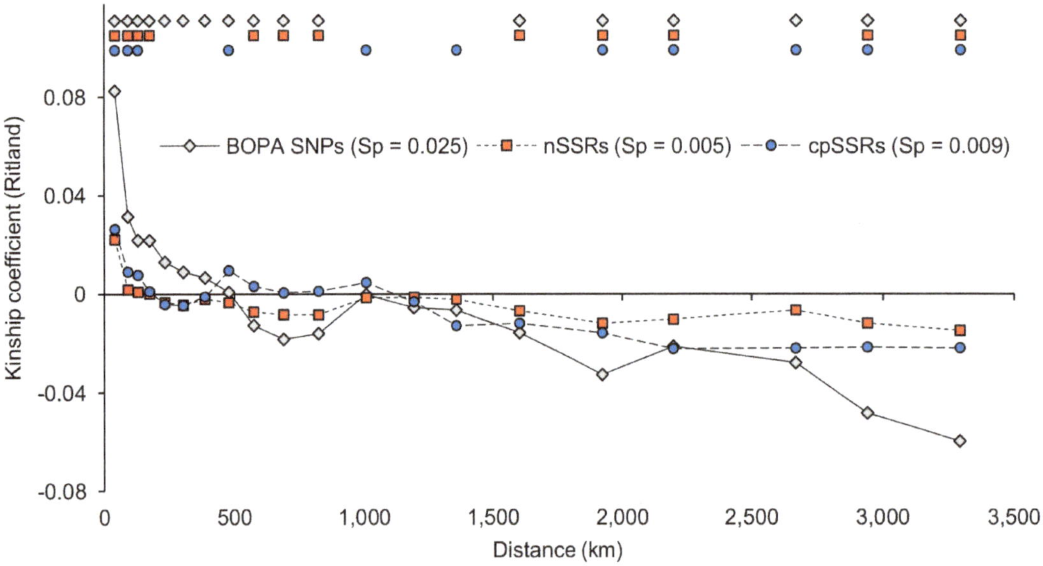

Figure 2. Spatial autocorrelation analysis profiles for wild barley accessions based on BOPA SNPs, nSSRs and cpSSRs. Geographic distances on the *x*-axis are the mean values of distance classes. The symbols at the top of the figure mark observations significantly larger or smaller ($P \leq 0.01$) than the average for distance classes. Values for the *Sp* statistic, calculated from the regression slope of the graph and the kinship coefficient of the first distance class [76], are also shown. Placing all three data sets on the same graph allows profiles to be compared. Increases in similarity at a distance class of around 1,000 km, and an earlier additional increase for cpSSRs at around 500 km, illustrate that a simple isolation-by-distance model is not sufficient to describe genetic variation in wild barley.

compared to the Eastern Mediterranean, in which latter region it can be postulated that a continuous presence of the taxon has led to the accumulation of genetic diversity (as shown in Figs. 4, 5A and B) there. The correspondence between past-present ecological niche modelling and our analysis of spatial genetic diversity in wild barley has important implications, as it supports the utility of niche modelling as a tool for identifying genetically diverse and potentially refugial areas. There are, however, other possible reasons why different levels of genetic diversity are observed across wild barley's range. The Eastern Mediterranean region (as represented, e.g., by altitudinal variation for the accessions included in the current study, see Fig. 5C) is, for example, particularly environmentally heterogeneous. This may have allowed more genetic variation to develop and accumulate there compared to Central Asia without recourse to an explanation based upon post-LGM macro-geographic range adjustment. We are not able to distinguish between these alternatives (or, indeed, to understand whether a combination of both range expansion and environmental heterogeneity are important) for determining the current pattern of spatial genetic diversity observed in wild barley. Nevertheless, one interesting feature of our data that deserves further exploration in this regard is the level of linkage disequilibrium (LD) between chromosome-position-mapped BOPA SNP markers in different parts of wild barley's range, as we relate below.

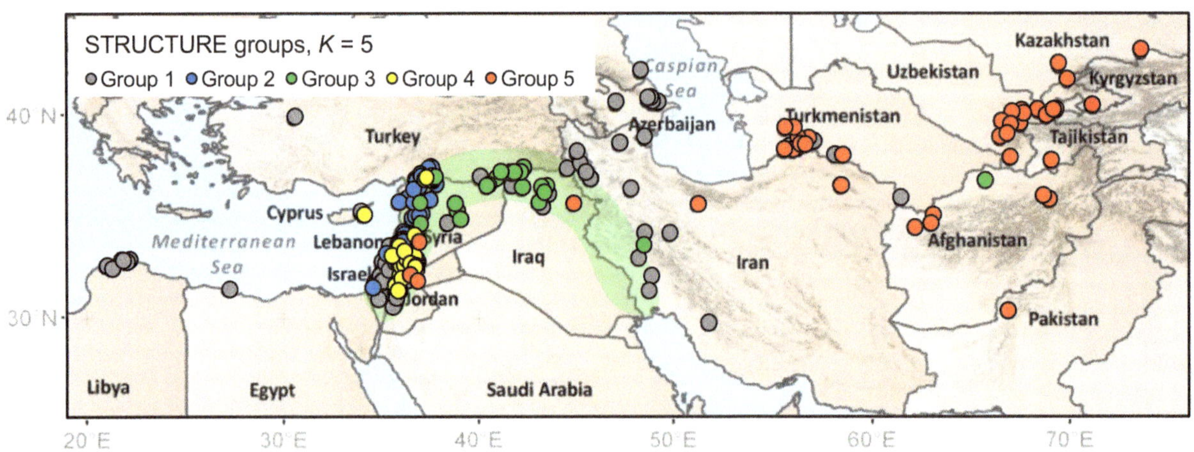

Figure 3. STRUCTURE group assignments for individual wild barley accessions. The results shown are based on $K = 5$ and for BOPA SNPs. The results for nSSRs (not shown) were similar. Results correspond with spatial autocorrelation analysis (Fig. 2) in describing a more complex genetic structure in wild barley than might be expected with a simple isolation-by-distance model.

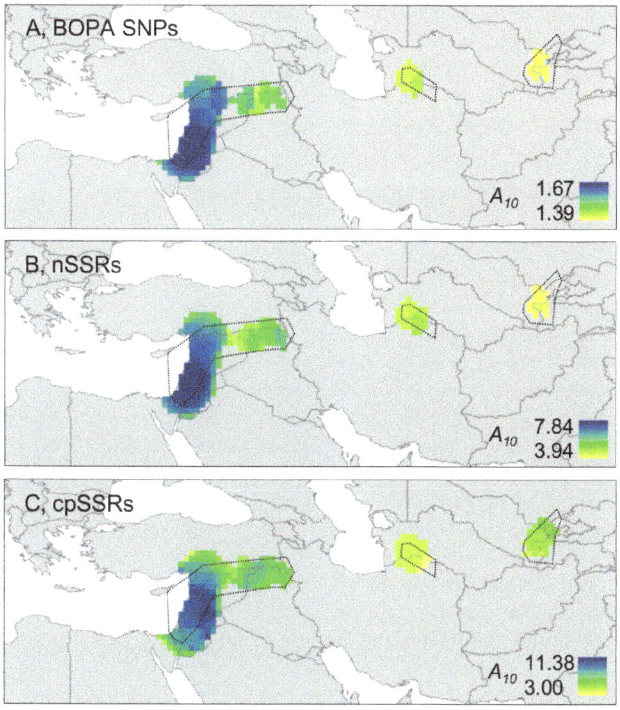

Figure 4. Allelic (A and B) and haplotype (C) richness (A_{10}) maps for wild barley. BOPA SNPs, nSSRs and cpSSRs all indicate the Eastern Mediterranean region as more diverse (highly diverse areas = dark blue) than Central Asia. As expected, nSSRs with high allelic diversity and cpSSRs with multiple haplotypes reveal relatively higher richness values within neighbourhoods (A_{10} as high as 7.84 and 11.38, respectively) than biallelic BOPA SNPs (maximum A_{10} = 1.67). Not all of the original sample range could be included in analysis because of the required minimum sampling intensity to calculate a standardised diversity value (see Materials and Methods; compare the current figure with Fig. 1). Accessions included in analysis in a particular geographic area are circumscribed by a dotted line. A, 2,426 from 2,505 BOPA SNPs used in calculations (SNPs excluded with ≥25% missing data in one or more grid cells); B, all 24 nSSRs used in calculations; C, all cpSSR haplotypes used in calculations.

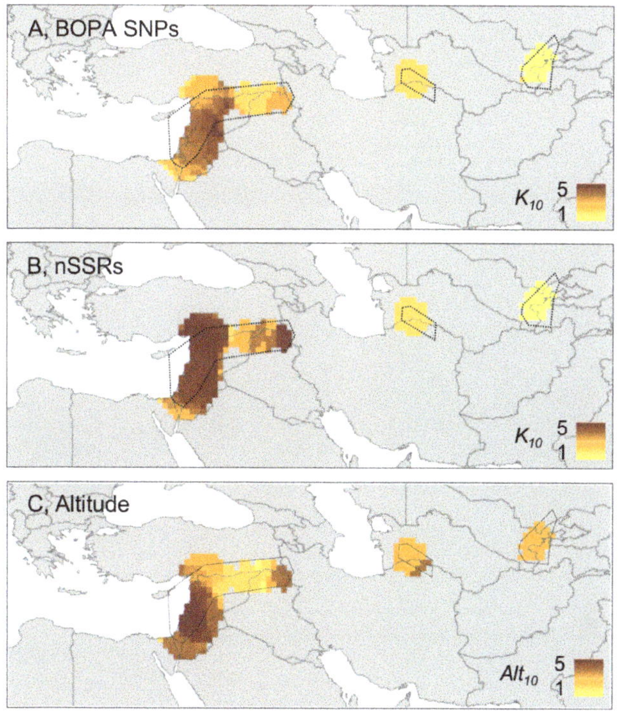

Figure 5. STRUCTURE group richness (A and B, K_{10}) and 'altitude richness' (C, Alt_{10}) maps for wild barley. A and B, richness estimates for BOPA SNPs and nSSRs, respectively, K = 5 in STRUCTURE analysis. Both marker sets indicate the Eastern Mediterranean region as more diverse (highly diverse areas = dark brown) than Central Asia. C, 'altitude richness' of wild barley sample sites, based on five altitude categories (<200 m, 200 to 600 m, 600 to 1,000 m, 1,000 to 1,400 m, >1,400 m). Altitude data provide an indication of environmental heterogeneity and were downloaded from WorldClim (www.worldclim.org/; values given in Table S1). Unlike the 19 bioclimatic variables used elsewhere in the current study, altitude data are actual values rather than interpolations from weather station records, so they are particularly appropriate for assessing real environmental heterogeneity [22,77]. Altitude richness estimates indicate sample points in the Eastern Mediterranean region as more diverse than those in Central Asia. Not all of the original sample range could be included in analyses because of the required minimum sampling intensity to calculate standardised diversity values (see Materials and Methods; compare the current figure with Figs. 1 and 3 [individual STRUCTURE K group assignments], see also Fig. 4). Accessions included in analyses in a particular geographic area are circumscribed by a dotted line. The analysis to generate 'altitude richness' was carried out in the same way as for STRUCTURE group richness, except 'altitude category' substituted for 'STRUCTURE group'.

To explore LD, we undertook a further analysis based on two sub-samples of our wild barley accessions taken to represent the Eastern Mediterranean and Central Asia regions, as shown in Figure 7. These sub-samples, each of 40 individuals, represent approximately balanced sets of material (as explained in the legend to Fig. 7) for LD comparison. Compared to the Eastern Mediterranean sub-sample, the Central Asian sub-sample is of lower genetic diversity, comes from a more uniform environment and has a much higher proportion of accessions collected from habitat established (apparently) since the LGM. For these sub-samples of accessions, we then calculated LD for pairs of chromosome-mapped BOPA SNPs falling into different centimorgan (cM) distance categories along each of barley's seven chromosomes. Estimates for LD were based on 487 SNPs (the number of mapped SNPs on each chromosome ranged from 57 to 80) with a minimum minor allele frequency of 0.1 in both sub-samples, while the chromosome distance interval for making pairwise comparisons was set at five cM (so comparisons for paired SNPs 0 to 5 cM apart, 5 to 10 cM apart, etc.). The level of LD was estimated with r^2 values (the squared correlation of allele frequencies [48,49]) using DNASP 5.00.07 [50] with all SNPs assigned homozygous status (i.e., no intra-locus component in analysis). Finally, once r^2 values were generated, they were compiled into mean values for chromosome distance categories (a minimum of 10 observations for a distance interval were required before assigning a mean value) for each chromosome, and then averaged across chromosomes, in EXCEL. Results were then expressed in graphical form comparing LD estimates across sub-samples (mean r^2 Central Asia sub-sample/mean r^2 Eastern Mediterranean sub-sample) (Fig. 8).

Our comparison of LD estimates indicates that at shorter chromosome distance intervals values of LD are relatively higher in Central Asia than in the Eastern Mediterranean, but that with increased distance along chromosomes values become more equal. Linkage disequilibrium values are difficult to interpret because of the many influencing factors, including population structure, selection pressures, mating patterns and changes in population size

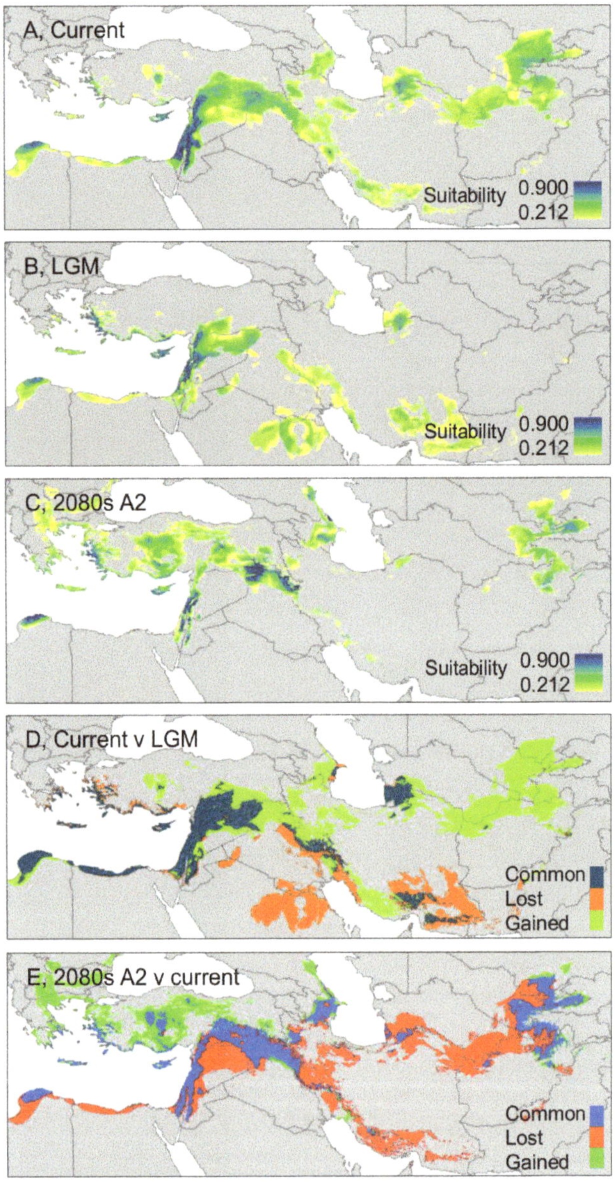

Figure 6. Potential wild barley distributions in current (A), past (B) and future (C) climates. Distributions are based on ecological niche modelling using MaxEnt (see Materials and Methods). D, differences between current and past modelled distributions, including areas lost and gained since the LGM. E, differences between future and current modelled distributions, including areas expected to be lost and gained by the 2080s. Note that past and future distribution maps take no account of rises or falls in sea levels or of other water bodies, and that these distributions are shown superimposed on current country boundaries.

and range [26,51–53]. In the absence of confounding factors, however, our observations are consistent with a relatively recent geographic expansion in Central Asia that has not allowed for enough time for recombination between proximate paired markers to result in equilibrium (convergence) between them [54–56]. Although interesting, we stress that this interpretation of the results of our LD analysis is speculative and must be treated with caution. For example, the selfing rate for wild barley can vary across populations (see [41] and [57] for Israeli and Jordanian populations, respectively), and if wild barley stands in Central Asia were more highly selfed [58], this could also explain why LD for proximate markers was higher there than in the Eastern Mediterranean. Clearly, more research on this topic is required.

Niche Modelling Suggests Small, But Statistically Significant, Losses of Genetic Diversity in Wild Barley Under Mid-Term Anthropogenic Climate Change

A comparison of present-day plant distributions with predictions for the 2080s is useful for devising responses to anthropogenic climate change [1,32]. For example, areas of predicted habitat loss may be targets for the collection of seed that can then be stored in gene banks. In addition, locations where habitat is likely to be retained may be priorities for *in situ* conservation measures [59,60]. Of most interest may be locations where habitat is predicted to be retained in geographic regions of general habitat loss. Our ecological niche modelling comparing the current distribution of wild barley with that predicted for the 2080s under the A2 emission scenario is shown in Figure 6 (C, E). The comparison suggests that suitable habitat will be lost in particular in large areas of Iran, northern Syria and in the border region of Afghanistan and Turkmenistan. At the same time, potential habitat will be gained most notably in parts of Turkey.

In order to calculate the possible losses in range-wide genetic diversity in wild barley associated with future habitat loss, we compared allelic richness at nSSRs for accessions predicted to be in shared future and current habitat ($N = 155$, i.e., excluding accessions in 'lost' habitat, as indicated in Table S1) with accessions in the current distribution ($N = 243$). We chose nSSRs as the estimator for this analysis because of their high allelic variability and hence sensitivity in describing diversity differences. Our analysis indicated a relatively small reduction in allelic richness under climate change (shared habitat, $A = 14.88$; current habitat, $A = 15.88$; estimates calculated in FSTAT 2.9.4 [61] and corrected by rarefaction to a sample size of 135 complete genotypes across all nSSRs). Although small, the difference was statistically significant ($P = 0.012$ based on a two-tailed t-test of individual locus allelic richness values undertaken in EXCEL). Further assessment, based on BOPA SNPs and cpSSRs, revealed a relatively modest 108 SNP alleles (4.6% of SNPs in the comparison) and four chloroplast haplotypes unique to 2080s 'lost' habitat. Our analysis therefore suggests that, overall, mid-term future losses in genetic diversity due to climate change are expected to be relatively low. A comparison of modelled current, past and future distributions (Fig. 6) suggests that in part this is because much predicted future habitat loss is in areas of putative post-LGM range expansion, where contemporary genetic variation is relatively low (e.g., in the border region of Afghanistan and Turkmenistan). On the other hand, shared future-current habitat includes much of the more genetically diverse putative LGM refugial regions. Of concern, though, could be future habitat loss in parts of northern Syria, where habitat is in common in the past-current comparison, but not in the future-current comparison (compare Figs. 6D, E).

It is important to consider a number of provisos when interpreting our findings. First, we have not considered in our analysis that existing wild barley populations in habitat that will be lost under climate change could migrate to (newly) environmentally-matched sites. Such migration will presumably be easier in areas with greater micro-geographic environmental heterogeneity, as the distances to be moved are then smaller. This would suggest migration is more feasible in the Eastern Mediterranean region (see Fig. 5C). Whether migrations are possible also depends on the level of human activity in wild barley's habitat. Human disturbance that provides opportunities for establishment could

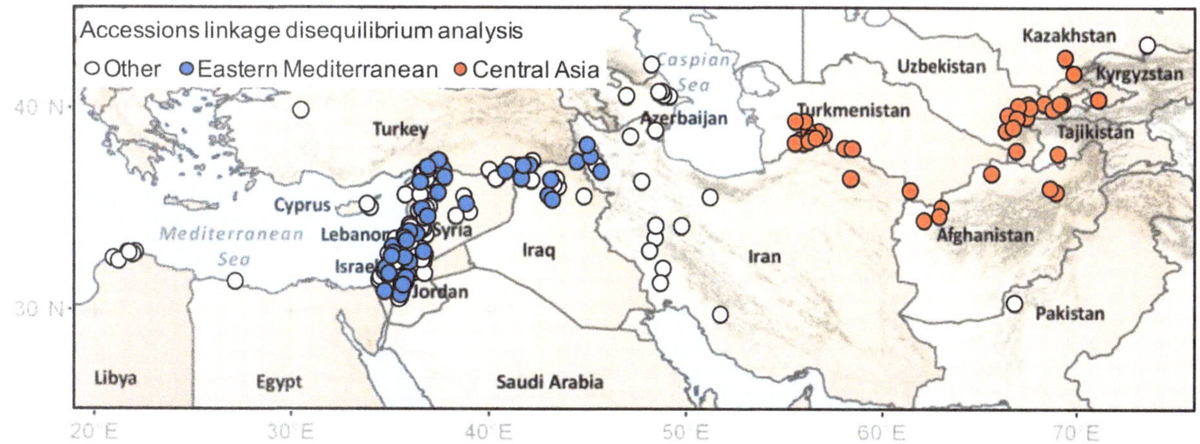

Figure 7. Locations of wild barley individuals sub-sampled from two regions for testing of linkage disequilibrium (LD). For LD assessment, forty accessions were chosen at random from the Eastern Mediterranean and Central Asia regions, across an approximately equal-dimensioned geographic area to minimise confounding sample size and dimensional effects in analysis (a significant issue in LD calculations [26]). In the case of the Eastern Mediterranean region, to ensure similar geographic coverage to Central Asia, sampling was extended eastward away from the coast below the peak of the Fertile Crescent into northwestern Iran. The coordinates of the accessions sampled for LD analysis are given in Table S1. Compared to the Eastern Mediterranean sub-sample, that from Central Asia had a less variable environment across accession collection sites (see Fig. 5C and the bioclimatic variables given in Table S1). Furthermore, climate modelling suggested a much greater proportion of accessions in the Central Asian sub-sample to be associated with range expansion since the LGM (see Fig. 6D and Table S1). Consistent with genetic diversity levels expressed on maps (Figs. 4, 5A, 5B), the latter sub-sample also had lower nuclear diversity according to FSTAT 2.9.4 [61] calculations (nSSR allelic richness for the Eastern Mediterranean sub-sample = 10.44, for Central Asia = 7.44, corrected by rarefaction to a sample size of 36 complete genotypes across all nSSRs; $P<0.001$ based on a two-tailed t-test of individual locus allelic richness values undertaken in EXCEL). Another factor that can confound LD comparisons is differences in allele frequency distributions between samples. We therefore tested allele frequency profiles for our two sub-samples, and found them to be similar (proportion of markers with a minimum minor allele frequency between 0.1 and 0.3 was 0.493 and 0.499 for the Eastern Mediterranean and Central Asia areas, respectively).

be beneficial, while modern agricultural practices that intensify crop production (excluding other plants from fields) and fragment wild habitat could be detrimental [62]. Second, in common with most other studies that compare present-day and potential future

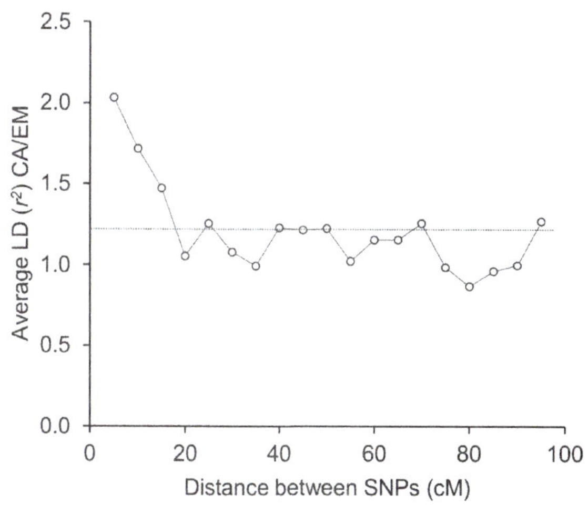

Figure 8. Comparison of mean linkage disequilibrium (LD) (r^2) values across all wild barley chromosomes. CA = Central Asia, EM = Eastern Mediterranean sub-samples. Comparisons are for BOPA SNPs at five cM intervals (SNPs 0 to 5 cM apart, 5 to 10 cM apart, etc.). The dotted line indicates the average value of all plotted comparisons. The graph indicates that when compared to longer pairwise SNP distances, LD estimates for shorter pairwise SNP distances are relatively higher in the Central Asia sub-sample than in the Eastern Mediterranean sub-sample. The difference between sub-samples appears to be lost after about 15 cM.

distributions to make conservation predictions (e.g., [5,63]), our analysis does not consider the possibilities for the local adaptation of plant populations to new climatic conditions. There is little relevant research on this topic for wild barley, which ideally requires multiple time-interval-based monitoring of wild stands [64]. Nevo et al. [65], however, did assess genetic variation and phenotypic traits in the same 10 natural stands of wild barley in Israel sampled first in 1980 and then again in 2008. The authors observed some change in the distribution of nSSR alleles and larger changes in flowering times (accessions sampled in 2008 flowered significantly earlier under greenhouse conditions). These changes could indicate responses to climate change, although other explanations are also possible. Interestingly, the changes in wild barley nSSR composition observed by Nevo et al. [65] were much smaller than those found in wild emmer wheat (*Triticum dicoccoides*) populations included in the same study and sampled at the same dates. This suggests different responses to climate change by different cereals in the Eastern Mediterranean region. Third and finally, our current assessment was based on genetic markers that are presumably (mostly) neutral with regard to phenotype, so we are not able to determine whether or not there will be important losses in functional genetic diversity under anthropogenic climate change.

Final Considerations

Our analyses are consistent with the view that climate change has played a role in determining the levels of present-day genetic variation observed in wild barley in different portions of its natural range. Our data support the utility of ecological niche modelling of current and past plant distributions for predicting geographic areas of high genetic diversity, and suggest limited future losses of genetic diversity in wild barley under mid-term future climate change. We have explored the use of chromosome-position-mapped SNPs for discriminating between different hypotheses to explain diversity patterns in wild barley, but more research is required on this topic, ideally using SNPs that have been physically positioned in the genome. With the promise soon of a complete genome sequence for barley (building on the current sequence assembly [66,67]), the physical distances between the SNPs used in the current study will soon be available. This will allow more formal LD analysis of possible range expansions and contractions in relation to climate change. Further research on wild barley should also explore the ensemble forecasting of distributions based on both the differences between GCM and the multiple statistical methods available for species modelling [68,69]. Modelling should also investigate the possible further downscaling of environmental data in predictions [70], which could provide greater accuracy [34,71].

The wider utility of past-present ecological niche modelling for locating centres of genetic diversity in the Fertile Crescent and Central Asia regions could be tested by examining the wild progenitors of other important cereals located there [8]. Molecular marker data sets are available for comparison purposes (e.g., for einkorn wheat [*Triticum monococcum*] and emmer wheat [72,73]), although more systematic assessments of genetic diversity are required based on fully geo-referenced samples. It would be interesting to model the distributions of different wild cereals at the time it is proposed that humans began to manage them significantly. How modelled distributions correspond with putative sites of first cultivation and first domestication [74] could then be explored. For example, modelling to understand distributions over the transition to the Younger Dryas (~12,000 years ago) would be useful. This was a relatively cold and unfavourable period for humans in the Fertile Crescent region that is believed to be associated with early cultivation events, leading eventually to domestications [75]. Distribution modelling of wild barley combined with geographically coincident sampling and genetic analysis of wild and landrace accessions of the taxon (and of other cereals) throughout the region will provide important insights into domestication processes [26].

Acknowledgments

We thank Shiaoman Chao, Nicola Cook, Linzi Jorgensen, Andrzej Kilian and Joy Roy who helped assemble the molecular marker data sets used in the current study, and Carlos Navarro and Denis Marechal who helped prepare future climate data. Dave Miller and Ian Williamson helped in the preparation of figures. Anonymous review greatly improved the manuscript. MvZ thanks the CGIAR Research Programme on Climate Change, Agriculture and Food Security (CCAFS) for support.

Author Contributions

Conceived and designed the experiments: JR MvZ IKD RW BS. Performed the experiments: JR IKD AB BS. Analyzed the data: JR MvZ IKD RW BS. Wrote the paper: JR MvZ IKD AB RW BS.

References

1. Warren R, VanDerWal J, Price J, Welbergen JA, Atkinson I, et al. (2013) Quantifying the benefit of early climate change mitigation in avoiding biodiversity loss. Nature Climate Change 3: 678–682.
2. Waltari E, Hijmans RJ, Peterson AT, Nyári ÁS, Perkins SL, et al. (2007) Locating Pleistocene refugia: comparing phylogeographic and ecological niche model predictions. PLoS ONE 2: e563. Available: www.plosone.org/
3. Provan J, Bennett KD (2008) Phylogeographic insights into cryptic glacial refugia. Trends in Ecology and Evolution 23: 564–571.
4. Nogués-Bravo D (2009) Predicting the past distribution of species climatic niches. Global Ecology and Biogeography 18: 521–531.
5. Thomas E, van Zonneveld M, Loo J, Hodgkin T, Galluzzi G, et al. (2012) Present spatial diversity patterns of *Theobroma cacao* L. in the neotropics reflect genetic differentiation in Pleistocene refugia followed by human-influenced dispersal. PLoS ONE 7: e47676. Available: www.plosone.org/
6. van Zonneveld M, Scheldeman X, Escribano P, Viruel MA, Van Damme P, et al. (2012) Mapping genetic diversity of cherimoya (*Annona cherimola* Mill.): application of spatial analysis for conservation and use of plant genetic resources. PLoS ONE 7: e29845. Available: www.plosone.org/
7. Newton AC, Flavell AJ, George TS, Leat P, Mullholland B, et al. (2011) Crops that feed the world 4. Barley: a resilient crop? Strengths and weaknesses in the context of food security. Food Security 3: 141–178.
8. Harlan JR (1975) Crops and man. Madison, Wisconsin, , USA: The American Society of Agronomy and the Crop Science Society of America.
9. Feuillet C, Langridge P, Waugh R (2008) Cereal breeding takes a walk on the wild side. Trends in Genetics 24: 24–32.
10. Nevo E, Chen G (2010) Drought and salt tolerances in wild relatives for wheat and barley improvement. Plant, Cell and Environment 33: 670–685.
11. Rostoks N, Mudie S, Cardle L, Russell J, Ramsay L, et al. (2005) Genome-wide SNP discovery and linkage analysis in barley based on genes responsive to abiotic stress. Molecular Genetics and Genomics 274: 515–527.
12. Ramsay L, Macaulay M, degli Ivanissevich S, MacLean K, Cardle L, et al. (2000) A simple sequence repeat-based linkage map of barley. Genetics 156: 1997–2005.
13. Provan J, Russell JR, Booth A, Powell W (1999) Polymorphic chloroplast simple sequence repeat primers for systematic and population studies in the genus *Hordeum*. Molecular Ecology 8: 505–511.
14. Elith J, Graham CH, Anderson RP, Dudík M, Ferrier S, et al. (2006) Novel methods improve prediction of species' distributions from occurrence data. Ecography 29: 129–151.
15. Hijmans RJ, Guarino L, Mathur P (2012) DIVA-GIS version 7.5 manual. California, , USA: University of California Davis.
16. Mitchell A (2005) The ESRI guide to GIS analysis, volume 2: spatial measurements and statistics. California, , USA: ESRI press.
17. Szpiech ZA, Jakobsson M, Rosenberg NA (2008) ADZE: a rarefaction approach for counting alleles private to combinations of populations. Bioinformatics 24: 2498–2504.
18. Groves CR, Game ET, Anderson MG, Cross M, Enquist C, et al. (2012) Incorporating climate change into systematic conservation planning. Biodiversity and Conservation 21: 1651–1671.
19. Steffenson BJ, Olivera P, Roy JA, Jin Y, Smith KP, et al. (2007) A walk on the wild side: mining wild wheat and barley collections for rust resistance genes. Australian Journal of Agricultural Research 58: 532–544.
20. Dai F, Nevo E, Wu D, Comadran J, Zhou M, et al. (2012) Tibet is one of the centers of domestication of cultivated barley. Proceedings of the National Academy of Sciences of the USA 109: 16969–16973.
21. Ren X, Nevo E, Sun D, Sun G (2013) Tibet as a potential domestication center of cultivated barley of China. PLoS ONE 8: e62700. Available: www.plosone.org/
22. Hijmans RJ, Cameron SE, Parra JL, Jones PG, Jarvis A (2005) Very high resolution interpolated climate surfaces for global land areas. International Journal of Climatology 25: 1965–1978.
23. Close TJ, Bhat PR, Lonardi S, Wu Y, Rostoks N, et al. (2009) Development and implementation of high-throughput SNP genotyping in barley. BioMed Central Genomics 10: 582. Available: www.biomedcentral.com
24. Roy JK, Smith KP, Muehlbauer GJ, Chao S, Close TJ, et al. (2010) Association mapping of spot blotch resistance in wild barley. Molecular Breeding 26: 243–256.
25. Moragues M, Comadran J, Waugh R, Milne I, Flavell AJ, et al. (2010) Effects of ascertainment bias and marker number on estimations of barley diversity from high throughput SNP genotype data. Trends in Applied Genetics 120: 1525–1534.
26. Russell JR, Dawson IK, Flavell A, Steffenson B, Weltzien E, et al. (2011) Analysis of >1000 single nucleotide polymorphisms in geographically matched samples of landrace and wild barley indicates secondary contact and chromosome-level differences in diversity around domestication genes. New Phytologist 191: 564–578.
27. Comadran J, Thomas W, van Eeuwijk F, Ceccarelli S, Grando S, et al. (2009) Patterns of genetic diversity and linkage disequilibrium in a highly structured *Hordeum vulgare* association-mapping population for the Mediterranean basin. Theoretical and Applied Genetics 119: 175–187.
28. Hardy OJ, Vekemans X (2002) SPAGeDi: a versatile computer program to analyse spatial genetic structure at the individual or population levels. Molecular Ecology Notes 2: 618–620.

29. Ritland K (1996) Estimators for pairwise relatedness and individual inbreeding coefficients. Genetical Research 67: 175–185.
30. Falush D, Stephens M, Pritchard JK (2007) Inference of population structure using multilocus genotype data: dominant markers and null alleles. Molecular Ecology Notes 7: 574–578.
31. Leberg PL (2002) Estimating allelic richness: effects of sample size and bottlenecks. Molecular Ecology 11: 2445–2449.
32. Davis AP, Gole TW, Baena S, Moat J (2012) The impact of climate change on indigenous arabica coffee (*Coffea arabica*): predicting future trends and identifying priorities. PLoS ONE 7: e47981. Available: www.plosone.org/
33. Elith J, Phillips SJ, Hastie T, Dudik M, Chee YE, et al. (2011) A statistical explanation of MaxEnt for ecologists. Diversity and Distributions 17: 43–57.
34. Aguirre-Gutiérrez J, Carvalheiro LG, Polce C, van Loon EE, Raes N, et al. (2013) Fit-for-purpose: species distribution model performance depends on evaluation criteria – Dutch hoverflies as a case study. PLoS ONE 8: e63708. Available: www.plosone.org/
35. Merow C, Smith MJ, Silander Jr JA (2013) A practical guide to MaxEnt for modeling species' distributions: what it does, and why inputs and settings matter. Ecography 36: 1058–1069.
36. van Zonneveld M, Jarvis A, Dvorak W, Lema G, Leibing C (2009) Climate change impact predictions on *Pinus patula* and *Pinus tecunumanii* populations in Mexico and Central America. Forest Ecology and Management 257: 1566–1576.
37. Phillips SJ, Dudík M (2008) Modeling of species distributions with Maxent: new extensions and a comprehensive evaluation. Ecography 31: 161–175.
38. Liu C, Berry PM, Dawson TP, Pearson RG (2005) Selecting thresholds of occurrence in the prediction of species distributions. Ecography 28: 385–393.
39. Ramirez J, Jarvis A (2010) Downscaling global circulation model outputs: the delta method. Decision and Policy Analysis Working Paper No. 1. Cali, Columbia: Centro Internacional de Agricultura Tropical (CIAT).
40. Vermeulen SJ, Challinor AJ, Thornton PK, Campbell BM, Eriyagama N, et al. (2013) Addressing uncertainty in adaptation planning for agriculture. Proceedings of the National Academy of Sciences of the USA 110: 8357–8362.
41. Brown AHD, Zohary D, Nevo E (1978) Outcrossing rates and heterozygosity in natural populations of *Hordeum spontaneum* Koch in Israel. Heredity 41: 49–62.
42. Petit RJ, Duminil J, Fineschi S, Hampe A, Salvini D, et al. (2005) Comparative organization of chloroplast, mitochondrial and nuclear diversity in plant populations. Molecular Ecology 14: 689–701.
43. Kozak KH, Graham CH, Wiens JJ (2008) Integrating GIS-based environmental data into evolutionary biology. Trends in Ecology and Evolution 23: 141–148.
44. Scheldeman X, van Zonneveld M (2010) Training manual on spatial analysis of plant diversity and distribution. Rome, Italy: Bioversity International.
45. van Etten J, Hijmans RJ (2010) A geospatial modelling approach integrating archaeobotany and genetics to trace the origin and dispersal of domesticated plants. PLoS ONE 5: e12060. Available: www.plosone.org/
46. Volis S, Mendlinger S, Turuspekov Y, Esnazarov U, Abugalieva S, et al. (2001) Allozyme variation in Turkmenian populations of wild barley, *Hordeum spontaneum* Koch. Annals of Botany 87: 435–446.
47. Fu Y-B, Horbach C (2012) Genetic diversity in a core subset of wild barley germplasm. Diversity 4: 239–257.
48. Hill WG, Robertson A (1968) Linkage disequilibrium in finite populations. Theoretical and Applied Genetics 38: 226–231.
49. Kelly JK (1997) A test of neutrality based on interlocus associations. Genetics 146: 1197–1206.
50. Librado P, Rozas J (2009) DNASP version 5, a software for comprehensive analysis of DNA polymorphism data. Bioinformatics 25: 1451–1452.
51. Slatkin M (2008) Linkage disequilibrium – understanding the evolutionary past and mapping the medical future. Nature Reviews Genetics 9: 477–485.
52. Excoffier L, Foll M, Petit RE (2009) Genetic consequences of range expansions. Annual Review of Ecology, Evolution, and Systematics 40:481–501.
53. Siol M, Wright SI, Barrett SCH (2010) The population genomics of plant adaptation. New Phytologist 188: 313–332.
54. McVean GAT (2002) A genealogical interpretation of linkage disequilibrium. Genetics 162: 987–991.
55. Caldwell KS, Russell J, Langridge P, Powell W (2006) Extreme population-dependent linkage disequilibrium detected in an inbreeding plant species, *Hordeum vulgare*. Genetics 172: 557–567.
56. Tenesa A, Navarro P, Hayes BJ, Duffy DL, Clarke GM, et al. (2007) Recent human effective population size estimated from linkage disequilibrium. Genome Research 17: 520–526.
57. Abdel-Ghani AH, Parzies HK, Omary A, Geiger HH (2004) Estimating the outcrossing rate of barley landraces and wild barley populations collected from ecologically different regions of Jordan. Theoretical and Applied Genetics 109: 588–595.
58. Morrell PL, Toleno DM, Lundy KE, Clegg MT (2006) Estimating the contribution of mutation, recombination and gene conversion in the generation of haplotypic diversity. Genetics 173: 1705–1723.
59. Keppel G, Van Niel KP, Wardell-Johnson GW, Yates CJ, Byrne M, et al. (2012) Refugia: identifying and understanding safe havens for biodiversity under climate change. Global Ecology and Biogeography 21: 393–404.
60. Shoo LP, Hoffmann AA, Garnett S, Pressey RL, Williams YM, et al. (2013) Making decisions to conserve species under climate change. Climatic Change 119: 239–246.
61. Goudet J (1995) FSTAT version 1.2: a computer program to calculate F statistics. Journal of Heredity 86: 485–486.
62. Bishaw Z, Struik PC, van Gastel AJG (2011) Wheat and barley seed system in Syria: farmers, varietal perceptions, seed sources and seed management. International Journal of Plant Production 5: 323–347.
63. Bálint M, Domisch S, Engelhardt CHM, Haase P, Lehrian S, et al. (2011) Cryptic biodiversity loss linked to global climate change. Nature Climate Change 1: 313–318.
64. Hansen MM, Olivieri I, Waller DM, Nielsen EE, the GeM Working Group (2012) Monitoring adaptive genetic responses to environmental change. Molecular Ecology 21: 1311–1329.
65. Nevo E, Fu Y-B, Pavlicek T, Khalifa S, Tavasi M, et al. (2012) Evolution of wild cereals during 28 years of global warming in Israel. Proceedings of the National Academy of Sciences of the USA 109: 3412–3415.
66. IBSC (The International Barley Genome Sequencing Consortium) (2012) A physical, genetic and functional sequence assembly of the barley genome. Nature 491: 711–717.
67. Mascher M, Richmond TA, Gerhardt DJ, Himmelbach A, Leah C, et al. (2013) Barley whole exome capture: a tool for genomic research in the genus *Hordeum* and beyond. The Plant Journal 76: 494–505.
68. Araujo MB, New M (2007) Ensemble forecasting of species distributions. Trends in Ecology and Evolution 22: 42–47.
69. Buisson L, Thuiller W, Casajus N, Lek S, Grenouillet G (2010) Uncertainty in ensemble forecasting of species distribution. Global Change Biology 16: 1145–1157.
70. Franklin J, Davis FW, Ikegami M, Syphard AD, Flint LE, et al. (2013) Modeling plant species distributions under future climates: how fine scale do climate projections need to be? Global Change Biology 19: 473–483.
71. Conlisk E, Syphard AD, Franklin J, Flint L, Flint A, et al. (2013) Uncertainty in assessing the impacts of global change with coupled dynamic species distribution and population models. Global Change Biology 19: 858–869.
72. Brown TA, Jones MK, Powell W, Allaby RG (2008) The complex origins of domesticated crops in the Fertile Crescent. Trends in Ecology and Evolution 24: 103–109.
73. Burger JC, Chapman MA, Burke JM (2008) Molecular insights into the evolution of crop plants. American Journal of Botany 95: 113–122.
74. Willcox G (2005) The distribution, natural habitats and availability of wild cereals in relation to their domestication in the Near East: multiple events, multiple centres. Vegetation History and Archaeobotany 14: 534–541.
75. Fuller DQ (2007) Contrasting patterns in crop domestication and domestication rates: recent archaeobotanical insights from the Old World. Annals of Botany 100: 903–924.
76. Vekemans X, Hardy OJ (2004) New insights from fine-scale spatial genetic structure analyses in plant populations. Molecular Ecology 13: 921–935.
77. Farr TG, Rosen PA, Caro E, Crippen R, Duren R, et al. (2007) The Shuttle Radar Topography Mission. Reviews of Geophysics 45: RG2004. doi: 10.1029/2005RG000183
78. Llaurens V, Castric V, Austerlitz F, Vekemans X (2008) High paternal diversity in the self-incompatible herb *Arabidopsis halleri* despite clonal reproduction and spatially restricted pollen dispersal. Molecular Ecology 17: 1577–1588.
79. Ismail SA, Ghazoul J, Ravikanth G, Shaanker RU, Kushalappa CG, et al. (2012) Does long-distance pollen dispersal preclude inbreeding in tropical trees? Fragmentation genetics of *Dysoxylum malabaricum* in an agro-forest landscape. Molecular Ecology 21: 5484–5496.

Confidence Intervals for Population Allele Frequencies: The General Case of Sampling from a Finite Diploid Population of Any Size

Tak Fung[1,2*], **Kevin Keenan**[3]

1 National University of Singapore, Department of Biological Sciences, Singapore, Singapore, 2 Queen's University Belfast, School of Biological Sciences, Belfast, Northern Ireland, United Kingdom, 3 Queen's University Belfast, Institute for Global Food Security, School of Biological Sciences, Belfast, Northern Ireland, United Kingdom

Abstract

The estimation of population allele frequencies using sample data forms a central component of studies in population genetics. These estimates can be used to test hypotheses on the evolutionary processes governing changes in genetic variation among populations. However, existing studies frequently do not account for sampling uncertainty in these estimates, thus compromising their utility. Incorporation of this uncertainty has been hindered by the lack of a method for constructing confidence intervals containing the population allele frequencies, for the general case of sampling from a finite diploid population of any size. In this study, we address this important knowledge gap by presenting a rigorous mathematical method to construct such confidence intervals. For a range of scenarios, the method is used to demonstrate that for a particular allele, in order to obtain accurate estimates within 0.05 of the population allele frequency with high probability ($\geq$95%), a sample size of >30 is often required. This analysis is augmented by an application of the method to empirical sample allele frequency data for two populations of the checkerspot butterfly (*Melitaea cinxia* L.), occupying meadows in Finland. For each population, the method is used to derive $\geq$98.3% confidence intervals for the population frequencies of three alleles. These intervals are then used to construct two joint $\geq$95% confidence regions, one for the set of three frequencies for each population. These regions are then used to derive a $\geq$95% confidence interval for Jost's *D*, a measure of genetic differentiation between the two populations. Overall, the results demonstrate the practical utility of the method with respect to informing sampling design and accounting for sampling uncertainty in studies of population genetics, important for scientific hypothesis-testing and also for risk-based natural resource management.

Editor: Guy Brock, University of Louisville, United States of America

Funding: TF is being supported by the National University of Singapore start-up grant WBS R-154-000-551-133. In addition, TF acknowledges past funding support by a Ph.D. studentship from the Beaufort Marine Research Award in Ecosystem Approach to Fisheries Management, carried out under the Sea Change Strategy and the Strategy for Science Technology and Innovation (2006–2013), with the support of the Marine Institute, funded under the Marine Research Sub-Programme of the Irish National Development Plan 2007–2013 (http://www.marine.ie/home/research/MIFunded/BeaufortAwards/). KK is being supported by a Ph.D. studentship from the Beaufort Marine Research Award in Fish Population Genetics, funded by the Irish Government under the Sea Change Strategy as above. The funders had no role in study design, data collection and analysis, decision to publish, or preparation of the manuscript.

Competing Interests: The authors have declared that no competing interests exist.

* E-mail: tfung01@qub.ac.uk

Introduction

Spatiotemporal patterns of genetic variation among populations are often used to test hypotheses about processes underlying the patterns, such as selection, migration and genetic drift (e.g., [1,2,3,4,5]). This genetic variation captures differences in genetic structure among populations, where the genetic structure of a population is determined by the distribution of alleles among individuals in the population. The allele distribution of a population is the result of a range of biological and environmental processes acting on the population and also on surrounding populations within the same geographical region, which result in non-random mixing of gametes among individuals in all populations. Patterns in allele distributions among populations are generally assessed by consideration of variations in allele frequencies (e.g., [6,7,8,9]).

Logistic and ethical constraints mean that in practice, a population is unlikely to be sampled in its entirety, such that the population allele frequencies have to be estimated using a subset of sampled individuals. For diploid organisms in a large population, it has been established that the frequency of an allele in a sample provides an unbiased estimate of the frequency in the population as a whole [10]. Thus, if samples of a given size are repeatedly taken from a large population, then the mean frequency of an allele in a sample converges to the population allele frequency as the number of samples increases. However, many studies only take a *single* sample from a population and present or use the resulting frequencies of alleles in the sample, without accounting for sampling uncertainty (e.g., [11,12,13,14,15,16,17,18]). Therefore, these studies implicitly assume that the sample allele frequencies are close to the population allele frequencies, which is by no means guaranteed. Interpretation of findings from these studies is therefore complicated by the potential for large sampling uncertainty. Ideally, in this single sample case, uncertainty bounds for the population allele frequencies would be quantified, based on the sample allele frequencies.

A recent study by Hale et al. [19] used computer simulations to draw samples from four diploid populations, with allele frequencies based on four empirical datasets. Subsequently, they used the sample data to derive the means, variances and ranges of allele frequencies in samples of varying size, as well as the means, variances and ranges of indicators that are a function of allele frequencies (specifically, heterozygosity and F_{ST}). Using these results, Hale et al. [19] concluded that a sample size of 30 is sufficient to accurately estimate population allele frequencies when using microsatellites. However, the authors did not quantify uncertainty bounds for the population allele frequencies using their sample data, such that their assessment of accuracy lacks a rigorous quantitative basis. Furthermore, for each sample size, sample frequencies were calculated using only 100 replicate samples, which may not give a good approximation of the true dispersion in the sample frequencies. This highlights a weakness of using a computational approach that lacks an underlying mathematical theory. Moreover, Hale et al. [19] did not consider the situation identified earlier, where one sample is taken and there is a need to quantify uncertainty of population allele frequencies using just the sample allele frequencies. For this situation, earlier studies have derived confidence intervals in order to capture uncertainty in the population allele frequencies. If a diploid population is of infinite size and is at Hardy-Weinberg equilibrium (HWE), then the frequency of an allele in a sample follows a binomial distribution [10]. Thus, Gillespie [20] proposed the use of the Wald confidence interval [21]. However, this interval only performs well for sufficiently large sample sizes – with small sample sizes, it is too short [22]. For the binomial distribution, other confidence intervals have been derived that do perform well for small sample sizes [22,23], notably the Clopper-Pearson interval that was derived eight decades ago [24]. However, these only apply in the limited cases when the population is at HWE. Weir [10] proposed a confidence interval that allows for deviations from HWE, analogous to the Wald confidence interval. However, like the latter, it is expected to perform well only for sufficiently large sample sizes [10]. This is problematic because the accuracy at a particular sample size is unknown, so "sufficiently large" cannot be rigorously quantified. Moreover, all the confidence intervals considered only apply to cases when the population can be assumed to be of infinite size, i.e. when the population size is much larger than the sample size. This does not reflect the range of scenarios encountered in empirical research (e.g., [13,25,26,27]).

In this study, we build on previous work by constructing confidence intervals for population allele frequencies for the general case where (i) the population is diploid, finite and can be of any size; (ii) the sample can take any size less than or equal to that of the population; and (iii) the population can deviate from HWE to any extent. These confidence intervals are guaranteed to contain the population allele frequency with a probability above a known threshold. The method derived for constructing these intervals is then used to calculate sample sizes required to achieve accurate estimates of population allele frequencies, under a range of scenarios. Here, accuracy is measured as the length of the confidence intervals. The sample sizes derived serve as a guide for determination of suitable sample sizes in future population genetic studies. In particular, we show that a sample size of 30 does not necessarily give accurate estimates, thus refining the conclusion of Hale et al. [19]. Lastly, we provide an example of how the method can be applied to microsatellite data for two populations of the checkerspot butterfly (*Melitaea cinxia* L.) [13], to derive confidence intervals for population allele frequencies and also for Jost's *D*, a measure of genetic differentiation between two populations that is a function of their population allele frequencies [9]. The data comes from a study [13] where it is unclear that the populations can be assumed to be of infinite sizes or at HWE. This example illustrates how the mathematical theory underlying our method can be used to quantify sampling uncertainty not only in population allele frequencies, but also in parameters that are functions of these frequencies, important for hypothesis-testing and also for natural resource management.

Overall, this study provides a rigorous mathematical quantification of sampling uncertainty in population allele frequencies, without the typically unrealistic constraints of assuming that a population has an infinite size and is at HWE. Thus, the results can be applied to a wide range of studies in population genetics.

Methods

In order to construct confidence intervals for the general case of taking samples of any size from a diploid population of any size (larger than or equal to the sample size) and with any degree of deviation from HWE, the sampling distribution of the allele frequencies in this general case is first exactly specified. This distribution is then used to derive formulae specifying confidence intervals that contain the population allele frequencies with probability above a known threshold. Using these formulae, confidence intervals are derived for a range of archetypal scenarios, which are then used to calculate sample sizes that permit accurate estimates of population allele frequencies in these scenarios. In addition, the formulae are applied to a real scenario where samples are taken from two butterfly populations [13], to construct confidence intervals for the population allele frequencies of these two populations. The intervals are then used to derive a corresponding confidence interval for Jost's *D* [9].

Derivation of sampling distribution of allele frequencies

Consider a population of M diploid individuals, from which a sample of N individuals is randomly drawn ($M \geq N$). At the locus of interest, there are $n > 1$ alleles, denoted by A_i, $i \in \{1, 2, \ldots, n\}$. Let the population allele frequencies be denoted by p_i, with the corresponding sample allele frequencies being denoted by $p_{i,N}$. Also, let P_{ij} be the frequency of individuals in the population with alleles A_i and A_j ($i, j \in \{1, 2, \ldots, n\}$), such that MP_{ij} is the number of corresponding individuals in the base population. p_i is related to P_{ij} by the formula $p_i = P_{ii} + \sum_{j=1, j\neq i}^{n} (P_{ij}/2)$. Here, P_{ii} is a measure of the homozygosity of individuals in the population with respect to allele A_i. $P_{ii} = p_i - \sum_{j=1, j\neq i}^{n} (P_{ij}/2)$, and thus $P_{ii} \leq p_i$. If $0 \leq p_i \leq 0.5$, then the minimum value of P_{ii} is 0, since all copies of allele A_i can be distributed among heterozygotes of allele A_i. However, if $p_i > 0.5$, then this is not possible – there must be at least one homozygote of allele A_i. In this case, the minimum number of homozygotes is realized when all heterozygotes have a copy of allele A_i, i.e. when $P_{ii} + \sum_{j=1, j\neq i}^{n} P_{ij} = 1$. Rearranging this for $\sum_{j=1, j\neq i}^{n} P_{ij}$ and substituting into $P_{ii} = p_i - \sum_{j=1, j\neq i}^{n} (P_{ij}/2)$ gives the minimum value of P_{ii} as $2p_i - 1$. Thus, overall, $\text{Max}\{0, 2p_i - 1\} \leq P_{ii} \leq p_i$.

The frequency of allele A_i in the sample of size N is given by $p_{i,N} = Y_{i,N}/(2N)$, where $Y_{i,N}$ is the number of copies of allele A_i in the sample. Thus, the probability distribution of $p_{i,N}$ is the same as the probability distribution of $Y_{i,N}$ except with the *x*-axis scaled by a factor $1/(2N)$. Denote the probability mass function (pmf) for $Y_{i,N}$ by $P(Y_{i,N} = y_{i,N})$. The range of $Y_{i,N}$ is $[0, 2N]$, so $P(Y_{i,N} = y_{i,N}) = 0$ for $y_{i,N}$ outside this range. Thus, for the following calculations, only $y_{i,N} \in [0, 2N]$ are considered. Now, $Y_{i,N} = 2X_{ii} + X_{i\cdot}$, where $X_{i\cdot} = \sum_{j=1, j\neq i}^{n} X_{ij}$ and X_{ij} is the number of individuals in the sample with alleles A_i and A_j. X_{ii}, $X_{i\cdot}$ and $Z_i = N - X_{ii} - X_{i\cdot}$ thus represent the number of individuals in the sample with two copies, one copy and no copies of allele A_i, respectively. X_{ii}, $X_{i\cdot}$ and Z_i follow a multivariate hypergeometric distribution with pmf given by:

$$P(X_{ii}=x_{ii},\, X_{i\cdot}=x_{i\cdot},\, Z_i=z_i) = \frac{\binom{MP_{ii}}{x_{ii}}\binom{M\sum_{j,j\neq i}^{n} P_{ij}}{x_{i\cdot}}\binom{M-MP_{ii}-M\sum_{j,j\neq i}^{n} P_{ij}}{z_i}}{\binom{M}{N}}, \tag{1}$$

where terms on the right-hand side in brackets are binomial coefficients. Since the number of individuals in the sample must equal N, $x_{ii}+x_{i\cdot}+z_i=N$. $P(Y_{i,N}=y_{i,N})$ is the sum of $P(X_{ii}=x_{ii}, X_{i\cdot}=x_{i\cdot}, Z_i=z_i)$ for all those biologically feasible combinations of x_{ii}, $x_{i\cdot}$ and z_i satisfying $y_{i,N}=2x_{ii}+x_{i\cdot}$. It will now be shown that this summation can be simplified to the sum of an expression that depends only on x_{ii} and $y_{i,N}$, over all x_{ii} between lower and upper bounds that only depend on $y_{i,N}$. This considerably eases calculation of $P(Y_{i,N}=y_{i,N})$. Firstly, the number of individuals in the sample with two copies, one copy or no copies of allele A_i cannot exceed the corresponding numbers in the sampled population. This gives rise to three inequalities:

$$x_{ii} \le MP_{ii}, \tag{2}$$

$$x_{i\cdot} \le M\sum_{j,j\neq i}^{n} P_{ij}, \tag{3}$$

$$N - x_{ii} - x_{i\cdot} \le M - MP_{ii} - M\sum_{j,j\neq i}^{n} P_{ij}. \tag{4}$$

Secondly, the number of individuals with two copies, one copy or no copies of allele A_i in the sample must be non-negative. This gives rise to three more inequalities:

$$x_{ii} \ge 0, \tag{5}$$

$$x_{i\cdot} \ge 0, \tag{6}$$

$$N - x_{ii} - x_{i\cdot} \ge 0. \tag{7}$$

Since $p_i = P_{ii} + \sum_{j=1, j\neq i}^{n}(P_{ij}/2)$ and $y_{i,N}=2x_{ii}+x_{i\cdot}$, then $\sum_{j=1, j\neq i}^{n} P_{ij} = 2(p_i - P_{ii})$ and $x_{i\cdot} = y_{i,N} - 2x_{ii}$. Thus, the inequalities (2)–(7) can be rearranged to obtain the double inequality:

$$\begin{aligned}&\mathrm{Max}\left\{\frac{y_{i,N}}{2} - Mp_i + MP_{ii},\, y_{i,N}-N,\, 0\right\} \le x_{ii} \le \\ &\mathrm{Min}\left\{MP_{ii},\, \frac{y_{i,N}}{2},\, M-N+y_{i,N}-2Mp_i+MP_{ii}\right\}.\end{aligned} \tag{8}$$

It is noted that $x_{i\cdot} = y_{i,N} - 2x_{ii}$, which means that $x_{i\cdot} \ge 0$ (inequality (6)) ensures that $x_{ii} \le y_{i,N}/2 \le N$, as required. Denote the upper and lower bounds in (8) by $L(y_{i,N})$ and $U(y_{i,N})$ respectively. Then $P(Y_{i,N}=y_{i,N})$ can be written as:

$$\begin{aligned}P(Y_{i,N}=y_{i,N}) &= \sum_{x_{ii}=\mathrm{ceiling}[L(y_{i,N})]}^{\mathrm{floor}[U(y_{i,N})]} P(X_{ii}=x_{ii},\, X_{i\cdot}=x_{i\cdot},\, Z_i=z_i) \\ &= \sum_{x_{ii}=\mathrm{ceiling}[L(y_{i,N})]}^{\mathrm{floor}[U(y_{i,N})]} P\begin{pmatrix} X_{ii}=x_{ii},\, X_{i\cdot}=y_{i,N}-2x_{ii}, \\ Z_i = N + x_{ii} - y_{i,N}\end{pmatrix},\end{aligned} \tag{9}$$

where z_i has been rewritten using $z_i = N - x_{ii} - x_{i\cdot}$ and $x_{i\cdot} = y_{i,N} - 2x_{ii}$, ceiling$[a]$ is the smallest integer larger than a, and floor$[a]$ is the largest integer smaller than a. The ceiling and floor functions are introduced to ensure that x_{ii} is an integer, which it must be to have a biological interpretation. $P(Y_{i,N}=y_{i,N})$ is equal to $P(p_{i,N}=y_{i,N}/(2N))$, the pmf for $p_{i,N}$, and thus defines the probability distribution of $p_{i,N}$. Equation (9) can be used to define the probability distribution of $p_{i,N}$ given only four parameters: M, p_i, P_{ii} and N.

Derivation of confidence intervals for sample allele frequencies

For given population size (M), population allele frequency (p_i), population frequency of homozygotes with allele A_i (P_{ii}) and sample size (N), the probability distribution for the sample allele frequency ($p_{i,N}$) can be calculated exactly using equation (9). The mean value of this distribution is:

$$\begin{aligned}E[p_{i,N}] &= E\left[\frac{Y_{i,N}}{2N}\right] = \frac{2E[X_{ii}]+E[X_{i\cdot}]}{2N} \\ &= \frac{2NP_{ii} + N\sum_{j=1, j\neq i}^{n} P_{ij}}{2N} = P_{ii} + \sum_{j=1, j\neq i}^{n}\frac{P_{ij}}{2} = p_i,\end{aligned} \tag{10}$$

as in the case of sampling from an infinite diploid population [10]. In equation (10), the expectation $E[X_{ij}] = NP_{ij}$ has been used, which follows from the fact that the X_{ij}'s, with $i, j \in \{1, 2, \ldots, n\}$, follow a multivariate hypergeometric distribution with parameters M, N and MP_{ij} [28]. In addition, it can be proved that the variance of $p_{i,N}$ is

$$\sigma^2[p_{i,N}] = \frac{(M-N)(p_i - 2p_i^2 + P_{ii})}{2(M-1)N} \tag{11}$$

(see File S1). The variance thus includes a standard finite correction factor $(M-N)/(M-1)$ that tends to 1 as $M\to\infty$, as required. Therefore, as $M\to\infty$, $\sigma^2[p_{i,N}] \to (p_i - 2p_i^2 + P_{ii})/(2N)$, the variance in the case of an infinite population size [10]. The cumulative distribution function (cdf) for $p_{i,N}$ is specified by:

$$\begin{aligned}P(p_{i,N} \le w) &= P\left(p_{i,N} \le \frac{y_{i,N}}{2N}\right) \\ &= P(Y_{i,N} \le y_{i,N}) = \sum_{k=0}^{y_{i,N}} P(Y_{i,N}=k),\end{aligned} \tag{12}$$

where $y_{i,N}$ is an integer in the interval [0, 2*N*]. This cdf can be calculated using equation (9) and is a function of p_i. To construct a CI for p_i given *M*, *N* and P_{ii}, consider testing the null hypothesis $H_0 : p_i = p_{i,0}$ against the alternative $H_1 : p_i \neq p_{i,0}$ at significance level α for an observed value of $Y_{i,N}$, denoted by $\hat{y}_{i,N}$. The null hypothesis is not rejected if using $p_i = p_{i,0}$, $\hat{y}_{i,N}$ falls within an acceptance region defined by $B_\alpha = [y_{i,N,\alpha/2}, y_{i,N,1-(\alpha/2)}]$, where $y_{i,N,\alpha/2}$ is the largest integer for which $P(Y_{i,N} < y_{i,N,\alpha/2}) \leq \alpha/2$ and $y_{i,N,1-(\alpha/2)}$ is the smallest integer for which $P(Y_{i,N} > y_{i,N,1-(\alpha/2)}) \leq \alpha/2$. B_α can be calculated using equation (12). One method of constructing a $\geq 100(1-\alpha)\%$ CI for p_i is to determine the set of values of $p_{i,0}$ for which the null hypothesis is not rejected at significance level α, denoted by $P_{i,H_0,\alpha} = \{p_{i,0} : \hat{y}_{i,N} \in B_\alpha\}$, and defining the CI as

$$[L_\alpha, U_\alpha] = \left[\text{Min}\left(P_{i,H_0,\alpha}\right), \text{Max}\left(P_{i,H_0,\alpha}\right)\right] \quad (13)$$

[29]. This hypothesis-testing approach corresponds to the "test-method" described by Talens [30], who applied it to construct CI's for parameters of the univariate hypergeometric distribution. If $P_{i,H_0,\alpha} = \varnothing$, the empty set, then the acceptance region needs to be extended to include $\hat{y}_{i,N}$ for at least one value of $p_{i,0}$. Thus, in this case, α is decreased until $\hat{y}_{i,N}$ is in the acceptance region for at least one $p_{i,0}$ value.

The CI specified by equation (13) is not an exact $100(1-\alpha)\%$ CI because the distribution for $Y_{i,N}$ is discrete and because there may be some $p_{i,0}$ values between $\text{Min}(P_{i,H_0,\alpha})$ and $\text{Max}(P_{i,H_0,\alpha})$ for which $\hat{y}_{i,N} \notin B_\alpha$, since $P(Y_{i,N} \leq y_{i,N})$ is not guaranteed to be a monotonic function of p_i (see equations (9) and (12)). It is noted that for the probability parameter in a binomial distribution, Clopper and Pearson [24] derived $\geq 100(1-\alpha)\%$ CI's using an analogous method. Thus, for the case of a large population size relative to the sample size ($M >> N$) and HWE, where $p_{i,N}$ approximately follows a binomial distribution, CI's for p_i derived using our method would be virtually the same as Clopper-Pearson CI's. This can be verified by explicitly calculating and comparing CI's using the two methods – for example, if $M = 1{,}000{,}000 >> N = 30$ and $\hat{y}_{i,N} = 10$, then both methods give the CI [0.083, 0.285] for p_i.

In deriving equation (13), it was assumed that P_{ii} is known. However, P_{ii} is generally unknown. In this case, a $\geq 100(1-\alpha)\%$ confidence region (CR) can be derived for p_i and P_{ii} in an analogous way, by considering the null hypothesis $H'_0 : p_i = p_{i,0}, P_{ii} = P_{ii,0}$. The CR can be derived by determining the set of vectors $(p_{i,0}, P_{ii,0})$ for which the null hypothesis is not rejected at significance level α. This set is denoted by $P_{i,H'_0,\alpha} = \{(p_{i,0}, P_{ii,0}) : \hat{y}_{i,N} \in B_\alpha\}$, where B_α is as defined before. Using the CR, $\geq 100(1-\alpha)\%$ CI's for p_i and P_{ii} can be defined as

$$[L'_{p_i,\alpha}, U'_{p_i,\alpha}] = \left[\text{Min}\left(P_{i,H'_0,\alpha,1}\right), \text{Max}\left(P_{i,H'_0,\alpha,1}\right)\right], \quad (14a)$$

and

$$[L'_{P_{ii},\alpha}, U'_{P_{ii},\alpha}] = \left[\text{Min}\left(P_{i,H'_0,\alpha,2}\right), \text{Max}\left(P_{i,H'_0,\alpha,2}\right)\right], \quad (14b)$$

where $P_{i,H'_0,\alpha,j}$ is the set of values consisting of the *j*th elements in the set of vectors $P_{i,H'_0,\alpha}$. In determining $P_{i,H'_0,\alpha}$, $p_{i,0}$ and $P_{ii,0}$ values over the biologically feasible ranges are tested. Since the number of copies of allele A_i in the population must be at least the number found in the sample, $\hat{y}_{i,N}/(2M) \leq p_{i,0}$. Also, the number of copies of all other alleles in the population must be at least the number found in the sample, $2N - \hat{y}_{i,N}$; this constrains the maximum value of $p_{i,0}$ according to $p_{i,0} \leq [2M - (2N - \hat{y}_{i,N})]/(2M) = 1 - [(2N - \hat{y}_{i,N})/(2M)]$. For a given value of $p_{i,0}$, $\text{Max}\{0, 2p_{i,0} - 1\} \leq P_{ii,0} \leq p_{i,0}$, as determined earlier. If $P_{i,H'_0,\alpha} = \varnothing$, then the acceptance region is extended to include $\hat{y}_{i,N}$ for at least one pair of values of $p_{i,0}$ and $P_{ii,0}$, by decreasing α.

CI's specified by equation (13) can be calculated given *M*, *N*, P_{ii} and α, whereas those specified by equations (14a) and (14b) can be calculated given just *M*, *N* and α. In this paper, computation of any CI's using these equations, incorporating the underlying formulae specified by equations (1), (8), (9) and (12), was carried out using the software package *Mathematica* v5.0 [31]. Supporting Webpage 1 provides *Mathematica* code for computation of the CI's and can be viewed at http://rpubs.com/kkeenan02/Fung-Keenan-Mathematica/. However, other software packages such as *MATLAB* [32] and *R* [33] could also be used to implement the formulae; indeed, an *R* version of the code is provided on Supporting Webpage 2 and can be viewed at http://rpubs.com/kkeenan02/Fung-Keenan-R/. All source code for the composition of Supporting Webpages 1 and 2, including the raw *Mathematica* and *R* code used, can be accessed at https://github.com/kkeenan02/Fung-Keenan2013/.

Determining minimum sample sizes for accurate estimation of population allele frequencies

In studies of population genetics, it is desirable to obtain a sample allele frequency close to the population allele frequency with high probability [19], i.e. a short CI of high probability. Thus, under three representative scenarios, we calculate the minimum sample sizes required to obtain $\geq 95\%$ CI's with lengths ≤ 0.2 and ≤ 0.1 across all possible values of the observed sample allele frequency, $\hat{p}_{i,N} = \hat{y}_{i,N}/(2N)$. These minimum sample sizes are denoted by $N_{\leq 0.2}$ and $N_{\leq 0.1}$ respectively. They are calculated by starting with $N = 10$, computing CI lengths for all possible values of $\hat{p}_{i,N}$ and then taking the maximum value. This is repeated for increasing *N* in increments of 10 until the maximum CI length becomes ≤ 0.1. The *N* value with maximum CI length closest to 0.2 is then chosen, and then increased or decreased as necessary to find $N_{\leq 0.2}$. $N_{\leq 0.1}$ is derived analogously. Maximum CI lengths of 0.2 and 0.1 represent small maximum absolute errors in the estimated allele frequency of 0.1 and 0.05 respectively, if the estimate is taken as the mid-point of the CI. The first scenario examined, Scenario 1, is sampling from a population of $M = 1{,}000$ at HWE. $M = 1{,}000$ is larger than or on the same order of magnitude as the upper bound of the range of size estimates for 15/24 (63%) species populations collated by Frankham et al. [25], covering mammals, birds, insects and plants. Thus, $M = 1{,}000$ is taken to represent a large population. Later, *M* is varied over two orders of magnitude to consider populations that range from small to very large. Since there is a HWE, $P_{ii} = p_i^2$. Also, in the sampled population, the number of homozygotes with allele *i*, MP_{ii}, must be an integer. This restricts the number of values that p_i can take to 11 equally spaced values in the interval [0, 1] (starting from 0).

Scenarios 2 and 3 represent situations where HWE does not hold, such that $P_{ii} \neq p_i^2$. In Scenario 2, P_{ii} is assumed to take its minimum value, which is $\text{Max}\{0, 2p_i - 1\}$, whereas in Scenario 3, P_{ii} is assumed to take its maximum value, which is p_i. In comparison with Scenario 1, p_i can now take a larger number of values – it can take the 2,001 equally spaced values in the interval [0, 1] in Scenario 2 and the 1,001 equally spaced values in the interval [0, 1] in Scenario 3. Since the variance of $p_{i,N}$ is an increasing function of P_{ii} (equation (11)), CI's derived using $p_{i,N}$ are expected to increase in length with P_{ii} as well. This means that CI lengths and minimum sample sizes ($N_{\leq 0.2}$ and $N_{\leq 0.1}$) derived for Scenarios 2 and 3 are expected to encompass the entire ranges of possible values.

In the three scenarios examined, P_{ii} is specified as a function of p_i, such that equation (13) can be used to calculate a CI for p_i given a value of $\hat{p}_{i,N}$. Calculation of this CI involves considering all possible values of p_i and determining under which values $\hat{p}_{i,N}$ falls within the corresponding acceptance region (as described above). An alternative scenario, Scenario 4, is that the relationship between P_{ii} and p_i is unknown. In this case, equation (14a) has to be used to calculate a CI for p_i given a value of $\hat{p}_{i,N}$, which involves considering all possible values of p_i and P_{ii}. This results in a considerable increase in computation time; for example, when sampling $N=30$ individuals from a population of $M=1{,}000$, 2,001 values of p_i need to be considered but 1,002,001 combinations of p_i and P_{ii} need to be considered, representing an increase in computational time by two orders of magnitude. However, for a given value of p_i, the maximum value of P_{ii} gives the highest variance for $p_{i,N}$ and is thus expected to maximize the length of the acceptance region within which $\hat{p}_{i,N}$ could fall within. Thus, CI's for p_i derived under Scenario 4 are expected to closely match those derived under the case of maximum homozygosity (Scenario 3). This can be verified by explicit calculation of the CI's – for example, given $M=100$, $N=30$ and $\hat{p}_{i,N}=15$, the CI's derived under Scenario 4 and Scenario 3 are [0.14, 0.4] and [0.15, 0.39], representing a difference in length of only 0.02. Similar results hold for $\hat{p}_{i,N}=10$ and 5. Therefore, results from Scenario 3 are used to approximate those for Scenario 4, obviating the need for long computational runtimes to explore all possible combinations of p_i and P_{ii}.

Lastly, from equation (11), the variance of $p_{i,N}$ is an increasing function of M. Thus, CI lengths for p_i are expected to increase with M, which would result in increases in $N_{\leq 0.2}$ and $N_{\leq 0.1}$. To test this explicitly, for a population at HWE and with $N=30$, maximum lengths for the $\geq 95\%$ CI for p_i (across all values of $\hat{p}_{i,N}$) are calculated for $M=100$, 250, 500, 750, 1,000, 2,500, 5,000, 7,500 and 10,000. This range corresponds to populations that are small to very large [25].

Application to an empirical data set for checkerspot butterflies

To demonstrate how the theory developed can be used in practice, it is applied to microsatellite data for samples from two populations of the checkerspot butterfly (*Melitaea cinxia* L.) occupying meadows on the Åland Islands in Finland [13]. Specifically, for the *CINX1* locus, $\geq 100(1-2\alpha)\%$ CI's are calculated for the frequencies of alleles A, B and C for the Prästö and Finström populations, using the corresponding sample allele frequencies (Table 1 of Palo et al. [13]). For each population, a CI is not calculated for the population frequency of the fourth and final allele D, because this is fixed by the population frequencies of the first three alleles. To achieve consistency with the notation used in our study, henceforth, alleles A, B, C and D are referred to as alleles A_1, A_2, A_3 and A_4 respectively. The sample size for the Prästö population is $N_1=53$, whereas that for the Finström population is $N_2=74$ [13]. p_{i,N_1} and q_{i,N_2}, $i \in \{1,2,3,4\}$, are used to denote the sample allele frequencies of A_i in the Prästö and Finström populations respectively. Similarly, p_i and q_i are used to denote the population allele frequencies of A_i in the two populations, respectively. The two populations consist of two and seven subpopulations respectively. Thus, technically, the two populations can be referred to as metapopulations, although this terminology is not used in our study for clarity. The subpopulations form part of a total of about 536 subpopulations on the Åland Islands, with an estimated total size ranging from 35,000 to at least 200,000 [13]. Thus, the size of each subpopulation is assumed to be $[(35{,}000+200{,}000)/2]/536=219$, such that the Prästö and Finström populations are assumed to have a size of $M_1=2\times 219=438$ and $M_2=7\times 219=1533$ respectively. The observed sample allele frequencies in the two populations, denoted by $\hat{p}_{i,N_1}$ and $\hat{q}_{i,N_2}$ respectively, are taken directly from [13]. These are used to calculate the observed number of copies of allele A_i in each population, denoted by $\hat{y}_{i,N_1}$ and $\hat{z}_{i,N_2}$ respectively, using the formulae $\hat{y}_{i,N_1}=2N_1\hat{p}_{i,N_1}$ and $\hat{z}_{i,N_2}=2N_2\hat{q}_{i,N_2}$. Due to rounding error in $\hat{p}_{i,N_1}$ and $\hat{q}_{i,N_2}$ values from [13], $\hat{y}_{i,N_1}$ and $\hat{z}_{i,N_2}$ had to be rounded to the nearest integer. Since the true homozygosity of each allele in each population is unknown [13], conservative CI's are calculated assuming P_{ii} takes its maximum value of p_i (this is expected to maximize the lengths of the CI's, as explained in the previous section). In addition, $\alpha=0.025/3$ is chosen, such that a $\geq 98.3\%$ CI is derived for each of the three population allele frequencies in each of the two populations. The reason for this choice of α is because for each population, using the Bonferroni Inequality [34], the cubic region defined by the three CI's can be taken as a $\geq 100(1-2\beta)\%$ confidence region (CR) for the three population allele frequencies, where $3\alpha \geq \beta$. The choice of $\alpha=0.025/3$ allows $\geq 95\%$ CR's to be derived for each set of three population allele frequencies in the two populations.

To demonstrate how the theory developed in this paper can be used to derive CI's for genetic indicators that are a function of the population allele frequencies, the $\geq 95\%$ CR's are used to calculate a $\geq 95\%$ CI for Jost's D for the *CINX1* locus and the Prästö and Finström butterfly populations. Jost's D is a measure of genetic distance between populations [9]. For the case of two populations, it is given by

$$D_{\text{Jost}}=2\left[1-\left(\frac{J_T}{J_S}\right)\right], \tag{15}$$

where, using the notation in this example,

$$\begin{aligned} J_T &= \sum_{i=1}^{4}\left(\frac{p_i+q_i}{2}\right)^2 = \frac{1}{4}(p_4+q_4)^2+\frac{1}{4}\sum_{i=1}^{3}(p_i+q_i)^2 \\ &= \frac{1}{4}\left(2-\sum_{i=1}^{3}p_i-\sum_{i=1}^{3}q_i\right)^2+\frac{1}{4}\sum_{i=1}^{3}(p_i+q_i)^2 \end{aligned} \tag{16}$$

and

$$\begin{aligned} J_S &= \frac{\sum_{i=1}^{4}p_i^2+\sum_{i=1}^{4}q_i^2}{2} = \frac{1}{2}(p_4^2+q_4^2)+\frac{1}{2}\sum_{i=1}^{3}(p_i^2+q_i^2) \\ &= \frac{1}{2}\left[\left(1-\sum_{i=1}^{3}p_i\right)^2+\left(1-\sum_{i=1}^{3}q_i\right)^2\right]+\frac{1}{2}\sum_{i=1}^{3}(p_i^2+q_i^2). \end{aligned} \tag{17}$$

A lower limit to a $\geq 95\%$ CI for D_{Jost} can be derived by minimizing the function in (15) given the constraints that the six population allele frequencies are contained within the two corresponding $\geq 95\%$ CR's, and the two constraints $1-\sum_{i=1}^{3}p_i=p_4\geq 0$ and $1-\sum_{i=1}^{3}q_i=q_4\geq 0$. This lower limit is denoted by L_D. Similarly, the upper limit to a $\geq 95\%$ CI for D_{Jost} can be derived by maximizing the function in (15) under the same constraints. This is denoted by U_D. In this study, the "Minimize" and "Maximize" functions in *Mathematica* v5.0 [31] are used to compute L_D and U_D, but corresponding functions may be used in other software packages, such as the "solnp" function in the "Rsolnp" *R* package. A $\geq$95% CI for D_{Jost} can then be defined as the interval $[L_D, U_D]$. Supporting Webpage 1 provides *Mathematica* code that can be used to calculate $[L_D, U_D]$ for the butterfly case study examined (http://rpubs.com/kkeenan02/Fung-Keenan-Mathematica/). Corresponding *R* code is presented on Supporting Webpage 2 (http://rpubs.com/kkeenan02/Fung-Keenan-R/).

Results

Maximum length of $\geq$95% confidence interval with increasing sample size

For Scenario 1, where the sampled diploid population of size $M = 1{,}000$ was at HWE, the maximum length of the $\geq$95% CI for the population frequency of allele A_i, p_i, considering all possible observed values of the sample allele frequency, $\hat{p}_{i,N}$, decreased non-linearly with sample size N (Figure 1). The minimum N required to achieve a maximum length ≤ 0.2, $N_{\leq 0.2}$, was 22, whereas the minimum N required to achieve a length ≤ 0.1, $N_{\leq 0.1}$, was 49 (Figure 1).

For given N and $\hat{p}_{i,N}$, the CI for p_i consists of all possible values of p_i for which $\hat{p}_{i,N}$ lies in the corresponding acceptance region, as described in *Methods*. This acceptance region is expected to increase with the variance of $p_{i,N}$, $\sigma^2[p_{i,N}]$. At HWE, the frequency of homozygotes of allele A_i, P_{ii}, is equal to p_i^2; thus, according to equation (11), $\sigma^2[p_{i,N}]$ takes its highest values at intermediate values of p_i. As a result, intermediate values of $\hat{p}_{i,N}$ are likely to fall within the acceptance regions of more values of p_i, such that the corresponding CI lengths are generally longer. Simulation results for Scenario 1 were broadly in agreement with these theoretical expectations (Figure 2).

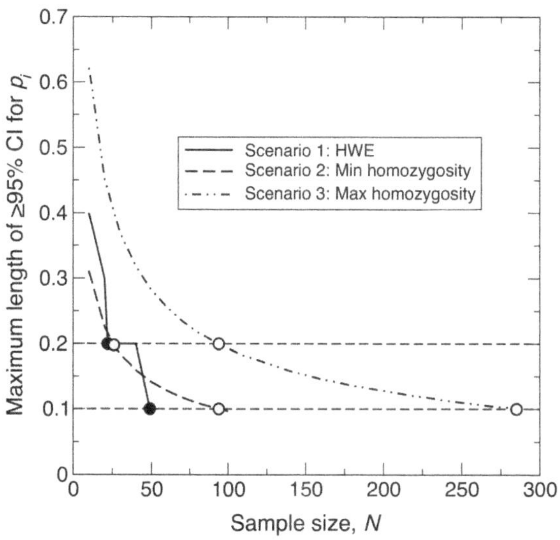

Figure 1. Change in maximum length of ≥95% confidence interval with increasing sample size. Graph showing how the maximum length of the $\geq$95% confidence interval (CI) for the population frequency of an allele A_i (p_i) changes with increasing sample size *N*, when sampling from a diploid population of size $M = 1{,}000$. For a given *N*, the maximum CI length was derived by calculating CI lengths for all possible values of the observed sample allele frequency and then taking the maximum length. The three curves correspond to three scenarios where the population is (1) at Hardy-Weinberg equilibrium (HWE), (2) attains its lowest homozygosity value with respect to A_i, and (3) attains its highest homozygosity value with respect to A_i. For each curve, the two filled circles represent the minimum *N* values required for the maximum CI length to reach values of ≤ 0.2 and ≤ 0.1. For visual guidance, the two dashed horizontal lines mark maximum CI lengths of 0.2 and 0.1. The exact values of *N* tested are described in *Methods*, and equation (13) was used to calculate the CI lengths.

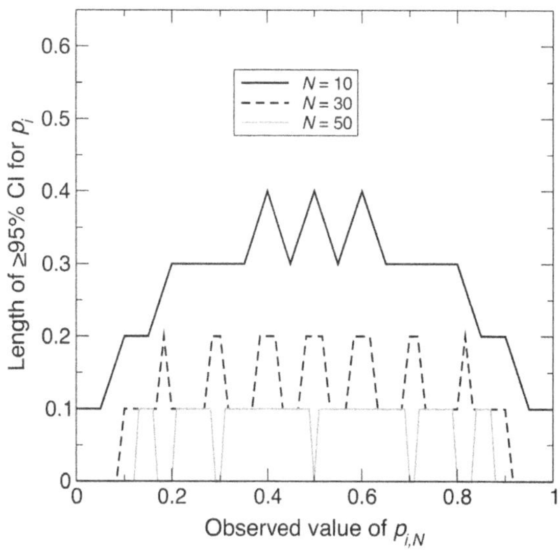

Figure 2. Change in length of $\geq$95% confidence interval across different observed values, under Hardy-Weinberg equilibrium. For a sample of size *N* taken from a population of size $M = 1{,}000$ at Hardy-Weinberg equilibrium, graph showing the length of the $\geq$95% confidence interval (CI) for the population frequency of an allele A_i (p_i) across all possible observed values of the sample allele frequency ($p_{i,N}$). The three curves correspond to $N = 10$, 30 and 50. Equation (13) was used to calculate the length of each CI.

In Scenario 2, where the population was no longer at HWE but attained its lowest P_{ii}, the maximum CI length also decreased non-linearly with increasing N (Figure 1). Compared with Scenario 1, $N_{\leq 0.2}$ was slightly larger and $N_{\leq 0.1}$ was larger by about a factor of two, taking values of 26 and 94 respectively (Figure 1). This is contrary to expectations that minimum homozygosity would give smaller $N_{\leq 0.2}$ and $N_{\leq 0.1}$ (see *Methods*). The reason is that although $\sigma^2[p_{i,N}]$ is smaller for given N and p_i compared with the case of HWE, which would be expected to result in fewer p_i values for which a given $\hat{p}_{i,N}$ falls within the corresponding acceptance regions and hence a shorter CI, there are more possible values of p_i (2,001 compared with 11 – see *Methods*), which increased the number of p_i values for which $\hat{p}_{i,N}$ falls within the corresponding acceptance regions. For Scenario 2, $P_{ii} = \text{Max}\{0, 2p_i - 1\}$, such that $\sigma^2[p_{i,N}]$ attains its highest values at p_i values close to 0.25 and 0.75 (equation (11)). Thus, for given N, values of $\hat{p}_{i,N}$ near 0.25 and 0.75 are expected to generally exhibit the longest CI lengths. Simulation results for Scenario 2 were broadly in agreement with these theoretical expectations (Figure 3).

For the last scenario, Scenario 3, the population attained its highest P_{ii}, representing the opposite extreme to Scenario 2. As for the previous two scenarios, the maximum CI length decreased non-linearly with increasing N (Figure 1), but this time, $N_{\leq 0.2}$ and $N_{\leq 0.1}$ took values of 94 and 285 respectively. These values were approximately four and six times as large as the corresponding values in Scenario 1 (Figure 1). In Scenario 3, $P_{ii} = p_i$, and

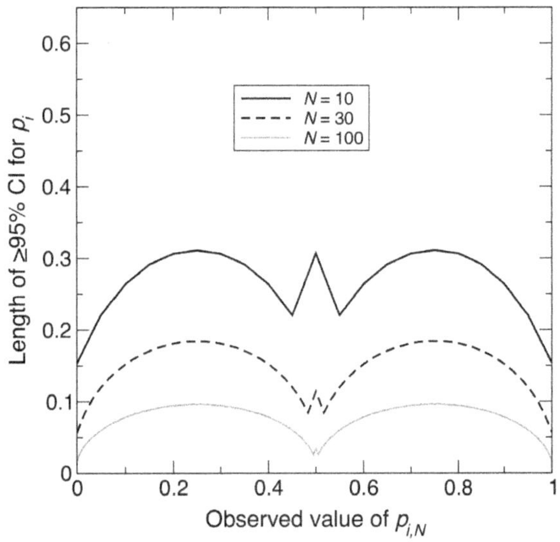

Figure 3. Change in length of $\geq$95% confidence interval across different observed values, under minimum homozygosity. For a sample of size N taken from a population of size $M = 1{,}000$ with the minimum homozygosity possible for an allele A_i, graph showing the length of the $\geq$95% confidence interval (CI) for the population frequency of A_i (p_i) across all possible observed values of the sample allele frequency ($p_{i,N}$). The three curves correspond to $N = 10$, 30 and 100. Equation (13) was used to calculate the length of each CI.

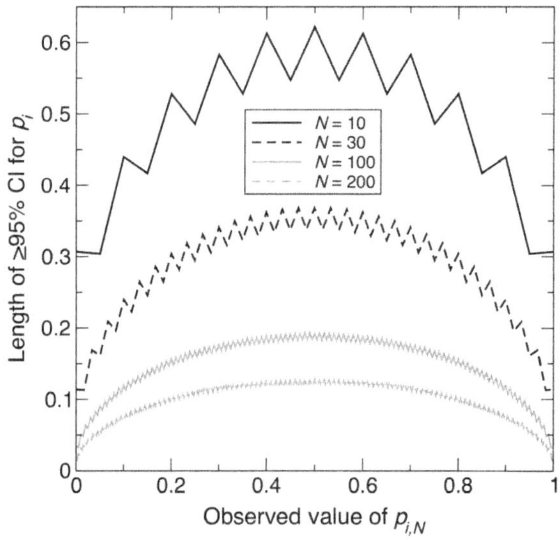

Figure 4. Change in length of $\geq$95% confidence interval across different observed values, under maximum homozygosity. For a sample of size N taken from a population of size $M = 1{,}000$ with the maximum homozygosity possible for an allele A_i, graph showing the length of the $\geq$95% confidence interval (CI) for the population frequency of A_i (p_i) across all possible observed values of the sample allele frequency ($p_{i,N}$). The four curves correspond to $N = 10$, 30, 100 and 200. Equation (13) was used to calculate the length of each CI.

equation (11) shows that intermediate values of p_i give the highest values of $\sigma^2[p_{i,N}]$. Therefore, as in Scenario 1, intermediate values of $\hat{p}_{i,N}$ are expected to generally exhibit the longest CI lengths for given N. Again, simulation results were consistent with these expectations (Figure 4).

Maximum length of $\geq$95% confidence interval with increasing population size

When taking a sample of size $N = 30$ from a diploid population at HWE, the maximum length of the $\geq$95% CI for p_i, across all $\hat{p}_{i,N}$, increased with population size M (Figure 5). As M was increased from a small value of 100 to 1,000, the maximum CI length remained the same at 0.2 (this is possible because there is nothing to prevent different values of M giving the same maximum CI length, using equation (13)). The maximum CI length increased when M was increased from 1,000 to a large value of 2,500, but only by 0.04. Thereafter, the length remained constant up to a very large value of $M = 7{,}500$, and only increased by a small amount of 0.02 with a further increase in M to 10,000. Thus, simulation results confirm the theoretical expectation that CI length increases with M (as explained in *Methods* – this expectation arises because $\sigma^2[p_{i,N}]$ increases with M, according to equation (11)). However, the increase in the CI length was modest as M was increased over two orders of magnitude.

$\geq$95% confidence interval for Jost's D between two checkerspot butterfly populations

For the Prästö checkerspot butterfly population, the $\geq$98.3% CI's for the population frequencies for three of the four alleles at the *CINX1* locus were computed using equation (13) and data from Palo et al. [13], as described in *Methods*. The three alleles are denoted by A_1, A_2 and A_3 respectively, with the population frequencies denoted by p_1, p_2 and p_3 respectively. The three corresponding $\geq$98.3% CI's derived were [0.002, 0.080], [0.598, 0.872] and [0.114, 0.381] respectively. Using the same methodology, for the Finström population, the $\geq$98.3% CI's for the population frequencies of A_1, A_2 and A_3 were calculated as[0.003, 0.109], [0.714, 0.914] and [0.003, 0.109] respectively.

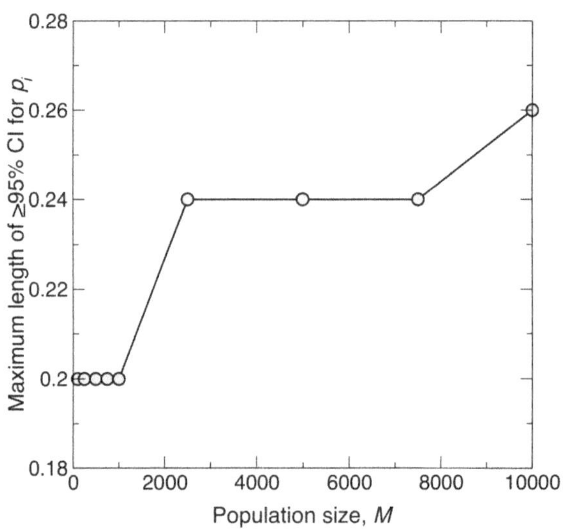

Figure 5. Change in maximum length of $\geq$95% confidence interval with increasing population size. Graph showing how the maximum length of the $\geq$95% confidence interval (CI) for the population frequency of an allele A_i (p_i) changes with increasing population size M, when taking samples of size $N = 30$. The population is at Hardy-Weinberg equilibrium. For a given M, the maximum CI length was derived by calculating CI lengths for all possible values of the observed sample allele frequency and then taking the maximum length. M values of 100, 250, 500, 750, 1,000, 2,500, 5,000, 7,500 and 10,000 were tested, as indicated by the filled circles.

The three $\geq$98.3% CI's for the Prästö population were used to form a cubic region, which corresponds to a $\geq$95% CR for A_1, A_2 and A_3 [34]. In the same way, the three $\geq$98.3% CI's for the Finström population were used to form a $\geq$95% CR for A_1, A_2 and A_3. Within these two CR's, the maximum and minimum values of D_{Jost} (D_{Jost} is given by equation (15)) were calculated and used to derive a $\geq$95% CI for D_{Jost}, as described in *Methods*. This CI for D_{Jost} was derived as $\left[2.34\times10^{-5}, 0.186\right]$. If D_{Jost} was calculated simply using the sample allele frequencies and equation (15), then only one value would have been obtained: 0.043.

Discussion

In scientific studies that use sample data to estimate unknown population parameters, the sampling uncertainty in the estimates needs to be quantified in order to make reliable inferences on population processes captured by the parameters. This forms an essential part of scientific hypothesis-testing. Therefore, in studies of population genetics, it is essential to quantify the sampling uncertainty of key population parameters used to infer past and present evolutionary processes. These include allele frequencies, which are often used to quantify genetic variation among populations, thereby allowing hypotheses on processes driving this variation to be tested (e.g., [1,2,3,4,5]). However, many studies do not include sampling uncertainty for allele frequencies, instead presenting and/or using single point estimates based on one sample per population (e.g., [11,12,13,14,15,16,17,18]). Thus, it is not possible to assess the accuracy of any inferences from these studies. This hinders not only the advance of scientific knowledge but also decision-making based on this knowledge, such as for sustainable management and conservation of natural resources. In this context, the work presented in this paper is valuable in that it provides a method of quantifying sampling uncertainty in allele frequencies for diploid populations, in the form of confidence intervals (CI's) containing true values with probability equal to or greater than a desired threshold.

The method presented pertains to the general case of a locus with n alleles, with a sample of size N taken from a population of size M and any degree of homozygosity with respect to the n alleles. In this case, the method allows construction of a CI for the population frequency of each allele, which can then be combined to create a joint confidence region (CR) for all the population allele frequencies at a given locus. It is noted that if more than one locus is considered simultaneously, then a joint CR for all population allele frequencies at all loci can be calculated by combining CI's in an analogous way. For the subcase of an infinite population size ($M\to\infty$) and a large sample size of $N\geq30$, Weir [10] had proposed an approximate $100(1-2\alpha)$% CI for the population allele frequency of an allele A_i, p_i. This is $[p_{i,N}-z_{1-\alpha}\hat{\sigma}[p_{i,N}], p_{i,N}-z_\alpha\hat{\sigma}[p_{i,N}]]$, where $p_{i,N}$ is the sample allele frequency; z_α satisfies $\Phi(z_\alpha)=\alpha$, with Φ being the cumulative distribution function (cdf) of the standard Normal distribution; and $\hat{\sigma}[p_{i,N}]$ is an estimate of the standard deviation of $p_{i,N}$ using sample data. $\hat{\sigma}[p_{i,N}]$ is specified by equation (11) with $p_{i,N}$ replacing p_i and $P_{ii,N}$, the sample frequency of homozygotes with allele A_i, replacing P_{ii}, the corresponding population frequency of homozygotes. However, the accuracy of this CI depends both on how close $\hat{\sigma}[p_{i,N}]$ is to the true standard deviation $\sigma[p_{i,N}]$ (specified by equation (11)) and how close the cdf of $p_{i,N}$ is to a Normal distribution with the same mean and variance. This accuracy has not been quantified [10] and thus, it is not known whether the CI actually contains p_i with a probability of at least $100(1-2\alpha)$, rendering its use problematic. In this study, we have rectified this problem by constructing a CI for p_i with probability coverage of at least $100(1-2\alpha)$, for the more general case where the population size can take any value larger than or equal to the sample size.

The method we derived was used to show that the sampling uncertainty in p_i, measured as the maximum length of the $\geq$95% CI for p_i across all possible values of the observed sample allele frequency $\hat{p}_{i,N}$, decreased non-linearly with N when sampling from a large population ($M=1{,}000$) under three archetypal scenarios. These three scenarios represent the cases where the population (1) is at Hardy-Weinberg equilibrium (HWE), (2) has the lowest value of P_{ii} and (3) has the highest value of P_{ii}. As expected from theory, for any given N, the maximum CI lengths for Scenario 3 were always greater than corresponding lengths in Scenarios 1 and 2. However, the maximum CI lengths for Scenario 2 was unexpectedly greater than that in Scenario 1 for some values of N, reflecting the greater possible number of values p_i can take in a finite population with minimum homozygosity compared to one at HWE. This illustrates how the finite size of a population can give opposite trends to those obtained under an assumption of infinite size. According to theory and simulations, Scenario 3 gives CI lengths that closely approximate those in the case where P_{ii} is unknown (see *Methods*). Thus, if P_{ii} is unknown, CI lengths derived under Scenario 3 should be used. This was the approach used in the application of our method to sample data for two butterfly populations, discussed further below. On the other hand, if there is evidence that the population is at HWE, then the shorter CI's derived under Scenario 1 could be used.

Under the three scenarios examined, the non-linear decreases in sampling uncertainty with increasing N are consistent with results from the simulation study of Hale et al. [19], who found that the average difference between $p_{i,N}$ and p_i also exhibited non-linear decreases with N. However, Hale et al. [19] did not use their simulation results to quantify sampling uncertainty for the realistic situation where only one sample is taken from a population; this situation was considered in our study. Furthermore, as mentioned in the Introduction, for given N, they only used 100 samples to numerically construct the distribution for $p_{i,N}$, resulting in an incomplete distribution that may not closely reflect the true distribution. This highlights a weakness of a simulation-based approach without a rigorous mathematical underpinning, which is present in our approach. Hale et al. [19] concluded that $N=25$–30 is sufficient to give accurate estimates of p_i, but this conclusion has to be interpreted in light of the limitations identified. Our results refine this conclusion by showing that across the three scenarios examined, $N=49$–285 is required to ensure that, with a high probability of ≥0.95, an estimate for p_i can be derived from any one sample that is within 0.05 of the true value; this corresponds to a CI of length ≤0.1. To ensure that the estimate is within 0.1 rather than 0.05, corresponding to a CI of length ≤0.2, our results show that $N=22$–94 is required. Thus, $N=30$ is not guaranteed to give "accurate" estimates of p_i under all or most scenarios, and N values up to 10 times larger could be required. Decreasing the population size M from 1,000 would help to decrease sampling uncertainty, but results showed that decreasing M over two orders of magnitude from 10,000 to 100 only resulted in modest decreases in the maximum CI length of ≤0.06, with no decrease when M was decreased from 1,000 to 100. Thus, the overall conclusion is that $N=30$ is often insufficient to guarantee accurate estimates of p_i, in the sense that p_i is within 0.05 or 0.1 of the estimate. Considering that alleles at highly polymorphic loci, such as microsatellites, often occur at population frequencies of <0.05 [35], it might be desirable to derive CI's for p_i that are of length <0.1. Thus, sample sizes even larger than the values found in our simulations might be required under some circumstances.

The application of our method to empirical data for two populations of the checkerspot butterfly [13] demonstrated how the underlying theory can be applied to construct joint $\geq 95\%$ CR's for the population frequencies of multiple alleles at a single locus. These CR's were then used to construct a $\geq 95\%$ CI for Jost's D, which measures genetic differentiation between the two populations. This illustrates how our method can be used to quantify sampling uncertainty in genetic indicators that are a function of population allele frequencies, thus facilitating hypothesis-testing and also risk-based natural resource management. In the example considered, the single point estimate of Jost's D using the sample allele frequencies was about four times lower than the upper bound of its $\geq 95\%$ CI. Thus, use of the single point estimate without accounting for sampling uncertainty could lead to misleading conclusions. The effectiveness of any management measures based on such conclusions would be compromised, hindering the achievement of objectives related to conservation or sustainable use. Our example therefore highlights the practical utility of the method that we have derived.

In conclusion, we have presented a rigorous mathematical method for quantifying sampling uncertainty in estimates of population allele frequencies, for a general case that has hitherto not been analyzed. In addition, we have demonstrated its practical application in informing sampling design and determining uncertainty in genetic indicators. Thus, the method derived advances both theory and practice, with broad implications for a range of disciplines, including: conservation genetics, evolutionary genetics, genetic epidemiology, genome wide association studies (GWAS), forensics and medical genetics. In particular, the method provides exact answers to the question of how many individuals need to be sampled from a population in order to achieve a given level of accuracy in estimates of population allele frequencies. This is a question that has rarely been studied before [19], despite its important practical implications. Previous studies have derived sample sizes required to sample, with high probability, at least one copy of all alleles at a locus above a given frequency [35,36,37], but these do not correspond to sample sizes required to achieve accurate estimates of population allele frequencies. Derivation of the latter requires explicit quantification of sampling uncertainty, as we have done in this study.

Possible future extensions

The CI's and CR's constructed using our method are conservative in the sense that they contain the true values with a probability equal to *or* greater than a desired threshold. This conservative property is useful in hypothesis-testing if there is a need to decrease the probability of obtaining a false positive below a certain threshold. However, researchers would ideally like to construct CI's and CR's containing true values with a known probability, not just with probability at or above a known threshold. Therefore, future research could attempt to tighten the intervals and regions that we have derived, ideally until they cover a known probability. For example, Cai and Krishnamoorthy [23] devised a method (using their "Combined Test" approach) that was shown to give shorter CI's for the probability parameters of both the binomial and univariate hypergeometric distributions, when compared with a hypothesis-testing approach analogous to the one used in this paper. If their method could be extended to the population allele frequency parameter for the more complex distribution used in this paper (equation (9)), based on a multivariate hypergeometric distribution (equation (1)), then this might result in tighter CI's for population allele frequencies when sampling from a finite diploid population.

In addition, the CI's derived in our paper were designed to quantify the uncertainty in population allele frequencies that arises from taking a random sample of a finite diploid population, which can exhibit any degree of relatedness. The "random" refers to equal probability of choosing individuals that may be of any type, which does not imply that once an individual of a particular type has been sampled, the next individual sampled is equally likely to belong to any of the types (i.e., does not imply individuals in the population or sample are unrelated). Relatedness among individuals in the population is implicitly included within the population frequencies of the different genotypes, and is thus accounted for when calculating the sampling distribution for the population frequency of an allele A_i, as specified by equation (1). However, this representation does not give an explicit quantification of the degree of relatedness among individuals in the population, for example using kinship coefficients. DeGiorgio and Rosenberg [38] used such coefficients to derive an unbiased estimator of heterozygosity ($H = 1 - \sum_{i=1}^{n} p_i^2$, for a locus with n alleles) in the case of sampling from a population of diploid individuals that could be related, and DeGiorgio et al. [39] extended these results to the case of individuals with arbitrary ploidy. In their calculations, the number of copies of allele A_i in each sampled individual k was treated as a random variable and the covariances of these variables were then related to the kinship coefficients. This is different to calculations in our paper, where the number of sampled individuals with alleles A_i and A_j was treated as a random variable (see *Methods*). The method in this paper might be revised by considering the number of copies of allele A_i for each sampled individual instead, following [38,39]. This could allow explicit quantification of relatedness in the context of deriving CI's for the population allele frequencies.

Acknowledgments

We would like to thank two anonymous reviewers for helpful and insightful comments, which have led to great improvements in this manuscript.

Author Contributions

Conceived and designed the experiments: TF KK. Performed the experiments: TF. Analyzed the data: TF KK. Contributed reagents/materials/analysis tools: TF KK. Wrote the paper: TF KK.

References

1. Bowcock AM, Kidd JR, Mountain JL, Hebert JM, Carotenuto L, et al. (1991) Drift, admixture, and selection in human evolution: A study with DNA polymorphisms. Proc Natl Acad Sci USA 88: 839–843.
2. Akey JM, Zhang G, Zhang K, Jin L, Shriver MD (2002) Interrogating a high-density SNP map for signatures of natural selection. Genome Res 12: 1805–1814.
3. Luikart G, England PR, Tallmon D, Jordan S, Taberlet P (2003) The power and promise of population genomics: from genotyping to genome typing. Nature 4: 981–994.
4. de Kovel CGF (2006) The power of allele frequency comparisons to detect the footprint of selection in natural and experimental situations. Genet Sel Evol 38: 3–23.

5. Friedlander SM, Herrmann AL, Lowry DP, Mepham ER, Lek M, et al. (2013) *ACTN3* allele frequency in humans covaries with global latitudinal gradient. PLoS ONE 8(1):e52282.
6. Nei M (1972) Genetic distance between populations. Am Nat 106: 283–292.
7. Nei M, Chesser RK (1983) Estimation of fixation indices and gene diversities. Ann Hum Genet, 47: 253–259.
8. Weir BS, Cockerham CC (1984) Estimating *F*-statistics for the analysis of population structure. Evolution 38: 1358–1370.
9. Jost L (2008) G_{ST} and its relatives do not measure differentiation. Mol Ecol 17: 4015–4026
10. Weir BS (1996) Genetic data analysis II. Sunderland, USA, Sinauer Associates. 445 p.
11. Eanes WF, Koehn RK (1978) An analysis of genetic structure in the Monarch butterfly, *Danaus plexippus* L. Evolution 32: 784–797.
12. Forbes SH, Hogg JT, Buchanan FC, Crawford AM, Allendorf FW (1995) Microsatellite evolution in congeneric mammals: domestic and bighorn sheep. Mol Biol and Evol 12: 1106–1113.
13. Palo J, Varvio S-L, Hanski I, Väinölä R (1995) Developing microsatellite markers for insect population structure: complex variation in a checkerspot butterfly. Hereditas 123: 295–300.
14. Luikart G, Allendorf FW, Cournuet J-M, Sherwin WB (1998) Distortion of allele frequency distributions provides a test for recent population bottlenecks. J Hered 89:238–247.
15. Rank NE, Dahlhoff EP (2002) Allele frequency shifts in response to climate change and physiological consequences of allozyme variation in a montane insect. Evolution 56: 2278–2289.
16. Hugnet C, Bentjen SA, Mealey KL (2004) Frequency of the mutant MDR1 allele associated with multidrug sensitivity in a sample of collies from France. J Vet Pharmacol Ther 27: 227–229.
17. Seider T, Fimmers R, Betz P, Lederer T (2010) Allele frequencies of the five miniSTR loci D1S1656, D2S441, D10S1248, D12S391 and D22S1045 in a German population sample. Forensic Sci Int Genet 4: e159–e160.
18. Petrejčíková E, Soták M, Bernasovská J, Bernasovský I, Rębała K, et al. (2011) Allele frequencies and population data for 11 Y-chromosome STRs in samples from Eastern Slovakia. Forensic Sci Int Genet 5: e53–e62.
19. Hale ML, Burg TM, Steeves TE (2012) Sampling for microsatellite-based population genetic studies: 25 to 30 individuals per population is enough to accurately estimate allele frequencies. PLoS ONE 7(9): e45170.
20. Gillespie JH (2004) Population genetics: a concise guide, second edition. Baltimore, USA, The John Hopkins University Press. 217 p.
21. Laplace PS (1812) Théorie analytique des probabilités. Paris, France, Courcier. 464 p.
22. Agresti A, Coull BA (1998) Approximate is better than "exact" for interval estimation of binomial proportions. Am Stat 52: 119–126.
23. Cai Y, Krishnamoorthy K (2005) A simple improved inferential method for some discrete distributions. Comput Stat Data Anal 48: 605–621.
24. Clopper CJ, Pearson ES (1934) The use of confidence or fiducial limits illustrated in the case of the binomial. Biometrika 26: 404–413.
25. Frankham R (1996) Relationship of genetic variation to population size in wildlife. Conserv Biol 10: 1500–1508.
26. Green DM (2003) The ecology of extinction: population fluctuation and decline in amphibians. Biol Conserv 111: 331–343.
27. Vredenberg VT, Knapp RA, Tunstall TS, Briggs CJ (2010) Dynamics of an emerging disease drive large-scale amphibian population extinctions. Proc Natl Acad Sci USA 107: 9689–9694.
28. Johnson NL, Kotz S, Balakrishnan N (1997) Discrete multivariate distributions. Hoboken, USA, Wiley-Blackwell. 328 p.
29. Cox DR, Hinkley DV (1974) Theoretical statistics. London, UK, Chapman and Hall. 511 p.
30. Talens E (2005) Statistical auditing and the AOQL-Method. Ridderkerk, The Netherlands, Labyrint Publications. 164 p.
31. Wolfram Research Inc. (2003) Mathematica Edition: Version 5.0. Champaign, Illinois, , USA, Wolfram Research Inc.
32. The MathWorks Inc. (2012) MATLAB version 8.0. Natick, Massachusetts, , USA, The MathWorks Inc.
33. R Development Core Team (2010) R: a language and environment for statistical computing. Vienna, Austria, R Foundation for Statistical Computing. 1731 p.
34. Rice JA (1995) Mathematical statistics and data analysis, second edition. Belmont, USA, Duxbury Press. 651 p.
35. Chakraborty R (1992) Sample size requirements for addressing the population genetic issues of forensic use of DNA typing. Hum Biol 64: 141–159.
36. Gregorius H-R (1980) The probability of losing an allele when diploid genotypes are sampled. Biometrics 36: 643–652.
37. Sjögren P, Wyöni P-I (1994) Conservation genetics and detection of rare alleles in finite populations. Conserv Biol 8: 267–270.
38. DeGiorgio M, Rosenberg NA (2009) An unbiased estimator of gene diversity in samples containing related individuals. Mol Biol Evol 26: 501–512.
39. DeGiorgio M, Jankovic I, Rosenberg NA (2010) Unbiased estimation of gene diversity in samples containing related individuals: exact variance and arbitrary ploidy. Genetics 186: 1367–1387.

9

Identifying Priority Areas for Conservation and Management in Diverse Tropical Forests

Karel Mokany[1]*, David A. Westcott[2], Soumya Prasad[3], Andrew J. Ford[2], Daniel J. Metcalfe[4]

1 Ecosystem Sciences, Commonwealth Scientific and Industrial Research Organisation, Canberra, ACT, Australia, 2 Ecosystem Sciences, Commonwealth Scientific and Industrial Research Organisation, Atherton, QLD, Australia, 3 Centre for Ecological Sciences, Indian Institute of Science, Bangalore, India, 4 Ecosystem Sciences, Commonwealth Scientific and Industrial Research Organisation, Dutton Park, QLD, Australia

Abstract

The high concentration of the world's species in tropical forests endows these systems with particular importance for retaining global biodiversity, yet it also presents significant challenges for ecology and conservation science. The vast number of rare and yet to be discovered species restricts the applicability of species-level modelling for tropical forests, while the capacity of community classification approaches to identify priorities for conservation and management is also limited. Here we assessed the degree to which macroecological modelling can overcome shortfalls in our knowledge of biodiversity in tropical forests and help identify priority areas for their conservation and management. We used 527 plant community survey plots in the Australian Wet Tropics to generate models and predictions of species richness, compositional dissimilarity, and community composition for all the 4,313 vascular plant species recorded across the region (>1.3 million communities (grid cells)). We then applied these predictions to identify areas of tropical forest likely to contain the greatest concentration of species, rare species, endemic species and primitive angiosperm families. Synthesising these alternative attributes of diversity into a single index of conservation value, we identified two areas within the Australian wet tropics that should be a high priority for future conservation actions: the Atherton Tablelands and Daintree rainforest. Our findings demonstrate the value of macroecological modelling in identifying priority areas for conservation and management actions within highly diverse systems, such as tropical forests.

Editor: Bruno Hérault, Cirad, France

Funding: This research was supported by a grant from the Australia-India Strategic Research Fund. The funders had no role in study design, data collection and analysis, decision to publish, or preparation of the manuscript.

Competing Interests: The authors have declared that no competing interests exist.

* E-mail: Karel.Mokany@csiro.au

Introduction

Tropical rainforests contain more than 60% of all known species, despite covering only 7% of the earth's surface [1,2]. The concentration of diversity in tropical forests endows these ecosystems with particular importance for retaining global biodiversity, yet also presents significant challenges for ecology and conservation science. Understanding patterns in tropical forest diversity and subsequently identifying priorities for conservation is severely hampered by the vast numbers of species present, many of which are rare (or yet to be discovered) and for most of which we have little information regarding their attributes or current distributions [3,4]. The combination of high diversity and major shortfalls in our knowledge of tropical forest biodiversity limits the applicability of commonly applied species-level modelling approaches [5] in improving our understanding of current patterns in diversity.

As a consequence of the challenges faced in understanding diversity in tropical forests, ecologists have often relied upon community-level approaches to improve our knowledge of these systems. This trend is typified in the Australian Wet Tropics, the largest area of tropical rainforest in Australia [6]. Here, a long history of ecological research has applied and refined structural and compositional classification of tropical plant diversity into forest "types", "categories", or "regional ecosystems" [7–12]. Across the entire Australian Wet Tropics, these forest categories have been mapped at a fine spatial resolution [12,13], and subsequently related to underlying environmental variables [14,15].

In the Australian Wet Tropics, simplifying the significant plant diversity into forest categories has enabled valuable research which has greatly improved our knowledge of these systems. For example, a number of studies have applied correlative modelling to project the distribution of different forest categories back in time, using reconstructed climate conditions, identifying possible rainforest refugia at the last glacial maximum [14,16–18]. Other research has projected the future distribution of forest categories under alternative scenarios of climate change, to identify the likely threat posed by global warming for these tropical forests [19,20]. Projections for the Australian Wet Tropics under both past and future climates in general show dramatic reductions in the spatial extent of most tropical rainforest categories as precipitation decreases [16].

Despite the insights provided, there are limits in the utility of a community classification approach to understand patterns and identify conservation priorities within diverse tropical forests. A community-level classification approach inherently ignores any variation in diversity within a community category, whilst assuming significant and uniform changes in diversity between categories [21]. Applying community-level categorical data to identify conservation and management priorities could therefore be heavily influenced by the number of categories applied, the relative differences between categories and the basis for the categorisation (e.g. structural v.s. compositional).

In moving beyond community classification approaches, macroecological modelling offers significant potential to both overcome shortfalls in our knowledge of biodiversity in tropical forests, while providing new insight into core patterns of diversity. Specifically, models of species richness (α-diversity) and pairwise compositional dissimilarity (β-diversity) can provide unique insight, by relating community diversity in a continuous manner to underlying environmental gradients [22,23]. Models of species richness and compositional dissimilarity already have demonstrated utility in improving our understanding of current diversity patterns within tropical forests [24,25]. These models of core elements of diversity can also be combined in a new approach (*Dynamic*FOAM), that predicts the composition of every community (i.e. grid cell) over a region, even where available knowledge is limited [26]. Application of these continuous macroecological modelling approaches provides significant potential to improve both our knowledge of the key drivers of tropical forest diversity and our capacity to identify priorities for conservation within these diverse systems.

Here we apply macroecological modelling to identify priority areas for conservation and management of plant diversity within the Australian Wet Tropics. We develop models of species richness and compositional dissimilarity for the 4,313 vascular plant species, then apply the *Dynamic*FOAM algorithm [26] to predict the composition of every community (grid cell) across the region (>1.3 million communities). We analyse these macroecological predictions in a number of novel ways to identify areas of high importance for conservation and management within these diverse tropical forests.

Materials and Methods

Spatial Information

Our study region (Fig. 1) was defined as the spatial domain including the Wet Tropics Bioregion in Queensland, Australia (from IBRA, 2012), and a 100 km buffer around it (approximately 14°39′S to 20°24′S and 143°54′E to 147°30′E). This region comprises an area of 87,365 km^2, of which 95.2% is natural vegetation (as classified by Queensland Land Use Mapping, 1999), while the area protected for conservation is 17.6% of the study region and 44.2% of the Wet Tropics Bioregion (as defined by CAPAD, 2010). Our analyses were carried out on a 250 m resolution spatial grid over this region, as defined by the Australian GEODATA 9 Second Digital Elevation Model – V3.

Biological Data

Plant community survey data over the study region were obtained from two sources: the Queensland Herbarium's CORVEG survey data [27], and; plant community surveys conducted by CSIRO. The methods applied by the Queensland Herbarium in surveying plant community composition are described in detail in Neldner *et al.* [27], and involved identifying all vascular plant species within a 10×50 m survey plot. In total, 517 Queensland Herbarium survey plots occurred within the study region, however, many of these plots contained species which were only identified to family or genus level, which is inappropriate for accurately quantifying species richness and compositional dissimilarity. We therefore removed those 298 survey plots where more than 10% of taxa that were not identified to the species level. There were no obvious spatial biases in the locations of the survey plots removed (Fig. S1). For the remaining 219 Queensland Herbarium survey plots, those taxa not identified to species level were removed from the analysis. This equated to an average of 1.97 records (range: 1–7) removed across 136 sites containing unidentified taxa. Again, there was no obvious spatial bias in the proportion of records removed from the retained Queensland Herbarium plots, though there was a slight trend for plots with more records to contain a larger proportion of unidentified taxa (Fig. S2).

Plant community surveys conducted by CSIRO between 2004 and 2012 followed a similar methodology to the Queensland Herbarium surveys, but were located in areas that would complement the Queensland Herbarium surveys, and were comprised of 20×50 m plots [28]. Survey data were obtained by CSIRO for 308 sites, and as with the Queensland Herbarium data, we removed those taxa not identified to species level. Almost all taxa in the CSIRO surveys were identified to species level (or to recognised HISPID names where specifies are yet to be formally described), therefore no sites needed to be removed due to coarse taxonomic identification.

Combining community surveys from the Queensland Herbarium and CSIRO gave a total of 527 community survey plots used to generate models of vascular plant species richness and compositional dissimilarity (Fig. 1). We standardised taxonomic nomenclature across these two data sets to be consistent with the Australian Plant Name Index. For both the Queensland Herbarium and CSIRO plant community surveys, we removed all non-native species, as defined by the Census of the Queensland Flora [29] and the Australian Plant Name Index. The analyses we present are therefore relevant to native plant biodiversity only.

For application of the *Dynamic*FOAM procedure (see below), we supplemented the community survey data described above with 104,831 plant species occurrence records, obtained from the Atlas of Living Australia on May 16, 2012. All plant occurrence records obtained were recent (recorded since 1990), spatially accurate (< 300 m uncertainty), and consistent with the taxonomic nomenclature applied for the plant community survey data. For the *Dynamic*FOAM procedure, we also included the records of the identified species from the 298 Queensland Herbarium survey plots that were excluded from the community modelling analyses.

Environmental Data

For the correlative modelling of plant community species richness and compositional dissimilarity, we utilised complete spatial environmental data across the study region. We applied the 250 m digital elevation model of the study region in ANUCLIM [30] to obtain climate data (precipitation, evaporation, temperature, radiation (adjusted for slope & aspect)) and bioclimatic variables (plant growth indices) across the spatial grid (averages for the period 1976 to 2005). We also applied a range of spatial geological (mean geological age, weathering index) and soil (depth, nutrients, bulk density, water holding capacity) data layers, obtained from various sources [31–34].

Scaling Community Diversity from Plot to Grid Cell

It is possible and commonly practised to generate models of species richness and compositional dissimilarity that are relevant to smaller ecological survey plots occurring within larger spatial grid cells. However, for the purposes of the present study, our objective was to generate predictions of species richness and compositional dissimilarity that are relevant to the spatial grain of the 250 m grid cells (62,500 m^2) rather than the much smaller area that the survey plots occupy (500–1,000 m^2). To transform our community diversity data so that they were relevant to the spatial grid cell

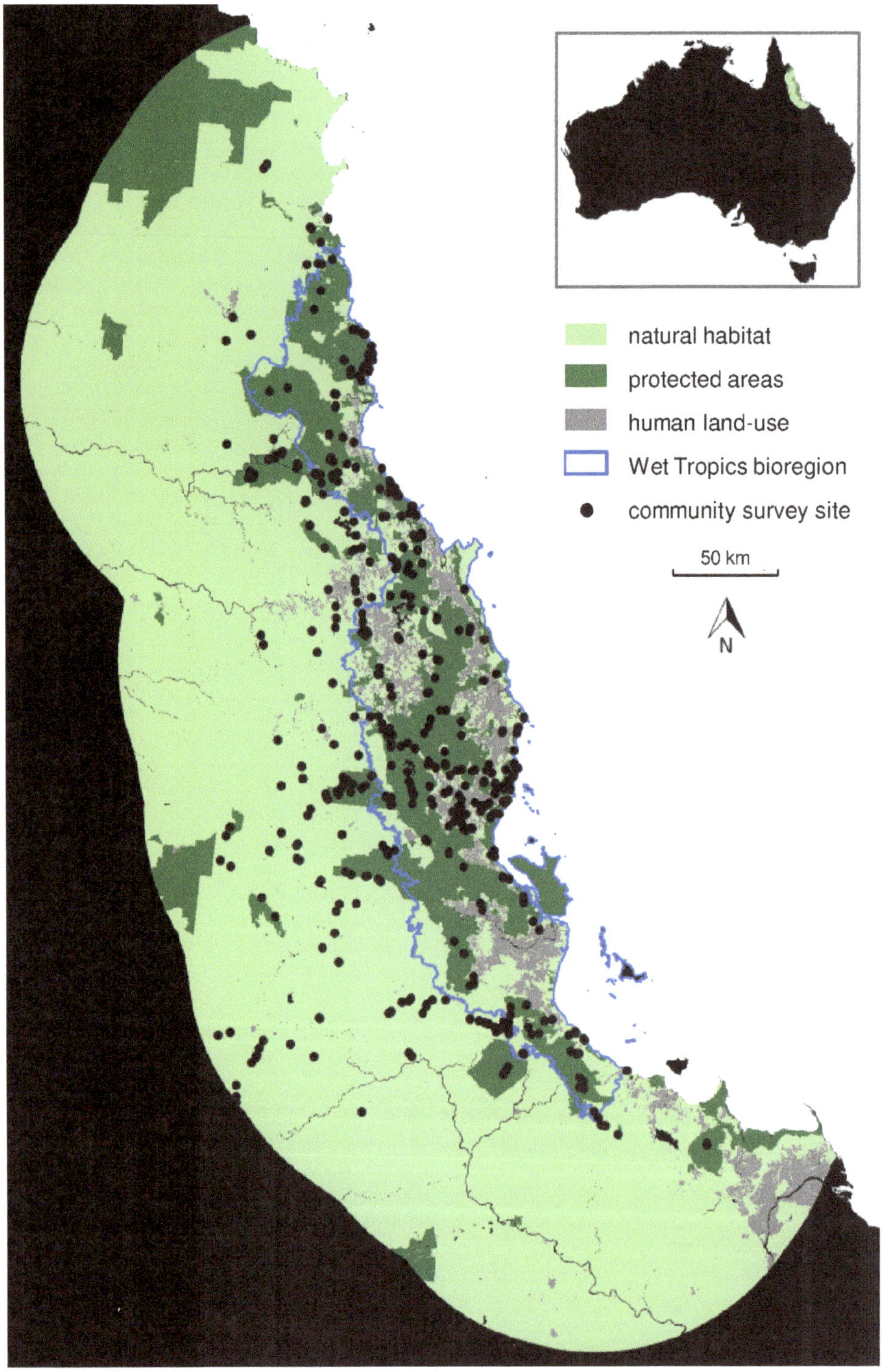

Figure 1. The study region. The study region, being the Wet Tropics Bioregion (blue line) and 100 km of surrounding area, in Queensland, Australia (inset). Protected areas (dark green), unprotected natural habitat (pale green) and modified habitat (grey) are shown, along with the locations of the 527 plant community surveys used in our analyses (black dots).

resolution, we applied a recently developed biodiversity scaling method [35].

Briefly, this approach applies the species-area power model ($S = cA^z$) to scale both species richness and compositional dissimilarity from small sample areas to larger areas. Under this approach, the species richness of a grid cell (S) is predicted from the observed richness of the community survey (c) and the area of the grid cell relative to the survey area (e.g. $A = 62{,}500\ m^2/500\ m^2$ for the Queensland herbarium data). To scale pair-wise compositional dissimilarity from the community surveys to the grid cells they occurred within, we first predict the number of species in common between the two grid cells i and j ($S_{com,ij}$) from the observed number of species in common between the two community surveys ($c_{com,ij}$) using the species area power relationship ($S_{com,ij} = c_{com,ij}A^{z_{com}}$). We then calculate the predicted Sorensen's compositional dissimilarity between the two grid cells ($\beta_{ij} = 1-[2S_{com,ij}/(S_i+S_j)]$) using the predicted species richness of each grid cell (S_i, S_j) and the predicted number of species in common between the two grid cells ($S_{com,ij}$). We applied a single scaling factor for species richness ($z = 0.25$) and for the number of species shared between a pair of communities ($z_{com} = 0.42$), as derived previously [35]. This scaling approach retains the underlying gradients in species richness and compositional dissimilarity that are observed across the community survey plots, but scales the absolute values so that they better represent those of the grid cells being modelled.

Species Richness Modelling

We generated a model of plant species richness using the Generalized Regression and Spatial Prediction package (GRASP [36]) in R [37]. GRASP applies a generalised additive modelling framework, and for the current implementation we used a Poisson link function with 3 degrees of freedom for each independent variable. We applied an interactive backward variable selection process, using a range of environmental variables that we hypothesised could be important in influencing plant community species richness in the study region. Candidate environmental variables were gradually omitted based on model Bayesian Information Criterion values, variable significance, and variable contribution to deviance reduction. Plant community species richness for each cell on the spatial grid of the region was predicted using the final model and the environmental variables for every grid cell.

Compositional Dissimilarity Modelling

We generated a model of pair-wise plant community compositional dissimilarity (Sorensen's dissimilarity) using Generalised Dissimilarity Modelling (GDM) which is an extension of matrix regression, designed to accommodate both the curvilinear relationship of observed compositional dissimilarity with increasing ecological distance between sites, and the variation in the rate of compositional turnover at different positions along environmental gradients [23]. We applied an interactive backward variable selection process, using a range of environmental variables that we hypothesised could be important in influencing plant community compositional dissimilarity in the study region. We applied custom written code in R [37] to implement a permutation test of variable and model significance. Using this procedure with 1,000 permutations, candidate environmental variables were gradually omitted based on variable significance and contribution to deviance reduction. We used the final model of compositional dissimilarity to generate spatially complete transformed environmental layers, which allow prediction of compositional dissimilarity between any pair of communities (i.e. pairs of cells on the spatial grid) [38].

Predicting Community Composition with *Dynamic*FOAM

We applied the *Dynamic*FOAM procedure to generate predictions of the composition of each community (i.e. grid cell) across the region. *Dynamic*FOAM is an optimisation algorithm that constructs species lists for each community which best meet the constraints of modelled estimates of the number of species present, the predicted dissimilarity in species composition between each pair of communities and any available data on the occurrences of specific species at specific sites [26]. When predicting the composition of all communities using *Dynamic*FOAM, it is possible to include a specified number of hypothetical species (e.g. undescribed species) [26], however, for the current implementation we applied only the 4,313 native species whose occurrences had been recorded within the study region. Given the stochastic nature of the *Dynamic*FOAM algorithm, we generated 10 replicate solutions, with the results presented here being summaries over these replicates.

Analysing Predictions of Community Composition

We applied a number of analytical approaches in examining our predictions of community composition across the study region. The median area of occurrence of species within each community (grid cell) was quantified by combining the predicted composition of each community and the predicted area of occurrence of each species over the study region. We also determined the predicted number of species within each community that were endemic to a circular area of radius 30 km centred on that community. We calculated the predicted number of "primitive" angiosperm families represented in each community by combining predicted community composition with family-level taxonomic affinity. Primitive angiosperm families were defined as phylogenetically near-basal, and included the families Austrobaileyaceae, Myristicaceae, Himantandraceae, Eupomatiaceae, Annonaceae, Atherospermataceae, Calycanthaceae, Hernandiaceae, Lauraceae, Monimiaceae, Winteraceae, Aristolochiaceae and Piperaceae [6,39]. Each of the above indices provide alternative measures of the conservation value of each community (grid cell), and many additional measures of conservation value could also have been obtained from our analyses. Here we simply demonstrate how these four alternative measures of conservation value can be synthesised into a single index of conservation value, by first normalising each to a 0–1 range (where 1 = highest conservation value), then calculating the mean value across all four normalised conservation attributes for each community.

Results

The model of plant species richness in the Australian Wet Tropics explained 40.7% of the total deviance (D^2), and included nine independent variables of climate and substrate (Table 1, Fig. 2). The strongest relationships with individual environmental variables were for species richness to increase with annual precipitation ($D^2 = 25.1\%$; $P<0.001$), soil plant available water holding capacity ($D^2 = 22.5\%$; $P<0.001$), C3 megaphyll plant growth index ($D^2 = 19.5\%$; $P<0.001$) and decrease with annual mean radiation ($D^2 = 21.9\%$; $P<0.001$) (Table 1, Fig. 2). Spatial projection of the species richness model showed moist coastal mountainous areas as having the greatest predicted species richness, while communities where predicted richness was lowest occurred in the drier inland areas in the south of the region (Fig. 3A).

The model of compositional dissimilarity for plant communities in the Australian Wet Tropics included seven independent climate and substrate variables, plus geographic distance, and explained 34.2% of the total deviance (Table 2, Fig. 4). The strongest relationships between compositional dissimilarity and individual environmental variables were for annual precipitation ($D^2 = 23.8\%$; $P<0.001$), annual mean radiation ($D^2 = 16.6\%$; $P<0.001$) and C3 megaphyll plant growth index ($D^2 = 16.7\%$; $P<0.001$) (Table 2, Fig. 4). In the full model with all eight variables, the amount of deviance in compositional dissimilarity explained solely by geographic distance was relatively low (lost D^2 when removed = 0.3%; Table 2, Fig. 4H).

We combined the model predictions of species richness for all 1,397,833 communities (grid cells) across the region with the model predictions of compositional dissimilarity between each pair

Table 1. Variable contribution to the Generalised Additive Model of plant species richness for the Australian Wet Tropics.

	All variable model	Single variable model
Variable	**% deviance lost**	**% deviance explained**
Annual mean radiation	0.8**	21.9**
Annual precipitation	0.4**	25.1**
Isothermality	0.4**	5.0**
Minimum temp. coldest period	2.5**	11.7**
C3 megaphyll growth index	3.4**	19.5**
C3 mesophyll growth index	3.4**	12.7**
Soil depth	5.0**	17.9**
Soil PAW holding capacity	9.0**	22.5**
Weathering index	6.1**	6.8**

**$P<0.001$; PAW = Plant available water; temp. = temperature.
For the full model of plant species richness (all variable model), the null deviance = 44,424, the residual deviance = 26,359, the residual degrees of freedom = 499, the deviance explained = 40.7% and the model P-value <0.001.

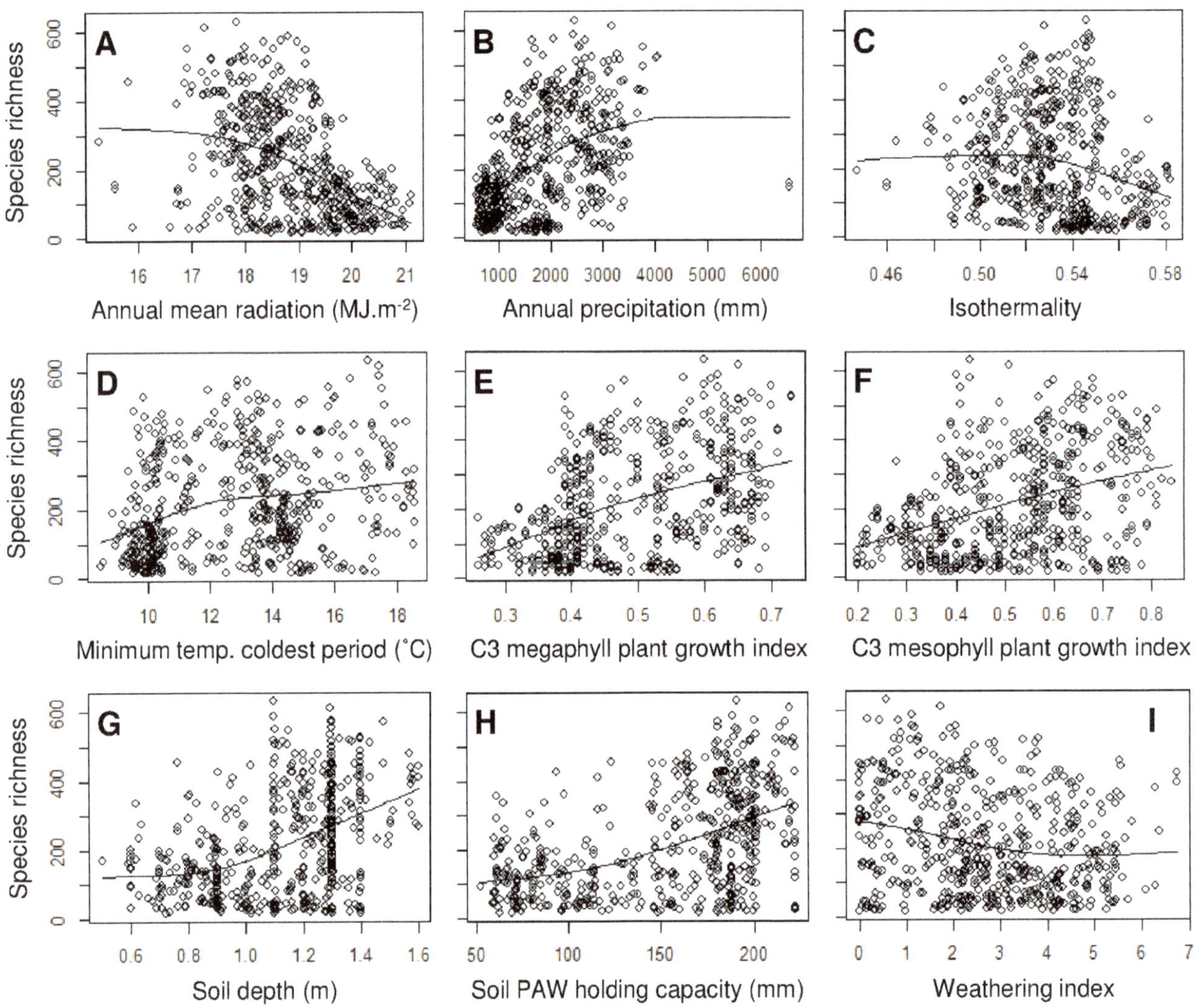

Figure 2. Species richness model response functions. Plant species richness as a function of each of the nine explanatory variables included in the model of plant species richness ((*A*)–(*I*)). The line shows the fitted relationship in the model, while observed data for the 527 community survey sites are shown by the open circles (richness scaled to grid cell area).

Table 2. Variable contribution to the Generalised Dissimilarity Model of pairwise community compositional dissimilarity for plants in the Australian Wet Tropics.

	All variable model	Single variable model
Variable	**% deviance lost**	**% deviance explained**
Annual precipitation	3.1**	23.8**
Maximum temp. warmest period	0.5*	3.4**
Annual mean radiation	0.6*	16.6**
C3 megaphyll growth index	0.5*	16.7**
C3 mesophyll growth index	0.7**	11.9**
Soil PAW holding capacity	2.0**	10.1**
Soil nutrient status	0.3*	4.8**
Geographic distance	0.3**	7.7**

*$P<0.01$;
**$P<0.001$; PAW = Plant available water; temp. = temperature.
For the full model of plant community compositional dissimilarity, the number of site pairs = 138,601, the intercept = 0.349, the null deviance = 48,010, the residual deviance = 31,568, the deviance explained = 34.2% and the model P-value <0.001.

of communities, to predict the current composition of every community using the *Dynamic*FOAM algorithm [26]. Across the ten replicate *Dynamic*FOAM solutions generated for the region, the proportion of species occurrences correctly predicted was 0.406 ($\pm 5.2\times10^{-3}$) with correctly predicted occurrences significantly ($P<0.01$) more frequent than random, while the mean absolute error in predicted Sorensen's dissimilarity between site pairs was 0.050 ($\pm 1.9\times10^{-4}$) (Fig. S3).

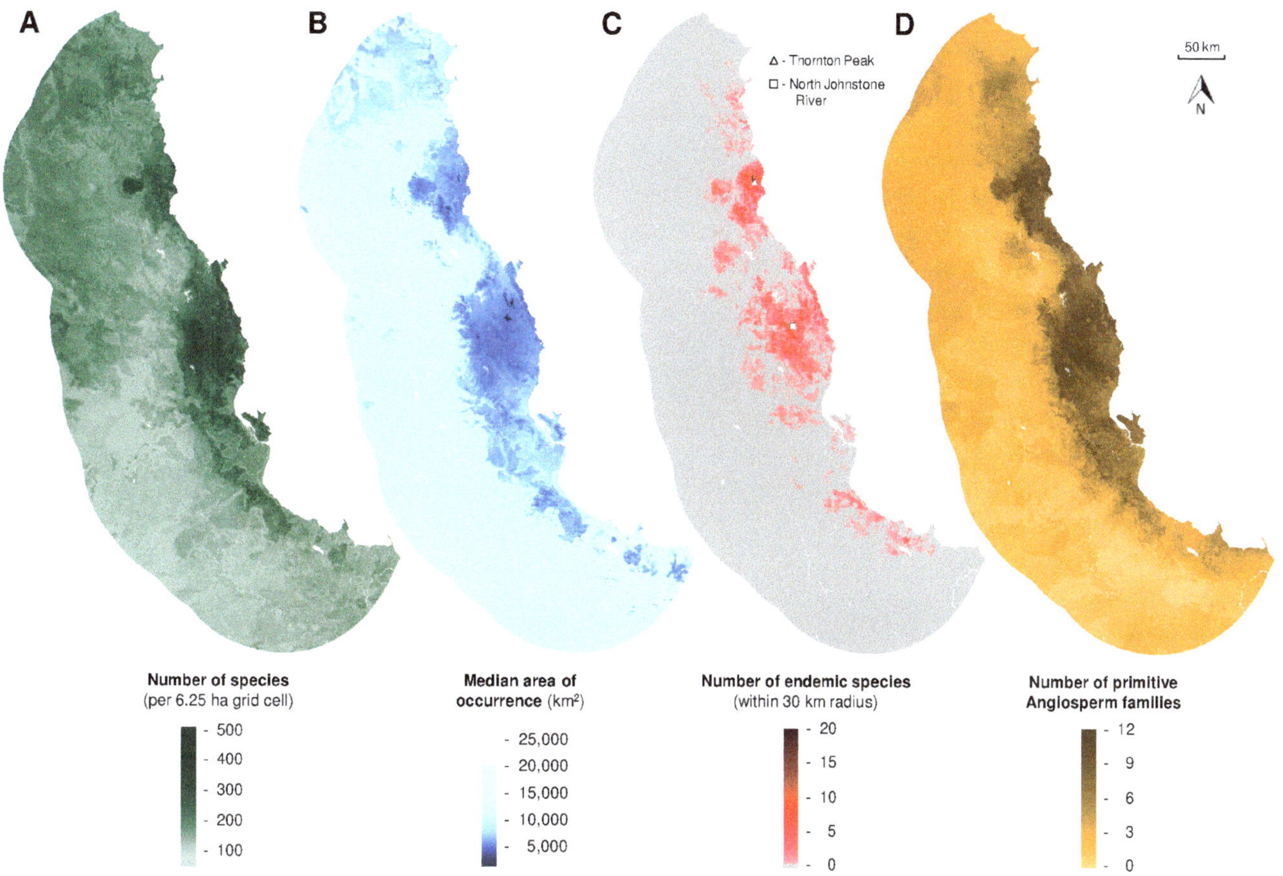

Figure 3. Predicted plant diversity attributes. Predictions across the study region of: (*A*) species richness; (*B*) median area of occurrence of species within each community; (*C*) the number of species present that are endemic to a circular area of radius 30 km centred on each community, and; (*D*) the number of primitive angiosperm families represented in each community (250 m grid cell).

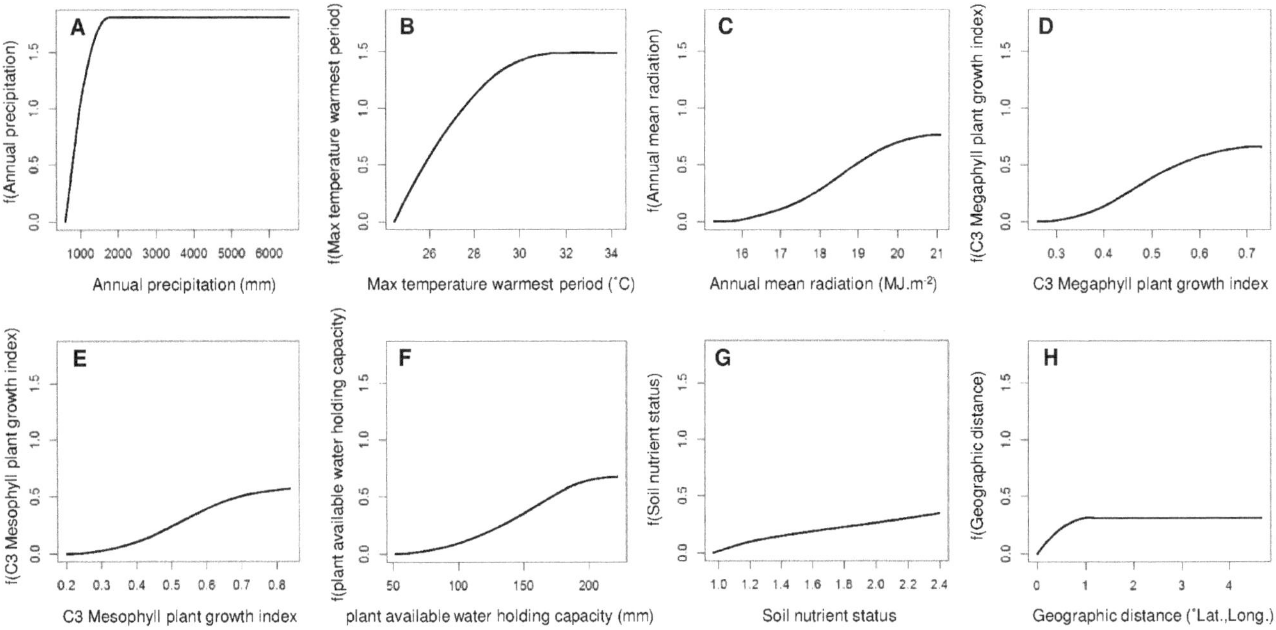

Figure 4. Compositional dissimilarity model response functions. GDM transformed environmental values as a function of the original environmental values for each of the eight explanatory variables included in the model of community compositional dissimilarity ((*A*) – (*H*)). The relative y-axis (transformed) range for the response of each variable indicates the relative strength of that variable in determining compositional dissimilarity, while the nonlinearity of the response indicates which sections of the environmental gradient have steeper predicted compositional dissimilarity.

Applying the predictions of the composition of every community (i.e. grid cell) across the region, we calculated three biodiversity metrics. First, we found that communities predicted to contain the rarest species (lowest median area of occurrence) were located in high altitude locations within the tropical forests (Fig. 3B). However, communities with the greatest predicted density of species endemic to the surrounding area of 30 km radius were concentrated in two locations: (1) the North Johnstone River in the Atherton Tablelands area, and (2) the Daintree rainforest around Thornton Peak (Fig. 3C). Finally, communities containing the largest number of "primitive" angiosperm families (described in [6,39]) were predicted to occur more evenly across the area of wet tropical forest (Fig. 3D). When the three attributes of plant biodiversity considered here (Fig. 3) were combined with species richness (Fig. 3A) in a single index of conservation value, habitat within two areas (Atherton Tablelands, Daintree rainforest) were predicted to have the greatest overall value for conserving plant diversity in the Australian Wet Tropics (Fig. 5).

Discussion

Drivers of Diversity

To identify conservation and management actions that are likely to best retain the diversity of tropical forests over time, we need to rapidly improve our understanding of current biodiversity in these systems. Here we have demonstrated how macroecological modelling can overcome shortfalls in our knowledge of diversity in tropical forests [40], providing valuable new insight into current patterns plus identifying priority areas for conservation and management. Although the Australian Wet Tropics are some of the world's best studied tropical rain forest, the same macro-ecological modelling approaches can be successfully applied in systems where biological information is much more limited, as the proportion of occurrences correctly predicted by the method is relatively insensitive to the amount of community survey data included in the analyses, at all but extremely low levels of input data [26].

For the Australian Wet Tropics, our continuous predictions of species richness, compositional dissimilarity and community composition for the 4,313 vascular plant species recorded in the region highlight which areas are likely to contain the greatest species richness (Fig. 3A), the most rare species (Fig. 3B), the greatest concentration of endemic species (Fig. 3C), and the greatest representation of "primitive" angiosperm families (Fig. 3D).These continuous predictions of plant diversity offer substantial advances over previous categorical assessment of vegetation patterns in the Australian Wet Tropics [12,13].

For both the models of species richness and compositional dissimilarity, annual precipitation was the most important individual predictor (Tables 1,2; Figs. 2B, 4A), supporting substantial previous research in both the Australian Wet Tropics and other tropical forests [4,15,41]. Outcomes for these tropical forests under climate change are therefore likely to strongly depend upon the nature of any concomitant changes in precipitation [16,42]. Other important explanatory variables were also shared by the macroecological models of species richness and compositional dissimilarity, including annual mean radiation and C3 megaphyll plant growth index (Tables 1,2; Fig. 2,4). Although both of these variables incorporate components of energy and temperature [30], they are also likely to be important through their relationship with moisture availability.

Spatial Projection of Plant Diversity

Here we have demonstrated how the predictions of the species richness and compositional dissimilarity models can be combined using *Dynamic*FOAM [26], to predict the composition of every community across the region (n = 1,397,833). These complete metacommunity predictions can then be analysed in numerous

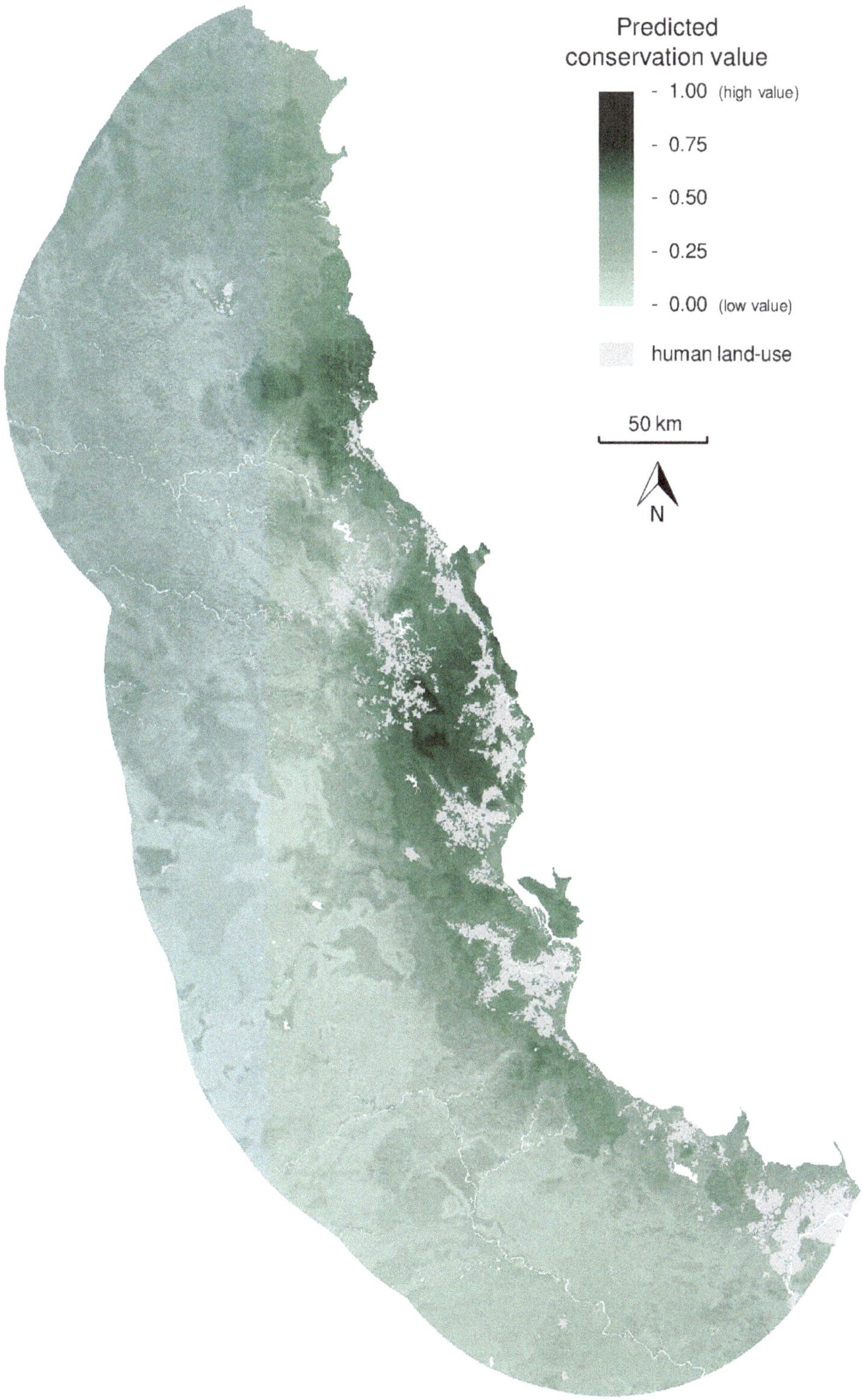

Figure 5. Predicted conservation value. Predictions of the overall conservation value for plant diversity of each community (grid cell) in the Australian Wet Tropics, as synthesised across the four different elements of diversity we considered (from Fig. 3).

ways to identify both major patterns in diversity and priorities for conservation or management. For example, in the Australian Wet Tropics, the communities predicted to contain the most rare species are not consistently those with the greatest species richness, but rather are areas with combinations of both moderately high species richness and high compositional dissimilarity to all other communities (Figs. 3A,B).

Another useful application of the predictions of community composition is in identifying areas of high species endemism. Areas that possess large numbers of endemic species are of particularly high value for conservation, through minimising the threat of extinction for the endemic species occurring there and because these areas may be important refuges for many other species under climate change [43,44]. Our analyses of species endemism within surrounding areas of 30 km radius identified two major centres of endemism for vascular plants: the North Johnstone River in the Atherton Tablelands, and; the Daintree rainforest around Thornton Peak (Fig. 3C). Interestingly, these predictions generally align with those of Nix & Switzer [14], who identified two large rainforest refugia in the same areas of the Australian Wet Tropics, by modelling the distribution of "rainforest", as a vegetation category, back to the last glacial maximum.

The Australian Wet Tropics flora has particular value globally for the large representation of "primitive" angiosperm families (13 out of 28) as described by Metcalfe and Ford [6,39]. Here we combined our predictions of plant community composition across the region with the identities of the 13 primitive angiosperm families, to quantify the predicted number of primitive angiosperm families occurring within each community (grid cell) (Fig. 3D). While the density of primitive angiosperm families corresponds roughly with overall plant species richness (Fig. 3A), communities in which primitive angiosperm families are best represented are not consistently those with the most species.

Conservation and Management Implications

Identifying priority areas for conservation and management often requires consideration of a range of biodiversity features [45]. These multiple features of biodiversity may be weighted in different ways for specific conservation applications, with a range of alternative algorithms and software available to achieve this [45,46]. Here we demonstrate how the predictions for multiple attributes of diversity emerging from our macroecological analyses (Fig. 3) can be simply synthesised into a single index of conservation value (Fig. 5). For more specific applications, other attributes of diversity could be additionally considered, the importance of specific diversity attributes could be weighted, or our predictions of diversity could be used in iterative assessment of alternative conservation/management options to incorporate complementarity between areas. Here, however, we simply demonstrate the potential for deriving a single measure of conservation value that is clearly composed of different elements of biodiversity that are perceived to be important.

For the Australian Wet Tropics, the Atherton Tablelands and Daintree rainforest areas were predicted to have the greatest overall conservation value for plant diversity (Fig. 5), possessing combinations of high species richness, many rare and endemic species, plus many primitive angiosperm families. The prediction of high conservation value for the Atherton Tablelands is of particular relevance for conservation and management, given the high levels of habitat loss already incurred in this area (Fig. 1). Indices of overall conservation value (Fig. 5) could be used to identify priorities for new protected areas in the region, capturing any high-value tropical forest currently outside the reserve estate. Our predicted conservation value could also be applied in targeting priority locations for management actions such as alien invasive species control measures or habitat restoration efforts, so that they best contribute to the maintenance of native plant diversity.

Future Directions

Here we have extended our knowledge of current patterns in plant diversity for the Australian Wet Tropics, through continuous modelling and prediction of species richness, compositional dissimilarity and complete metacommunity composition. We have applied these predictions to identify current priority areas for conservation and management, however, a thorough assessment of alternative management scenarios to retain biodiversity over time requires a more dynamic modelling approach. Indeed, the models and predictions of community composition generated here could form the basis for semi-mechanistic macroecological modelling to assess spatiotemporally explicit outcomes for biodiversity under specified climate and management scenarios [47]. Such semi-mechanistic modelling can incorporate dynamic dispersal processes and could be applied to help identify adaptive management actions under climate change [48]. Improved projections for plant diversity under alternative climate and management scenarios could also contribute to improving our understanding of likely outcomes for the diverse fauna that inhabit these tropical forests [49,50].

Supporting Information

Figure S1 The proportion of QLD Herbarium Corveg survey sites that contain different proportions of taxa not identified to species level, and the spatial distribution of the retained and omitted data across the Australian Wet Tropics.

Figure S2 The proportion of records omitted from the retained QLD Herbarium survey sites as a function of the total number of taxa recorded, and the spatial distribution of sites.

Figure S3 The difference between pairwise dissimilarity from the *Dynamic*FOAM predicted community compositions for the Australian Wet Tropics and both the pairwise dissimilarity predicted by the Generalised Dissimilarity Model and the observed dissimilarity from the community survey sites.

Acknowledgments

We thank the Queensland Herbarium and K.J Williams for providing biological and environmental data. The CSIRO survey data were collected under permits issued by the Queensland State Government.

Author Contributions

Conceived and designed the experiments: KM DAW SP. Performed the experiments: KM AJF DJM. Analyzed the data: KM. Wrote the paper: KM DAW SP AJF DJM.

References

1. Dirzo R, Raven PH (2003) Global state of biodiversity and loss. Annual Review of Environment and Resources 28: 137–167.
2. Bradshaw CJA, Sodhi NS, Brook BW (2008) Tropical turmoil: a biodiversity tragedy in progress. Frontiers in Ecology and the Environment 7: 79–87.
3. Lomolino M (2004) Conservation biogeography. In: Lomolino M, Heaney L, editors. Frontiers of Biogeography: new directions in the geography of nature. Sunderland, Massachusetts: Sinauer Associates. 293–296.
4. Carson WP, Schnitzer SA, editors (2008) Tropical Forest Community Ecology. Chichester: Wiley-Blackwell. 536 p.
5. Elith J, Leathwick JR (2009) Species Distribution Models: Ecological Explanation and Prediction Across Space and Time. Annual Review of Ecology Evolution and Systematics 40: 677–697.
6. Metcalfe DJ, Ford AJ (2009) Floristics and Plant Biodiversity of the Rainforests of the Wet Tropics. In: Stork N, Turton SM, editors. Living in a Dynamic Tropical Forest Landscape: Blackwell Publishing. 123–132.

7. Webb LJ (1959) A Physiognomic Classification of Australian Rain Forests. Journal of Ecology 47: 551–570.
8. Webb LJ (1968) Environmental Relationships of the Structural Types of Australian Rain Forest Vegetation. Ecology 49: 296–311.
9. Williams W, Tracey J (1984) Network Analysis of Northern Queensland Tropical Rainforests. Australian Journal of Botany 32: 109–116.
10. Tracey J (1982) The vegetation of the humid tropical region of North Queensland. Melbourne: CSIRO Press.
11. Sattler PS, Williams RD (1999) The Conservation Status of Queensland's Bioregional Ecosystems. Brisbane: Environmental Protection Agency.
12. Stanton P, Stanton D (2005) Vegetation of the Wet Tropics of Queensland Bioregion. Cairns: Wet Tropics Management Authority.
13. Tracey J, Webb L (1975) Vegetation of the Humid Tropical Region of North Queensland. Indooroopilly, Australia: CSIRO.
14. Nix H, Switzer M (1991) Rainforest animals: atlas of vertebrates endemic to Australia's wet tropics. Canberra: Australian National Parks and Wildlife Service.
15. Hilbert DW, Van Den Muyzenberg J (1999) Using an artificial neural network to characterize the relative suitability of environments for forest types in a complex tropical vegetation mosaic. Diversity and Distributions 5: 263–274.
16. Hilbert DW, Graham A, Hopkins MS (2007) Glacial and interglacial refugia within a long-term rainforest refugium: The wet tropics bioregion of NE Queensland, Australia. Palaeogeography Palaeoclimatology Palaeoecology 251: 104–118.
17. VanDerWal J, Shoo LP, Williams SE (2009) New approaches to understanding late Quaternary climate fluctuations and refugial dynamics in Australian wet tropical rain forests. Journal of Biogeography 36: 291–301.
18. Graham CH, VanDerWal J, Phillips SJ, Moritz C, Williams SE (2010) Dynamic refugia and species persistence: tracking spatial shifts in habitat through time. Ecography 33: 1062–1069.
19. Hilbert DW, Ostendorf B, Hopkins MS (2001) Sensitivity of tropical forests to climate change in the humid tropics of north Queensland. Austral Ecology 26: 590–603.
20. Ostendorf B, Hilbert DW, Hopkins MS (2001) The effect of climate change on tropical rainforest vegetation pattern. Ecological Modelling 145: 211–224.
21. Ferrier S, Guisan A (2006) Spatial modelling of biodiversity at the community level. Journal of Applied Ecology 43: 393–404.
22. Currie DJ (2001) Projected effects of climate change on patterns of vertebrate and tree species richness in the conterminous United States. Ecosystems 4: 216–225.
23. Ferrier S, Manion G, Elith J, Richardson K (2007) Using generalized dissimilarity modelling to analyse and predict patterns of beta diversity in regional biodiversity assessment. Diversity & Distributions 13: 252–264.
24. Ter Steege H, Pitman N, Sabatier D, Castellanos H, Van Der Hout P, et al. (2003) A spatial model of tree α-diversity and tree density for the Amazon. Biodiversity & Conservation 12: 2255–2277.
25. Jones MM, Ferrier S, Condit R, Manion G, Aguilar S, et al. (2013) Strong congruence in tree and fern community turnover in response to soils and climate in central Panama. Journal of Ecology 101: 506–516.
26. Mokany K, Harwood TD, Overton JM, Barker GM, Ferrier S (2011) Combining a- and B-diversity models to fill gaps in our knowledge of biodiversity. Ecology Letters 14: 1043–1051.
27. Neldner VJ, Wilson BA, Thompson EJ, Dillewaard HA (2012) Methodology for Survey and Mapping of Regional Ecosystems and Vegetation Communities in Queensland. Brisbane: Queensland Herbarium, Queensland Department of Science, Information Technology, Innovation and the Arts. 124 p.
28. Mayfield MM, Dwyer JM, Chalmandrier L, Wells J, Bonser SP, et al. (2013) Differences in forest plant functional trait distributions across land use and productivity gradients. American Journal of Botany: (in press).
29. Bostock PD, Holland AE (2010) Census of the Queensland Flora 2010. Brisbane: Queensland Herbarium, Department of Environment and Resource Management.
30. Xu T, Hutchinson M (2010) ANUClim Version 6.1 User Guide. Canberra: Fenner School of Environment and Society, The Australian National University.
31. McKenzie N, Jacquier D, Cresswell H (2000) Estimation of Soil Properties Using the Atlas of Australian Soils. Canberra: CSIRO.
32. Western A, McKenzie N (2004) Soil hydrological properties of Australia. Melbourne: CRC for Catchment Hydrology.
33. Raymond O (2009) New digital geological map of Australia. AusGeo News 93: 3.
34. Wilford J (2012) A weathering intensity index for the Australian continent using airborne gamma-ray spectrometry and digital terrain analysis. Geoderma 183–184: 124–142.
35. Mokany K, Jones MM, Harwood TD (2013) Scaling pairwise β-diversity and α-diversity with area. Journal of Biogeography: 40: 2299–2309.
36. Lehmann A, Overton JM, Leathwick JR (2003) GRASP: generalized regression analysis and spatial prediction. Ecological Modelling 160: 165–183.
37. R Development Core Team (2012) R: A Language and Environment for Statistical Computing. Vienna, Austria: R Foundation for Statistical Computing.
38. Overton J, Barker G, Price R (2009) Estimating and conserving patterns of invertebrate diversity: a test case of New Zealand land snails. Diversity & Distributions 15: 731–741.
39. Metcalfe DJ, Ford AJ (2009) A re-evaluation of Queensland's Wet Tropics based on "primitive" plants. Pacific Conservation Biology 15: 80–86.
40. Mokany K, Ferrier S (2011) Predicting impacts of climate change on biodiversity: a role for semi-mechanistic community-level modelling. Diversity & Distributions 17: 374–380.
41. Engelbrecht BMJ, Comita LS, Condit R, Kursar TA, Tyree MT, et al. (2007) Drought sensitivity shapes species distribution patterns in tropical forests. Nature 447: 80–82.
42. Malhi Y, Aragao LEOC, Galbraith D, Huntingford C, Fisher R, et al. (2009) Exploring the likelihood and mechanism of a climate-change-induced dieback of the Amazon rainforest. Proceedings of the National Academy of Sciences 106: 20610–20615.
43. Moritz C, Richardson KS, Ferrier S, Monteith GB, Stanisic J, et al. (2001) Biogeographical concordance and efficiency of taxon indicators for establishing conservation priority in a tropical rainforest biota. Proceedings of the Royal Society of London Series B: Biological Sciences 268: 1875–1881.
44. Carnaval AC, Hickerson MJ, Haddad CFB, Rodrigues MT, Moritz C (2009) Stability Predicts Genetic Diversity in the Brazilian Atlantic Forest Hotspot. Science 323: 785–789.
45. Moilanen A (2007) Landscape Zonation, benefit functions and target-based planning: Unifying reserve selection strategies. Biological Conservation 134: 571–579.
46. Ball IR, Possingham HP, Watts M (2009) Marxan and relatives: Software for spatial conservation prioritisation. In: Moilanen A, Wilson K, Possingham H, editors. Spatial conservation prioritisation: Quantitative methods and computational tools. Oxford: Oxford University Press. 185–195.
47. Mokany K, Harwood TD, Williams KJ, Ferrier S (2012) Dynamic macroecology and the future for biodiversity. Global Change Biology 18: 3149–3159.
48. Mokany K, Harwood TD, Ferrier S (2013) Comparing habitat configuration strategies for retaining biodiversity under climate change. Journal of Applied Ecology 50: 519–527.
49. Williams SE, Bolitho EE, Fox S (2003) Climate change in Australian tropical rainforests: an impending environmental catastrophe. Proceedings of the Royal Society of London Series B-Biological Sciences 270: 1887–1892.
50. Shoo LP, Williams SE, Hero J-M (2005) Climate warming and the rainforest birds of the Australian Wet Tropics: Using abundance data as a sensitive predictor of change in total population size. Biological Conservation 125: 335–343.

10

High-Level Genetic Diversity and Complex Population Structure of Siberian Apricot (*Prunus sibirica* L.) in China as Revealed by Nuclear SSR Markers

Zhe Wang[1], Ming Kang[2], Huabo Liu[1], Jiao Gao[1], Zhengdong Zhang[1], Yingyue Li[1], Rongling Wu[1], Xiaoming Pang[1]*

1 National Engineering Laboratory for Tree Breeding, Key Laboratory of Genetics and Breeding in Forest Trees and Ornamental Plants, Ministry of Education, Center for Computational Biology, College of Biological Sciences and Biotechnology, Beijing Forestry University, Beijing, China, **2** South China Botanical Garden, Chinese Academy of Sciences, Guangzhou, China

Abstract

Siberian apricot (*Prunus sibirica* L.), an ecologically and economically important tree species with a high degree of tolerance to a variety of extreme environmental conditions, is widely distributed across the mountains of northeastern and northern China, eastern and southeastern regions of Mongolia, Eastern Siberia, and the Maritime Territory of Russia. However, few studies have examined the genetic diversity and population structure of this species. Using 31 nuclear microsatellites, we investigated the level of genetic diversity and population structure of Siberian apricot sampled from 22 populations across China. The number of alleles per locus ranged from 5 to 33, with an average of 19.323 alleles. The observed heterozygosity and expected heterozygosity ranged from 0.037 to 0.874 and 0.040 to 0.924 with average values of 0.639 and 0.774, respectively. A STRUCTURE-based analysis clustered all of the populations into four genetic clusters. Significant genetic differentiation was observed between all population pairs. A hierarchical analysis of molecular variance attributed about 94% of the variation to within populations. No significant difference was detected between the wild and semi-wild groups, indicating that recent cultivation practices have had little impact on the genetic diversity of Siberian apricot. The Mantel test showed that the genetic distance among the populations was not significantly correlated with geographic distance ($r = 0.4651$, $p = 0.9940$). Our study represents the most comprehensive investigation of the genetic diversity and population structure of Siberian apricot in China to date, and it provides valuable information for the collection of genetic resources for the breeding of Siberian apricot and related species.

Editor: David D. Fang, USDA-ARS-SRRC, United States of America

Funding: This work was supported by grants from the Program for New Century Excellent Talents in University by the Ministry of Education, China (NCET-10-0223), and the Starting Funds from the Office for Science and Technology of Beijing Forestry University, China (2008-03), to X.M.P. The funders had no role in study design, data collection and analysis, decision to publish, or preparation of the manuscript.

Competing Interests: The authors have declared that no competing interests exist.

* E-mail: xmpang@bjfu.edu.cn

Introduction

Siberian apricot (*Prunus sibirica* L.), an ecologically and economically important tree species, is widely distributed across the mountainous areas of northern and northeastern China, eastern Siberian, and Mongolia [1]. It can adapt to a variety of harsh environmental conditions, including cold stress, drought stress, and reduced soil fertility, making it one of the primary choices for controlling desertification in northern and northwestern China. Siberian apricot almond is not only a traditional dry food, but also an important raw material for food, cosmetics, and biodiesel manufacturing. Thus, Siberian apricot is important to the income of farmers in these areas [2,3].

In recent decades, almond products have become increasingly popular on the domestic and international market. Consequently, many almond processing plants have been established around the major areas of production in China. However, Siberian apricot resources are declining due to backward management patterns and deterioration of the natural environment [4]. Furthermore, diseases and insect pests such as awning caterpillar (*Malacosoma neustria testacea Motsch*) and leaf roller (*Adoxophyes honmai*) have made the originally fragile natural environment even worse [5]. Despite the hardiness of Siberian apricot, its flowers will wither if a late frost hits during flowering, and this can cause a serious reduction in yield or no yield at all. Therefore, there is an urgent need to develop a Siberian apricot cultivar with increased tolerance to both abiotic and biotic stresses. The success of breeding programs is based on the knowledge and availability of genetic variability for efficient selection [6]. However, Siberian apricot, as a building block for breeding programs, has not been extensively studied in China until now.

Increased knowledge of the genetic diversity and population structure of Siberian apricot in China will provide the basis for protecting, utilizing, and improving our resources. Therefore, an assessment of the extent and nature of the genetic variation in Siberian apricot is important for breeding and genetic resource conservation programs. Traditionally, genetic diversity has been assessed based on morphological characteristics, which are often influenced by the environmental conditions. With the advent of molecular markers, including restriction fragment length polymorphisms, amplified fragment length polymorphisms, simple sequence repeats (SSRs), and single nucleotide polymorphisms, much progress has been made in understanding the genetic diversity and population structure of various species [7–10]. Among these markers, SSRs have been the first choice for the study of genetic diversity and population structure owing to their desirable genetic attributes, including high numbers of polymorphisms, wide genomic distribution, co-dominant inheritance, and high degree of reproducibility [11,12]. Nuclear SSR makers have also proven to be very useful for the evaluation of genetic diversity in apricot [13,14]: they have been employed to investigate the genetic diversity of Siberian apricot in the Yan Mountains of China [15]. However, a comprehensive analysis of Siberian apricot genetic diversity and its population structure in China at the DNA level is lacking.

In this study, 31 nuclear SSR loci developed previously for this species [16] were used to analyze the genetic diversity and structure of Siberian apricot populations in China. The objectives of the study were to provide a complete picture of the organization of genetic diversity of Siberian apricot populations in China and to reveal the origin of the genetic variation in Siberian apricot populations.

Materials and Methods

Sampling

A total of 672 individuals of Siberian apricot representing 22 populations were collected throughout the areas of distribution in China (Table 1). A total of 25 to 32 individuals were sampled for each population, and the coordinate of each tree was recorded using a global positioning system. The distance between any two individuals at each location was >50 m. The 22 populations were from 21 sampling locations (P5 and P17 were from the same region) across 18 longitudes in the east-west direction and across 6 latitudes in the north-south direction. The highest altitude of the locations was 1,334 m (P20), while the lowest altitude was 87 m (P1). Daqing (P14) had a minimum altitude gap of only 3 m, while Weichang (P20) had a maximum altitude gap of 271 m. The sampled populations were divided into six groups according to their geographical locations. The Yan Mountains group (G1) included P7, P8, P9, P10, P11, and P12; the Greater Khingan Mountains group (G2) included P18, P19, P20, P21, and P6; the Western Liaoning Hills group included (G3) P1, P2, P3, P4, P5, P17, and P22; the Northeast Plain group (G4) included P13 and P14; the Linkou group (G5) included P15; and the Daqingshan Mountain group (G6) included P16 (Figure 1). No specific permits were required for this field study. All sampling locations were public space where anyone can enter and collect forest products, regardless of ownership. In addition, the field study did not involve endangered or protected species.

In China, Siberian apricot has been cultivated for decades in an experimental forest. Currently, the main method of propagation is to sow seeds collected from the immediate area or near the region without selection. Three such populations were collected and designated as semi-wild type. All other populations were from the wild (Table 1). Young leaves were collected and placed immediately in Ziploc bags preloaded with colored silica gel to dry them and preserve them for DNA extraction.

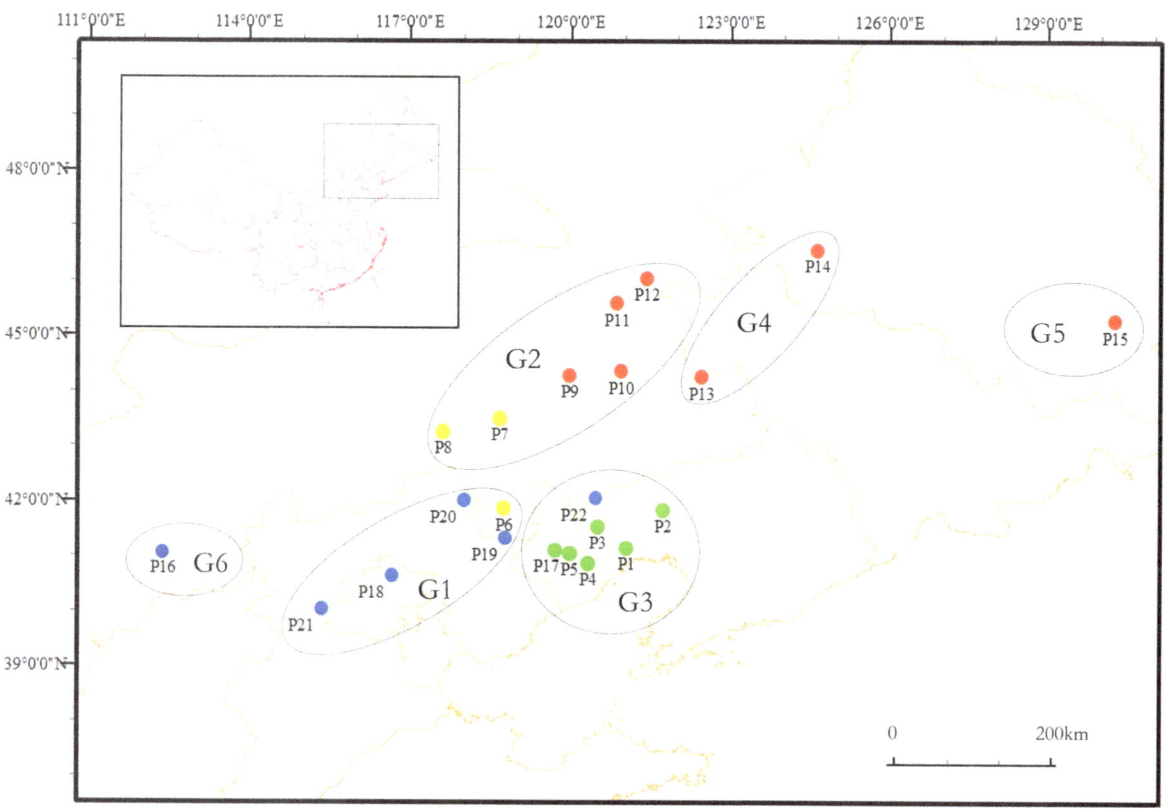

Figure 1. Geographic distribution of the analysed Siberian apricot sampling in China. The image was generated by the software ArcGIS (ESRI, Redlands, CA, USA). Green dots represent genetic cluster C1, yellow dots represent genetic cluster C2, red dots represent genetic cluster C3, blue dots represent genetic cluster C4.

Table 1. Summary of Siberian apricot sampling locations around China.

Population	Population ID	Sample size	Locality	Elevation (m)	Origin
Jinzhou, Liaoning	P1	31	N41°09′ E121°03′	87~226	Wild
Fuxin, Liaoning	P2	30	N41°50′ E121°44′	384~456	Semi-wild
Chaoyang, Liaoning	P3	30	N41°32′ E120°30′	361~533	Wild
Huludao, Liaoning	P4	31	N40°52′ E120°19′	237~267	Wild
Kazuo, Liaoning	P5	30	N41°03′ E119°58′	530~736	Wild
Jinshan, Inner Mongolia	P6	31	N41°52′ E118°46′	998~1126	Wild
Daban, Inner Mongolia	P7	32	N43°30′ E118°41′	681~780	Semi-wild
Jingpeng, Inner Mongolia	P8	29	N43°15′ E117°38′	1092~1275	Wild
Tianshan, Inner Mongolia	P9	32	N44°17′ E119°58′	486~519	Wild
Lubei, Inner Mongolia	P10	32	N44°21′ E120°56′	301~319	Wild
Tuliemaodu, Inner Mongolia	P11	32	N45°35′ E120°52′	477~564	Wild
Keqinzhongqi, Inner Mongolia	P12	32	N46°02′ E121°26′	475~584	Wild
Baicheng, Jilin	P13	32	N44°15′ E122°27′	158~167	Wild
Daqing, Heilongjiang	P14	32	N46°34′ E124°39′	147~150	Wild
Linkou, Heilongjiang	P15	30	N45°17′ E130°17′	255~316	Wild
Wulancabu, Inner Mongolia	P16	31	N41°04′ E112°22′	157~195	Wild
Kazuo, Liaoning	P17	30	N41°06′ E119°43′	317~364	Semi-wild
Huairou, Beijing	P18	30	N40°38′ E116°41′	426~507	Wild
Pingquan, Hebei	P19	29	N41°19′ E118°47′	641~733	Wild
Weichang, Hebei	P20	31	N42°01′ E118°01′	1063~1334	Wild
Zhuolu, Hebei	P21	30	N40°02′ E115°22′	1149~1222	Wild
Chifeng, Inner Mongolia	P22	25	N42°03′ E120°27′	693~821	Wild

Microsatellite DNA Analysis

Total genomic DNA was extracted from dry leaves collected from all localities using a modified version of the cetyl trimethylammonium bromide method [17]. The quality and concentration of the extracted DNA was determined by 1% agarose gel electrophoresis and ultraviolet spectrophotometry.

Thirty-one microsatellite loci were employed to study the genetic diversity on wild Siberian apricot accessions including 23 recently developed in Siberian apricot [16,18], one from apricot (*Prunus armeniaca* L.) [19] and seven from peach (*Prunus persica* L.) [20–22] (Table S1). The forward primer of each pair was tagged with a section of the universal M13 sequence (5′-TGTAAAAC-GACGGCCAGT-3′) during synthesis. Amplification was performed in a 10-μL reaction mixture containing 1 μL of DNA template (10 ng/μL), 5 μL of 2X Taq mix, 0.4 μL of the forward primer (1 μM), 1.6 μL of the reverse primer, 1.6 μL of M13 primer (1 μM) with a fluorescent label (FAM, HEX, ROX, or TAMRA), and 1.4 μL of ddH_2O. The reaction conditions were: 94°C for 5 min, followed by 30 cycles of 94°C for 30 s, 55°C for 30 s, and 72°C for 30 s, followed by 8 cycles of 94°C for 30 s, 53°C for 40 s, and 72°C for 30 s, with a final extension at 72°C for 10 min. The products were separated in an ABI 3730×L DNA Analyzer using GeneScan-500LIZ as an internal marker (Applied Biosystems, Foster City, CA, USA). The amplicon fragments were sized using Gene-Marker 1.75 software (SoftGenetics LLC, State College, PA, USA). All rare alleles and private alleles were re-amplified. For the alleles from the homozygous loci, the purified PCR products were sent to sequence. For the alleles from the heterozygous loci, the targeted fragments were separated, cloned and sequenced following the protocol by Chen et al [23]. These sequences were compared with target fragments to distinguish whether they were non-specific amplifications.

Data Analysis

FLEXBIN was used for automated binning of the microsatellite raw data [24], and the Excel Microsatellite Toolkit [25] was employed to convert the size data into various formats for further analysis. The level of genetic diversity was estimated using GENALEX software version 6.41 [26] with the following statistics: number of alleles (Na), effective number of alleles (Ne), Shannon's Information Index (I), observed heterozygosity (Ho), expected heterozygosity (He) [27], and F-statistics calculations (F_{IS}, F_{IT}, and F_{ST}).

Clustering based on a Bayesian model was used to evaluate the genetic structures of the Siberian apricot populations with the software package STRUCTURE [28] in its extended version 2.3.3 [29,30]. The admixture model and independent allelic frequencies were employed to analyze the data set without prior population information. The length of the burn-in period and number of MCMC reps after burn-in were set to 25,000 and 100,000, respectively. These steps were used to determine the ancestry value, which estimates the proportion of an individual's genome that originated from a given genetic group. The algorithm was run ten times for each K value, from 1 to 22. Using an ad hoc quantity constructed from the second-order rate of change of the likelihood function with respect to K (ΔK), the distribution of ΔK showed a clear peak at the true value of K [31].

The observed genetic variation among and within the populations and genetic groups was characterized by an analysis of molecular variance (AMOVA) using ARLEQUIN version 3.5

Table 2. Diversity indices of 31 nuclear microsatellite loci from data of 672 individuals.

Locus	Na	Ne	I	Ho	He	Fis	Fit	Fst	Nm	Rare alleles
PSL1	16	4.652	1.837	0.600	0.785	0.181	0.238	0.070	3.334	6
PSL3	7	1.553	0.781	0.308	0.356	0.073	0.137	0.069	3.361	3
PSL6	26	9.485	2.628	0.689	0.895	0.179	0.228	0.059	3.952	7
PSL7	23	10.513	2.577	0.807	0.905	0.047	0.113	0.069	3.376	8
PSL8	11	3.378	1.342	0.617	0.704	0.036	0.121	0.089	2.567	7
PSL10	19	4.744	1.928	0.733	0.789	-0.049	0.074	0.118	1.869	9
PSL11	24	11.934	2.681	0.790	0.916	0.077	0.154	0.083	2.758	7
PSL12	21	9.535	2.529	0.816	0.895	0.034	0.092	0.060	3.929	4
PSL13	20	3.528	1.816	0.654	0.717	0.028	0.086	0.059	3.970	11
PSL14	12	2.261	1.295	0.518	0.558	0.034	0.074	0.042	5.765	4
PSL16	24	2.924	1.899	0.313	0.658	0.485	0.523	0.074	3.108	7
PSL18	27	12.249	2.765	0.781	0.918	0.094	0.150	0.062	3.808	10
A1-10	20	6.345	2.293	0.388	0.842	0.513	0.548	0.071	3.253	6
A3-9	8	3.907	1.470	0.388	0.744	0.421	0.484	0.109	2.045	3
A3-66	27	11.305	2.770	0.770	0.912	0.103	0.157	0.060	3.931	6
H1-7	23	10.389	2.606	0.838	0.904	0.000	0.072	0.072	3.205	6
H1-11	17	5.955	2.154	0.711	0.832	0.076	0.148	0.079	2.931	5
H1-77	33	13.653	2.987	0.874	0.927	0.016	0.056	0.041	5.891	7
H1-87	23	10.363	2.589	0.798	0.904	0.047	0.118	0.075	3.080	7
H2-11	14	4.725	1.806	0.814	0.788	-0.078	-0.033	0.042	5.641	5
H2-27	12	3.222	1.536	0.653	0.690	-0.030	0.052	0.080	2.887	4
H2-77	31	13.117	2.864	0.475	0.924	0.455	0.497	0.077	2.981	11
H2-79	19	5.487	2.119	0.779	0.818	-0.010	0.050	0.060	3.940	7
BPPCT 002	29	7.198	2.352	0.799	0.861	0.004	0.073	0.069	3.374	16
BPPCT 030	12	5.825	1.945	0.731	0.828	0.024	0.121	0.100	2.253	3
CPPCT 6	17	9.249	2.386	0.831	0.892	0.005	0.068	0.063	3.714	4
UDP96-001	5	1.041	0.119	0.037	0.040	0.030	0.059	0.029	8.244	3
UDP96-005	22	4.872	2.140	0.388	0.795	0.474	0.515	0.079	2.923	10
UDP96-010	12	1.844	0.975	0.437	0.458	-0.019	0.043	0.061	3.859	7
UDP98-412	19	9.794	2.523	0.856	0.898	-0.026	0.049	0.073	3.180	2
ssrPaCITA 15	26	5.871	2.210	0.621	0.830	0.212	0.288	0.097	2.317	12
Mean	19.323	6.804	2.062	0.639	0.774	0.111	0.173	0.071	3.595	6.677
Total	599									207

Na: Number of Different Alleles; Ne: Number of Effective Alleles; I: Shannon's Information Index; Ho: Observed Heterozygosity; He: Expected Heterozygosity; F_{IS}: inbreeding coefficient; F_{IT}: over inbreeding coefficient; F_{ST}: fixation index; Nm: Gene Flow.

[32]. This analysis subdivided the 22 populations into two different origin groups, six geographical groups and K groups. Three hierarchical divisions were identified based on the genetic variance: within populations, among populations within groups, and among groups using a nonparametric permutation procedure incorporating 10,000 iterations. In addition, we tested all of the loci for deviations from Hardy-Weinberg equilibrium (HWE) using ARLEQUIN version 3.5 [33] with 100,000,000 steps in the Markov chain [34] and 100,000 dememorization steps. We selected F_{ST} and R_{ST} to calculate the genetic differentiation of all population pairs. The values of F_{ST} and R_{ST} were calculated using FSTAT version 2.9.3 [35] and ARLEQUIN version 3.5 [33], respectively. To examine the effect of geographic distance on genetic structure, correlations between the pairwise genetic distances, represented by $F_{ST}/(1-F_{ST})$ estimates [36], and pairwise geographic distances among 19 wild populations, which were calculated according to the latitude and longitude of each site with Vincenty's formula (http://www.movable-type.co.uk/scripts/latlong- vincenty.html), were tested using the Mantel test implemented by Isolation By Distance Web Service version 3.23 (http://ibdws.sdsu.edu/~ibdws/) [37,38]. We also employed Monmonier's maximum difference algorithm to highlight geographical features corresponding to pronounced genetic discontinuity using BARRIER version 2.2 [39].

Results

Genetic Diversity Among the Loci

A total of 31 microsatellite loci were used to genotype 672 individuals of Siberian apricot (Table 2). The genetic profiles detected 599 alleles, which ranged between 5 and 33 per locus, 207 of which were rare alleles with a frequency below 1%. The

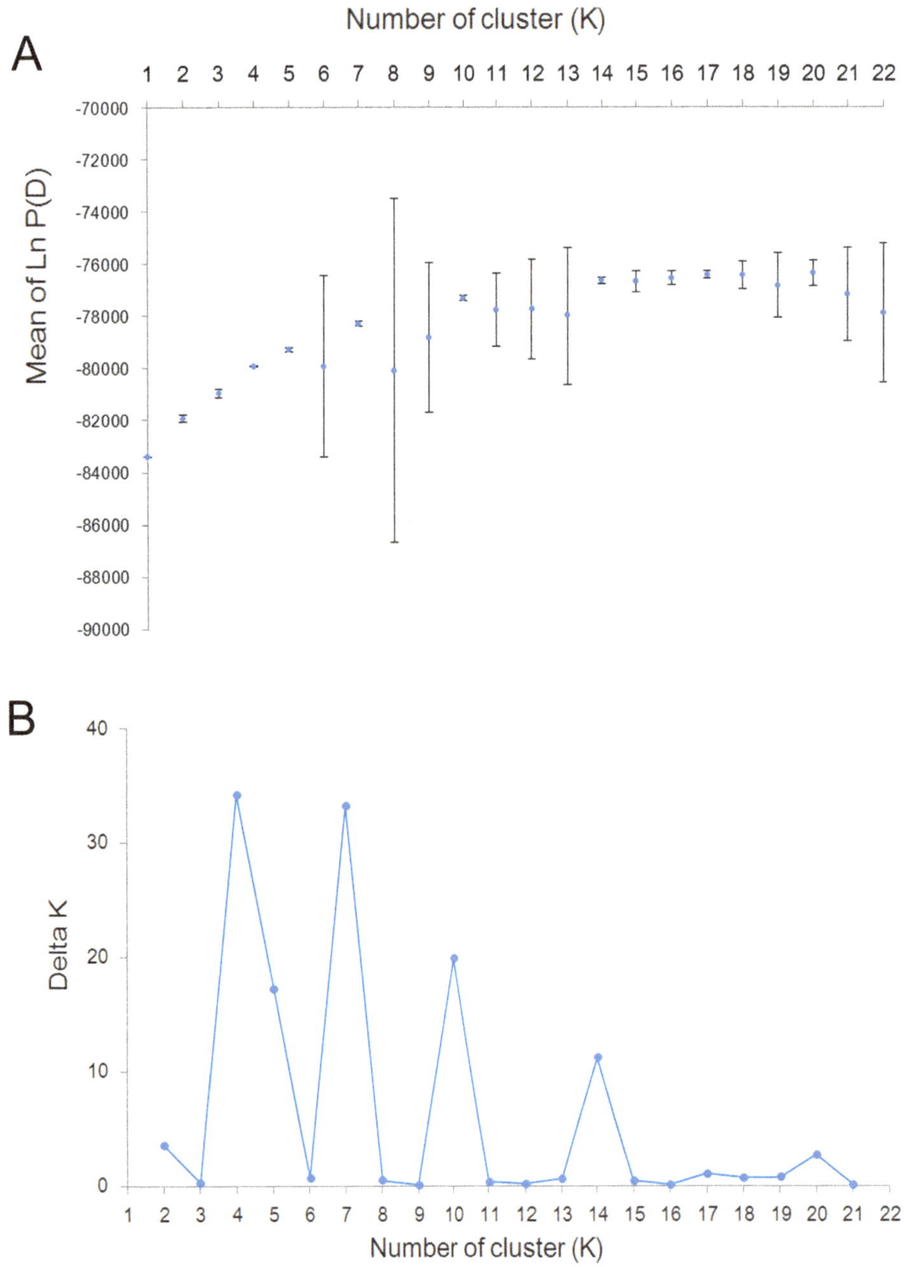

Figure 2. Plot of the Ln P(D) ± SD and delta K (ΔK). The mean of Ln P(D) was based on ten repetitions for each K value.

sequencing results showed that all the rare alleles were true alleles with the sequences containing expected microsatellites. The Ne in the total samples was 6.804 (range, 1.041 to 13.653). I ranged from 0.119 to 2.987, with an average of 2.062. The Ho in the total sample was 0.639, which deviated from the He (0.774). Genetic subdivision using F-statistics estimated a moderate inbreeding coefficient (0.173) and an F_{ST} value of 0.071 across all loci, indicating moderate genetic differentiation among the sites. The values of gene flow ranged from 1.869 to 8.224, with an average of 3.595. Most loci conformed to HWE and no population had a particularly large number of loci that deviated from HWE.

Genetic Structure of the Siberian Apricot Samples

The genetic structure of the Siberian apricot samples was investigated by a Bayesian-based population assignment analysis using STRUCTURE [28]. Our results show a clear maximum for ΔK at K = 4 (Figure 2B), in which all individuals were classified into four different clusters. About 80% individuals belonged to each genetic cluster, which showed strong ancestry values with an average >0.90 (Table S2). Regarding the genetic cluster 1 (C1) which included P1, P2, P3, P4, P5 and P17, only 14 individuals (7.4%) showed ancestry values <0.60. Eighteen individuals which were from the locations belonged to other genetic clusters. These individuals corresponded to two accessions from P9, two accessions from P10, two accessions from P11, six accessions from P12, one accession from P18, and five accessions from P19

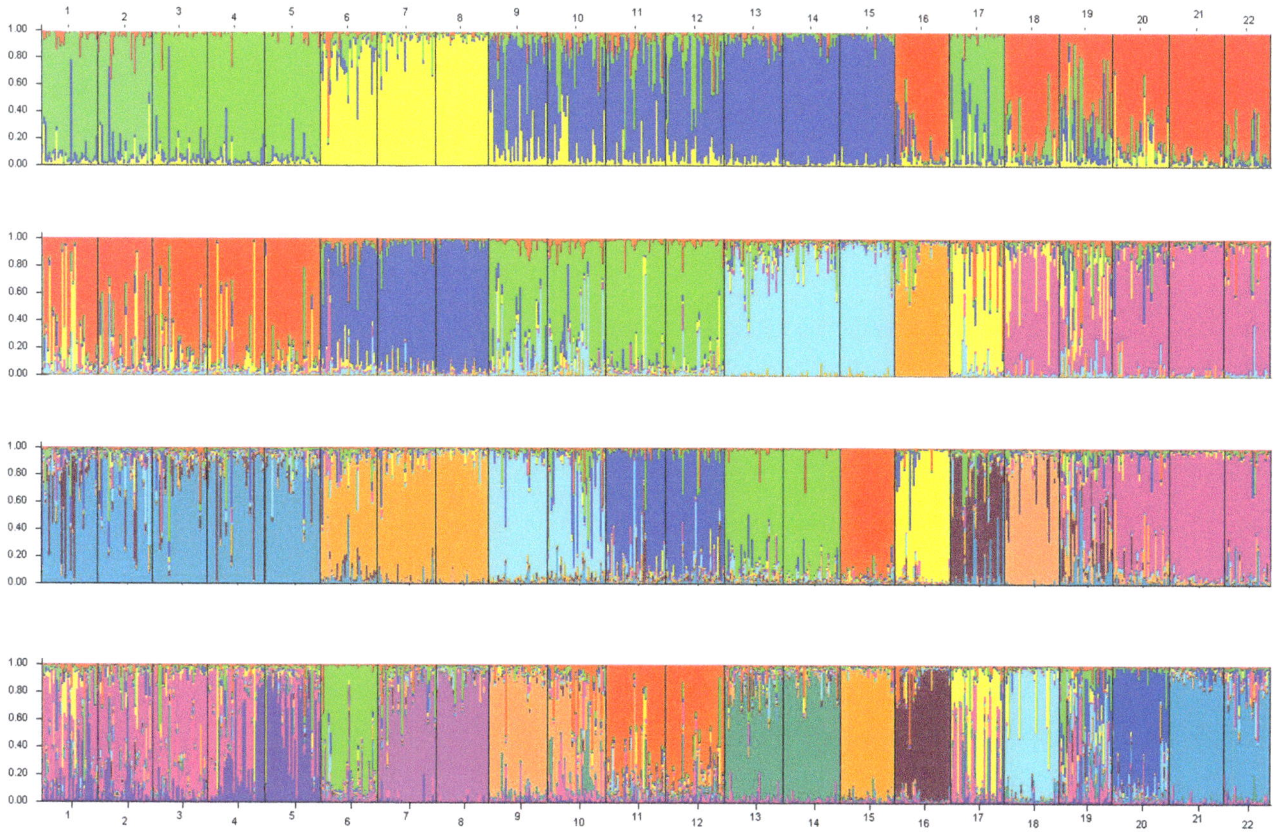

Figure 3. Clustering of 22 Siberian apricot populations. Each individual is shown as a vertical line divided into segments representing the estimated membership proportion in the four, seven, ten and fourteen ancestral genetic clusters inferred with STRUCTURE.

(Figure 3 and Table S2). The genetic cluster 2 (C2) consisted of P6, P7 and P8, and only 13 individuals (13.1%) showed ancestry values <0.60. We found 10 individuals were from the locations belonged to other genetic clusters for C2. These individuals corresponded to two accessions from P2, one accession from P9, three accessions from P10, one accession from P11, one accession from P19, and two accessions from P20. Within the genetic cluster 3 (C3), which contained population P9, P10, P11, P12, P13, P14 and P15, only 25 individuals (11.6%) showed ancestry values < 0.60, and 11 individuals were from the locations belonged to other genetic clusters. These individuals correlated to one accession from each of the population P1, P2, P3 and P6, five accessions from P17, and two accessions from P19 (Figure 3 and Table S2); All remaining populations including P16, P18, P19, P20, P21 and P22 were clustered into the genetic cluster 4 (C4), among which 16 individuals (9.5%) showed ancestry values <0.60. And only two individuals were from P6 which belonged to C2 (Figure 3 and Table S2).

At the same time, the second largest ΔK at K = 7 was much larger than the remaining values. In addition, two clear peaks were observed at K = 10 and 14 (Figure 2B). When K = 7, P16 and P17 were separated into two new genetic clusters from genetic cluster C1 and genetic cluster C2, while genetic cluster C3 was divided into two genetic clusters. On the basis of seven genetic clusters, the 3th, 4th and 7th genetic clusters, were all split into two genetic clusters while K = 10. When k = 14, the 1st and the 2nd genetic clusters were further divided into two genetic clusters, and the 10th genetic cluster was divided into three detailed genetic clusters on the basis of the clustering of ten genetic clusters (Figure 3).

Genetic Diversity Among the Siberian Apricot Populations

The population genetic parameters used in this study are summarized in Table 3. The highest degree of genetic diversity occurred in P4 (Ne = 6.084, Ho = 0.672, and He = 0.753), P10 (Ne = 5.709, Ho = 0.705, and He = 0.763), and P17 (Ne = 5.700, Ho = 0.666, and He = 0.769), while the diversity was lowest in P8 (Ne = 4.155, Ho = 0.592, and He = 0.674), P16 (Ne = 3.544, Ho = 0.564, and He = 0.632), and P21 (Ne = 4.163, Ho = 0.558, and He = 0.658). P16 and P21 were unique populations with3 and 4 more private alleles respectively, although they had the lowest genetic diversity.

The value of Na for the wild genotypes was significantly higher than that for the semi-wild genotypes (Table 3). The number of private alleles in the wild genotypes was far greater than that in the semi-wild genotypes. These differences could be associated with the huge disparities in sample size. The values of Ho and He for the wild genotypes were almost equal to the values for the semi-wild genotypes.

The Ho and He values in genetic cluster C3 were slightly larger than those in the other genetic clusters (Table 3) whereas genetic cluster C4 was the lowest Ho value; however, regardless of whether the individuals were considered to be wild or semi-wild, and regardless of whether they belonged to which genetic cluster, the Ho value was significantly lower than the He value. This result is in agreement with the high value of the fixation index, suggesting a deficit of heterozygotes with regard to the expectations of HWE.

Table 3. Genetic diversity estimations in wild and semi-wild groups, genetic clusters and all populations.

Pop	Sample size	Na	Ne	Ho	He	F	Private alleles
C1	182	16.645	6.583	0.653	0.768	0.147	32
C2	92	11.903	5.221	0.642	0.721	0.106	5
C3	222	15.806	6.630	0.668	0.780	0.141	29
C4	176	15.258	5.609	0.583	0.733	0.194	26
Wild	580	18.968	6.767	0.638	0.773	0.171	152
Semi-wild	92	16.694	6.607	0.643	0.770	0.161	11
Pop1	**31**	10.742	5.351	0.661	0.733	0.087	7
Pop2	**30**	10.355	5.465	0.641	0.736	0.122	3
Pop3	**30**	10.129	5.370	0.618	0.710	0.122	1
Pop4	**31**	11.226	6.084	0.672	0.753	0.123	6
Pop5	**30**	9.806	5.708	0.650	0.744	0.137	3
Pop6	**31**	8.742	4.609	0.671	0.704	0.047	2
Pop7	**32**	8.774	4.524	0.631	0.702	0.101	2
Pop8	**29**	7.419	4.155	0.621	0.674	0.076	1
Pop9	**32**	10.226	5.099	0.652	0.734	0.115	3
Pop10	**32**	10.484	5.709	0.705	0.763	0.071	3
Pop11	**32**	10.032	5.572	0.666	0.750	0.116	1
Pop12	**32**	9.935	5.165	0.657	0.738	0.104	0
Pop13	**32**	9.871	5.436	0.652	0.736	0.117	2
Pop14	**32**	9.806	5.114	0.661	0.725	0.086	0
Pop15	**30**	8.290	4.921	0.683	0.740	0.082	4
Pop16	**31**	7.613	3.544	0.564	0.632	0.099	3
Pop17	**30**	11.000	5.700	0.666	0.769	0.132	6
Pop18	**30**	8.226	4.648	0.592	0.678	0.122	1
Pop19	**29**	10.000	5.264	0.641	0.734	0.121	3
Pop20	**31**	8.968	4.766	0.575	0.693	0.181	2
Pop21	**30**	7.710	4.163	0.558	0.658	0.143	4
Pop22	**25**	8.645	4.556	0.548	0.669	0.171	2

Na: Number of Different Alleles; Ne: Number of Effective Alleles; Ho: Observed Heterozygosity; He: Expected Heterozygosity; F: Inbreeding coefficient.

A comparison of private alleles between the wild (152) and semi-wild (11) populations showed a significant difference between them (Table 3). When all populations were considered, P1 contained the most private alleles (7); no private alleles were found in P12 and P14.

AMOVA

Our AMOVA revealed that a low percentage of variation was divided among natural populations, different origins, geographical distribution, and genetic clusters, respectively (Table 4). About 94% of the variation was attributed to differences within populations in all variance partitions. A hierarchical AMOVA of the four genetic clusters using STRUCTURE revealed that 1.87% of the variance was distributed among them, and it produced the largest F_{ST} value (0.06008). Seven genetic clusters revealed the highest percentage of variation (3.48%) among them, and it produced the second largest F_{ST} value (0.06002). With the populations grouped according to their geographical origin, a lower percentage of variation (1.98%) could be explained by geographic factors. When the populations were grouped according to their origin, a negative percentage of variation was detected among the groups.

Genetic and Geographic Relatedness

The pairwise genetic differentiation values (F_{ST} and R_{ST}) calculated for the 22 populations showed genetic differentiation between each population (Table 5). All of the F_{ST} values were significantly different from 0 in all pairwise comparisons between the 22 populations ($p<0.01$). The lowest values of F_{ST} were observed between P1-P2, P1-P3, P1-P4, P1-P17, P2-P4, P4-P5, P4-P17, P5-P17, P11-P12 and P13-P14. Populations from different genetic clusters appeared to be more differentiated from each other, corresponding well to the classification of the genetic cluster. Of the R_{ST} values, 10 (8 of which were from paired populations coming from different clusters) were not significantly different from 0. The pairwise genetic differentiation values (F_{ST}) between the four genetic clusters showed a higher genetic differentiation between three population pairs (C1-C2, C1-C4, and C3-C4) (Table S3).

The highest F_{ST} values were observed for pairwise comparisons between P16 and other populations. The geographic distance matrix ranged from 35.8 to 1,526.5 km, based on the latitude and longitude values of all 19 wild populations. Rousset's genetic distance values [$F_{ST}/(1-F_{ST})$] [36] indicated that the most closely related Siberian apricot populations were P11 and P12, even though the geographical distance between them was not the closest. The greatest geographic distance (1,526.5 km) was between P15 and P16; however, this pairing did not have the largest Rousset's distance (0.123). The largest Rousset's distance (0.156) was between P16 and P22. The Mantel test (Figure 4) showed that genetic distance was not significantly correlated with geographic distance ($r=0.4651$, $p=0.9940$).

The Identification of Genetic Barriers

A genetic barrier prediction analysis using Monmonier's maximum difference algorithm identified three putative barriers when all populations were included (Figure 5A). The first barrier separated the western peripheral population P16 from all other populations. The second predicted barrier separated population P22, which was located in the center of the distribution areas. The third predicted barrier separated population P20,. When only the 19 wild populations were included (Figure 5B), the first barriers separated P16, similar to the result obtained when all of the populations were included. The second predicted barrier separated P20 and P22 from the other populations. There was a gap between P20 and P22 that could be associated with each other. The second and the third predicted barriers together separated P8 from the other populations.

Discussion

Genetic Diversity of Siberian Apricot in China

Heterozygosity is an important measurement of gene diversity [40]. In our study, a relatively high level of genetic diversity was detected at microsatellite loci in Siberian apricot; the mean Ho and He values were 0.639 and 0.774, respectively. Similar values were reported for populations of Siberian apricot in the Yan Mountains (Ho = 0.668, He = 0.788) [15]. Fewer polymorphisms have been reported for apricot (*Prunus armenica* L.; Ho = 0.615, He = 0.621) [41]. The genetic diversity of Chinese wild almond (*Amygdalus nana* L.; Ho = 0.339, He = 0.219) is reportedly even lower [42]. Ferrer et al. [43] found that the number of loci and populations included in studies might affect estimates of genetic diversity. In our study, the number of loci and samples was larger

Table 4. Analysis of molecular variance from microsatellite data using Arlequin version 3.5.

Source of variation	d.f.	Sum of squares	Variance components	Percentage of variation	Fixation Index
Variance partition[a]					
Among populations	21	956.633	**0.58429 Va**	5.59	$F_{ST}=0.05592$
Within populations	1322	13039.582	**9.86353 Vb**	94.41	
Total	1343	13996.215	10.44782		
Variance partition[b]					
Among groups	1	24.409	−0.07047 Va	−0.68	$F_{ST}=0.05111$
Among populations within groups	20	932.224	**0.60174 Vb**	5.79	$F_{SC}=0.05750$
Within populations	1322	13039.582	**9.86353 Vc**	94.89	$F_{CT}=-0.00678$
Total	1343	13996.215	10.39480		
Variance partition[c]					
Among groups	5	390.503	**0.20726 Va**	1.98	$F_{ST}=0.05968$
Among populations within groups	16	566.130	**0.41874 Vb**	3.99	$F_{SC}=0.04072$
within populations	1322	13039.582	**9.86353 Vc**	94.03	$F_{CT}=0.01976$
Total	1343	13996.215	10.48953		
Variance partition[d]					
Among groups	3	301.894	**0.19642 Va**	1.87	$F_{ST}=0.06008$
Among populations within groups	18	654.739	**0.43402 Vb**	4.14	$F_{SC}=0.04215$
within populations	1322	13039.582	**9.86353 Vc**	93.99	$F_{CT}=0.01872$
Total	1343	13996.215	10.49397		
Variance partition[e]					
Among groups	6	511.653	**0.30467 Va**	3.48	$F_{ST}=0.06002$
Among populations within groups	15	444.980	**0.32512 Vb**	2.43	$F_{SC}=0.03191$
within populations	1322	13039.582	**9.86353 Vc**	94.10	$F_{CT}=0.02903$
Total	1343	13996.215	10.49332		
Variance partition[f]					
Among groups	9	612.562	**0.30269 Va**	2.89	$F_{ST}=0.05844$
Among populations within groups	12	344.071	**0.30951 Vb**	2.95	$F_{SC}=0.03042$
within populations	1322	13039.582	**9.86353 Vc**	94.16	$F_{CT}=0.02889$
Total	1343	13996.215	10.47573		
Variance partition[g]					
Among groups	13	746.937	**0.33358 Va**	3.19	$F_{ST}=0.05737$
Among populations within groups	9	209.696	**0.26671 Vb**	2.55	$F_{SC}=0.02633$
within populations	1322	13039.582	**9.86353 Vc**	94.26	$F_{CT}=0.03188$
Total	1343	13996.215	10.46381		

The hierarchical analysis included 22 sampling populations in China including within populations, among populations within groups and among groups.
[a]The first analysis included all populations as one hierarchical group.
[b]The second analysis included two different origin groups.
[c]The third analysis included six geographical groups.
[d]The fourth analysis included four genetic clusters.
[e]The fifth analysis included seven genetic subclusters.
[f]The fifth analysis included ten genetic subclusters.
[g]The fifth analysis included fourteen genetic subclustersF_{ST} variance among coefficient of individual relative to the total variance.
F_{SC} variance among subpopulations within groups.
F_{CT} variance among groups relative to the total variance.

than in the aforementioned studies. The geographic range of the species and species characteristics (e.g., long-lived, outcrossing, and wind-pollinated) also influenced the genetic diversity, and high heterozygosity could be favorable in long-lived plants growing in arid zones. Indeed, Siberian apricot is long-lived, wind-pollinated, self-incompatible, and distributed across a wide area with a harsh environment, which may be one cause of the high level of genetic diversity and high number of alleles of per loci detected in Siberian apricot populations. We have found many morphological variations in our field investigation, such as double petals apricot, green sepal apricot, big flower apricot, late flowering apricot, heart-shaped apricot, sweet benevolence apricot and so on, which have not been reported previously. Among the populations, P16 and P21 had the lowest level of genetic diversity (Table 3). P16 is

Table 5. Pairwise F_{ST} values (below diagonal) and Pairwise R_{ST} values (above diagonal) between 22 populations.

	P1	P2	P3	P4	P5	P6	P7	P8	P9	P10	P11	P12	P13	P14	P15	P16	P17	P18	P19	P20	P21	P22
P1	0	0.033	0.015	0.038	0.024	0.095	0.038	0.021	0.045	0.085	0.049	0.042	0.048	0.043	0.090	0.106	0.058	0.097	0.023	0.080	0.009ns	0.007ns
P2	0.012	0	0.049	0.075	0.036	0.099	0.068	0.047	0.054	0.070	0.076	0.061	0.087	0.058	0.068	0.115	0.044	0.102	0.055	0.083	0.034	0.004ns
P3	0.018	0.021	0	0.057	0.035	0.124	0.081	0.057	0.054	0.096	0.044	0.055	0.071	0.061	0.074	0.101	0.082	0.129	0.011ns	0.091	0.015ns	0.017ns
P4	0.017	0.019	0.024	0	0.025	0.090	0.060	0.044	0.047	0.082	0.051	0.048	0.076	0.063	0.122	0.100	0.067	0.120	0.066	0.086	0.045	0.040
P5	0.026	0.024	0.034	0.015	0	0.075	0.052	0.021	0.045	0.053	0.056	0.037	0.081	0.037	0.091	0.061	0.043	0.072	0.039	0.063	0.025	0.003ns
P6	0.035	0.039	0.052	0.042	0.048	0	0.077	0.042	0.037	0.048	0.089	0.072	0.121	0.056	0.115	0.084	0.030	0.064	0.122	0.020	0.100	0.073
P7	0.051	0.057	0.060	0.052	0.053	0.031	0	0.027	0.055	0.072	0.054	0.021	0.052	0.025	0.109	0.097	0.039	0.082	0.075	0.079	0.051	0.052
P8	0.064	0.064	0.074	0.062	0.060	0.038	0.041	0	0.043	0.051	0.054	0.023	0.039	0.018	0.090	0.068	0.031	0.040	0.052	0.044	0.036	0.021
P9	0.044	0.047	0.058	0.043	0.037	0.051	0.040	0.059	0	0.059	0.033	0.043	0.089	0.045	0.092	0.063	0.020	0.120	0.059	0.038	0.055	0.027
P10	0.027	0.028	0.030	0.022	0.027	0.038	0.042	0.052	0.025	0	0.076	0.061	0.111	0.041	0.061	0.046	0.047	0.075	0.114	0.012ns	0.079	0.047
P11	0.029	0.039	0.045	0.031	0.042	0.049	0.051	0.069	0.039	0.026	0	0.054	0.073	0.041	0.085	0.076	0.046	0.136	0.047	0.072	0.034	0.036
P12	0.031	0.031	0.043	0.033	0.040	0.042	0.050	0.065	0.046	0.029	0.016	0	0.046	0.019	0.101	0.065	0.038	0.076	0.056	0.050	0.058	0.048
P13	0.028	0.037	0.040	0.029	0.045	0.046	0.056	0.072	0.044	0.023	0.038	0.037	0	0.034	0.083	0.112	0.091	0.082	0.077	0.108	0.052	0.067
P14	0.052	0.054	0.060	0.040	0.052	0.067	0.064	0.089	0.054	0.035	0.055	0.052	0.021	0	0.062	0.060	0.048	0.046	0.068	0.042	0.040	0.026
P15	0.056	0.059	0.058	0.048	0.064	0.064	0.082	0.084	0.069	0.050	0.058	0.054	0.047	0.057	0	0.083	0.076	0.096	0.091	0.085	0.060	0.059
P16	0.098	0.086	0.112	0.090	0.099	0.096	0.081	0.106	0.097	0.082	0.087	0.088	0.088	0.102	0.109	0	0.044	0.091	0.109	0.050	0.101	0.069
P17	0.017	0.024	0.027	0.014	0.019	0.037	0.044	0.062	0.031	0.023	0.027	0.031	0.023	0.033	0.047	0.089	0	0.082	0.066	0.040	0.061	0.034
P18	0.058	0.060	0.074	0.061	0.070	0.066	0.071	0.077	0.072	0.058	0.067	0.070	0.054	0.074	0.092	0.104	0.053	0	0.130	0.093	0.098	0.075
P19	0.025	0.024	0.041	0.035	0.040	0.029	0.048	0.063	0.046	0.030	0.033	0.032	0.034	0.050	0.062	0.080	0.026	0.038	0	0.100	0.010ns	0.036
P20	0.040	0.042	0.055	0.057	0.063	0.049	0.055	0.069	0.065	0.044	0.054	0.060	0.061	0.083	0.092	0.102	0.053	0.059	0.030	0	0.088	0.057
P21	0.066	0.068	0.080	0.079	0.083	0.077	0.080	0.092	0.085	0.069	0.081	0.088	0.074	0.103	0.110	0.102	0.078	0.052	0.046	0.050	0	0.009ns
P22	0.075	0.068	0.084	0.090	0.089	0.083	0.095	0.118	0.101	0.074	0.086	0.090	0.082	0.103	0.109	0.135	0.080	0.069	0.048	0.048	0.065	0

Significant values at the 1% nominal level are bolded.

located in the western edge of the distribution area whereas P21 is located 1,200 m above sea level at the southwestern edge of the Yan Mountains. The marginal distribution would reduce the opportunity to communicate with other populations and lead to a low level of genetic diversity.

The Ho value was lower than the He value at all 31 loci (Table 2), indicating a deficiency of heterozygotes at these loci. A heterozygote deficiency was also observed at the population level (Table 3). Similar findings related to heterozygote deficiency have been observed in other trees [44–46]. In *Cinnamomum insularimontanum* Hyata (*Lauraceae*) from southern Korea, a heterozygote deficiency was explained as a process of partial selfing rather than the presence of null alleles or a temporal Wahlund effect. A deficiency of heterozygotes in the tropical species *Sextonia rubra* (Mez) van der Werff was explained as an effect of biparental breeding due to limited pollen dispersal among relatives [45]. In flowering dogwood trees, a deficiency in heterozygotes was explained as the result of half-sibling mating occurring over a small geographical area [46]. The seed-setting rate by self-pollination in Siberian apricot is very low; such trees usually exhibit self-incompatibility. Thus, the deficiency of heterozygotes in Siberian apricot in our study may be the result of low levels of inbreeding. Further research on the mating system, pollen dispersal, and seeds in Siberian apricot populations is needed to infer the precise cause of the deficiency in heterozygotes.

Genetic Structure of Siberian Apricot

An AMOVA revealed that genetic variation within populations accounted for about 94.4% of the total (Table 4). Outcrossing woody plants tend to be more genetically diverse and have less genetic differentiation among populations [32]. The percentage of genetic variation within populations of Siberian apricot in the Yan Mountains was shown to be up to 96% [15]. The negative percentage of variation detected among wild and semi-wild groups suggests that there is no significant difference between them. Furthermore, the values of Ne, He,Ho and F in the semi-wild population were similar to those in the wild population. This indicates that the sources of semi-wild populations might be selected randomly from the seeds of wild populations, and that recent cultivation practices have had little impact on the genetic diversity of Siberian apricot. The relatively low values of Na in the semi-wild group might be due to the small sample size.

The use of R- and F-statistics when estimating genetic differentiation assumes a stepwise-mutation model (SMM) and an infinite-allele model, respectively. R-statistics was developed to take into account the high homoplasy inherent in microsatellite markers [47]. However, several analyses of population structure have reached the conclusion that many microsatellite loci do not fit an SMM process [48–50]. Balloux et al. [51] showed that microsatellites could mutate following a fairly strict SMM model. De Andrés et al. [52] also chose F_{ST} instead of R_{ST} when calculating the genetic differentiation among grapevine populations. Compared with R_{ST}, F_{ST} was more consistent with our other analysis. Though all the genetic differentiation between pairwise populations was significant, the lowest values of F_{ST} still could be observed between P1-P2, P1-P3, P1-P4, P1-P17, P2-P3, P2-P4, P4-P5, P4-P17, P5-P17, P11-P12 and P13-P14. An UPGMA dendrogram based on Nei's unbiased genetic distance showed these population pairs had the shortest genetic distance

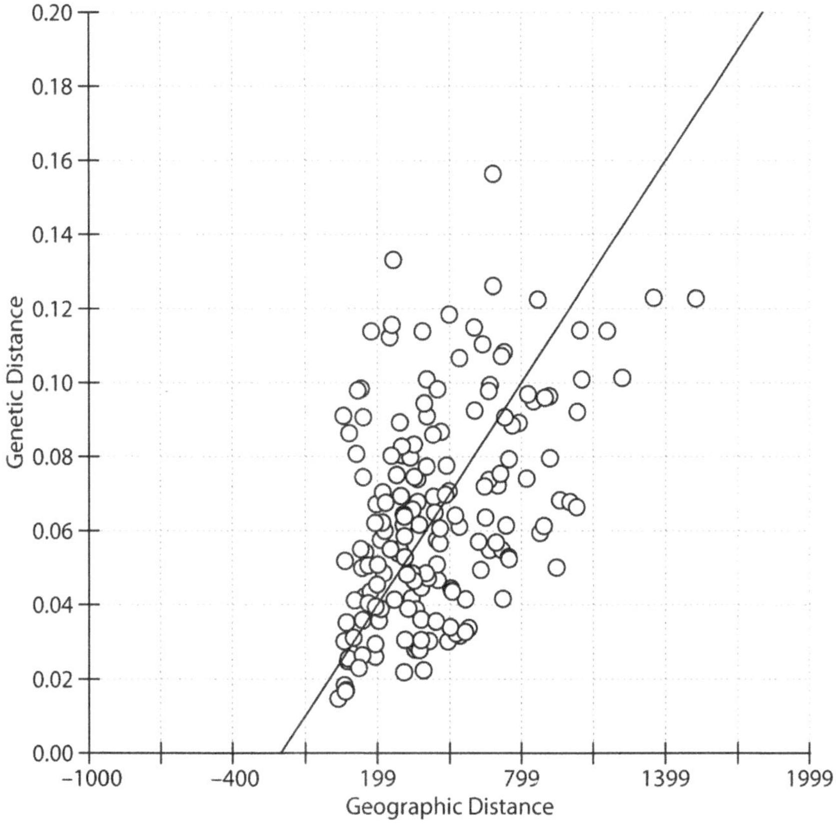

Figure 4. Correlation between genetic distance and geographic distance for Siberian apricot populations. Genetic distance is represented by pairwise $F_{ST}/(1-F_{ST})$ estimates among populations, which is regressed against the geographic distance. The RMA regression line overlays the scatterplot.

(Figure S1). In addition, the clustering analysis showed these pairwise populations with low F_{ST} value clustered into a genetic cluster. All pairwise populations with the lowest genetic differentiation were from the same region (Figure 1), except P13-P14. We did not find variation in the Siberian apricot trees around P13 and P14, indicating that they are isolated populations. The distance from P13 to P14 is about 300 km, which is far enough that the two populations have little chance to exchange genes. The F_{ST} value between P13 and P14 (Table 5) is not significantly different, suggesting a low degree of genetic differentiation between them. It may be that a long time ago human activity severed the continuity of their distribution, but that the later development of a similar environment at the two sampling locations (the eastern edge of the Greater Khingan Mountains and western edge of the Northeast Plain) guided the evolution of the two populations in the same direction. Isolated populations cannot communicate with outside populations, which may increase the chance of inbreeding. The relatively high positive value of F (Table 3) supports this possibility.

STRUCTURE has been successfully used in a large variety of population genetic studies, including in studies of genetic structure, the distinguishing of breeds, and the detection of hybrids between cultivated and wild assortments [53–55]. In general, two models are used to identify the true optimum number of subsets (K) in STRUCTURE. The first model, described by Pritchard et al. [28], is based on the probability Pr(X|K) (called Ln P[D] in STRUCTURE), and the K value that provides the maximum Ln P(D) value is selected as the optimum number of subsets [56]. Evanno et al. [31] found that in many cases the estimated Ln P(D) does not help visualize the correct number of clusters (K). They recommended using an ad hoc statistic, ΔK, based on the rate of change in the log probability of data between successive K values evaluated by STRUCTURE to more accurately detect the real number of clusters [57,58]. However, Vigouroux et al. [59] pointed out that the ΔK method of Evanno et al. [31] always favored K = 2 in the main structure analysis. When large datasets are analyzed, a convergence problem for the Gibbs sampler algorithm used in STRUCTURE may occur [60,61]. Recently, Jacobs et al. [62] grouped populations by maximizing the allocation of genetic diversity among subgroups (i.e., maximizing the F_{ST} values). This provided a new means of identifying the true optimum number of subsets. The result of AMOVA showed that the maximum F_{ST} value (0.06008) when all populations were grouped into four genetic clusters (Table 4).

In this study, STRUCTURE identified four main genetic groups (clusters) (Table S2). All genetic clusters showed high average ancestry values, as compared to their own clusters. The populations from G3 were almost all clustered into genetic cluster C1, while the populations from G1 were clustered into genetic cluster C4 except P6. However, based on the ancestry values of all of the individuals (Table S2), we found that a high number of individuals from P9 (2 individuals), P10 (2 individuals), P11 (2 individuals), P12 (6 individuals), P18 (1 individual), and P19 (5 individuals), belonged to the other genetic clusters, which were clustered in genetic cluster C1. Similar results were also found in the other genetic clusters. Siberian apricot reproduces mainly by seeds from ripe and dehiscent fruits. A natural gene flow over such distance could not be possible, and one putative explanation could be dispersal by some kinds of rodent and human actions. The

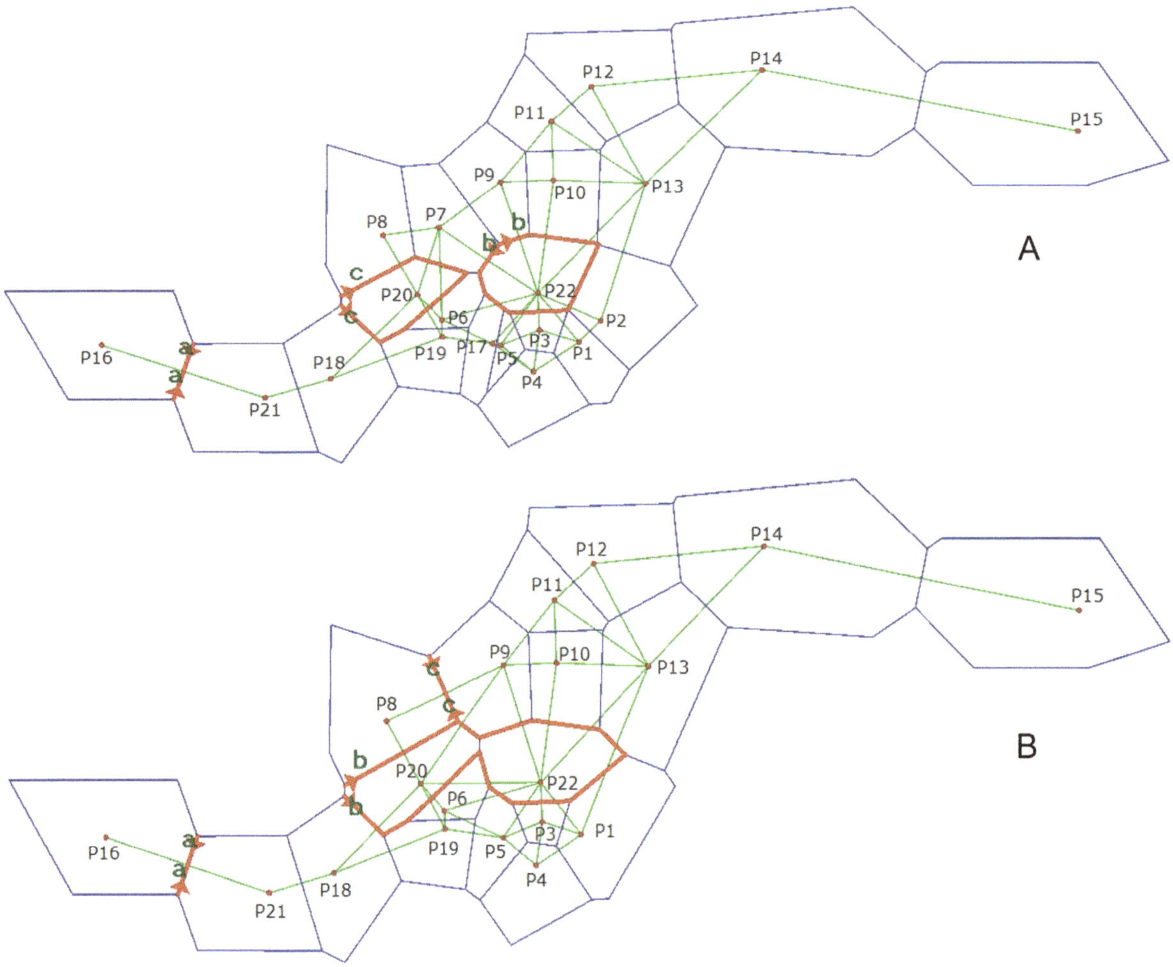

Figure 5. Genetic barriers predicted by BARRIER (version 2.2). The genetic barriers are shown in bold lines with arrows. A: genetic barrier predication using all populations; B: genetic barrier predication using all wild populations.

Korean field mouse (*Apodemus peninsulae*), whitebellied rat (*Niviventer confucianus*), striped field mouse (*Apodemus agrarius*), and other rodents feed and store Siberian apricot seeds, which makes long-distance gene flow possible and improves the level of genetic diversity.

According to the results of our structure analysis, most of the populations that were geographically close were generally clustered into the same cluster. An analysis based on the Mantel test (Figure 4) showed that the genetic distance was not significantly correlated with the geographic distance ($r=0.4651$, $p=0.9940$), suggesting that geographic distance is not the principal factor influencing genetic differentiation in Siberian apricot. The distance between P5 and P17 was <17 km; however, the populations were not clustered into the same cluster when $K>4$. Furthermore, significant genetic differentiation was detected between them (Table 5), suggesting that the seeds from the semi-wild population in P17 was not from the local.

P16 was separated by the first predicted barriers, regardless of whether the three semi-wild populations were included or not (Figure 5). However, the second and third predicted barriers produced different results. If semi-wild populations were excluded, P22 had an exchange with P20 (belonging to genetic cluster C4) while P8 did not exchange with other populations. It is possible that the seed resources of P7 were from P8, because they were from the same genetic cluster when $K=14$, and the seed resources of P17 were from genetic cluster C1 and genetic cluster C3. Most of the populations with low-level genetic diversity (Table 3) were separated from other populations, indicating that the barrier was an important factor influencing genetic diversity. Further investigation into how these genetic barriers are related to geographic or other factors is needed.

Conclusions

Our studies show a relatively high level of genetic diversity among Siberian apricot populations in China. However, a significant deficiency in heterozygotes was detected at the locus and population levels, which may be the result of low-level inbreeding. Our structure analysis clustered all of the populations into four genetic clusters. There was no significant difference between the wild and semi-wild groups, indicating that recent cultivation practices have had little impact on the genetic diversity of Siberian apricot. Our study represents the most comprehensive investigation of the genetic diversity and population structure of Siberian apricot in China and will provide valuable information for the collection of genetic resources for the breeding of Siberian apricot and related species.

Acknowledgments

We wish to thank Aohan Forestry Bureau, Inner Mongolia Autonomous Region and Chengde Forestry Bureau in China for their help in collecting the samples.

Author Contributions

Conceived and designed the experiments: RW XP. Performed the experiments: HL ZW YL JG ZZ. Analyzed the data: ZW MK. Contributed reagents/materials/analysis tools: ZW. Wrote the paper: ZW XP RW.

References

1. Wang LB (2011) Resource Investigation and Distribution Pattern of Three Armeniaca Species. Forest Resources Management 5: 65–70.
2. Gumus M, Kasifoglu S (2010) Performance and emission evaluation of a compression ignition engine using a biodiesel (apricot seed kernel oil methyl ester) and its blends with diesel fuel. Biomass and bioenergy 34: 134–139.
3. Wang LB (2012) Evaluation of Siberian Apricot (*Prunus sibirica* L.) Germplasm Variability for Biodiesel Properties. Journal of the American Oil Chemists' Society 89: 1743–1747.
4. Li XF, Liu MY, Guo XY (2005) Introduction on sustainable management and development of wild Siberian apricot. Inner Mongolia Forestry Investigation and Design 28: 21–22.
5. Wang ZM, Feng LJ, Feng CF, Liu LQ (2001) A diseases and insect survey report of Siberian apricot in Cayouzhongqi. Inner Mongolia Forestry Investigation and Design 24: 35–36.
6. Ali M, Rajewski J, Baenziger P, Gill K, Eskridge K, et al. (2008) Assessment of genetic diversity and relationship among a collection of US sweet sorghum germplasm by SSR markers. Molecular Breeding 21: 497–509.
7. Bovine HapMap C, Gibbs RA, Taylor JF, Van Tassell CP, Barendse W, et al. (2009) Genome-wide survey of SNP variation uncovers the genetic structure of cattle breeds. Science 324: 528–532.
8. Hagen S, Khadari B, Lambert P, Audergon JM (2002) Genetic diversity in apricot revealed by AFLP markers: species and cultivar comparisons. Theor Appl Genet 105: 298–305.
9. Sánchez-Pérez R, Martínez-Gómez P, Dicenta F, Egea J, Ruiz D (2006) Level and transmission of genetic heterozygosity in apricot (*Prunus armeniaca* L.) explored using simple sequence repeat markers. Genetic Resources and Crop Evolution 53: 763–770.
10. Vicente Md, Truco M, Egea J, Burgos L, Arús P (1998) RFLP variability in apricot (*Prunus armeniaca* L.). Plant Breeding 117: 153–158.
11. Kijas J, Fowler J, Thomas M (1995) An evaluation of sequence tagged microsatellite site markers for genetic analysis within *Citrus* and related species. Genome 38: 349–355.
12. Du Q, Wang B, Wei Z, Zhang D, Li B (2012) Genetic diversity and population structure of Chinese White poplar (*Populus tomentosa*) revealed by SSR markers. J Hered 103: 853–862.
13. Hormaza JI (2002) Molecular characterization and similarity relationships among apricot (*Prunus armeniaca* L.) genotypes using simple sequence repeats. Theor Appl Genet 104: 321–328.
14. Zhebentyayeva TN, Reighard GL, Gorina VM, Abbott AG (2003) Simple sequence repeat (SSR) analysis for assessment of genetic variability in apricot germplasm. Theor Appl Genet 106: 435–444.
15. Liu HB, Wang Z, Liu J, Ma LY, Wang SQ, et al. (2012) Genetic Diversity and Genetic Structure of Siberian Apricot Populations in the Yan Mountains. Scientia Silvae Sinicae 48: 68–74.
16. Liu HB, Liu J, Wang Z, Ma LY, Wang SQ, et al. (2013) Development and Characterization of Microsatellite Markers in *Prunus sibirica* (*Rosaceae*). Applications in Plant Sciences 1: 1200074.
17. Doyle JJ (1987) A rapid DNA isolation procedure for small quantities of fresh leaf tissue. Phytochem bull 19: 11–15.
18. Wang Z, Liu HB, Liu J, Li YY, Wu RL, et al. (2014) Mining new microsatellite markers for Siberian apricot (Prunus sibirica L.) from SSR-enriched genomic library. Scientia Horticulturae 166: 65–69.
19. Lopes M, Sefc K, Laimer M, Da Câmara Machado A (2002) Identification of microsatellite loci in apricot. Molecular Ecology Notes 2: 24–26.
20. Aranzana M, Garcia-Mas J, Carbo J, Arús P (2002) Development and variability analysis of microsatellite markers in peach. Plant Breeding 121: 87–92.
21. Dirlewanger E, Cosson P, Tavaud M, Aranzana J, Poizat C, et al. (2002) Development of microsatellite markers in peach [Prunus persica (L.) Batsch] and their use in genetic diversity analysis in peach and sweet cherry (Prunus avium L.). Theor Appl Genet 105: 127–138.
22. Testolin R, Marrazzo T, Cipriani G, Quarta R, Verde I, et al. (2000) Microsatellite DNA in peach (*Prunus persica* L. Batsch) and its use in fingerprinting and testing the genetic origin of cultivars. Genome 43: 512–520.
23. Chen Q, Zhang L, Yuan Z, Yan Z, Zheng Y, et al. (2008) Empirical verification of heterogeneous DNA fragments generated from wheat genome-specific SSR primers. Canadian Journal of Plant Science 88: 1065–1071.
24. Amos W, Hoffman J, Frodsham A, Zhang L, Best S, et al. (2007) Automated binning of microsatellite alleles: problems and solutions. Molecular Ecology Notes 7: 10–14.
25. Park SDE (2001) Trypanotolerance in West African cattle and the population genetic effects of selection. Ph D thesis, University of Dublin.
26. Peakall R, Smouse PE (2006) Genalex 6: genetic analysis in Excel. Population genetic software for teaching and research. Molecular Ecology Notes 6: 288–295.
27. Nei M (1973) Analysis of gene diversity in subdivided populations. Proc Natl Acad Sci U S A 70: 3321–3323.
28. Pritchard JK, Stephens M, Donnelly P (2000) Inference of population structure using multilocus genotype data. Genetics 155: 945–959.
29. Pritchard JK, Donnelly P (2001) Case-control studies of association in structured or admixed populations. Theor Popul Biol 60: 227–237.
30. Hubisz MJ, Falush D, Stephens M, Pritchard JK (2009) Inferring weak population structure with the assistance of sample group information. Mol Ecol Resour 9: 1322–1332.
31. Evanno G, Regnaut S, Goudet J (2005) Detecting the number of clusters of individuals using the software Structure: a simulation study. Mol Ecol 14: 2611–2620.
32. Hamrick J, Godt M (1996) Effects of life history traits on genetic diversity in plant species. Philosophical Transactions of the Royal Society of London Series B: Biological Sciences 351: 1291–1298.
33. Excoffier L, Lischer HE (2010) Arlequin suite ver 3.5: a new series of programs to perform population genetics analyses under Linux and Windows. Mol Ecol Resour 10: 564–567.
34. Guo SW, Thompson EA (1992) Performing the exact test of Hardy-Weinberg proportion for multiple alleles. Biometrics 48: 361–372.
35. Goudet J (2001) FSTAT, a program to estimate and test gene diversities and fixation indices (version 2.9. 3).
36. Rousset F (1997) Genetic differentiation and estimation of gene flow from F-statistics under isolation by distance. Genetics 145: 1219–1228.
37. Jensen JL, Bohonak AJ, Kelley ST (2005) Isolation by distance, web service. BMC Genet 6: 13.
38. Mantel N (1967) Ranking procedures for arbitrarily restricted observation. Biometrics 23: 65–78.
39. Manni F, Guerard E, Heyer E (2004) Geographic patterns of (genetic, morphologic, linguistic) variation: how barriers can be detected by using Monmonier's algorithm. Hum Biol 76: 173–190.
40. Slatkin M, Barton NH (1989) A comparison of three indirect methods for estimating average levels of gene flow. Evolution: 1349–1368.
41. Donoso J, Aros D, Meneses C, Narváez C, Infante R, et al. (2008) Genetic relationships in apricot (*Prunus armeniaca* L.) using SSR markers and their implications for breeding. Journal of Food Agriculture & Environment 6: 378–382.
42. Tahan O, Geng Y, Zeng L, Dong S, Chen F, et al. (2009) Assessment of genetic diversity and population structure of Chinese wild almond, *Amygdalus nana*, using EST-and genomic SSRs. Biochemical Systematics and Ecology 37: 146–153.
43. Ferrer MM, Eguiarte LE, Montana C (2004) Genetic structure and outcrossing rates in *Flourensia cernua* (*Asteraceae*) growing at different densities in the South-western Chihuahuan Desert. Ann Bot 94: 419–426.
44. Chung MY, Nason JD, Epperson BK, Chung MG (2003) Temporal aspects of the fine-scale genetic structure in a population of *Cinnamomum insularimontanum* (*Lauraceae*). Heredity (Edinb) 90: 98–106.
45. Veron V, Caron H, Degen B (2005) Gene flow and mating system of the tropical tree *Sextonia rubra*. Silvae genetica 54: 275–280.

46. Hadziabdic D, Wang X, Wadl PA, Rinehart TA, Ownley BH, et al. (2012) Genetic diversity of flowering dogwood in the Great Smoky Mountains National Park. Tree Genetics & Genomes 8: 855–871.
47. Slatkin M (1995) A measure of population subdivision based on microsatellite allele frequencies. Genetics 139: 457–462.
48. Estoup A, Garnery L, Solignac M, Cornuet JM (1995) Microsatellite variation in honey bee (*Apis mellifera* L.) populations: hierarchical genetic structure and test of the infinite allele and stepwise mutation models. Genetics 140: 679–695.
49. Goodman SJ (1998) Patterns of extensive genetic differentiation and variation among European harbor seals (*Phoca vitulina vitulina*) revealed using microsatellite DNA polymorphisms. Mol Biol Evol 15: 104–118.
50. Ross KG, Krieger MJ, Shoemaker DD, Vargo EL, Keller L (1997) Hierarchical analysis of genetic structure in native fire ant populations: results from three classes of molecular markers. Genetics 147: 643–655.
51. Balloux F, Brunner H, Lugon-Moulin N, Hausser J, Goudet J (2000) Microsatellites can be misleading: an empirical and simulation study. Evolution 54: 1414–1422.
52. De Andres MT, Benito A, Perez-Rivera G, Ocete R, Lopez MA, et al. (2012) Genetic diversity of wild grapevine populations in Spain and their genetic relationships with cultivated grapevines. Mol Ecol 21: 800–816.
53. Pineiro R, Fuertes Aguilar J, Munt DD, Nieto Feliner G (2007) Ecology matters: Atlantic-Mediterranean disjunction in the sand-dune shrub Armeria pungens (*Plumbaginaceae*). Mol Ecol 16: 2155–2171.
54. Coart E, Vekemans X, Smulders MJ, Wagner I, Van Huylenbroeck J, et al. (2003) Genetic variation in the endangered wild apple (*Malus sylvestris* (L.) Mill.) in Belgium as revealed by amplified fragment length polymorphism and microsatellite markers. Mol Ecol 12: 845–857.
55. Koopman WJ, Li Y, Coart E,van de Weg WE, Vosman B, et al. (2007) Linked vs. unlinked markers: multilocus microsatellite haplotype-sharing as a tool to estimate gene flow and introgression. Molecular ecology 16: 243–256.
56. López-Gartner G, Cortina H, McCouch SR, Moncada MDP (2009) Analysis of genetic structure in a sample of coffee (*Coffea arabica* L.) using fluorescent SSR markers. Tree Genetics & Genomes 5: 435–446.
57. Krutovsky KV, Clair JBS, Saich R, Hipkins VD, Neale DB (2009) Estimation of population structure in coastal Douglas-fir [*Pseudotsuga menziesii* (Mirb.) Franco var. *menziesii*] using allozyme and microsatellite markers. Tree Genetics & Genomes 5: 641–658.
58. D'Hoop BB, Paulo MJ, Kowitwanich K, Sengers M, Visser RG, et al. (2010) Population structure and linkage disequilibrium unravelled in tetraploid potato. Theor Appl Genet 121: 1151–1170.
59. Vigouroux Y, Glaubitz JC, Matsuoka Y, Goodman MM, Sanchez GJ, et al. (2008) Population structure and genetic diversity of New World maize races assessed by DNA microsatellites. Am J Bot 95: 1240–1253.
60. Rosenberg NA, Pritchard JK, Weber JL, Cann HM, Kidd KK, et al. (2002) Genetic structure of human populations. Science 298: 2381–2385.
61. Corander J, Marttinen P, Siren J, Tang J (2008) Enhanced Bayesian modelling in BAPS software for learning genetic structures of populations. BMC Bioinformatics 9: 539.
62. Jacobs MM, Smulders MJ, van den Berg RG, Vosman B (2011) What's in a name; Genetic structure in *Solanum* section *Petota* studied using population-genetic tools. BMC evolutionary biology 11: 42.

11

Hidden Diversity in Sardines: Genetic and Morphological Evidence for Cryptic Species in the Goldstripe Sardinella, *Sardinella gibbosa* (Bleeker, 1849)

Rey C. Thomas Jr.[1]*, Demian A. Willette[2], Kent E. Carpenter[3], Mudjekeewis D. Santos[1]

1 Genetic Fingerprinting Laboratory, National Fisheries Research and Development Institute, Quezon City, Metro Manila, Philippines, **2** Department of Ecology and Evolutionary Biology, University of California Los Angeles, Los Angeles, California, United States of America, **3** Department of Biological Sciences, Old Dominion University, Norfolk, Virginia, United States of America

Abstract

Cryptic species continue to be uncovered in many fish taxa, posing challenges for fisheries conservation and management. In *Sardinella gibbosa*, previous investigations revealed subtle intra-species variations, resulting in numerous synonyms and a controversial taxonomy for this sardine. Here, we tested for cryptic diversity within *S. gibbosa* using genetic data from two mitochondrial and one nuclear gene regions of 248 individuals of *S. gibbosa*, collected from eight locations across the Philippine archipelago. Deep genetic divergence and subsequent clustering was consistent across both mitochondrial and nuclear markers. Clade distribution is geographically limited: Clade 1 is widely distributed in the central Philippines, while Clade 2 is limited to the northernmost sampling site. In addition, morphometric analyses revealed a unique head shape that characterized each genetic clade. Hence, both genetic and morphological evidence strongly suggests a hidden diversity within this common and commercially-important sardine.

Editor: Erik Sotka, College of Charleston, United States of America

Funding: This project was funded mainly through a National Fisheries Research and Development Institute grant to Mudjekeewis Santos, Demian Willette and Eunice Bognot. This research was partly supported by a grant from the National Science Foundation (NSF OISE-0730256) to Kent Carpenter and Paul Barber. The funders had no role in study design, data collection and analysis, decision to publish, or preparation of the manuscript.

Competing Interests: The authors have declared that no competing interests exist.

* E-mail: reycthomas@gmail.com

Introduction

Accurately defining the limits of a species is essential to studying its biology, ecology, conservation, and management [1–3]. Species are typically categorized according to gross morphology [4]. However, species diversity can be masked by a lack of obvious morphological differences between cryptic species [5]. Cryptic species are morphologically similar but genetically distinct lineages, and are often overlooked by species identification using gross morphology alone [6]. Genetic differentiation can be used to distinguish morphologically similar lineages [7–9], in which the genotypic-clustering species definition is utilized to identify cryptic species [10–12]. Cryptic species are widely distributed across different taxa and geographies in the marine realm [13–14], occurring either in allopatry [15,16] or sympatry [8,17] as sister species [16], or a result of convergent evolution [18]. In the Indo-West Pacific, cryptic species of widely distributed reef fishes contribute significantly to overall marine biodiversity [19]. Thus, inaccurate species delimitation would overlook cryptic species and underestimate biodiversity, a fact that can lead to flawed conservation and management strategies [5,20].

Marine small pelagic fishes comprise the majority of world's landed fish catch [21], and members of the family Clupeidae contribute significantly to this volume [22]. The Clupeidae reach their highest diversity in the Indo-West Pacific [23], a region also proposed to be the geographic origin of this family [24]. Within the Indo-West Pacific, the goldstripe sardinella *Sardinella gibbosa* (Bleeker, 1849) is among the most abundant and widespread marine pelagic species. It has a distribution that extends from the East African coast to Taiwan, the Philippines, Indonesia and Northern Australia [23]. Historically, it is among the most abundant and commercially important species in the Indo-West Pacific sardine fishery [25,26]. In particular, the goldstripe sardinella is the second most abundant sardine occurring in the Philippine archipelago [27]. Migratory patterns of *S. gibbosa* have been correlated with the availability and seasonality of planktonic prey in the environment [26,28,29]. Biological data from the coast of India suggests a peak spawning period that lasts from early March towards the end of May [26,30]. However, two distinct length and age groups have been observed [31], and variations in the number of scale striations have been observed in *S. gibbosa* off the South African coast [32]. Morphological classification of this sardine has been complicated by subtle intra-species variations, leading to several recorded synonyms for *S. gibbosa* [23]. Type specimens, now considered synonyms of *S. gibbosa*, were previously known as *Clupea immaculata* (Southern Japan and China) [33], *Fimbriclupea dactyolepis* (Northwest Australia) [34], and *Sardinella taiwanensis* (Taiwan) [35]. Such subtle biological and morphological differences documented in *S. gibbosa* may hint of hidden diversity within the sardine.

Table 1. Sampling information for *S. gibbosa.*

Location	*N*	Collection date
Sta. Ana, Cagayan (CAG)	50	June, 2012
Atimonan, Quezon (QUE)	30	July, 2011
Balayan Bay, Batangas (BAT)	22	November, 2011
Manila Bay, Manila (MNL)	30	May, 2012
Tacloban, Leyte (LEY)	30	July, 2011
Ilo-Ilo City, Ilo-Ilo (ILO)	30	May, 2011
Banate, Ilo-Ilo (BAN)	32	April, 2013
Puerto Princesa, Palawan (PAL)	24	December, 2012
Kudat, Sabah (KUD)	21	March, 2011
Yilan County, Taiwan (YIL)	1	April, 2011
Nha Trang, Vietnam (NTR)	26	October, 2011
Phu Quouc, Vietnam (PQU)	26	October, 2011
Songkla, Thailand (SON)	4	October, 2011
Surat Thani, Thailand (SUR)	17	October, 2011
Koh Samui, Thailand (KOH)	8	October, 2011
Trang Province, Thailand (TRA)	27	October, 2011
Total	378	

The objective of this work has been to explore the possible cryptic species within the goldstripe sardinella in the Philippine archipelago by examining molecular and morphometric data. We investigated the occurrence of cryptic species using genotypic clustering for both mitochondrial and nuclear markers. We also used morphometric variations to test for subtle morphological differentiation between the genetically-partitioned groups. Both genetic and morphological evidence provide strong support for an unexpected multiple-species complex within the common and commercially-important sardine *S. gibbosa* in the Philippine archipelago.

Materials and Methods

Sampling

A total of 378 individuals of *S. gibbosa* were collected from sixteen fish markets across the Philippines, Taiwan, Malaysia, Vietnam and Thailand (Table 1). Body coloration, morphometric, and meristic characters were recorded for frozen then thawed samples. Tissue samples and voucher specimens preserved in absolute ethanol were stored at the National Fisheries Research and Development Institute, Quezon City, Philippines.

PCR amplification

Genomic DNA was extracted from muscle tissue samples using either modified Chelex® (Bio-Rad, Hercules, CA) DNA extraction protocol [36], or salting-out method [37]. Approximately 540 bp of the ribosomal 16S gene region are amplified by polymerase chain reaction (PCR) using the primers 16Sar (5′-CGCCTGTTTATCAAAAACAT-3′) and 16Sbr (5′-CCGGTCTGAACTCAGATCACGT-3′) [38]. Additional mitochondrial and nuclear DNA sequences were obtained only for Philippine collections due to mounting sequencing costs and limited time available for this study. The primers CRA (5′-TTCCACCTCTAACTCCCAAAGCTAG-3′) and CRE (5′-CCTGAAGTAGGAACCAGATG-3′) were used to amplify 560 bp of mitochondrial DNA control region [39].

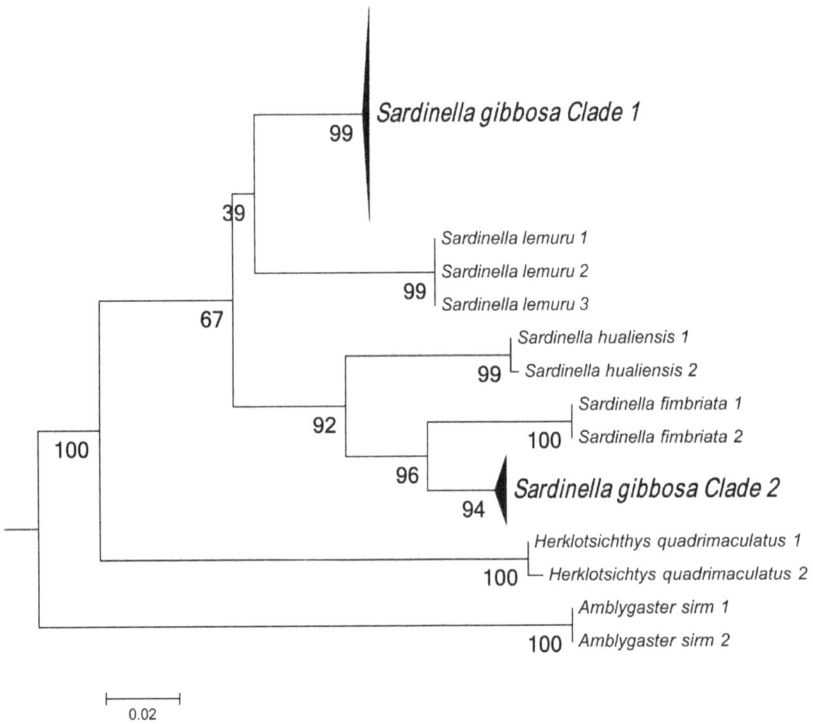

Figure 1. Maximum-likelihood (ML) tree inferred using 16S rRNA sequences. Bootstrap support values were calculated using 1,000 replicates. Sequences of *S. hualiensis* obtained from GenBank were JN580490.1 and JN580479.1.

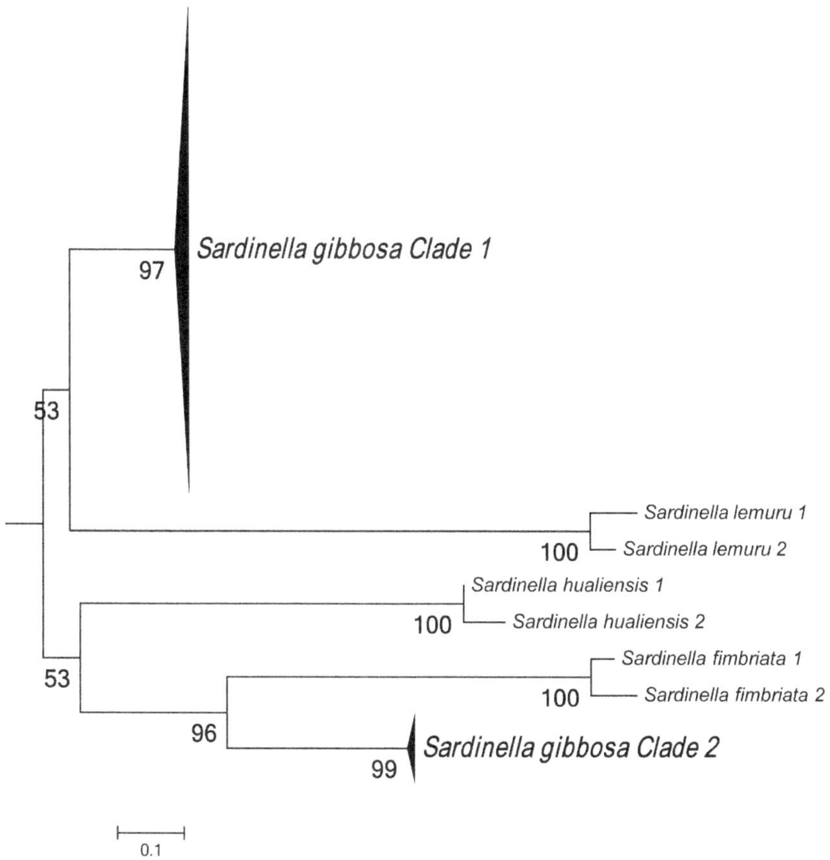

Figure 2. Maximum-likelihood (ML) tree generated from mitochondrial control region sequences. Bootstrap support values were calculated using 1,000 replicates.

Nuclear DNA was obtained using the primers S7RPEX1F (5′-TGGCCTCTTCCTTGGCCGTC-3′) and S7RPEX2R (5′-AACTCGTCTGGCTTTTCGCC-3′) to amplify 700 bp of the 1st intron of the ribosomal S7 gene [40]. The total volume of the reaction mixture to amplify mtDNA gene regions is 25 µl; consisting of 13.5 µl of nuclease-free water, 2.5 µl of 10x PCR buffer, 2.5 µl of 10 mM dNTP, 2.0 µl of 25 mM MgCl2, 1.0 µl of 5x BSA, 1.25 µl of 10 mM of both primers, and 0.125 µl of *Taq*

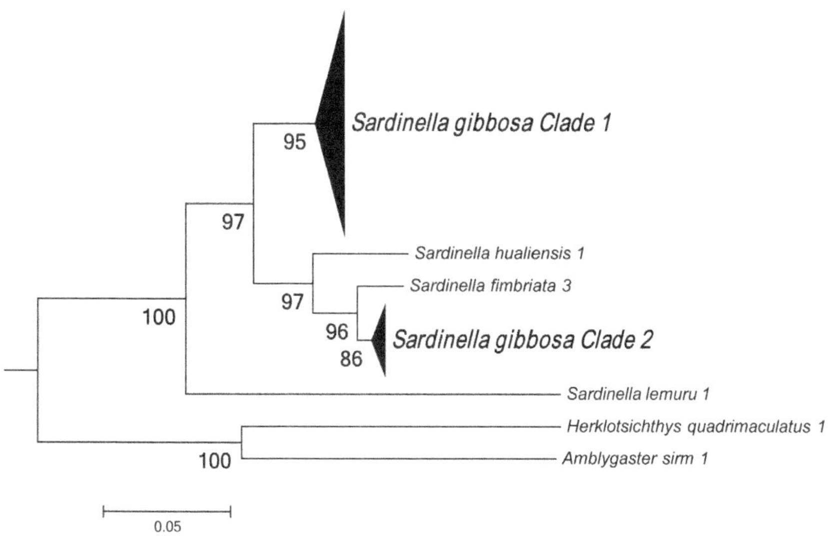

Figure 3. Maximum-likelihood (ML) tree generate from the 1st intron of ribosomal S7 sequences. Bootstrap support values were calculated using 1,000 replicates.

Table 2. Pairwise genetic distances calculated for each taxa.

16S

Species	1	2	3	4	5	6	7
Sardinella gibbosa Clade 1		0.03	0.05	0.04	0.03	0.08	0.07
Sardinella gibbosa Clade 2	0.13		0.02	0.03	0.04	0.09	0.10
Sardinella fimbriata	0.22	0.08		0.04	0.07	0.13	0.11
Sardinella hualiensis	0.15	0.11	0.16		0.04	0.14	0.14
Sardinella lemuru	0.11	0.18	0.27	0.16		0.09	0.06
Amblygaster sirm	0.33	0.36	0.50	0.50	0.37		0.12
Herklotsichthys quadrimaculatus	0.30	0.37	0.39	0.47	0.25	0.47	

Control region

Species	1	2	3	4	5
Sardinella gibbosa Clade 1		0.07	0.08	0.07	0.06
Sardinella gibbosa Clade 2	0.40		0.09	0.12	0.10
Sardinella fimbriata	0.52	0.52		0.20	0.12
Sardinella hualiensis	0.43	0.64	1.00		0.10
Sardinella lemuru	0.40	0.61	0.75	0.66	

S7 intron

Species	1	2	3	4	5	6	7
Sardinella gibbosa Clade 1		0.02	0.03	0.10	0.03	0.13	0.24
Sardinella gibbosa Clade 2	0.10		0.01	0.09	0.02	0.15	0.26
Sardinella fimbriata	0.10	0.03		0.11	0.02	0.17	0.27
Sardinella hualiensis	0.11	0.07	0.08		0.09	0.27	0.53
Sardinella lemuru	0.34	0.35	0.38	0.37		0.16	0.48
Amblygaster sirm	0.51	0.57	0.63	0.80	0.60		0.09
Herklotsichthys quadrimaculatus	0.62	0.72	0.77	1.00	0.95	0.41	

The standard error of mean is shown in the upper right diagonals.

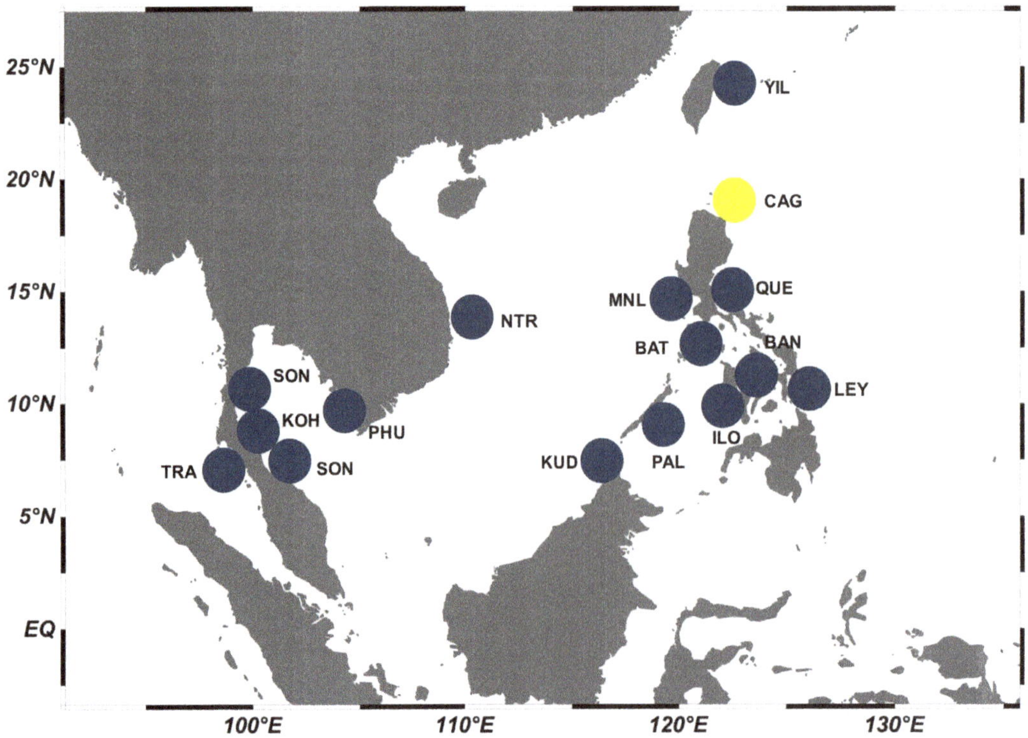

Figure 4. Sampling sites and clade distribution for *S. gibbosa*. Each color represents the cryptic clades. Clade 1 is represented in blue, clade 2 is shaded in yellow.

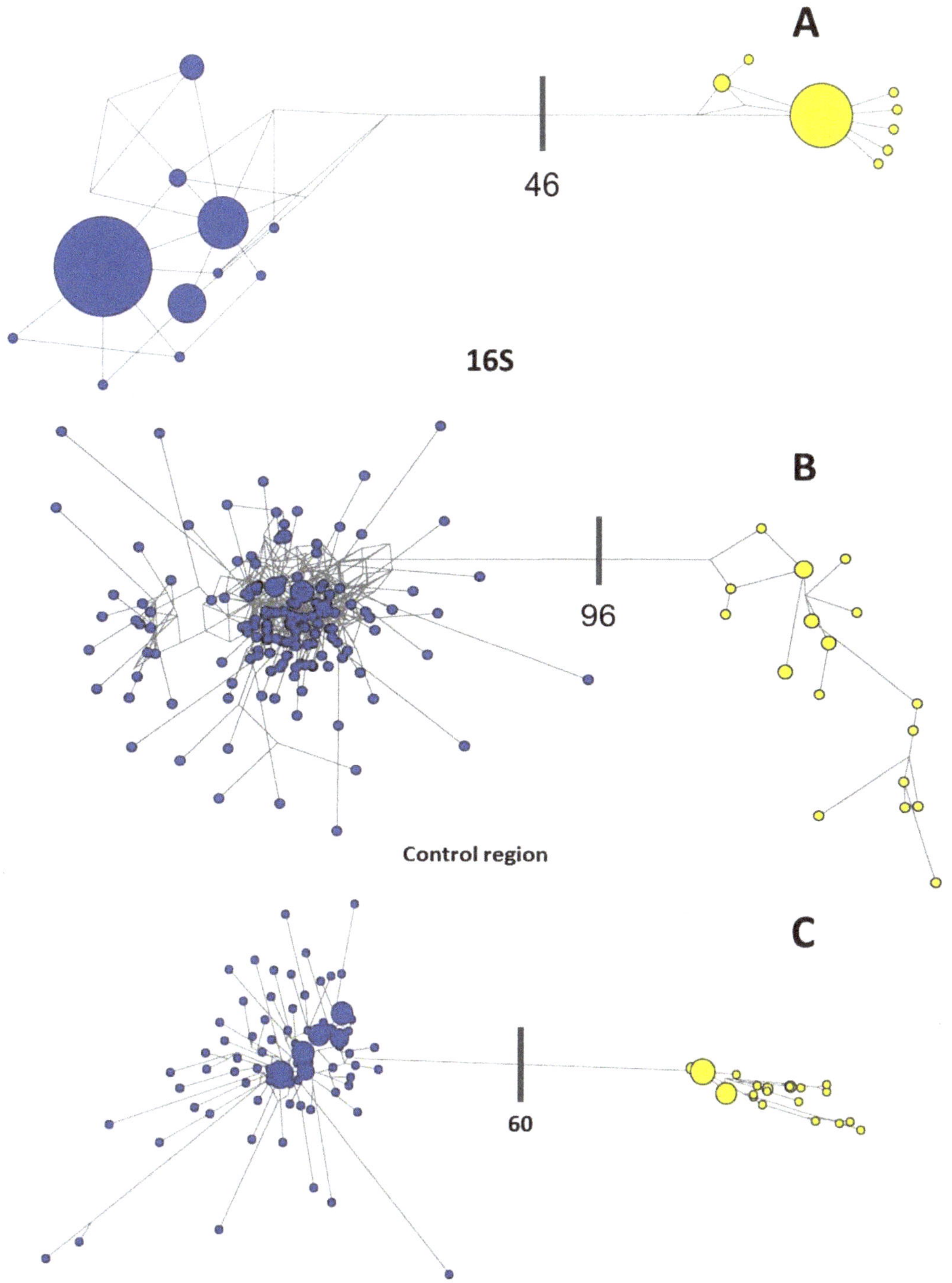

Figure 5. Median-joining network inferred for 16S (A) and control region (B) and S7. Each pie represents a haplotype proportional to its frequency. Thick bars represents step changes between Clade 1 (blue) and Clade 2 (yellow) which is labeled according to the number of base-pair difference.

DNA polymerase. Thermal cycling conditions consisted of an initial denaturation of 94°C for 10 min followed by 38 cycles of DNA denaturing at 94°C for 30 s, primer annealing at 45°C for 45 s, and sequence extension at 72°C for 45 s, ending with a final extension of 72°C for 10 min. For the ribosomal S7 intron, we utilized PCR conditions as previously described [40]. Successful PCR products were purified using ExoSAP-IT® (USB Corp, Cleveland, OH). The reaction mixture consisted of 2 µl of ExoSAP-IT and 22 µl of PCR product, and eventually incubated at 37°C for 15 min followed by another 15 min at 80°C to inactivate the enzyme. Purified PCR products were sent to either Macrogen, Inc. Korea or UC-Berkeley for DNA sequencing.

Sequence data was deposited on the public domain database GenBank [Accession numbers pending].

Phylogenetic reconstruction

Sequences were assembled in Geneious v5.4 [41] and aligned using MUSCLE v3.8.31 [42]. A best-fit nucleotide substitution model was determined using jMODELTEST v2 [43,44]. Phylogenetic analysis using maximum likelihood (ML) criteria was inferred from MEGA v5.2.1 [45] using the best-fit nucleotide substitution models, namely, Kimura-2-Parameter (K2P) for 16S, three parameter model (TPM) for control region and Hasegawa-Kishino-Yano (HKY) for the S7 intron. Also included in the analysis for outgroup comparison were the closely related taxa, namely, *Sardinella fimbriata, S. hualiensis, S. lemuru, Herkoltsichthys quadrimaculatus* and *Amblygaster sirm* sequences. Further, such species have overlapping geographic distribution with *S. gibbosa* throughout the Indo-West Pacific [23]. However, we excluded *Amblygaster sirm* and *Herklotsichthys quadrimaculatus* as outgroups for control region dataset since they are highly divergent and inclusion of these taxa created large indels in sequence alignment. Allelic state of the nuclear S7 intron was estimated using PHASE v2.1 [46,47] as implemented in DnaSP v5.0 [48]. The phylogenetic network was inferred using the median-joining network implemented in NETWORK v4.6 [49] using the default settings.

Morphological analysis

To complement genetic data, variability within *S. gibbosa* from 10 individuals per site was quantified by morphometric measurements representing the head shape. Measurements (in mm) obtained using a Vernier caliper were standard length, snout length (tip of snout to eye), head length (tip of snout to edge of operculum), eye diameter (horizontal diameter), upper jaw length, and post-orbital length (right edge of eye to end of operculum). All measurements were converted into ratios to represent proportion with respect to standard length. A principal component analysis implemented in PC-ORD v4.10 [50] was performed on natural log-transformed ratios which separated morphological variations into linear combinations of variables that describe overall head shape. In addition, analysis of similarities (ANOSIM) and similarity of percentage analysis (SIMPER) were conducted on log-transformed morphometric ratios in Primer v5.2.4 [51] to determine the percentage contribution of morphometric ratios to the overall variations in head shape.

Results

Maximum-likelihood analysis of 16S rRNA sequences support the existence of two species within *S. gibbosa* (Figure 1). In concordance with 16S data, mitochondrial control region sequences revealed similar clustering (Figure 2). Clustering for both markers exhibited monophyletic clades with high bootstrap support. In addition, nuclear DNA sequences of the first intron of S7 gene revealed a deep divergence between Clade 1 and Clade 2 (Figure 3). None of the phylogenetic analyses indicated that the two morphotypes initially identified as *S. gibbosa* are sister species. Consistent across examined gene regions, genetic distances calculated for both mitochondrial and nuclear gene regions exhibited divergence comparable to species-level differentiation (Table 2). Clade 1 is broadly distributed across the collection sites except at the northernmost locations (Figure 4). In contrast, Clade 2 is geographically restricted to this one northernmost site in Cagayan Province. Further, the single sample from Yilan County, Taiwan did not cluster with Clade 2, despite the site's close proximity with Cagayan Province. However, the current dataset for Taiwan, Vietnam, Thailand and Malaysia are only limited to the mitochondrial 16S gene region. Nevertheless, median joining network for all three markers, at least for Philippine sites, revealed numerous base-pair mutations between Clades 1 and 2 (Figure 5).

All specimens exhibited the diagnostic characters for *S. gibbosa*, including the dark spot at dorsal fin origin. However, head shape and pigmentation of both lower and upper jaws differ between the two clades identified using genetic markers (Figure 6). Principal component analysis (PCA) revealed strong morphometric differentiation in head dimensions (Figure 7). The first four principal components (PC) account for 95.42% of overall variance (PC1 – 48.74%; PC2 – 21.27%; PC3 – 14.94%; PC4 - 10.79%) (Table 3). PC1 was highly correlated with variance in upper jaw length, eye diameter, and post-orbital length, respectively. On the other hand, PC's 2 through 4 were associated with differences in the ratios of head length, upper jaw, and eye diameter. In concordance with the genetic clades, a surprisingly similar clustering was observed in plots of the principal components (Figure 7). Scatter plots for PC1 and PC3 separated collections from Quezon Province into a distinct cluster. Such grouping might indicate a unique sub-population or subspecies in Clade 1. A shorter snout and upper jaw with respect to head length in individuals from the four clades accounted for such clustering in principal component analysis.

Multivariate analysis of morphometric ratios using ANOSIM showed significant variations between head shape of the three

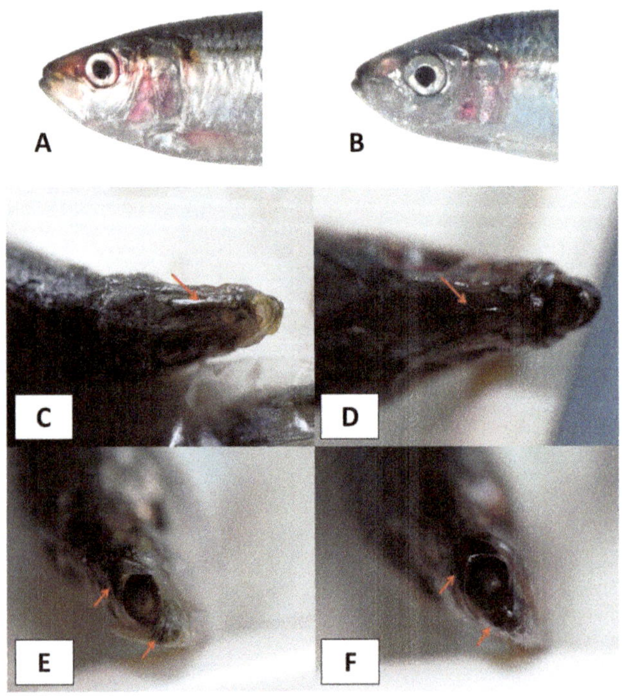

Figure 6. Photos of *Sardinella gibbosa* representing Clade 1 (A) and Clade 2 (B). Between-clade color differences in median frontal line of the head (C–D) and pigmentation of the upper and lower jaw (E–F).

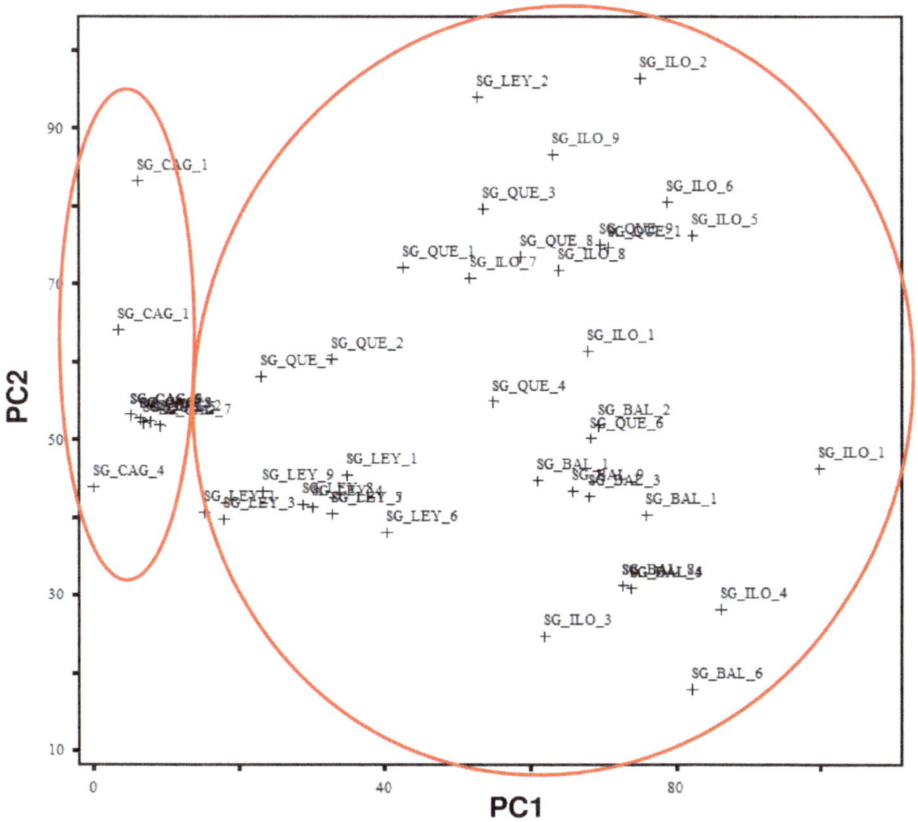
PC2
PC1
90
70
50
30
10
0
40
80
SG_LEY_2
SG_ILO_2
SG_ILO_9
SG_CAG_1
SG_QUE_3
SG_ILO_6
SG_ILO_5
SG_QUE_1
SG_QUE_8
SG_ILO_7
SG_ILO_8
SG_ILO_1
SG_QUE_2
SG_QUE_4
SG_BAL_2
SG_QUE_6
SG_CAG_4
SG_LEY_1
SG_LEY_9
SG_BAL_3
SG_BAL_1
SG_LEY_5
SG_LEY_6
SG_ILO_4
SG_ILO_3
SG_BAL_6

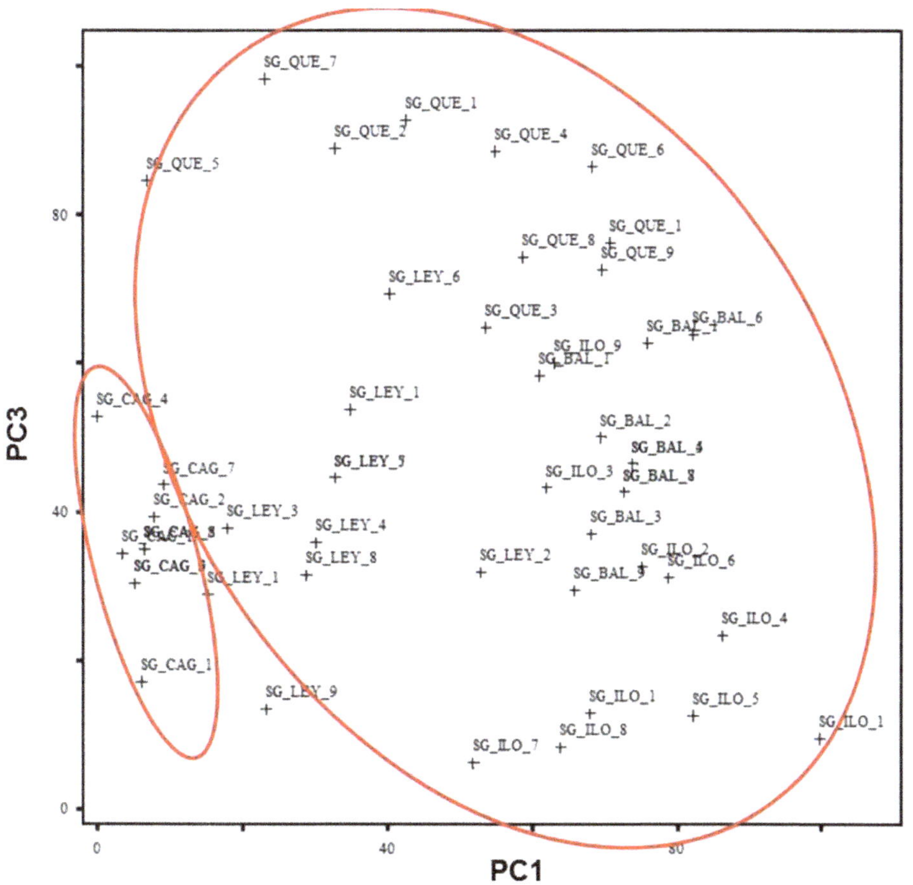
PC3
PC1
80
40
0
0
40
SG_QUE_7
SG_QUE_1
SG_QUE_4
SG_QUE_6
SG_QUE_5
SG_QUE_8
SG_QUE_9
SG_LEY_6
SG_QUE_3
SG_BAL_6
SG_ILO_9
SG_CAG_4
SG_LEY_1
SG_BAL_2
SG_CAG_7
SG_LEY_5
SG_ILO_3
SG_BAL_8
SG_CAG_2
SG_LEY_3
SG_LEY_4
SG_BAL_3
SG_LEY_8
SG_LEY_2
SG_ILO_6
SG_ILO_4
SG_CAG_1
SG_LEY_9
SG_ILO_1
SG_ILO_5
SG_ILO_8
SG_ILO_7

Figure 7. Plots of the first versus the second (A) and the first versus the third (B) principal components (PC) of 60 nominal specimens (n = 10 per population).

genotypic clades (R = 0.486; p<0.01). Differences in snout, post-orbital, and head lengths distinguished the two clades of *S. gibbosa* (Table 4). Strong differentiation in head and upper jaw length accounted for 52.26% of variation between Clade 1 and Clade 2. On the other hand, variations in eye, snout, and post-orbital length contributed 47.74% of overall difference between the two clades. Clade 1 had shorter heads with respect to standard length. Lastly, Clade 2 individuals had the lowest post-orbital length, and subsequently a shorter operculum.

In addition to morphometric difference, Clade 2 has a distinct black pigmentation on the edge of mouth and frontal line between the nostrils (Figure 6). Similar blackish coloration has been observed in the caudal fins of Clade 2. Further, a leaner body characterized Clade 2 in contrast with the more rounded shape of Clade 1 (data not shown). Lastly, Clade 2 lacked the gold stripe across the lateral body wall which characterized Clade 1 (Figure 6).

Discussion

Many discoveries of cryptic species have been based on prior observations of subtle behavioral, biological, or morphological intra-species variations [5]. However, phenotypic differentiation does not necessarily complement genotypic divergence [9], as evident in the lack of congruence between genetics and diagnostic morphological characters [6,10,52]. In extreme cases, morphological variations are randomly shared among genetically distinct lineages within a cryptic species complex [11,53]. To avoid the inconsistency between genetics and morphology, the straightforward approach is to identify cryptic species using multi-locus genetic data [54,55], a method that can help avoid the pitfalls of morphological species delimitation [12]. Cryptic species identified through genetic clustering can then be bolstered by support from additional morphological or biological traits.

In *S. gibbosa*, molecular evidence from both mitochondrial and nuclear DNA strongly supports two cryptic species. Clade 1 is widely distributed throughout the Philippine islands, while Clade 2 is geographically restricted to the Cagayan Province. Such allopatric distribution has been reported to occur in other cryptic species [15,16]. In addition, the two clades exhibited the same clustering for all markers (Figures 1–3), a finding consistent with the 'genotypic-clustering' species definition [12], and substantiated by agreement between multi-locus genotypic data [56,57]. Likewise, a lack of reciprocal monophyly in Clades 1 and 2 showed that the two lineages are different and not sister species. Clades 1 and 2 were paraphyletic with each other, a phylogenetic pattern that commonly occurs among cryptic species [58,59]. In addition, the 10–40% genetic distances between cryptic clades are comparable to species-level differences (Table 2) [60]. In some pairs, genetic distances for 16S rRNA and S7 intron of *S. gibbosa* exceed levels typically distinguishing closely-related species [60]. It is also interesting to note that Clades 1 and 2 do not have shared haplotypes in both the conserved 16S rRNA and the polymorphic mitochondrial control region (Figure 5). Consistent patterns in both maternally and bi-parentally inherited genetic markers demonstrate a lack of gene flow between the two cryptic species. Such patterns fall within the framework of the general species concept for two unique species [1]. Hence, genetic information from both mitochondrial and nuclear DNA presents solid evidence for two biologically distinct species.

Distinct morphometric variations in head shape characterized both Clade 1 and 2 of the *S. gibbosa* species. Multivariate analysis of head measurements revealed clustering comparable to the genetic clades. Similar clustering due to head shape variations have characterized sub-species within a sardine [23,61], a result later confirmed by molecular evidence from mitochondrial data [62]. Morphological differences between closely related sardines are often characterized by slight differences in measurements or meristic counts, resulting in an ambiguous and often controversial taxonomic status [23]. For instance, the sister sardine species *Sardinella tawilis* and *Sardinella hualiensis* share diagnostic characters; and excluding habitat preference, differ only in head length and lower gillraker count [63]. However, intra-species morphological variations in sardines are often presumed to be an artifact of localized adaptations to environment, due to a lack of support from significant genetic differentiation between morphological forms [64,65]. In contrast, the morphological disparity between the cryptic clades of *S. gibbosa* complements genetic divergence (Figure 6), and thus is not a mere localized ecological adaptation. Lastly, clustering in PCA due to head shape in Clades 1 and 2 of *S. gibbosa* falls within the phenetic-clustering species delimitation, as

Table 3. Summary of principal component analysis (eigenvalues and eigenvectors) calculated from 5 morphometric ratios describing head shape.

	PC 1	PC 2	PC 3	PC 4	PC 5
Eigenvalue	2.440	1.063	0.747	0.540	0.213
% variance	48.74	21.27	14.94	10.79	4.27
Eigenvector					
head	0.403	**−0.126**	**0.875**	0.101	−0.213
snout	−0.130	**0.893**	0.228	**−0.366**	−0.011
eye	**−0.515**	0.123	**0.291**	**0.651**	**0.459**
postorbital	**−0.453**	**−0.408**	**0.306**	**−0.652**	**0.331**
upper jaw	**−0.591**	−0.077	0.058	0.082	**−0.796**

Loadings with absolute values >0.3 are shown in bold.

Table 4. Summary of 5 morphometric measurements representing variations in head shape.

	Morphometric ratio (% *Ls*)		Clade 1 vs. Clade 2 (% difference)	
Morphometric measurement	**Clade 1**	**Clade 2**	**Contribution**	**Cumulative**
head length	22.33	24.27	32.00	32.00
upper jaw	40.48	34.62	20.25	52.26
eye	26.82	23.08	18.78	71.04
snout	30.31	30.06	15.74	86.78
post-orbital	40.58	38.46	13.22	100.00

Morphometric ratio refers to each character value as a percentage of standard length (% *Ls*). Contribution and cumulative differences calculated using SIMPER describes, in percentage, each morphometric character's contribution to head shape variation between Clade 1 and Clade 2.

there is a lack of intermediate forms in between the two clades [12,66].

Combined, genetic and morphological data reveal a hidden diversity in a common and commercially important sardine. Our findings expand the previous investigations on the biology, ecology, and morphology of *S. gibbosa* that alluded to a cryptic diversity [25,31]. Discovery of new fish species in the Northern Philippines [67,68] including a sardine beyond its previously known distribution [69], suggests that this region harbors undocumented and unique fauna. Such a pattern presents the possibility that Clade 2 might be a new species. Alternatively, Clades 1 and 2 might be previously documented synonyms of *S. gibbosa* [23]. Based on geographic proximity and morphological similarity, the most likely candidate synonym is *S. taiwanensis* [35]; however, further scrutiny of type specimens is necessary for validation. Nevertheless, the findings in this study demonstrate that a combination of both morphological and genetic data is essential to assess diversity in taxonomically ambiguous sardines. Here, strong evidence of two ecologically similar, but genetically and morphologically distinct species warrants appropriate management strategies for separate sardine fisheries.

Acknowledgments

The authors thank the researchers at Genetic Fingerprinting Laboratory of NFRDI for their superb support. Sample collection was done with the assistance of the regional offices of Philippine Bureau of Fisheries and Aquatic Resources through the National Stock Assessment Program. Samples were collected and verified with the help of Eunice Bognot of BFAR-NFRDI. Additional feedbacks on the paper were contributed by Dr. Mary Beth Maningas of the University of Santo Tomas. We would like to extend our utmost thanks to Benedict Maralit and Sheryll Hipolito of Tokyo University of Marine Science and Technology, for gathering archived literature from Japan. Also, we would like to recognize Mr. Fumiaki Nishihara for his excellent work on translating Japanese texts.

Author Contributions

Conceived and designed the experiments: RCT MDS DAW. Performed the experiments: RCT. Analyzed the data: RCT DAW. Contributed reagents/materials/analysis tools: MDS DAW RCT KEC. Wrote the paper: RCT DAW MDS KEC.

References

1. Haudorf B (2011) Progress towards a general species concept. Evolution 65: 923–931.
2. De Quieroz K (2007) Species concepts and species delimitation. Systematic Biology 56: 879–886.
3. Hey J, Waples R, Arnold M, Butlin R, Harrison R (2003) Understanding and confronting species uncertainty in biology and conservation. Trends in Ecology and Evolution 18:597–603.
4. Mayr E (1963) Animal species and evolution. Oxford University Press, London.
5. Bickford D, Lohman D, Sodhi N, Ng P, Meier R, et al. (2007) Cryptic species as a window on diversity and conservation. Trends in Ecology and Evolution 22: 148–155.
6. Knowlton N (1993) Sibling species in the sea. Annual Review of Ecology and Systematics 24: 189–216.
7. Piggott M, Chao N, Beheregaray L (2011) Three fishes in one: cryptic species in an Amazonian floodplain forest specialist. Biological Journal of the Linnaean Society 102: 391–403.
8. Boissin E, Feral J, Chenuil A (2008) Defining reproductively isolated units in a cryptic and syntopic species complex using mitochondrial and nuclear markers: the brooding brittle star, *Amphipholis squamata* (Ophiuroidea). Molecular Ecology 17: 1732–1744.
9. Knowlton N (2000) Molecular genetic analyses of species boundaries in the sea. Hydrobiologia 420: 73–90.
10. Carolan J, Murray T, Fitzpatrick U, Crossley J, Schmidt H, et al. (2012) Colour patterns do not diagnose species: quantitative evaluation of a DNA barcoded cryptic bumblebee complex. PLOS-ONE 7: e29251.
11. Hebert P, Penton E, Burns J, Janzen D, Hallwachs W (2004) Ten species in one: DNA barcoding reveals cryptic species in the neotropical skipper butterfly *Astraptes fulgator*. Proceedings of the National Academy of Sciences USA 101: 14812–14817.
12. Mallet J (1995) A species definition for the modern synthesis. Trends in Ecology and Evolution 10: 294–299.
13. Beheregaray L, Caccone A (2007) Cryptic biodiversity in a changing world. Journal of Biology 6: 9.
14. Pfenninger M, Schwenk K (2007) Cryptic animal species are homogenously distributed among taxa and biogeographic regions. BMC Evolutionary Biology 7: 121.
15. Ross K, Gotzek D, Ascunce M, Shomeaker D (2010) Species delimitation: a case study in a problematic ant taxon. Systematic Biology 59: 162–184.
16. Brown D, Brennman R, Koepfli K, Pollinger JP, Mila B, et al. (2007) Extensive population genetic structure in the giraffe. BMC Biology 5, 57.
17. Stuart B, Inger R, Voris H (2006) High level of cryptic species diversity revealed by sympatric lineages of Southeast Asian forest frogs. Biology Letters 2: 470.
18. Goodman S, Maminirina C, Weyeneth N, Bradman H, Christidis L, et al. (2009) The use of molecular and morphological characters to resolve the taxonomic identity of cryptic species: the case of the *Miniopterus manavi* (Chiroptera, Miniopteridae). Zoologica Scripta 38: 339–363.
19. Hubert N, Meyer C, Bruggermann H, Guerin F, Komeno R, et al. (2012) Cryptic diversity in Indo-Pacific coral-reef fishes revealed by DNA-barcoding provides new support to the centre-of-overlap hypothesis. PLOS ONE 7: e28987
20. Trontelj P, Fiser C (2009) Cryptic species diversity should not be trivialized. Systematics and Biodiversity 7:1–3.
21. Atarhouch T, Ruber L, Gonzales E, Albert E, Rami M, et al. (2006) Signature of an early genetic bottleneck in a population of Moroccan sardines (*Sardina pilchardus*). Molecular Phylogenetics and Evolution 39: 373–383.
22. FAO. © 2006-2013. National Aquaculture Sector Overview. Philippines. National Aquaculture Sector Overview Fact Sheets. Text by Paclibare, J.O. In: *FAO Fisheries and Aquaculture Department* [online]. Rome. Updated 1 February 2005. [Cited 13 February 2013].
23. Whitehead P (1985) Clupeioid fishes of the world (Suborder Clupeoidei): An annotated & illustrated catalogue of the herrings, sardines, pilchards, sprats, shads, anchovies & wolf-herrings. FAO Fisheries Synopsis vol 7, Rome.

24. Lavoue S, Miya M, Musikasinthorn P, Chen W, Nishida M (2012) Mitogenomic evidence for an Indo-West Pacific origin of the Clupeioidei (Teleostei: Clupeiformes). PLoS ONE 8(2): e56485.
25. Okera W (1974) Morphometrics, 'condition' & gonad development of the East African *Sardinella gibbosa* (Bleeker) & *Sardinella albella* (Valenciennes). Journal of Fish Biology 6: 801–812.
26. Nair R (1973) Indian sardines (their biology & fishery). CSIR Zoological Monograph (2). CSIR, New Delhi.
27. Willette D, Bognot E, Mutia M, Santos M (2011) Biology and ecology of sardines in the Philippines: a review. BFAR Technical Paper Series 13: 1–20.
28. Nyunja J, Mavuti K, Wakwabi E (2002) Trophic ecology of *Sardinella gibbosa* (Pisces: Clupeidae) & *Atherinomorous lacunosus* (Pisces: Atherenidae) in Mtwapa Creek & Wasini Channel, Kenya. Western Indian Ocean Journal of Marine Science 1: 181–189.
29. Lazarus S (1975) Observations on the food & feeding habits of *Sardinella gibbosa* from Vizhinjam. Indian Journal of Fisheries 24: 107–112.
30. Sekharan K (1969) Spawning concentrations of the sardine, *Sardinella gibbosa* (Bleeker), off the north Andhra coast in March-April 1969. Indian Journal of Fisheries 16: 156–160.
31. Dutt S (1961) Biometric studies on *Sardinella* spp. of Waltair coast. J Zoo Soc India 13: 78–89.
32. Okera W (1970) An Analysis of the features of *Sardinella gibbosa* (Bleeker) scales, with special reference to the problem of age determination. Journal of the East Africa Natural History Society and National Museum 28: 1–14.
33. Kishinouye K (1908) Notes on the natural history of the sardine (*Clupea melanosticta* Schlegel). Journal of the Imperial Fisheries Bureau 14(3): 71–105, Pls. 13–21.
34. Whitley G (1940) Illustrations of some Australian fishes. Australian Zoologist 9: 397–428m Pls. 30–31.
35. Raja B, Hiyama Y (1969) Studies on the systematics and biometrics of a few Indo-Pacific sardines. Records of Oceanographic Works in Japan 10(1): 75–103.
36. Walsh P, Metzger A, Higuchi R (1991) Chelex 100 as a medium for simple extraction of DNA for PCR based typing from forensic material. Biotechniques 10: 506–513.
37. Bruford M, Hanotte O, Brookfield J, Burke T (1998) Multilocus and single-locus DNA fingerprinting. In: Molecular Genetic Analysis of Populations: A Practical Approach 2nd ed, (ed Hoelzel A), pp. 287–336. IRL Press, Oxford, UK.
38. Palumbi S (1996) Nucleic acids II: the polymerase chain reaction. In Hills DM, Moritz C, Mable BK (eds) Molecular Systematics, Sinauer, Sunderland.
39. Lee W, Howell W, Kocher T (1995) Structure & evolution of teleost mitochondrial control regions. Journal of Molecular Evolution 41: 54–66.
40. Chow S, Hazama K (1998) Universal PCR primers for S7 ribosomal protein gene introns in fish. Molecular Ecology 7: 1247–1263.
41. Geneious version 5.4 created by Biomatters. Available: http://www.geneious.com
42. Edgar R (2004) MUSCLE: Multiple sequence alignment with high accuracy and high throughput. Nucleic Acid Research 32 (5), 1792–1797.
43. Darriba D, Taboada G, Doallo R, Posada D (2012) jModelTest 2: more models, new heuristics and parallel computing. Nature Methods 9: 772.
44. Guindon S, Gascuel O (2003) A simple, fast and accurate method to estimate large phylogenies by maximum-likelihood. Systematic Biology 52: 696–704.
45. Tamura K, Peterson D, Stecher G, Nei M, Kumar S (2011) MEGA5: Molecular evolutionary genetics analysis using maximum-likelihood, evolutionary distance and maximum parsimony method. Molecular Biology and Evolution 28: 2731–2739.
46. Stephens M, Donnelly P (2003) A comparison of Bayesian methods for haplotype reconstruction from population genotype data. American Journal of Human Genetics 73: 1162–1169.
47. Stephens M, Donnelly P (2001) A new statistical method for haplotype reconstruction from population data. American Journal of Human Genetics 68: 978–989.
48. Librado P, Rozas J (2009) DnaSP v5: As software for comprehensive analysis of DNA polymorphism data. Bioinformatics 25: 1451–1452.
49. Bandelt H, Forster P, Rohl A (1999) Median-joining networks for inferring intraspecific phylogenies. Molecular Biology and Evolution 16: 37–48.
50. McCune B, Mefford M (1999) Multivariate analysis of ecological data v4.10. MjM Software, Gleneden Beach, Oregon.
51. Clarke K, Warwick R (2001) Change in marine communities: an approach to statistical analysis and interpretation. Primer-E Ltd. Plymouth.
52. Carolan J, Murray T, Fitzpatrick U, Crossley J, Schmidt H, et al. (2012) Colour patterns do not diagnose species: Quantitative evaluation of a DNA barcoded cryptic bumblebee complex. PLoS-ONE 7(1): e29251.
53. Colborn J, Crabtree R, Shaklee J, Pfeiler E, Bowen B (2001) The evolutionary enigma of bonefishes (*Albula spp.*): cryptic species and ancient separations in a globally distributed shorefish. Evolution 55: 807–820.
54. Matthews L, Anker A (2009) Molecular phylogeny reveals extensive ancient and ongoing radiations in a snapping shrimp species complex (Crustacea, Alpheidae, *Alpheus armillatus*). Molecular Phylogenetics and Evolution 50: 268–281.
55. Ladner J, Palumbi S (2012) Extensive sympatry, cryptic diversity and introgression throughout the geographic distribution of two coral species complexes. Molecular Ecology 21: 2224–2238.
56. Niemiller M, Near T, Fitzpatrick B (2012) Delimiting species using multilocus data: diagnosing cryptic diversity in the southern cavefish, *Typhlichthys subterraneus* (Teleostei: Amblyopsidae). Evolution 66–3: 846–866.
57. Dupuis J, Roe A, Sperling F (2012) Multilocus species delimitation in closely related animals and fungi: one marker is not enough. Molecular Ecology 21: 4422–4436.
58. Shaffer H, Thomson R (2007) Delimiting species in recent radiations. Systematic Biology 56: 896–907.
59. Funk D, Omland K (2003) Species-level paraphyly and polyphyly: Frequency, causes and consequences, with insights from animal mitochondrial DNA. Annual Reviews in Ecology and Systematics 34: 397–423.
60. Lavoue S, Sullivan J, Hopkins C (2003) Phylogenetic utility of the first two introns of the S7 ribosomal protein gene in African electric fishes (Mormyroidea: Teleostei) and congruence with other molecular markers. Biological Journal of the Linnean Society 78: 273.
61. Silva A (2003) Morphometric variation among sardine (*Sardina pilchardus*) populations from the northeastern Atlantic and the western Mediterranean. ICES J Mar Sci 60: 1352–1360.
62. Atarhouch T, Ruber L, Gonzales E, Albert E, Rami M, et al. (2006) Signature of an early genetic bottleneck in a population of Moroccan sardines (*Sardina pilchardus*). Molecular Phylogenetics and Evolution 39: 373–383.
63. Willette D, Carpenter K, Santos M (2014)Evolution of the freshwater sardinella, *Sardinella tawilis* (Clupeiformes: Clupeidae), in Lake Taal, Philippines and identification of its marine sister-species, *Sardinella hualiensis*. Bulletin of Marine Science 90: In press.
64. Kinsey S, Orsoy T, Bert T, Mahmoudi B (1994) Population structure of the Spanish sardine *Sardinella aurita*: natural morphological variation in a genetically homogenous population. Marine Biology 118: 309–317.
65. Hedgecock D, Hutchinson E, Li G, Sly F, Nelson K (1989) Genetic and morphometric variation in the Pacific sardine, *Sardinops sagax caerulea*: Comparisons and contrasts with historical data and with variability in the Northern anchovy, *Engraulis mordax*. Fishery Bulletin 87: 653–671.
66. Sokal R, Crovello T (1970) The biological species concept: A critical evaluation. The American Naturalist 104: 127–153.
67. Gill A, Williams J (2011) Description of two new species of Pseudochrominae from northern Palawan and Mindoro, Philippine Islands (Teleostei: Perciformes: Pseudochromidae). Zootaxa 3140: 49–59.
68. Gill A, Shao-K, Chen J (1995) *Pseudochromis cerasina*, a new species of pseudochromine dottyback fish from the West Pacific (Perciformes: Pseudochromidae). Proceedings of the Biological Society of Washington 117:17–22.
69. Willette D, Santos M, Aragon M (2011) First report of the Taiwan sardinella *Sardinella hualiensis* (Clupeiformes: Clupeidae) in the Philippines. Journal of Fish Biology 79: 2087–2094.

Microsatellite Marker Analysis Reveals the Complex Phylogeographic History of *Rhododendron ferrugineum* (Ericaceae) in the Pyrenees

Olivia Charrier[1,2]*, Pierre Dupont[1,2], André Pornon[1,2], Nathalie Escaravage[1,2]

1 University Toulouse III Paul Sabatier, Lab Evolution & Diversite Biologique EDB, F-31062 Toulouse, France, **2** CNRS, EDB, UMR 5174, F-31062 Toulouse, France

Abstract

Genetic variation within plant species is determined by a number of factors such as reproductive mode, breeding system, life history traits and climatic events. In alpine regions, plants experience heterogenic abiotic conditions that influence the population's genetic structure. The aim of this study was to investigate the genetic structure and phylogeographic history of the subalpine shrub Rhododendron ferrugineum across the Pyrenees and the links between the populations in the Pyrenees, the Alps and Jura Mountains. We used 27 microsatellite markers to genotype 645 samples from 29 Pyrenean populations, three from the Alps and one from the Jura Mountains. These data were used to estimate population genetics statistics such as allelic richness, observed heterozygosity, expected heterozygosity, fixation index, inbreeding coefficient and number of migrants. Genetic diversity was found to be higher in the Alps than in the Pyrenees suggesting colonization waves from the Alps to the Pyrenees. Two separate genetic lineages were found in both the Alps and Pyrenees, with a substructure of five genetic clusters in the Pyrenees where a loss of genetic diversity was noted. The strong differentiation among clusters is maintained by low gene flow across populations. Moreover, some populations showed higher genetic diversity than others and presented rare alleles that may indicate the presence of alpine refugia. Two lineages of R. ferrugineum have colonized the Pyrenees from the Alps. Then, during glaciation events R. ferrugineum survived in the Pyrenees in different refugia such as lowland refugia at the eastern part of the chain and nunataks at high elevations leading to a clustered genetic pattern.

Editor: Giovanni G. Vendramin, CNR, Italy

Funding: Funding provided by UniversitÃ© Paul Sabatier Toulouse - CNRS. The funders had no role in study design, data collection and analysis, decision to publish, or preparation of the manuscript.

Competing Interests: The authors have declared that no competing interests exist.

* E-mail: olcharrier@gmail.com

Introduction

The maintenance of genetic diversity within populations is essential for the conservation of species since it can enable adaptation to future environmental changes [1], [2]. With high genetic diversity, it is more likely that some individuals will possess alleles that could allow the population to adapt to a new habitat [3]. High genetic diversity may be maintained via gene flow among individuals and among populations. In recent years, many studies have shown strong impacts of global change on species' genetic diversity [4]. Because climate warming is greater at high than at low elevations [5] alpine biodiversity hotspots that are home to many endemic and rare species may be particularly threatened by climate change [6]. Thus, to predict how these biodiversity hotspots will deal with climate changes, it is important to understand the genetic patterns of the species and their changes over time [7].

Genetic structure and genetic divergence within and among plant populations result from a number of contemporary and historical factors acting at various temporal and spatial scales. Alpine habitats are typically characterized by small spatial changes in environmental variables such as topography, temperature, snow cover, soil moisture or bedrock [8] favorable to phenotypic differentiation and local genetic variability over short geographical distances. In addition to abiotic factors, life history traits can also influence genetic features of populations. For instance, selfing perennials and clonal species generally exhibit lower levels of genetic diversity and higher levels of population differentiation than outcrossing perennials [9]. In contrast, mixed mating systems maintain a high level of genetic diversity and may be advantageous for plants in variable alpine environments [6]. The combined effects of both life history traits and small-scale abiotic heterogeneity may result in local genetic adaptations and high genetic differentiation of populations, notably where gene exchanges between populations are restricted as is often the case in mountain habitats [10]. Therefore, plant-community fragmentation pattern, steep climatic gradients leading to asynchronous flowering phenology or disjoint pollinator communities [7], and topographical barriers may hamper gene flow (through seed and pollen movements) and may increase differentiation between populations.

At large scales, species genetic patterns are often shaped by past climate-driven range dynamics [11]. In temperate regions, species ranges have changed dramatically over the Pleistocene with its successions of glaciations and warming periods. In mountains, periodic advances and retreats of glaciers have, on several occasions, forced species to descend to lowland and peripheral refugia and/or to survive in isolated ice-free areas above glaciers (nunataks). Genetic isolation, bottlenecks and potential allopatric evolution of populations in refugia have been followed by repeated founding events during population expansions, which have led to loss of genetic diversity and heterozygosity [11]. Additionally, populations located at eroding margins of species distribution ranges following climate warming could have persisted in suitable isolated lowland habitats [12]. It has been suggested [13–14] that such small populations isolated for so long could display low within-population genetic diversity (as a result of genetic drift) but have disproportionally high levels of between-population differentiation and regional genetic diversity and distinctiveness [15].

The genetic pattern and phylogeography of many species [16], [17] have been investigated in various mountainous regions, but to date, only a very limited number of studies have focused on the Pyrenees [18–20]. The Pyrenees are a mountain range stretching across the isthmus that lies between the Iberian Peninsula and the rest of the European continent. Acting as a biogeographical barrier during postglacial expansions, they are a limit for numerous Iberian and northern species. However, lower lands at the eastern and western extremities of the mountain range have been corridors for migration of several plant and animal species [21,22]. Due to oceanic influences to the west and Mediterranean influences to the east of the chain, the climate varies greatly along a longitudinal gradient in the Pyrenees (Météo France data). This bioclimatic pattern potentially allowed some species to survive glaciations in various refugia along the chain. Studies have suggested that species have survived glaciations [23–25] in two main refugia - one in south central and another in the eastern Pyrenees - thanks to mild Mediterranean and oceanic climatic influences. However both geological and biological data [26] demonstrate that during glacial periods the ice sheet did not completely cover the highlands and rarely descended below 1000 m a.s.l except for big glaciers pushed down into large valleys and to the plains [23]. These nunataks and unglaciated peripheral area could have provided refugia for some species [18] and determined several recolonizing routes sculpturing a specific and complex genetic structure in the Pyrenees.

Here we investigate the genetic structure and phylogeographical history of the alpine shrub *Rhododendron ferrugineum* (Ericaceae) in the Pyrenees. This species is insect pollinated and reproduces both sexually (mixed mating system) and vegetatively through layering. It is widely distributed at high elevation, on non-calcareous bedrock. The aim of the present study is to understand to what extent the phylogeographical history of the species has shaped its present genetic patterns. The Asian origin of the genus *Rhododendron* [27] implies an east/west migration from Asia to Europe. In France, this would lead to the Alps being colonized by *R. ferrugineum* before the Pyrenees, with the Pyrenean populations most likely derived from populations from the Alps. After the migration to the Pyrenees, the species had to survive successive glaciation events. Different hypotheses of glacial refugia and postglacial colonization and their consequences on the genetic patterns of *R. ferrugineum* may be proposed: 1. The species could have survived glaciation episodes in one or both of the two main refugia at the eastern and south central part of the Pyrenees (HYP 1; [24]). Thus, the highest genetic diversity should be found in one or both of these areas and the genetic pattern would be shaped by founder effects i.e. gradual reduction of genomic variability along the colonizing routes from main refugia. This would have occurred because colonization usually involves only a fraction of the genetic diversity present in refugial areas [28]. 2. The species could also have survived in nunataks along the mountain chain (HYP 2). Hypothetically, high elevation nunatak populations would have experienced long periods of isolation, inbreeding and genetic drift [18]. This pattern would lead to a patchy distribution of genetic clusters with low levels of genetic diversity and high population differentiation; 3. A pattern with less differentiated clusters than in the previous hypothesis may be observed if the species had survived in large lowland refugia in which extensive gene flow had occurred. The recolonization of highland areas from lowland refugia could have left small, isolated and marginal rear-edge populations like those currently observed at low altitude along the chain (HYP 3). Indeed, the strong abiotic mountain heterogeneity may have allowed remnant populations to match suitable ecological conditions at intermediate or low altitudes. Moreover, the long life span of the clones [29,30] could have permitted their survival even with low seedling recruitment. Such rear-edge populations should exhibit more rare alleles and higher among-population genetic diversity than populations recently established from high altitude source populations [15]. These three hypotheses are not exclusive and it is conceivable that the genetic pattern observed may result from the different mechanisms at work successively or simultaneously within the bioclimatic regions of the Pyrenees inducing diverse phylogenetic patterns along the chain.

Materials and Methods

Species studied

Sampling of the species of interest was conducted in accordance to national and international guidelines (authorization from the ONF for the populations BetL, BetH, LecL, PraL, PuiH, from the Parc National des Pyrénées for the population TroH, from the Conseil général des Pyrénées-Orientales for population BouH, and no permissions were needed for the other sites).

Rhododendron ferrugineum L. (Ericaceae) is an evergreen shrub with a mean height of 70 cm that dominates subalpine landscapes in the Alps and Pyrenees between 1500 to 2200 m a.s.l, on north-to-west facing slopes, sometimes reaching 90–100% of the vegetation cover. It is also found in the Jura Mountains, in the Northern Apennines and in the Apuan Alps [31]. In the Pyrenees, infrequent smaller and isolated populations can be found lower down, between 900–1500 m a.s.l. A single population of some scattered individuals remains at low altitude (1500 m a.s.l.) in the French Jura Mountains. *Rhododendron ferrugineum* is a clonal, self-compatible [32] mass-flowering species (producing until 3000 flowers per m^2; [33]) pollinated by honeybees and bumblebees [32,34].

Study sites and sample collection

We sampled 29 populations across the Pyrenees, the population in the Jura Mountains and three populations in the Alps (Figure 1, Table 1) ranging from 1070 m to 2200 m a.s.l. The populations were divided into three groups according to their elevation: low altitude (hereafter L; from 1000 to 1450 m a.s.l), intermediate altitude (hereafter I; from 1450 to 1900 m a.s.l.), high altitude (hereafter H; from 1900 to 2200 m a.s.l.). In the Pyrenees, populations were sampled along a longitudinal transect through three regions: Languedoc-Roussillon (eastern part), Midi-Pyrénées (central part) and Aquitaine (western part). Due to calcareous bedrock, populations in the western part of the Pyrenees are rare and we only sampled one. An estimation of population size is given in Table 1 [<1 ha, 1–10 ha, 10–20 ha, 20–50 ha, 50–100 ha, >100 ha). The mean geographical distance between populations from Pyrenees *vs.* Alps, Pyrenees *vs* Jura Mountains and Alps *vs.* Jura Mountains was 900 km (ranging from 515 to 1430 km), 695 km (from 590 to 830 km) and 335 km (from 165 to 640 km) respectively. The distances between populations across the

Pyrenees ranged from 3 to 370 km with an average of 126 km. In each of the 33 populations (except for the very small JuraL and LapL populations), young leaves of twenty individuals were sampled (a total of 645 individuals sampled), separated by at least 5 m in order to avoid sampling clonal individuals, were dried in silica gel in the field and then stored at −20°C.

DNA extraction and microsatellite markers

For each individual sampled, 20 mg of leaf tissue was ground with two tungsten beads (diameter 3 mm) in a sterile 2 ml Eppendorf tube at 30 Hz for 90 s using TissueLyser II (Qiagen). Then, the quality and amount of DNA obtained from the DNeasy-Plant Mini Kit (Qiagen, Courtaboeuf, France) following the manufacturer's instructions was verified using a Nanodrop ND 1000 spectrophotometer (Peqlab, Erlangen, Germany).

We used nuclear microsatellites as suitable molecular markers to reconstruct regional phylogeographical patterns [18]. Chloroplasts (personal observation) or other types of genetic markers such as AFLP proved to be little informative in this species [35]. The 27 polymorphic microsatellite markers (Table S1) were previously developed [36,37] using pyrosequencing technologies (454 FLX Titanium, Roche Applied Science, Meylan, France). Forward primers were labeled with fluorochromes (Eurofins MGW Operon, Courtaboeuf, France) and were used in six multiplexed PCRs, optimized using Multiplex Manager 1.1 [38].

Amplifications were carried out in a 10 µl mix containing 10 ng of template DNA, 0.7 × Qiagen Multiplex PCR Master Mix, 10 µM of each primer and RNase free water. Cycling conditions were 15 min at 95°C, 30 × (30 s at 95°C, 90 s at 56°C, 45 s at 72°C) and 30 min at 60°C. Each set of reactions included a negative (water) and a positive (known genotype) control. Fragment lengths were read on an ABI 3730 sequencer (Applied Biosystems, Courtaboeuf, France) and scored with Genemapper version 4.0 software with the GeneScan – 600 LIZ (Applied Biosystems, Courtaboeuf, France) as internal size standard.

Data analysis

Genetic diversity. All 645 individuals were tested for clonality with GenAlEx 6.5 software [39]. The presence of null alleles at each locus was examined using Micro-Checker software [40], and tests for departures from Hardy-Weinberg equilibrium (HWE) were tested by Fisher's exact tests using Genepop 3.4 [41]. We used GenAlEx 6.5 software to calculate for each locus and population the number of alleles, the allelic richness (Ar), the observed (H_O) and expected heterozygosity (H_E) and the inbreeding coefficient (F_{IS}).

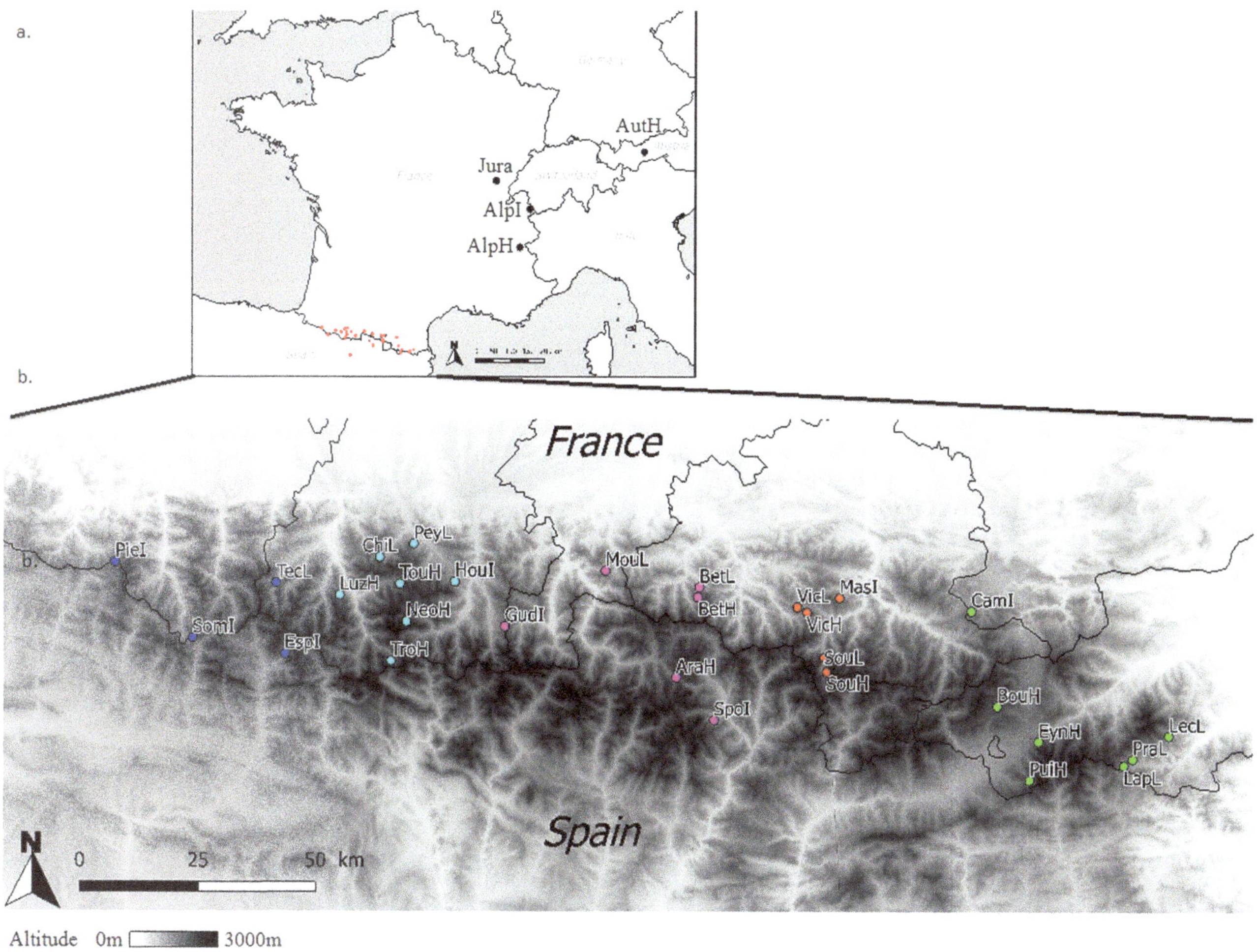

Figure 1. Location of the sampled populations. a. Location of the sampled *Rhododendron ferrugineum* populations in the Alps, Jura Mountains and Pyrenees, b. detailed map of the Pyrenean populations (the colors correspond to the five genetic clusters).

Table 1. Geographical information and genetic diversity of *Rhododendron ferrugineum* populations.

Moutain range	Sites	Code population	Location	Population size (ha)	Altitude (m)	n	Ar	Private allele	H_O	H_E	F_{IS}
Alps	La Vormaine	AlpI	45°02'07"N 06°24'19"E	>100	1560	20	3.70	1	0.49	0.49	**0.04**
	Col du Lautaret	AlpH	46°00'08"N 06°57'18"E	5–10	2100	20	4.30	10	0.59	0.57	**−0.01**
	Parscherkofel	AutH	47°12'32"N 11°27'18"E	50–100	2000	20	2.89	2	0.41	0.39	**−0.03**
Jura		JuraL	42°57'N 05°44'E	<1	1300	10	2.67	0	0.48	0.44	**−0.06**
Eastern Pyrenees	Léca	LecL	42°28'07"N 02°31'40"E	<1	1250	20	2.37	1	0.34	0.30	−0.11
	Prats de Mollo	PraL	42°25'21"N 02°24'38"E	<1	1450	20	2.44	1	0.31	0.32	**0.02**
	Lapreste	LapL	42°24'34"N 02°22'51"E	<1	1230	15	2.44	0	0.42	0.34	−0.21
	Eyne	EynH	42°28'27"N 02°07'05"E	5–10	1915	20	2.78	0	0.38	0.35	**−0.06**
	Puigmal	PuiH	42°23'26"N 2°05'0"E	5–10	2045	20	2.74	0	0.39	0.34	−0.10
	Lac des Bouillouses	BouH	42°33'31"N 01°59'42"E	50–100	2005	20	2.56	0	0.37	0.34	**−0.05**
Central Pyrenees	Camurac	CamI	42°46'30"N 01°55'45"E	10–20	1600	20	2.92	0	0.42	0.38	−0.09
	Massat	MasI	42°49'18"N 01°30'75"E	<1	1500	20	2.59	0	0.41	0.40	**0.0007**
	Soulcem	SouH	42°39'28"N 01°27'27"E	50–100	2035	20	2.63	0	0.41	0.39	**−0.01**
	Soulcem	SouL	42°41'28"N 01°27'07"E	<1	1240	20	2.74	0	0.37	0.39	0.08
	Etang de Lhers	VicH	42°47'36"N 01°24'16"E	20–50	1940	20	2.74	0	0.38	0.35	−0.07
	Etang de Lhers	VicL	42°48'23"N 01°22'31"E	1–5	1275	20	2.44	0	0.36	0.32	−0.09
	Espot	SpoI	42°33'47"N 01°05'36"E	<1	1515	20	4.18	0	0.35	0.43	0.10
	Bethmale	BetL	42°51'42"N 01°04'02"E	1–5	1385	20	2.67	0	0.40	0.37	**−0.06**
	Bethmale	BetH	42°50'24"N 01°03'42"E	50–100	1920	20	2.92	0	0.40	0.37	**−0.03**
	Val d'Aran	AraH	42°39'45"N 00°58'56"E	<1	2075	20	2.26	0	0.29	0.29	**0.05**
	Le Mourtis	MouL	42°54'32"N 00°46'31"E	5–10	1440	20	2.59	0	0.33	0.33	**0.04**
	Peyragudes	GudI	42°47'42"N 00°27'14"E	5–10	1650	20	2.67	0	0.39	0.37	**−0.03**
	Hourquette d'ancizan	HouI	42°53'57"N 00°18'13"E	5–10	1565	20	2.74	0	0.42	0.40	**−0.04**
	Néouvielle	NeoH	42°48'N 00°09'E	50–100	2200	20	2.56	0	0.47	0.44	**−0.03**
	Le Peyras	PeyL	42°59'15"N 00°10'46"E	5–10	1220	20	2.33	0	0.36	0.36	**0.03**
	Col du Tourmalet	TouH	42°53'57"N 00°07'54"E	5–10	1950	20	2.81	1	0.42	0.39	**−0.04**
	Cirque de Troumouse	TroH	42°43'41"N 00°05'40"E	5–10	2080	20	2.44	0	0.39	0.38	**0.02**
	Chiroulet	ChiL	42°57'38"N 00°04'17"E	1–5	1240	20	2.70	0	0.38	0.37	**0.02**
	Luz-Ardiden	LuzH	42°52'45"N 00°03'34"W	10–20	1905	20	2.63	0	0.37	0.37	**0.02**
	Huesca	EspI	42°45'08"N 00°14'18"W	>100	1640	20	2.27	0	0.20	0.30	0.32
	Lac du Tech	TecL	42°54'40"N 00°15'32"W	<1	1240	20	2.41	0	0.39	0.34	−0.11
	Col du Somport	SomI	42°47'41"N 00°31'38"W	<1	1645	20	2.26	0	0.33	0.37	0.16
Western Pyrenees	Col de la Pierre Saint Martin	PieI	42°58'07"N 00°45'52"W	<1	1830	20	2.00	0	0.28	0.29	**0.09**

Altitudinal classes are given in the code population (L: low, I: intermediate, H: high altitude).
n = sample size, Ar = allelic richness, H_O = observed heterozigosity, H_E = expected heterozygosity, F_{IS} = within population coefficient of inbreeding (non-significant values are in bold, P<0.01).

Population structure and differentiation. The pairwise F_{ST} (level of genetic differentiation; [42]) among populations were calculated with Genodive v2.0 [43]. Then, the statistical significance of the pairwise F_{ST} values was assessed by 1000 random permutations of individuals across populations with Genetix 4.05 [44].

To analyze the genetic structure among populations, we used Structure software (ver. 2.3.3; [45]) in order to determine the most likely number of homogenous clusters in the sample (K). The admixture model was applied with 50 runs for each K value from 1 to "number of populations +1", each run comprising a burn-in period of 20 000 iterations followed by 100 000 iterations. The optimal value of *K* was evaluated by considering the highest mean likelihood value of *K*, *i.e.*, L (*K*), as well as the ΔK method [46]. These results were computed and visualized in the online interface Structure Harvester [47].

To evaluate the distribution of genetic variance within and among populations, and among groups when populations were nested into genetic clusters, altitude (low, intermediate and high altitude) or population size (according to the six population size groups in Table 1) we performed analyses of molecular variance (AMOVA; [48]) using Genodive v2.0. The significance of the results was tested using 999 random permutations of the data.

Mantel tests [49] were performed ("ncf" package implemented in R, R Development Core Team, 2008) to test for isolation-by-distance (IBD; divergence due to drift and mutation proportionally increasing with geographical distance), investigating the correlation between matrices of genetic (F_{ST}) and geographical (km) distances. Mantel test were also used to determine wheter elevational separation was related to genetic differentiation. Calculation of the gene flow on the basis of rare alleles was conducted [50] and the number of migrants (N_m) was corrected for the number of samples.

Results

A total of 153 alleles at 27 microsatellite loci were detected across 645 individuals at all sites (Table S2). No clones were identified among the individuals sampled as revealed by the test of clonality.

Significant departures from HWE across loci were detected in 12 out of 33 (or 36%) sites. The RF87P2 marker was the only locus that showed no departures from HWE. Most of the loci showed significant departures from HWE in two to seven populations, except RF14P3, RF47P1 (11 populations each), RF56P1 (13 populations) and RF96P2 (31 populations). These instances likely reflect occasional departures from random mating rather than the presence of null alleles. Indeed, the probability of null alleles was low for all loci, according to Oosterhout's method [40], mean estimation of null alleles was −0.031 (ranging from −0.3675 to 0.1352). No differences in the results were observed when including the microsatellite marker RF96P2 or not, so we decided to keep it in the analyses.

Genetic diversity and allelic richness among populations

The number of alleles ranged from 2 to 15 (mean 6.25) alleles per locus. A total of 16 private alleles (restricted to a single population; Table 1) was detected (10.45%), mostly in AlpH (10), AutH (2), AlpI (1) and also in some Pyrenean populations (TouH, PraL, LapL had one private allele each). Allelic richness (Ar) varied across sites (2–4.30). The highest Ar were observed in the Alps (AlpI = 3.70 and AlpH = 4.30) and in one of the southernmost populations of the central Pyrenees (SpoI = 4.18), while the lowest Ar were observed at the western end of the Pyrenees (PieI = 2.00, SomI = 2.26, EspI = 2.27) and in the central Pyrenees (AraH = 2.26). Ar, H_O and H_E were 4.6, 0.37 and 0.35 respectively in the Pyrenees, and 5.2, 0.5 and 0.45 in the Alps. Expected (H_E) and observed heterozygosity (H_O) ranged from 0.20 to 0.59 and from 0.29 to 0.56 respectively for all sites. H_E was high in the Alps and Jura Mountains (ranging from 0.39 to 0.57), and in some Pyrenean populations (SpoI and NeoH, 0.43 and 0.44 respectively) and the lowest in the easternmost (LecL, PraL, AraH) and westernmost (EspI, PieI) Pyrenean populations. These latter populations were small and isolated (<1 ha) except the large and central population EspI (>100 ha). Significant heterozygosity excesses (F_{IS} coefficients ranging from −0.07 to −0.21) were detected in three eastern (LecL, LapL, PuiH) and four central populations of the Pyrenees (CamI, VicH, VicL, TecL). In contrast, significant heterozygosity deficiency was found in four populations with particularly high values of F_{IS} coefficients in SomI (0.16) and EspI (0.31). Because the frequencies of null alleles did not appear to be significant, the highest F_{IS} values could be due to inbreeding.

Genetic differentiation and structuration among populations

The F_{ST} values (Table S3) were significantly different from zero for all population pairs, except for three pairwise comparisons: BetH/BetL and SouH/SouL that were geographically really close (less than 3 km) and CamI/EynH (35 km apart). The mean F_{ST} value was relatively low (0.267) and most of the highest values (>0.33) were found in the Pyrenees between eastern and western populations, reaching 0.45 for LecL/AraH. The population from the Jura Mountains showed a highly variable level of F_{ST} (from 0.25 to 0.42).

To infer the relationship between populations we used the clustering algorithm implemented in the program Structure. Comparing all 33 populations (K: 1 to 34), ΔK as a function of K reached a peak at K = 2 and then reached a second peak at K = 6 (Figure 2a). At K = 2, populations from the Alps, Jura and the eastern Pyrenees were assigned to the same cluster while populations from western and central Pyrenees were clearly grouped apart. Individuals from AlpI showed a similar proportion of membership to both genetic clusters. At K = 6, populations from the Alps and Jura were assigned to one cluster (CL1) and those of the Pyrenees were grouped in 5 clusters (CL2 to CL6) along a longitudinal gradient. Except between CL5 and CL4 or CL6, there was low admixture between the other Pyrenean clusters.

The analysis of molecular variance (AMOVA; Table 2) based on F_{ST} showed that as expected for polymorphic loci such as microsatellites, the within-population variance component was rather high (>66%, P = 0.001). The among variation component (F_{ST}) ranged from 9% to 31% according to the grouping and was statistically significant in all combinations tested (P = 0.001). At a superior hierarchical level, the genetic clusters could explain a significant part of the total genetic variance (24.4%, P = 0.001) whereas it was not significant when grouping populations into altitudinal groups (P = 0.956) or population size (P = 0.222).

Allelic richness (Ar) was highest in CL1 (Alps and Jura, 5.2) and lowest in CL6 (western Pyrenees, 2.9) and ranged from 3.2 to 3.8 in the other clusters (Table 3). F_{ST} values were all significantly different among clusters (Table 4). The highest F_{ST} values were found between CL2 (eastern Pyrenees) and CL5 or CL6 (western Pyrenees, 0.29 and 0.31 respectively). The F_{ST} values between CL1 and the Pyrenean clusters were quite low (ranging from 0.16 to 0.20).

There was a positive relationship (Mantel test) between genetic distance and geographical distance matrices across populations located in the Pyrenees (R = 0.784, P = 0.001). The increase of isolation with geographical distance was consistent with the previous results (see Structure analyses). When including populations from the Alps and Jura, the Mantel test was still positive (R = 0.26) but not significant (P = 0.383), confirming the lack of strong differentiation among the populations from the Alps and Pyrenees. At the Pyrenean scale, the correlation among elevational distances and FST was slightly positive but not significant (R = 0.092, P = 0.06).

Gene flow among populations

Gene flow among all studied populations (Alps, Pyrenees and Jura) was low and reached a mean value of Nm = 0.72 migrants per generation. The values reached 2.88 for all Pyrenean populations but 2.10 in the eastern lineage (CL2) and 5.9 in the western lineage (CL3 to CL6).

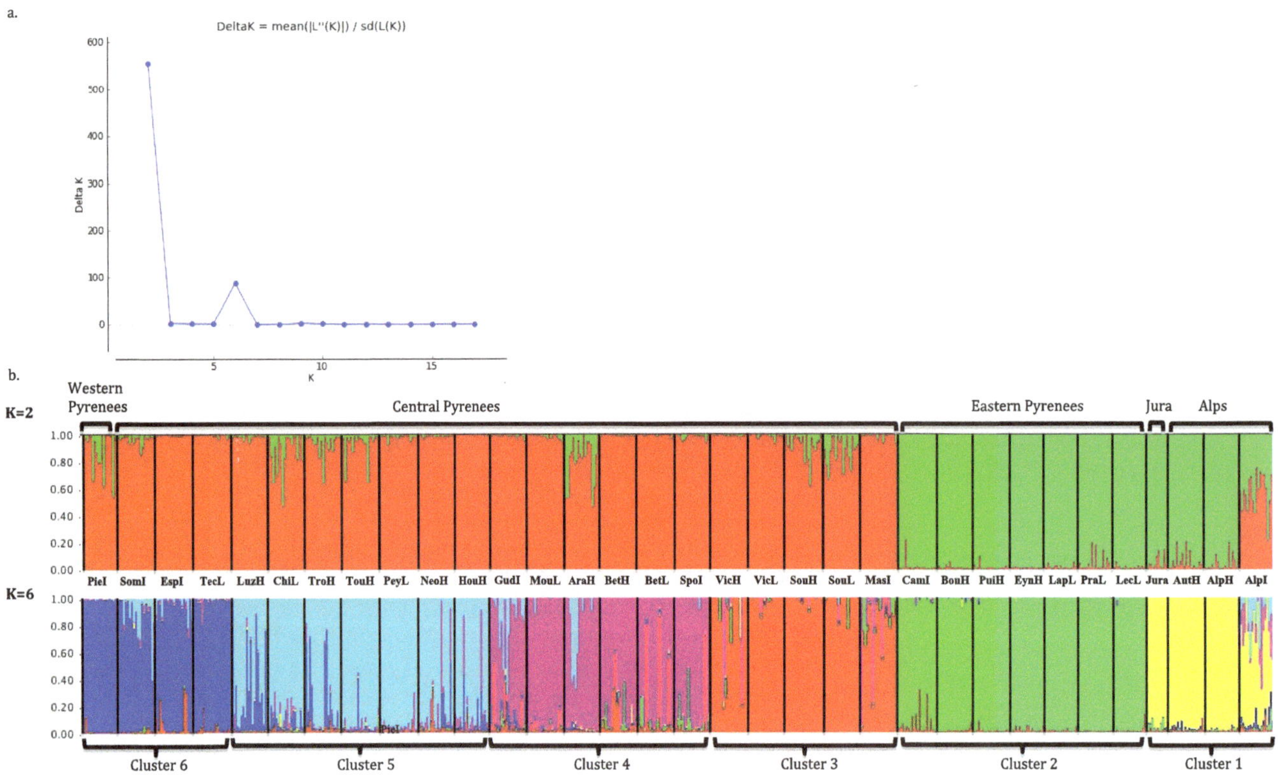

Figure 2. Genetic structure of the 33 populations. a. Plot of Delta K according to K. b. STRUCTURE clustering results obtained at K = 2 and K = 6. Each individual is represented by a thin bar corresponding to the sum of assignment probabilities to the K cluster. Black bars separate populations.

Discussion

Genetic diversity in R. ferrugineum populations

Microsatellite analysis revealed a relatively high level of genetic variability in *R. ferrugineum* populations in the Pyrenees (Ar: 2–4.18, H_E: 0.29–0.44) but lower than in the Alps and the Apennine populations (Ar: 2.71–5.57, H_E: 0.37–0.68; [31]) or in other long-lived species found in the Pyrenees (*Pinus uncinata*: H_E: 0.83–0.99, (25); *Quercus petraea*: H_E: 0.76–0.84, [51]). Although clonality and selfing are known to decrease genetic variability [52], the long life span of individuals [29,30] and the strong life-time inbreeding depression eliminating most selfed individuals (personal data) may contribute to maintaining relatively high genetic diversity even in small *R. ferrugineum* populations. Accordingly, most of the populations studied showed a F_{IS} close to zero and a great proportion of the genetic variation was found within populations as observed in other *R. ferrugineum* populations [31] and other alpine species [19], [53], [18]. However, selfing has probably contributed to individual recruitment in populations harboring higher heterozygote deficiency (SomI, PieI, EspI, SpoI). Such high heterozygote deficiency has also been reported in the Apennines and the Alps [31]. Surprisingly, significant heterozygote excess

Table 2. Partitioning of the observed microsatellite variation (AMOVA) based on F_{ST} of *Rhododendron ferrugineum* populations.

Source of Variation	SSD	d.f.	MS	Var-comp	% Variation	P-value
Within Population	4355.120	548	7.947	7.947	0.662	0.001
Among Population	732.618	24	30.526	1.135	0.095	0.001
Among Genetic cluster	1461.242	4	365.310	2.929	0.244	0.001
Within Population	4354.038	548	7.945	7.945	0.690	0.001
Among Population	2103.864	26	80.918	3.659	0.310	0.001
Among Altitude	93.988	2	46.994	−0.167	0.00	0.956
Within Population	4366.874	548	7.894	7.894	0.685	0.001
Among Population	2079.120	23	83.837	3.372	0.293	0.001
Among Population size	106.877	5	56.922	0.258	0.022	0.222

Table 3. Genetic diversity of *Rhododendron ferrugineum* per genetic cluster.

	Ar	H_O	H_E	F_{IS}
Cluster 1	5.630	0.498	0.584	0.148
Cluster 2	3.630	0.375	0.364	−0.030
Cluster 3	3.185	0.386	0.406	0.048
Cluster 4	3.778	0.360	0.390	0.078
Cluster 5	3.481	0.402	0.429	0.062
Cluster 6	2.926	0.315	0.384	0.179

Ar = allelic richness, H_O = observed heterozigosity, H_E = expected heterozygosity, F_{IS} = within population coefficient of inbreeding.

Table 4. F_{ST} values among genetic clusters (p-values for all pairs<0.001).

	Cluster 1	Cluster 2	Cluster 3	Cluster 4	Cluster 5	Cluster 6
Cluster 1	—	0.203	0.171	0.177	0.165	0.166
Cluster 2		—	0.237	0.275	0.297	0.311
Cluster 3			—	0.101	0.153	0.230
Cluster 4				—	0.108	0.235
Cluster 5					—	0.129
Cluster 6						—

occurred mostly in eastern (LapL, LecL, PuiH), and central populations (CamI, VicH, VicL) and in one of the westernmost populations (TecL).

Alpine origin of the two Pyrenean *R. ferrugineum* lineages

The *Rhododendron* genus is known to originate from Asia [27] and colonized Europe from east to west. Our hypothesis was that migration from the Alps to the Pyrenees occurred with a loss of genetic diversity along the colonization routes. In our study, we identified two lineages in the Alps: the first, found only in the northern population (AlpI), was related to western Pyrenean populations. This lineage would possibly be found in more populations with more extensive sampling in the Alps. The second lineage, related to the eastern and central Pyrenees, was found in the same population AlpI but also in south and central-eastern populations from the Alps and in the Jura Mountains. At K = 6, populations from the Jura Mountains and the Alps are found in the same cluster, that may indicate the occurrence of recent gene flow between the Alps and the Jura Mountains, possibly in low land refugia during glaciations. It would be interesting to complete our data by a large sampling of populations from the Alps and the Jura Mountains in order to detail the European phylogenetic pattern of the species. Compared to the Alpine populations, Pyrenean populations were characterized by a substantial loss of genetic diversity (Ar, H_E, H_O) suggesting bottlenecks during the migration events [54], [16]. Cold, dry steppe/tundra occupied most of the land between the Alps and the Pyrenees during the last maximum glaciation [55] and could have allowed migration of Alpine species such as *R. ferrugineum* either from the southwestern Alps [56], [57] or from northern Alpine populations through the French Massif Central which is considered to be a key area in the colonization of the Pyrenees by Alpine species [58].

Genetic pattern and phylogeographical history of *R. ferrugineum* in the Pyrenees

We found that genetic diversity of *R. ferrugineum* was not correlated to altitude or population size; instead there was a sharp longitudinal split between the two genetic lineages. Other studies in alpine ecosystems similarly found no changes in genetic diversity depending on population size or along altitudinal gradients [59]. Along environmental gradients, several parameters such as overlap of generations or recruitment frequency may influence intrapopulation genetic variation and gene flow among populations may be sufficient to maintain relatively high genetic diversity even in the smallest populations.

The longitudinal genetic pattern seemed to result from the migration of two Alpine lineages and from the establishment of populations in refugia along the Pyrenean chain due to glaciation events. The strong differentiation between the eastern and western Pyrenean lineages and the low level of gene flow observed between the two suggested that very few admixtures occurred since the two lineages had become established. This may result from topographical barriers (high summits, glaciated areas), seed dispersion by gravity and short pollinator travels hampering gene flow among populations and genetic homogenization. An east-west genetic separation had also been observed in other Pyrenean species (i.e. *Papaver alpinum*, [57], *Trifolium alpinum*, [19]). The populations of the eastern lineage were little differentiated even though they were distant from one another (35 km between the two undifferentiated populations CamI and EynH), were of very different sizes, grew in contrasting environments, and experienced a low level of among-population current gene flow. Very small and isolated populations (PraL, LecL, LapL) showed a relatively high level of intrapopulation heterozygosity (H_E, H_O) and exhibited private alleles (PraL, LapL) despite relatively lower allelic richness likely due to the loss of rare alleles by genetic drift, common in such small populations [60]. These populations survive in rather unusual ecological conditions for this species, i.e. on north facing slopes in deep, shaded valleys and mainly beneath beech forest (*Fagus sylvatica*) or sometimes in the vicinity of Mediterranean *Quercus ilex* forests. These data suggested that they were more likely to be rear-edge remnant populations [12] left there after the recolonization of highland areas (HYP 3) rather than the result of a secondary colonization of lowland areas by high altitude individuals. Together, these findings and the absence of clear founder effects suggested that the eastern populations within CL2 have experienced recent and extensive gene flow. Such intense gene exchanges could have occurred in large lowland refugia during the cold periods of the Pleistocene and possibly later in genetically connected populations growing over high altitude plateaus, which are typical of eastern Pyrenees landscapes. These large high altitude populations could have then become fragmented possibly during the last 30 ky when climate became warmer and drier [26].

The four well separated and differentiated genetic clusters nested in the western Pyrenean lineage appeared as a sub-structure of the east/west configuration and presumably diverged more recently. Overall, we did not observe a gradient of genetic variability loss among these clusters as expected from a range-expansion wave [11]. Rather, we observed a consistent genetic heterozygosity and allelic richness and a significant isolation by distance (Mantel test) among them. Wright's isolation by distance model suggests that gene flow is reduced as geographical distance increases. This may explain the very low level of admixture between CL3/CL6. The patchy distribution of often homogenous and highly distinct clusters enclosing a similar level of within- and between-population genomic variability is compatible with the hypothesis of the survival of the species in numerous Pyrenean refugia as proposed for other Pyrenean species [19],[18].

Moreover, the level of differentiation, the sharp genetic boundaries between most clusters and the number of migrants per generation suggested a long isolation of populations and the action of genetic drift in several high altitude refugia (HYP 2) rather than the survival in large and genetically connected populations in low elevation unglaciated areas. The higher genetic diversity in SpoI likely accounts for such refugia as well as TouH (having one private allele) where genetic variability was greater than in other populations. However, the relatively higher level of admixture between CL5 and CL4 or CL6 suggested that they had recently exchanged genes. Because the admixture concerned mainly isolated and sometimes very distant (AraH and CL5 populations) populations at high and intermediate altitude, contemporary gene exchanges among them through pollen or seed dispersal was very unlikely. We suspected that, at least in this part of the Pyrenees, longitudinal gene exchanges could have occurred possibly in lowland refugia across Pyrenean piedmonts before the post-glacial recolonization of high altitude area through valleys (HYP 1). Such a recolonizing pattern has been observed in sessile oak [51] and suggested for *Aster pyrenaeus* [20]. Thus, the phylogeographical pattern in central and western Pyrenees could result both from population isolation and differentiation in nunataks and gene exchange in some lowland refugia. Among clusters, CL6 showed the lowest genetic variability (EspI, SomI and PieI) and genetic diversity gradually decreasing in this cluster along the sequence TecL, SomI and PieI. Thus, the genetic pattern of CL6 could be chiefly related to founder events and the loss of genetic diversity and heterozygosity [11] during westward expansion of the species.

Implications for conservation

The longitudinal genetic structuration found in *R. ferrugineum* and particularly the east/west separation between the two main genetic lineages has already been observed in other studies [57], [19]. This suggests that many other alpine species may have been subject to similar events as *R. ferrugineum* and, nowadays, could present a comparable genetic pattern. We state that these data now should be usefully considered in management projects for Pyrenean species particularly those having an extended east/west distribution. According to the European Economic Community laws (92/43/EEC habitat directives, 1992) *R. ferrugineum* heathlands must be locally preserved due to their high ecological and patrimonial value. We think that, henceforth, in aim to preserve genetic diversity of the species, the genetic pattern here revealed should be taken into consideration to choose sites that have to be protected. The eastern lowland isolated populations (LecL, PraL, LapL) grow in very specific habitat and present rare alleles. According to the "rear edge population hypothesis", they harbor all together the bulk of species' genetic diversity [12]. Scientific programs have to be developed to follow the population dynamic and the genetic diversity evolution of these very original populations and take an inventory of other currently unknown rear-edge populations. Moreover, studying population dynamics at the edge of a species' range is crucial to understand the response of species to environmental changes and their susceptibility to extinction [31]. Also the populations TouH (presenting a rare allele) and SpoI (having a high allelic richness) should be considered as populations of interest for conservation. However, other parameters should be taken into account such as surrounding plant community or pollinator diversity. For example, it is known that the number of bumblebees species, which are one of the main pollinators of *R. ferrugineum*, is high in the eastern Pyrenees, especially close to population EynH [61]. Consequently, it could be an interesting population to monitor.

In conclusion, the Pyrenees are usually considered to only be a latitudinal barrier to northward or southward species expansions. In agreement with previous findings [18,19], our study of the subalpine shrub *R. ferrugineum* highlights a much more complex role of the Pyrenees in species genetic pattern and biogeography. The species showed two genetically isolated lineages, which have probably colonized the Pyrenees from the Alps and Jura Mountains. Our findings suggest that the eastern lineage most likely survived in large lowland refugia and that low-altitude small and isolated rear-edge populations have survived until now after the upward recolonization of highland areas agreeing with our HYP 3. Extensive gene flow may have occurred in lowland refugia as well as among populations growing in high altitude plateaus. The sub-structure of the central and western lineage suggests that the species mostly survived glaciations in isolated nunataks (HYP 2) but that some lowland refugia could also have played a role locally (HYP 1). Our study provides a useful genetic reference framework to select *R. ferrugineum* heathlands in management projects in the Pyrenees and points out priority populations for protection.

Supporting Information

Table S1 Characteristics of 27 polymorphic microsatellite markers in *Rhododendron ferrugineum.*

Table S2 Genetic parameters for the 27 microsatellite markers (non-significant values are in bold, P<0.01).

Table S3 Pairwise F_{ST} comparisons between all sampling sites above the diagonal as calculated by Genetix. Non-significant F_{ST} are in bold.

Acknowledgments

We thank L. Villar, C. Gallet, A. Delestrade, R. Douzet, J. Wagner and B. Doche for providing samples of *R. ferrugineum* from Spain, the Alps and the Jura Mountains. We also thank P. Winterton for editing the manuscript and the interns for assistance in the field and in the laboratory especially T. Fort, P-JG Malé and G. Besnard for helpful advice. This work was carried out in the lab EDB, part of the "Laboratoire d'Excellence (LABEX) entitled TULIP (ANR-10-LABX-41).

Author Contributions

Performed the experiments: OC PD AP NE. Analyzed the data: OC PD AP NE. Contributed reagents/materials/analysis tools: OC PD AP NE. Wrote the paper: OC PD AP NE.

References

1. Frankel OH, Brown AHD, Burdon JJ (1995) The conservation of plant biodiversity. Cambridge University press.
2. Freeman S, Herron JC (2007) Evolutionary analysis. 4th Edition. Pearson Education, Inc.
3. Barrett RDH, Schluter D (2007) Adaptation from standing genetic variation. Trends Ecol Evol. 23 (1), doi:10.1016/j.tree.2007.09.008.
4. Pauls SU, Nowak C, Balint M, Pfenninger M (2013) The impact of global climate change on genetic diversity within populations and species. Mol. Ecol. 22(4), 925–946.
5. Beniston M (1997) Variations of snow depth and duration in the Swiss Alps over the last 50 years: Links to changes in large-scale forcings. Climate Change. 36, 281–300.
6. Körner C (2003). Alpine plant life, 2nd edition, Springer, Heidelberg.

7. Hahn T, Kettle CJ, Ghazoul J, Frei ER, Matter P, et al. (2012) Patterns of genetic variation across altitude in three plant species of semi-dry grasslands. PLoS ONE. 7(8): e41608, doi:10.1371/journal.pone.0041608.
8. Hadado TT, Rau D, Bitocchi E, Papa R (2010) Adaptation and diversity along an altitudinal gradient in Ethiopian barley (Hordeum vulgare L.) landraces revealed by molecular analysis. BMC Plant Biology. 10:121
9. Hamrick JL, Godt MJW (1996) Effects of life history traits on genetic diversity in plant source: philosophical transactions. Biological Sciences. 351(1345), 1291–1298.
10. Ægisdóttir HH, Kuss P, Stöcklin J (2009) Isolated populations of a rare alpine plant show high genetic diversity and considerable population differentiation. AoB. 104 (7): 1313–1322.
11. Hewitt GM (2000) The genetic legacy of the quaternary ice ages. Nature, 405, 907–913.
12. Hampe A, Petit RJ (2005) Conserving biodiversity under climate change: the rear edge matters. Ecol. Let. 8, 461–467.
13. Aitken SN, Libby WJ (1994) Evolution of the pigmy-forest edaphic subspecies of Pinus contorta across an ecological staircase. Evol. 48 (4), 1009–1019.
14. Ledig FT (2000) Founder effects and the genetic structure of Coulter pine. J. Hered. 91, 307–315.
15. Hensen I, Kilian C, Wagner V, Durka W, Pusch J, et al. (2010) Low genetic variability and strong differentiation among isolated populations of the rare steppe grass Stipa capillata L. in central Europe. Plant Biology. 12, 526–536.
16. Stehlik I, Schneller JJ, Bachmann K (2002) Immigration and in situ glacial survival of the low-alpine Erinus alpinus (Scrophulariaceae). Biol. J. Linn. Soc. 77, 87–103.
17. Byars SG, Parsons Y, Hoffmann AA (2009) Effect of altitude on the genetic structure of an Alpine grass, Poa hiemata. AoB. 103, 885–899, doi:10.1093/aob/mcp018.
18. Segarra-Moragues JG, Palop-Esteban M, Gonzalez-Candelas F, Catalan P (2007) Nunatak survival vs. tabula rasa in the Central Pyrenees: a study on the endemic plant species *Borderea pyrenaica* (Dioscoreaceae). J. Biogeogr. 34, 1893–1906.
19. Lauga B, Malaval S, Largier G, Regnault-Roger C (2009) Two lineages of Trifolium alpinum (Fabaceae) in the Pyrenees: evidence from random amplified polymorphic DNA (RAPD) markers. Acta Bot. Gallica. 156(3), 317–330.
20. Escaravage N, Cambecèdes J, Largier G, Pornon A (2011) Conservation genetics of the rare Pyreneo-Cantabrian endemic Aster pyrenaeus (Asteraceae). AoB Plants, plr029 doi: 10.1093/aobpla/plr029.
21. Serrat D, Ventura J (1993) Glaciers of the Pyrenees, Spain and France, U.S. Geological survey professional paper. 1386- E- 2.
22. Petit RJ, Aguinagalde I, de Beaulieu JL, Bittkau C, Brewer S, et al. (2003) Glacial Refugia: Hotspots but not melting pots of genetic diversity. Science. 300(5625), 1563–5.
23. Jeannel R (1946) Le peuplement des Pyrénées. Revue française d'entomologie. 14 (1) 53–104.
24. Medail F, Diadema K (2009) Glacial refugia influence plant diversity patterns in the Mediterranean Basin. J. Biogeogr. 36, 1333–1345.
25. Dzialuk A, Muchewicz E, Boratynski A, Montserrat JM, Boratyńska K, et al. (2009) Genetic variation of Pinus uncinata (Pinaceae) in the Pyrenees determined with cpSSR markers. Plant Syst Evol. 277, 197–205.
26. Jalut G, Monserrat MJ, Fontugne M, Delibrias G, Vilaplana JM, et al. (1992) Glacial to interglacial vegetation changes in the northern and southern Pyrenees: deglaciation, vegetation cover and chronology. Quaternary ScienceReviews. 11, 449–480.
27. Goetsch L, Eckert AJ, Hall BD (2005) The molecular systematics of Rhododendron (Ericaceae): a phylogeny based upon RPB2 gene sequences. American Society of Plant Taxonomists, 30: 616–626.
28. Taberlet P, Fumagalli L, Wust-Saucy AG, Cosson JF (1998) Comparative phylogeography and postglacial colonization routes in Europe. Mol. Ecol. 7, 453–464.
29. Escaravage N, Questiau S, Pornon A, Doche B, Taberlet P (1998) Clonal diversity in a Rhododendron ferrugineum L. (Ericaceae) population inferred from AFLP markers. Mol. Ecol. 7, 975–982.
30. Pornon A, Escaravage N, Thomas P, Taberlet P (2000) Dynamics of genotypic structure in clonal Rhododendron ferrugineum (Ericaceae) populations. Mol. Ecol. 9, 1099–1111.
31. Bruni I, De Mattia F, Labra M, Grassi F, Fluch S, et al. (2012) Genetic variability of relict Rhododendron ferrugineum L. populations in the Northern Apennines with some inferences for a conservation strategy. Plant Biosystems, 146:1, 24–32.
32. Escaravage N, Pornon A, Doche B, Tillbottraud I (1997) Breeding system in an alpine species: Rhododendron ferrugineum (Ericaceae) in the French northern Alps. Can. J. Bot. 75, 736–743.
33. Delmas CEL, Escaravage N, Pornon A (2013) Massive floral display affects insect visits but not pollinator-mediated pollen transfer in Rhododendron ferrugineum. Plant Biology, in press doi: 10.111/plb.12039.
34. Escaravage N, Wagner J (2004) Pollination effectiveness and pollen dispersal in a Rhododendron ferrugineum (Ericaceae) population. Plant Biology. 6, 000A ± 000J.
35. Wolf PG, Doche B, Gielly L, Taberlet P (2004) Genetic structure of Rhododendron ferrugineum at a wide range of spatial scales. J. Hered. 95(4), 301–308.
36. Delmas CEL, Lhuillier E, Pornon A, Escaravage N (2011) Isolation and characterization of microsatellite loci in Rhododendron ferrugineum (Ericaceae) using pyrosequencing technology. Am. J. Bot. e120–e122.
37. Charrier O, Delmas CEL, Pornon A, Escaravage N (2012) Development of 18 microsatellite markers in Rhododendron ferrugineum (Ericaceae) for investigating genetic structure at margins. Conservation Genetics Resources, doi 10.1007/s12686-012-9831-2.
38. Holleley C, Geerts P (2009) Multiplex Manager 1.0: A cross-platform computer program that plans and optimizes multiplex PCR. BioTechniques. 46: 511–517.
39. Peakall R, Smouse PE (2006) GENALEX 6: genetic analysis in Excel. Population genetic software for teaching and research. Mol. Ecol. Notes. 6, 288–295.
40. Van Oosterhout C, Hutchinson WF, Wills DPM, Shipley P (2004) Micro-checker: software for identifying and correcting genotyping errors in microsatellite data. Mol. Ecol. Notes. 4(3), 535–538.
41. Raymond M, Rousset F (1995) GENEPOP (version 1.2): population genetics software for exact tests and ecumenicism. J. Hered. 86, 248–249.
42. Weir BS, Cockerham CC (1984) Estimating F-Statistics for the analysis of population structure. Evolution. 38(6), 1358–1370.
43. Meirmans PG, Van Tienderen PH (2004) GENOTYPE and GENODIVE: two programs for the analysis of genetic diversity of asexual organisms. Mol. Ecol. Notes. 4(4), 792–794.
44. Belkhir K, Borsa P, Chikhi L, Raufaste N, Bonhomme F (2004)GENETIX 4.0.Laboratoire Génome, Populations, Interactions, CNRS UMR 5000, Université de Montpellier II, Montpellier(France).
45. Pritchard JK, Stephens M, Donnelly PJ (2000) Inference of population structure using multilocus genotype data. Genetics. 155, 945–959.
46. Evanno G, Regnaut S, Goudet J (2005) Detecting the number of clusters of individuals using the software STRUCTURE: a simulation study. Mol. Ecol. 14, 2611–2620.
47. Earl DA, vonHoldt BM (2012) STRUCTURE HARVESTER: a website and program for visualising STRUCTURE output and implementing the Evanno method. Conservation Genetics Resources. 4, 359–361.
48. Excoffier L, Smouse PE, Quattro JM (1992) Analysis of Molecular Variance inferred from metric distances among DNA haplotypes: Application to human mitochondrial DNA restriction data. Genetics, 131, 479–491.
49. Mantel N (1967) The detection of disease clustering and a generalized regression approach. Cancer Res. 27, 209–220.
50. Barton NH, Slatkin M (1986) A quasi-equilibrium theory of the distribution of rare alleles in a subdivided population. Heredity. 56, 409–416.
51. Alberto F, Niort J, Derory J, Lepais O, Vitalis R, et al. (2010) Population differentiation of sessile oak at the altitudinal front of migration in the French Pyrenees. Mol. Ecol. 19, 2626–2639.
52. Koelling VA, Hamrick JL, Mauricio R (2012) Genetic diversity and structure in two species of Leavenworthia with self-incompatible and self-compatible populations. Heredity. 106 (2), 310–318.
53. Holderegger R, Stehlik I, Abbott RJ (2002) Molecular analysis of the Pleistocene history of Saxifraga oppositifolia in the Alps. Mol. Ecol. 11, 1409–1418.
54. Demesure B, Sodzi N, Petit RJ (1995) A set of universal primers for amplification of polymorphic noncoding regions of mitochondrial and chloroplast DNA in plants. Mol. Ecol. 4, 129–131.
55. Burga CA, Perret R (1998) Vegetation und Klima der Schweiz seit dem jüngeren Eiszeitalter. Ott, Thun.
56. Schönswetter P, Stehlik I, Holderegger R, Tribsch A (2005) Molecular evidence for glacial refugia of mountain plants in the European Alps. Mol. Ecol. 14, 3547–3555.
57. Kropf M, Comes HP, Kadereit JW (2006) Long-distance dispersal vs vicariance: the origin and genetic diversity of alpine plants in the Spanish Sierra Nevada. New Phytol. 172, 169–184.
58. Braun-Blanquet J (1948) Les souches préglaciaires de la flore Pyrénéennes. Collectana Botanica. 2, 1–23.
59. Ohsawa T, Ide Y (2007) Global patterns of genetic variation in plant species along vertical and horizontal gradients on mountains. GEB, DOI: 10.1111/j.1466-8238.2007.00357.x.
60. Lesica P, Allendorf FW (1995) When are peripheral populations valuable for conservation? Conserv. Biol. 9(4), 753–760.
61. Iserbyt S, Durieux E-A, Rasmont P (2008) The remarkable diversity of bumblebees (Hymenoptera: Apidae: Bombus) in the Eyne Valley (France, Pyrénées-Orientales). Annales de la Société entomologique de France (n.s.), 44 (2): 211–241.

13

Unexpectedly Low Rangewide Population Genetic Structure of the Imperiled Eastern Box Turtle *Terrapene c. carolina*

Steven J. A. Kimble[1]*, O. E. Rhodes Jr.[2], Rod N. Williams[1]

1 Department of Forestry and Natural Resources, Purdue University, West Lafayette, Indiana, United States of America, **2** Savannah River Ecology Laboratory, Aiken, South Carolina, United States of America

Abstract

Rangewide studies of genetic parameters can elucidate patterns and processes that operate only over large geographic scales. Herein, we present a rangewide population genetic assessment of the eastern box turtle *Terrapene c. carolina*, a species that is in steep decline across its range. To inform conservation planning for this species, we address the hypothesis that disruptions to demographic and movement parameters associated with the decline of the eastern box turtle has resulted in distinctive genetic signatures in the form of low genetic diversity, high population structuring, and decreased gene flow. We used microsatellite genotype data from (n = 799) individuals from across the species range to perform two Bayesian population assignment approaches, two methods for comparing historical and contemporary migration among populations, an evaluation of isolation by distance, and a method for detecting barriers to gene flow. Both Bayesian methods of population assignment indicated that there are two populations rangewide, both of which have maintained high levels of genetic diversity ($H_O = 0.756$). Evidence of isolation by distance was detected in this species at a spatial scale of 300 – 500 km, and the Appalachian Mountains were identified as the primary barrier to gene flow across the species range. We also found evidence for historical but not contemporary migration between populations. Our prediction of many, highly structured populations across the range was not supported. This may point to cryptic contemporary gene flow, which might in turn be explained by the presence of rare transients in populations. However these data may be influenced by historical signatures of genetic connectivity because individuals of this species can be long-lived.

Editor: Adam Stow, Macquarie University, Australia

Funding: Support for this project was provided by the Indiana Department of Natural Resources, Division of Fish and Wildlife, Wildlife Diversity Section, State Wildlife Improvement Grant E2-08-WDS15; Indiana Division of Forestry Grant E-9-6-A558; and the Department of Forestry and Natural Resources, Purdue University. The funders had no role in study design, data collection and analysis, decision to publish, or preparation of the manuscript.

Competing Interests: The authors have declared that no competing interests exist.

* E-mail: sjkimble@gmail.com

Introduction

The quantification of genetic parameters of species is basic to understanding their natural history. This is especially important in declining species in need of appropriate conservation approaches [1]. Species in decline, however, have by definition undergone demographic reductions [2], which may confound our ability to differentiate between natural and anthropogenically induced changes in genetic parameters. For example, when studying long-lived species investigators must distinguish between pre-disturbance (and therefore presumably stable) and anthropogenically induced genetic patterns to inform management strategies [3,4]. Furthermore, population genetic studies involving declining species are often confined to drawing broad management conclusions from limited data on a few individuals or populations, resulting in deviations from analytical assumptions which can negatively affect the reliability of results [5].

Wide-ranging species of conservation concern, whose interpopulation migration patterns are influenced by habitat fragmentation, also could be expected to exhibit genetic evidence of disruption to large- and small-scale movement behaviors [6]. Such species are often characterized by high dispersal (intrapopulation movement) and migration (interpopulation movement) habits, and thus are expected to be vulnerable to habitat fragmentation [7,8]. Unfortunately, the genetic attributes of many declining, wide-ranging species are poorly studied over large spatial scales, despite the fact that many are marked for conservation management planning.

A genetic population is commonly defined in the field as a group of conspecifics that are genetically similar and are to varying degrees separated from other populations of conspecifics and are likely to be more locally adapted [9]. Populations of long-lived, wide-ranging species should be managed at geographic scales that are appropriate for conservation planning at multiple levels of biological resolution [10,11]. Management plans conducted at geographic scales significantly smaller than that of the population may fail to incorporate mechanisms that maintain genetic diversity [12], while those conducted at scales larger than that of the population may lead to the loss of locally adapted genes [13]. For example, anthropogenic interpopulation movement of genotypes may introduce alleles that are locally maladaptive in the receiving population [13,14]. In addition, rangewide approaches are tremendously useful for identification of significant management units, i.e., genetic populations for which management plans should be made for species or populations of conservation concern [1,15] even those with low population differentiation [11]. Determination of the appropriate scale for management of turtles, in particular, is of paramount importance [16] because approximately 40% of Chelonians worldwide are considered endangered or vulnerable [16,17]. Simply removing local sources of extrinsic stressors, without regard to range-wide reservoirs of genetic resources, may not be sufficient mitigation against loss of genetic diversity because future threats such as climate change and novel disease may overwhelm genetically depauperate species [4,16].

The eastern box turtle *Terrapene c. carolina* is a declining species whose conservation plans are difficult to develop because of the complexities of studying a species that is cryptic, long-lived (>100 years in the wild; N. Karraker, unpublished data; [18]), and whose generations overlap. It is a terrestrial turtle species that historically ranged across much of the eastern United States [19], but which has suffered substantial demographic declines [19–22], likely due to some combination of habitat destruction and fragmentation, road mortality, collection, and disease [19]. Understanding the effects of demographic declines on patterns of genetic diversity and structure in box turtles is a necessary precursor to developing management strategies for their protection, but previous studies suggest that such studies should be conducted at scales larger than single states [23–25].

Our intent here was to explore range-wide genetic patterns of the eastern box turtle as a means to inform conservation planning for this species and provide a model for other species with similar traits and demographic histories. To accomplish this we first tested the hypothesis that genetic isolation by distance is significant across the species range due to the limited migration ability of this species. Second, we tested the hypothesis that habitat loss and fragmentation has formed multiple, geographically discrete and genetically differentiated populations across the species range. Because habitat reduction increases distance among patches of suitable habitat, we also tested a third hypothesis that increasing isolation due to habitat fragmentation has caused a reduction in the number of migrants among populations.

Methods

Sample collection

We conducted searches for eastern box turtles via visual encounter by car and on foot across much of their range (Figure 1). We sampled at multiple geographic scales to determine at what spatial extent populations occur. To this end, we sampled at eight sites in Indiana (a state whose forests are heavily fragmented by agriculture; [26]), in the four states surrounding Indiana (Illinois, Kentucky, Ohio, and Michigan), and from another nine states across the species range. We intentionally avoided sampling in the south and southwestern parts of the range where the eastern box turtle is sympatric with two other subspecies of *T. carolina* [25] to avoid confounding results with alleles from different subspecies. For each individual we recorded UTM coordinates, morphometric data, sex, activity, and any unusual markings or signs of injury or disease. We took tissue samples for genetic analysis, usually blood (~10µL), following the protocol of Kimble and Williams [27]. We assigned each turtle a unique number and filed a corresponding pattern of notches into the marginal scutes [28] to identify recaptures. These notches were subsequently sealed with surgical adhesive. Individuals were processed as quickly as possible and released immediately at the point of capture.

All animals we handled were done so in accordance with the Purdue Animal Care and Use Protocol 07-037 and amendments thereto. For all animals we handled we obtained all relevant permissions and permits from the appropriate government agencies, land trusts, and property managers before sampling began. We sampled in the Chattahoochee National Forest area of Georgia under permits from the USDA Forest Service Chattahoochee-Oconee National Forest office and Georgia Department of Natural Resources, Wildlife Resources Division; in Illinois under a permit from the Illinois Department of Natural Resources; in Indiana under permits from the Indiana Department of Natural Resources, Fish and Wildlife Division and Division of Nature Preserves and under permissions from NICHES Land Trust and Wabash College; in New York under a permit from the New York Department of Environmental Conservation, Division of Fish, Wildlife and Marine Resources; in North Carolina under a permit from the North Carolina Wildlife Resources Commission, Division of Wildlife Management; and in Ohio under the Ohio Department of Natural Resources, Division of Wildlife. Samples from other states were collected by collaborators working under their own permits.

Laboratory

We digested tissue samples using a modified proteinase K protocol and extracted DNA with a phenol-chloroform-isoamyl alcohol [29]. We resuspended purified DNA in 50 µL of TLE (10 mM tris, 0.1 mM EDTA, pH 8.0) and quantified DNA concentration on a spectrophotometer (NanoDrop 8000, Thermo Scientific, Wilmington, DE). We then diluted all DNA samples to 20 ng/µL in pure water prior to PCR.

We carried out PCR using 11 microsatellite loci developed specifically for the eastern box turtle [30]. We combined the 11 loci into three multiplexes and two singletons (Table 1). All reactions contained 60 ng DNA template, 10 mM tris-HCl, 0.05 mg/mL BSA, 50 mM KCl, 0.9 mM $MgCl_2$, 0.2 mM of each dNTP, 0.3 U of Taq polymerase, and multiplex-specific concentration of end-labeled fluorescent primers in a total reaction volume of 10 µL. We analyzed all PCR products on an automatic sequencer (ABI 3730XL, Applied Biosystems, Foster City, CA). We automatically scored genotypes with GeneMapper (version 3.7, Applied Biosystems, Foster City, CA) and checked each call manually at least twice. We reamplified ~10% of all genotypes to ensure repeatability and reamplified any that disagreed a third time. We used Cervus (version 3.0.3; [31]) to check for accidentally duplicated samples and in the case of duplication removed all but the first sample taken from an individual.

Statistical analysis

To test for the presence of null alleles and large allelic dropout, we used Microchecker (version 2.2.3; [32]). To assess how well our dataset and each population met Hardy-Weinberg equilibrium (HWE) assumptions, we used the web version of Genepop (version 4.0.10; [33]) to test for a significant deficit of heterozygotes using an exact method [34] with default parameters. We used a Markov chain Monte Carlo (MCMC) method [35] to assess significant deviations from HWE. We quantified deviations from Hardy-Weinberg equilibrium (HWE) by estimating F_{IS} in Genepop with 10,000 iterations.

Isolation by distance

We evaluated rangewide isolation by distance by using a Mantel test [36] performed in Alleles in Space (Ais, version 1.0; [37]) to determine whether a significant correlation existed between pairwise matrices of Nei's [38] genetic distances and geographic distances matrices. We log_{10}-transformed the geographic distance matrix to meet the assumption of normality [39] and used 1,000

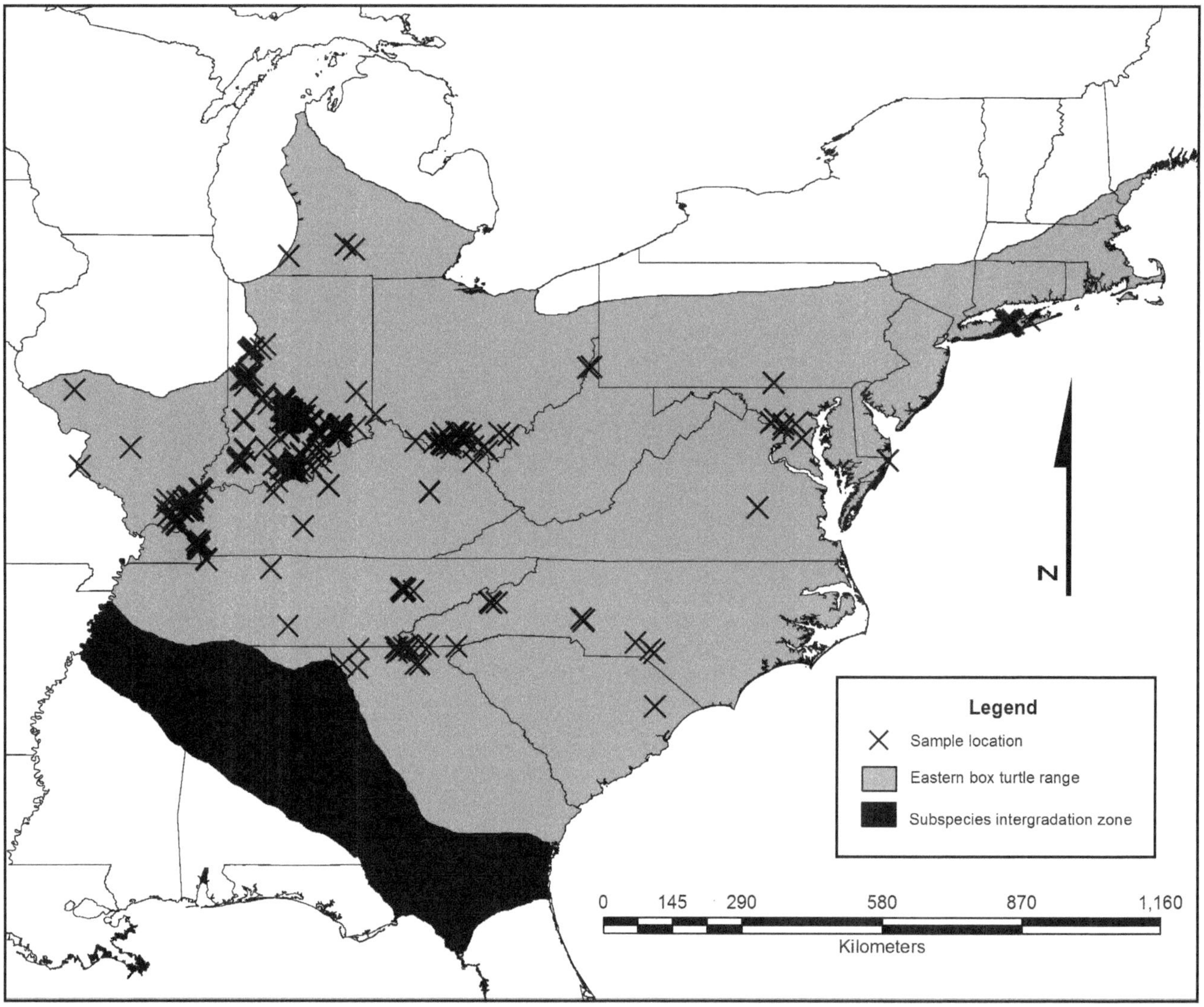

Figure 1. Locations of rangewide samples collected from *Terrapene c. carolina*. Due to the resolution, some marks represent more than one sample. Range data after Dodd (2001).

Table 1. Microsatellite PCR multiplex parameters for the eastern box turtle, *Terrapene. c. carolina*.

Multiplex	Locus	T_A (°C)	Primer concentration (μM)	Fluorophore
I	TCC_di_045	58.0	0.3	6-FAM
I	TCC_tetra_070	58.0	0.3	NED
I	TCC_di_082[1]	58.0	0.3	6-FAM
II	TCC_di_189	60.0	0.6	6-FAM
II	TCC_tetra_043	60.0	0.6	HEX
II	TCC_tetra_012/342	60.0	0.3	NED
III	TCC_di_366	63.0	0.5	6-FAM
III	TCC_di_318	63.0	0.3	HEX
III	TCC_di_352	63.0	0.3	NED
IV	TCC_di_345	63.0	0.3	6-FAM
V	TCC_di_300	57.0	0.3	6-FAM

[1]TCC_di_082 was PCR amplified separately and pooled with Multiplex I before genotyping.

randomized replicates to assess significance. To estimate the geographic scale at which IBD may begin to operate, we also performed a complementary spatial autocorrelation analysis in AIS, using 10 classes of equal distance and 1,000 replicates to assess significance.

Determination of population number, individual assignments, and barriers

To improve analytical robustness against violations of software-specific assumptions, a common approach in population genetics [40], we used two Bayesian analyses to infer the size, shape, and individual membership of populations. The first approach was in GENELAND (version 4.0.0; [41]) in program R [42] without spatial priors. We set GENELAND to search for the most likely number of populations (*k*) from 1–25 over 1,000,000 MCMC iterations with thinning every 100th iteration and a burn-in of 50,000. We assumed uncorrelated allele frequencies to avoid artifacts of uneven geographic sampling [43] and the potential for overestimation of *k* associated with the use of correlated frequency allele models [44]. We set the Poisson process maximum to 800 and the maximum number of nuclei allowed for the Poisson-Voronoi tessellation to 2,400 [41,45]. We ran the MCMC 10 times for each value of *k* and used the highest mean probability density value as the inferred *k* value. We also estimated the probability of population assignment for each individual.

To corroborate results, we used the Bayesian algorithm STRUCTURE (version 2.3; [46]). We performed 10 independent runs for each value of *k* from 1 to 25 for 1,000,000 iterations, including a 50,000 iteration burn-in. We ran the model with default settings except we used an admixture model, and with allele frequencies uncorrelated (as with GENELAND). To visualize and infer the most likely value of *k* we used STRUCTURE HARVESTER (version 0.6.92; [47]), which employs the Δk method of Evanno et al [48]. STRUCTURE also assigns a probability value to each the population assignment of each individual. For both GENELAND and STRUCTURE we analyzed each population individually for sub-structuring. We compared the results of both algorithms and used the population assignment with the highest confidence per individual for further analyses. Finally, we compared the individuals assigned to each population by both GENELAND and STRUCTURE for agreement. We tested resulting genetic populations for violations of HWE and dropped loci with relatively high null allele estimates as necessary.

To estimate the location of specific natural or anthropogenic barriers to gene flow which might have contributed to population structure identified across the species range, we used the maximum difference algorithm of Monmonier [49], which has recently been applied to landscape genetics [45,50–52]. This process builds a connectivity network of the sample locations using Delaunay triangulation, and then estimates the barriers among them by following contiguous connectivity links between samples that represent the highest genetic distances. We used this method in AIS using genetic distances corrected for geography ("pseudo-slopes"). We set the number of barriers to *k* - 1, with *k* being the number of populations identified by GENELAND and STRUCTURE.

Previous studies in the population genetics of *Terrapene* species suggest that populations may operate at geographic scales greater than most management jurisdictions, such as state or national parks, and perhaps even larger than a single state [4,23,25]. If this is the case, managers of local sites cannot manage for the entire population but instead must work with the genetic reality of the individuals under their control. To this end, we also report results for each management unit in which we sampled at least nine individuals (Table S1).

Quantification of migration among populations

We used the genetic population clusters identified by the Bayesian analyses to detect recent migration among them using BAYESASS (version 3; [53]), a genotype-based Bayesian platform appropriate even for populations that do not meet assumptions of HWE. We ran 1,000,000 iterations with a burn-in of 100,000 and sampling intervals of 100. We set the mixing parameters for the MCMC chain to 0.02 for migration, 0.1 for inbreeding, and 0.05 for allele frequencies so that the resulting migration parameter swapping acceptance rates were between 20% and 40% as suggested by the authors. We ran the analysis five times to check for convergence and visualized chain mixing, convergence, and burn-in values in TRACER (version 1.5; [54]).

We used MIGRATE (version 3.1.1; [55]) to estimate historical migration rates (among populations identified by the Bayesian clustering analyses). Under the maximum likelihood (ML) framework, we used a Brownian motion model of mutation, suitable for microsatellite data that likely do not adhere to a strictly stepwise model of mutation. We used five independent replicates of 10 short chains 10,000 iterations in length, three long chains 100,000 iterations in length, and four heated static chains at temperatures 1.0, 1.5, 3.0 and 10,000.

Results

Sample collection and laboratory

We collected tissue samples from 1,603 wild eastern box turtles from across much of the species range (Figure 1). We successfully resolved all quality control disagreements among PCR amplifications and between scorers and excluded all individuals with more than three loci for which genotype data could not be resolved (n = 45). Thus, the final data set included 1,558 individuals. For all genetic analyses, we randomly selected 24 individuals from the two locations where we sampled deeply for other purposes (the Hardwood Ecosystem Experiment in south-central Indiana: n = 627; and Oak Ridge, Tennessee: n = 182), to even the sampling distribution across sites. This resulted in a total of 799 individuals from which data were used for all analyses of genetic structure, gene flow and IBD.

Statistical analysis

Mean allelic richness from the sample sites ranged from 7.6 to 33.6 (Table S1). When all 799 samples were pooled, there was a significant deviation from Hardy-Weinberg equilibrium across all 11 loci ($\chi^2 = \infty$, df = 22, $p<0.001$). MICROCHECKER estimated the potential presence of numerous null alleles at low frequencies (Table S1). All but two of the management unit populations (e.g., Patuxent Wildlife Research Center in Maryland) were also significantly out of equilibrium, but these two have low sample sizes (9 and 12). A few loci suffered from heterozygote deficiencies in some populations (Table S2).

Isolation by distance

Isolation by distance was significant across the species range (Figure 2; $r = 0.13$, $p<0.001$). Additionally, the spatial autocorrelation analysis estimated that the pairwise genetic distance in the 300 – 500 km distance class ($A_y = 0.795$) exceeded the mean pairwise genetic distance for all pairwise distances ($A_y = 0.793$). This indicates that significant isolation begins to operate at this geographic distance and that panmixia operates at geographic scales less than this threshold [56,57].

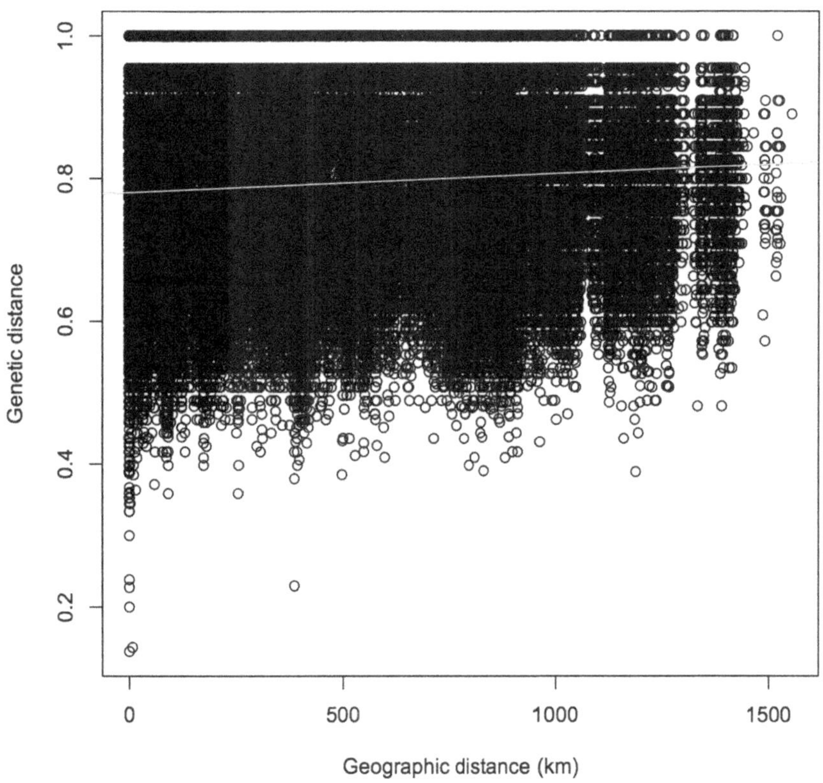

Figure 2. Correlogram of genetic isolation by geographic distance between pairs of eastern box turtle *Terrapene c. carolina* individuals across its range.

Determination of population number, individual assignments, and barriers

The nonspatial model GENELAND reported $k=2$ in 7 of 10 runs. One of the runs reporting $k=2$ also had the highest mean of probability density value and so we used this as the point estimate of the number of populations. Secondary runs on each of the two clusters revealed no substantial substructuring, i.e., population assignment probabilities were low. The final map (Figure 3) shows the population boundary roughly following the spine of the Appalachian Mountains. The nonspatial model STRUCTURE also indicated that $k=2$ according to the point value estimation method of Evanno et al [48]. More than 85% of individuals were assigned a population assignment probability of >0.9. Secondary analyses on these two clusters also showed no substantial substructuring, i.e., all Δk values were low. Individuals with admixed ancestry did not tend to cluster along the boundary between the Western and Eastern populations, a signature of a zone of interbreeding. STRUCTURE and GENELAND assigned 95.6% of individuals to the same two population groups.

There was a nearly universal lack of Hardy-Weinberg equilibrium, despite the fact that the Bayesian methods for population delineation construct populations by minimizing deviations from HWE and linkage. This might be caused by technical reasons (e.g., null alleles; [58]), or by the violation of HWE assumptions that are not appropriate for box turtles [9]). These deviations may be explained by the high polymorphisms at our loci, which range from 16 to 83 alleles per locus (mean: 36.1) and which resulted in the detection of many rare alleles. As allelic richness increases at a locus, the likelihood that they will be found as homozygotes declines in a finite sample size of the population, increasing the likelihood of deviations from HWE [9]. Furthermore, while H_E and H_O estimates vary by locus in each population, mean H_E and H_O estimates are similar across populations which suggests that departures from HWE are artifacts of the algorithms and do not compromise the interpretations of our data (Table S2). It has also been demonstrated that distinguishing between deviations from HWE and the presence of null alleles can be difficult, and the common tactic of excluding "problematic" loci may result in a loss of the most informative loci [59].

Quantification of migration among populations

Because both STRUCTURE and GENELAND were strongly concordant, we coded AIS to find one Monmonier barrier between them. The resulting line was largely congruent with the map that GENELAND generated, drawing the barrier along a line running north to south from Pennsylvania through Maryland, Virginia, North Carolina and South Carolina. We refer to the two resulting populations hereafter as "Western" and "Eastern". Though both GENELAND and STRUCTURE operate by minimizing linkage disequilibrium and departures from HWE, both populations were also significantly out of HWE (Table S1). Both the Western and Eastern populations had three loci with a null allele rate 0.10 – 0.12 but exclusion of these from both the Bayesian approaches and the test for HWE had not substantial effect on the results.

BAYESASS gave no evidence for recent immigration between populations as estimates included zero in their credible intervals. By contrast, MIGRATE estimated a historical migration rate of 15.2 migrants/generation from the Western population into the Eastern, and 17.4 migrants/generation from East to West.

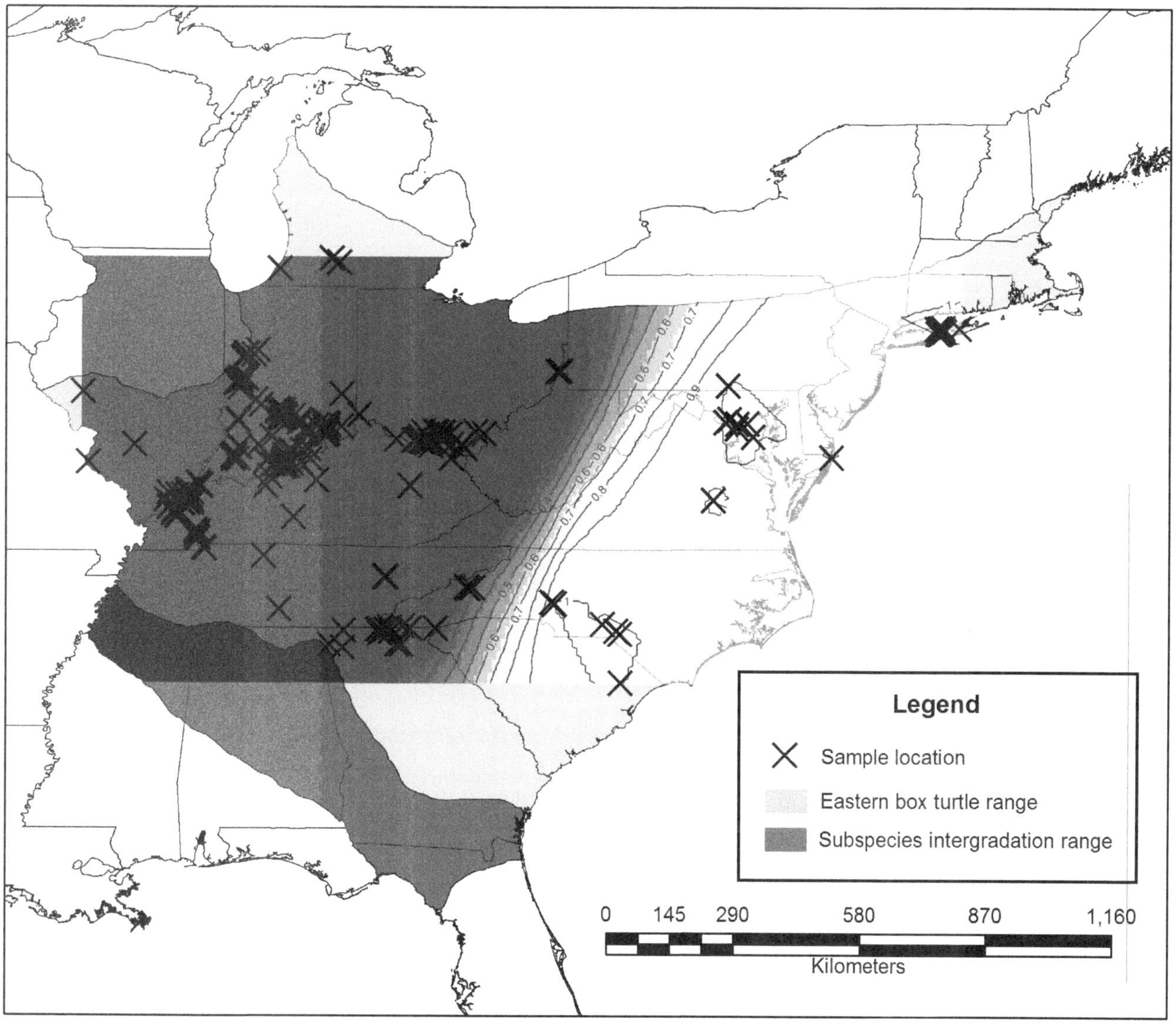

Figure 3. Map of the two populations found rangewide in the eastern box turtle *Terrapene. c. carolina*. The probability map of an individual turtle belonging to the Western population was generated by GENELAND and increases with darker shades of gray. The probability of an individual belonging to the Eastern population would be proportionally opposite. Note that the border between the populations follows the Appalachian Mountains.

Discussion

Life history traits of *Terrapene* species make the formulation of clear hypotheses about the geographic patterns of genetic populations difficult. Much of the natural history evidence for *Terrapene* suggests that members of this genus should maintain population structure at relatively small geographic scales. For example, one *T. ornata* individual was recaptured 27 times over ten years within 7.6 m of its initial capture location [60] and home range size for *T. carolina* has been reported between 0.02 and 187.67 ha [19,61]. Observational data suggest that juvenile dispersal may be very short distances, approximately 100 m or fewer [62,63]. Ultimately, evidence of low dispersal and highly conserved adult home ranges suggests that members of this genus should display a high degree of population structure at small spatial scales.

Alternatively, there is some evidence to the contrary. First, the existence of transient adult box turtles has been hypothesized due to the lack of recapture, despite intense effort spanning decades [20,21,64]. Furthermore, individuals with transient behavior have been observed in two radiotelemetry studies (*T. c. triunguis*, [70]; S. Kimble, *T. c. carolina*, unpublished data.) where adult males traveled roughly linear paths that were many times longer than the width of the standard home range (~10 km over two active seasons, [65]; ~7 km over 1 active season, S. Kimble, unpublished data), and have been observed mating along the way [65]. Second, *T. carolina* box turtles are not closely tied to bodies of water [19], and thus may suffer from fewer geographic constraints on gene flow than do aquatic turtles. Third, while little is known about juvenile *Terrapene* dispersal [19], parent-offspring pairs have been found up to 27.1 km apart [S. Kimble, unpublished data] suggesting some mechanism for higher gene flow than is currently appreciated.

Our data support the latter theory by indicating that there are only two populations across most of the range of the eastern box turtle. Furthermore the Appalachian Mountains may act or have acted as a barrier to gene flow at the continental scale, although eastern box turtles are currently known to inhabit all but the highest altitudes in North Carolina [19]. The finding that populations operate at large geographic scales is supported by previous work in the *Terrapene* species complex. In *T. ornata*, individuals from two sites 120 km apart were found to be genetically panmictic [66]. In *Terrapene c. carolina*, Marsack and Swanson [24] found that individuals separated by 30 to 70 km in southwestern Michigan also constituted a single population. Hagood [23] and Butler and colleagues [25] documented low genetic structure in *T. c. carolina* across larger distances of 160 km and 250 km, respectively. These studies suggest that the approach taken in this study, evaluating population genetic patterns at the scale of the species range, is the appropriate approach for the eastern box turtle.

Our data also indicate that isolation by distance is operating over relatively large spatial scales across the range of this species. The spatial autocorrelation analysis demonstrated that at approximately 300 – 500 km, mean pairwise genetic distances begin to exceed the average for the entire data set, suggesting a geographic extent at which populations operate. The same analysis by Hagood [23] in *T. c. carolina* returned a similar result of 450 to 650 km, which is approximately the distance from the Appalachian Mountains to the edge of the species range. These results, combined with the GENELAND and STRUCTURE results, describe a species that though apparently highly philopatric, has (or had) high gene flow across vast areas. Though box turtles are reported throughout much of Appalachia [19], cryptic barriers to gene flow such as terrain and elevation may cause subtle population barriers in turtles [67] that may result in the Appalachians serving as a modest barrier.

Overall, we detected little evidence that habitat fragmentation is so far affecting population genetic structure in eastern box turtles. The shape and scope of the populations appear to be more consistent with a historical landscape than with current patterns of landscape fragmentation (Fig 1). Individuals from as far distant as eastern Tennessee and southwestern Michigan were assigned to the same population and few private alleles were detected from samples across these geographically distant populations (Table S1). The exception was possibly a signal of incipient decline: we did find evidence that migration between the two rangewide populations has recently been reduced or eliminated, a result expected in the presence of increasing habitat fragmentation. However, generation times can be long in *Terrapene*, with longevity in the wild at least 100 years (N. Karraker, unpublished data; [18]). As few as three generations could have passed since European settlers started clearing large swaths of forests in the range of the eastern box turtle [68], suggesting the idea that the genetic signatures we see in *Terrapene* box turtles may be largely historical. The loss of migration events between the two populations may be the first signal of reduced gene flow.

Furthermore, though we do not know what historical levels were, genetic diversity appears to remain high. Allelic richness and observed heterozygosity are high (Table S1). Long-lived species that have experienced demographic declines yet retain high genetic diversity include the Nile crocodile [69] and the harpy eagle [70]. However, while the former has experienced population growth under the protection of a CITES listing, the latter has not exhibited any such rebound. To date, all long-term demographic studies of *Terrapene* have documented declines [20–22], suggesting that box turtles require more active management to further prevent declines and extirpations.

This work represents the first rangewide study of population genetics patterns in the *Terrapene carolina* complex and provides insights into the ecology of a subspecies that appears to have only begun to exhibit the effects of habitat fragmentation within the last few generations. Indeed, our data indicate that genetic diversity remains high across the range of this *Terrapene subspecies*, yet the detection of historical gene flow and the lack of recent immigration events are signatures of recently reduced population connectivity. The loss of genetic diversity is a major threat to chelonian species worldwide [16,17]. However, in long-lived species with overlapping generations, signatures of genetic loss may be masked for decades or even centuries [71]. Turtles are some of the longest-living vertebrates on the planet and yet many have suffered severe demographic declines, necessitating immediate management plans.

Acknowledgments

We wish to thank the many individuals who generously helped collect eastern box turtle tissue samples from across the species range, including M. Allender, M. Baragona, J. Beane, K. Buhlmann, N. Burgmeier, J. Butler, D. Carlson, V. Clarkston, K. Creely, M. Cook, M. Cross, T. Despot, A. Durso, N. Engbrecht, E. Estabrook, J. Faller, S. Foertmeyer, A. Garcia, B. Geboy, S. Hagood, J. Hall, K. Hanauer, C. Hennessy, P. Henry, A. Hoffman, J. Greathouse, T. Green, J. Groves, L. & M. Jackson, T. Jedele, B. Johnson, S. Johnson, L. Keener-Eck, J. Kissel, A. Krainyk, S. Klueh, V. Kinney, N. Levitte, K. Lilly, J. MacNeil, J. Mitchell, T. Mitchell, K. Norris, J. Petranka, H. Powell, K. Powers, S. Price, J. Richards, J. Riegel, S. Ritchie, J. Shuey, B. Shaffer, P. Spinks, G. Stephens, C. Szwed, C. Tabaka, B. Thompson, B. Tomson, C-C. Tsai, T. Tuberville, M. Turnquist, B. Weigel, K. Westerman, M. Wildnauer, L. Woody, and the staff of the Woodlands Nature Station at Land Between the Lakes National Recreation Area. We wish to thank all members of the Williams lab for valuable help and willingness in improving the manuscript. R. Burke, G. Dharmarajan, K. Dodd, J. Fike, R. Howard, G. Nyberg, N. Karraker, S. Klueh, M. Kremer, M. Lannoo, N. Lichti, Z. Olson, K. Smith, B. Pauli, M. Pochon, J. Robb, and E. Latch also gave important support for the completion of this project.

Author Contributions

Conceived and designed the experiments: SJAK OER RNW. Performed the experiments: SJAK. Contributed reagents/materials/analysis tools: OER RNW. Wrote the paper: SJAK. Conceived of the ideas: SJAK OER RNW. Collected and analyzed the data: SJAK.

References

1. Frankham R, Ballou JD, Briscoe DA (2002) Introduction to Conservation Genetics. Cambridge University Press. Cambridge. 642 p.
2. Whitlock MC, McCauley DE (1999) Indirect measures of gene flow and migration: $F_{ST}\neq 1/(4Nm+1)$. Heredity 82: 117–125.
3. Avise JC, Ball RM, Arnold J (1988) Current versus historical population sizes in vertebrate species with high gene flow: a comparison based on mitochondrial DNA lineages and inbreeding theory for neutral mutations. Mol Biol Evol 5: 331–344.

4. Kuo C-H, Janzen FJ (2004) Genetic effects of a persistent bottleneck on a natural population of Ornate Box Turtles (*Terrapene ornata*). Conserv Genet 5: 425–437.
5. Lowe WH, Allendorf FW (2010) What can genetics tell us about population connectivity? Mol Ecol 19: 3038–3051.
6. Ewers RM, Didham RK (2006) Confounding factor in the detection of species responses to habitat fragmentation. Biol Rev 81: 117–142.
7. Steen DA, Gibbs JP (2004) Effects of roads on the structure of freshwater turtle populations. Conserv Biol 18: 1143–1148.
8. Aresco MJ (2005) Mitigation measures to reduce highway mortality of turtles and other herpetofauna at a north Florida lake. J Wildl Manage 69: 549–560.
9. Hartl DL, Clark AG (2007) Principles of Population Genetics, 4th edn. Sinauer Associates, Sunderland, Massachusetts.
10. Rousset F (1997) Genetic differentiation and estimation of gene flow from *F*-statistics under isolation by distance. Genetics 145: 1219–1228.
11. Pearse DE, Arndt AD, Valenzuela N, Miller BA, Cantarelli V, et al. (2006) Estimating population structure under nonequilibrium conditions in a conservation context: continent-wide population genetics of the giant Amazon river turtle, *Podocnemis expansa* (Chelonia; Podocnemididae). Mol Ecol 15: 985–1006.
12. Nei M, Maruyama T, Chakraborty R (1975) The bottleneck effect and genetic variability in populations. Evol 29: 1–10.
13. Lesica O, Allendorf FW (1999) Ecological genetics and the restoration of plant communities: mix or match? Restor Ecol 7: 42–50.
14. McKay JK, Christian CE, Harrison S, Rice KJ (2005) "How local is local? – A review of practical and conceptual issues in the genetics of restoration. Restor Ecol 13:432–440.
15. Waples RS (1995) Evolutionary significant units and the conservation of biological diversity under the Endangered Species Act. Amer Fish Soc Symp 17:8–27.
16. Alacs EA, Janzen FJ, Scribner KT (2007) Genetic issues in freshwater turtle and tortoise conservation. Chel Res Mon 4: 107–123.
17. Gibbons JW, Scott DE, Ryan TJ, Buhlmann KA, Tuberville TD, et al. (2000) The global decline of reptiles, déjà vu amphibians. Biosci 50: 653–666.
18. Cook RP, Brothernon DK, Behler JL (2010) Inventory of amphibians and reptiles at the Williams Floyd Estate, Fire Island National Seashore. Natural Resouce Report NPS/NCBN/NRTR-2010/380. U.S. Department of the Interior, National Park Service, Fort Collins, Colorado.
19. Dodd CK (2001) North American Box Turtles: A Natural History. University of Oklahoma Press, Norman, Oklahoma. 231 p.
20. Stickel LF (1978) Changes in a box turtle population during three decades. Copeia 1978: 221–225.
21. Williams EC, Parker WS (1987) A long-term study of a box turtle (*Terrapene carolina*) population at Allee Memorial Woods, Indiana, with emphasis on survivorship. Herpetol 43: 328–335.
22. Hall RJ, Henry PFP, Bunck CM (1999) Fifty-year trends in a box turtle population in Maryland. Biol Cons 88: 165–172.
23. Hagood S (2009) Genetic differentiation of selected Eastern Box Turtle (*Terrapene carolina carolina*) populations in fragmented habitats, and a comparison of road-based mortality rates to population size. PhD Dissertation, University of Maryland.
24. Marsack K, Swanson BJ (2009) A genetic analysis of the impact of generation time and road-based habitat fragmentation on Eastern box turtles (*Terrapene c. carolina*). Copeia 4: 647–652.
25. Butler JM, Dodd Jr CK, Aresco M, Austin JD (2011) Morphological and molecular evidence indicates that the Gulf Coast box turtle (*Terrapene carolina major*) is not a distinct evolutionary lineage in the Florida Panhandle. Biol J Linn Soc Lond 102: 889–901.
26. Miller BK (1993) Wildlife trends in Indiana. Hoosier Farmland Wildlife Notes, 1: 2.
27. Kimble SJA, Williams RN (2012). Temporal variance in hematologic and plasma biochemical reference intervals for free-ranging eastern box turtle (*Terrapene carolina carolina*). J Wildl Dis 48: 799–802.
28. Ernst CH, Barbour RW, Hershey MF (1974) A new coding system for hardshelled turtles. J Ky Acad Sci 35: 27–28.
29. Sambrook J, Russell DW (2001) Molecular Cloning: A Laboratory Manual, 3rd edn. Cold Springs Harbor Press, Cold Springs Harbor, New York. 2344 p.
30. Kimble SJA, Fike JA, Rhodes Jr OE, Williams RN (2011) Identification of 12 polymorphic microsatellite loci for the eastern box turtle (*Terrapene carolina carolina*). Conserv Genet Resour 3: 65–67.
31. Kalinowski ST, Taper ML, Marshall TC (2007) Revising how the computer program Cervus accommodates genotyping error increases success in paternity assignment. Mol Ecol 16: 1099–1006.
32. Van Oosterhout C, Hutchinson WF, Wills DPM, Shipley P (2004) Micro-checker: software for identifying and correcting genotyping errors in microsatellites. Mol Ecol Notes 4: 535–538.
33. Raymond M, Rousset F (1995a) Genepop (version 1.2): Population genetics software for exact tests and ecuminicism. J Hered 86: 248–249.
34. Raymond M, Rousset F (1995b) An exact test for population differentiation. Evol 49: 1280–1283.
35. Guo SW, Thompson EA (1992) Performing the exact test of Hardy-Weinberg proportion for multiple alleles. Biometrics 48: 361–372.
36. Mantel N (1967) Detection of disease clustering and a generalized regression approach. Cancer Res 27: 209–220.
37. Miller MP (2005) Alleles in Space (Ais): computer software for the joint analysis of interindividual spatial and genetic information. J Hered 96: 722–724.
38. Nei M (1972) Genetic distance between populations. Am Nat 106: 283–292.
39. Dharmarajan G, Beasley JC, Fike JA, Rhodes Jr OE (2009) Population genetic structure of raccoons (*Procyon lotor*) inhabiting a highly fragmented landscape. Can J Zool, 87: 814–827.
40. Latch EK, Scognamillo DG, Fike JA, Chamberlain MJ, Rhodes Jr OE (2008) Deciphering ecological barriers to North American River Otter (*Lontra canadensis*) gene flow in the Louisiana landscape. J Hered 99: 265–274.
41. Guillot G, Mortier F, Estoup A (2005) Geneland: a computer package for landscape genetics. Mol Ecol Notes 5: 712–715.
42. R Development Core Team (2009) R: A language and environment for statistical computing. R Foundation for Statistical Computing, Vienna, Austria. http://www.r-project.org.
43. Serre D, Pääbo S (2004) Evidence for gradients of human genetic diversity within and among continents. Genome Res 14: 1679–1685.
44. Falush D, Stephens M, Pritchard JK (2003) Inference of population structure using multilocus genotype data: linked loci and correlated allele frequencies. Genet 164: 1567–1587.
45. Dupanloup I, Schneider S, Excoffier L (2002) A simulated annealing approach to define the genetic structure of populations. Mol Ecol 11: 2571–2581.
46. Pritchard JK, Stephens M, Donnelly P (2000) Inference of population structure using multilocus genotype data. Genet 155: 945–959.
47. Earl DA, vonHoldt BM (2012) Structure Harvester: a website and program for visualizing Structure output and implementing the Evanno method. Conserv Genet Resour 4: 359–361.
48. Evanno G, Regnaut S, Goudet J (2005) Detecting the number of clusters of individuals using the software Structure: a simulation study. Mol Ecol 14: 2611–2620.
49. Monmonier MS (1973) Maximum-difference barriers: an alternative numerical regionalization method. Geo Anal 3: 245–261.
50. Manel S, Schwartz L, Luikart L, Taberlet P (2003) Landscape genetics: combining landscape ecology and population genetics. Trends Ecol Evol 18: 189–197.
51. Palmé AE, Su Q, Rautenberg A, Manni F, Lascoux M (2003) Postglacial recolonization and cpDNA variation of silver birch, *Betula pendula*. Mol Ecol 12: 201–212.
52. Tsai Y-H E, Manos PS (2010) Host density drives the postglacial migration of the tree parasite, *Epifagus virginiana*. Proc Natl Acad Sci USA 107: 17035–17040.
53. Wilson GA, Rannala B (2003) Bayesian inference of recent migration rates using multilocus genotypes. Genet 163: 1177–1191.
54. Rambaut A, Drummond AJ (2007) Tracer v.5, available at beast.bio.ed.ac.uk/Tracer.
55. Beerli P, Felsenstein J (2001) Maximum likelihood estimation of a migration matrix and effective population sizes in *n* subpopulations by using a coalescent approach. Proc Natl Acad Sci USA 98: 4563–4568.
56. Wright S (1943) Isolation by distance. Genet 25: 114–138.
57. Sokal RR, Wartenberg DE (1983) A test of spatial autocorrelation analysis using an isolation-by-distance model. Genet 105: 219–237.
58. Lemer S, Rochel E, Planes S (2011) Correction method for null alleles in species with variable microsatellite flanking regions, a case study of the black-lipped pearl oyster *Pinctada margaritifera*. J Hered 102: 243–246.
59. Dharamarajan GD, Beatty WS, Rhodes OE Jr (2013) Heterozygote deficiencies caused by a Wahlund effect: dispelling unfounded expectations. J Wild Manag 77:226–234.
60. Metcalf E, Metcalf AL (1970) Observations on ornate box turtles (*Terrapene ornata ornata* Agassiz). Trans Kans Acad Sci 73: 96–117.
61. Currylow AF, MacGowan BJ, Williams RN (2012) Short-term forest management effects on a long-lived ectotherm. Plos One 7(7): e40473.
62. Madden R (1975) Home range, movements, and orientation in the eastern box turtle *Terrapene carolina carolina*. PhD dissertation, City University of New York.
63. Burke RL, Capitano W (2012) Eastern box turtle, *Terrapene carolina*, neonate overwintering ecology on Long Island.New York. Chel Conserv Biol 10: 256–259.
64. Schwartz CW, Schwartz ER (1974) The three-toed box turtle in central Missouri: its population, home range, and movements. Part I. Missouri Department of Conservation, Jefferson City, Missouri.
65. Kiester AR, Schwartz CW, Schwartz ER (1982) Promotion of gene flow by transient individuals in an otherwise sedentary population of box turtles (*Terrapene carolina triunguis*). Evol 36: 617–619.
66. Richtsmeier RJ, Bernstein NP, Demastes JW, Black RW (2008). Migration, gene flow, and genetic diversity within and among Iowa populations of Ornate Box Turtles (*Terrapene ornata ornata*). Chel Conserv Biol 7: 3–11.
67. Latch EK, Boarman WI, Walde A, Fleischer RC (2011). Fine-scale analysis reveals cryptic landscape genetic structure in desert tortoises. Plos One 6: e27794.
68. Steyaert LT, Knox RG (2008) Reconstructed historical land cover and biophysical parameters for studies of land-atmosphere interactions within the eastern United States. J Geophys Res 113: 1–27.
69. Bishop JM, Leslie AJ, Bourquin SL, O'Ryan C (2009) Reduced effective population size in an overexploited population of the Nile crocodile (*Crocodylus niloticus*) Biol Cons 142:2335–2341.

70. Lerner HR, Johnson JA, Lindsay AR, Kiff LF, Mindell DP (2009) It's not too late for the harpy eagle (*Harpia harpyja*): high levels of genetic diversity and differentiation can fuel conservation programs. PLOS One 4:e7336.

71. Victory ER, Glaubitz JC, Rhodes Jr OE, Woeste KE (2006). Genetic homogeneity in *Jugland nigra* (Juglandaceae) at nuclear microsatellites. Am J Bot 93: 118–126.

14

Maintenance of Genetic Diversity in an Introduced Island Population of Guanacos after Seven Decades and Two Severe Demographic Bottlenecks: Implications for Camelid Conservation

Benito A. González[1]☯, Pablo Orozco-terWengel[2]☯, Rainer von Borries[3], Warren E. Johnson[4], William L. Franklin[5], Juan C. Marín[6]*

1 Laboratorio de Ecología de Vida Silvestre, Facultad de Ciencias Forestales y de la Conservación de la Naturaleza, Universidad de Chile, Santiago, Chile, 2 School of Bioscience, Cardiff University, Cardiff, Glamorgan, Wales, United Kingdom, 3 Facultad de Medicina, Universidad de Chile, Santiago, Chile, 4 Smithsonian Conservation Biology Institute, Smithsonian Institution, Front Royal, Virginia, United States of America, 5 Department of Natural Resource Ecology and Management, Iowa State University, Ames, Iowa, United States of America, 6 Laboratorio de Genómica y Biodiversidad, Facultad de Ciencias, Universidad del Bío-Bío, Chillán, Chile

Abstract

Fifteen guanacos were introduced to Staats Island in the Falklands/Malvinas archipelago from Patagonia in the 1930s. Twenty five years later, the population was culled from 300 to 10–20 individuals, but quickly rebounded to a population of almost 400 animals that today retain the genetic signature of the founding event and later bottleneck. The goals of this study were to (i) make a genetic assessment of this island population through comparisons with mainland populations and simulations, and (ii) assess the likely source-population of the introduced guanacos. Genetic variation was estimated from 513 bp of mitochondrial DNA sequence and 15 microsatellite loci among 154 guanacos collected from eight localities, including the adjacent mainland and the islands of Tierra del Fuego and Staats Island. Of the 23 haplotypes observed among our samples, the Staats Island population only contained three haplotypes, all of which were shared with the coastal Monte Leon population in southern Patagonia. Mitochondrial DNA and microsatellite variations on Staats Island were comparable to most mainland populations and greater than those observed on Tierra del Fuego. Patterns of genetic structure suggest that the Staats Island guanaco population was founded with animals from southern Patagonia (as opposed to northern Patagonia or Tierra del Fuego), but that effective reductions in population size lasted only a few generations and that surviving animals were a random sample of the pre-bottleneck genetic variation.

Editor: Taras K. Oleksyk, Univerity of Puerto Rico at Mayaguez, United States of America

Funding: This work was partially funded by Falkland Research Expeditions, Hidden Corners Inc., Turner Foundation, University of South Florida, and the Comisión Nacional de Investigación, Ciencia y Tecnología (CONICYT, Post-Doctoral grants N° 3050046). The funders had no role in study design, data collection and analysis, decision to publish, or preparation of the manuscript.

Competing Interests: The authors have the following commercial funder and/or competing interest: Hidden Corners Inc., who partially funded the expeditions to Falkland/Malvinas archipelago. This does not alter the authors' adherence to all the PLOS ONE policies on sharing data and materials.

* E-mail: jcmarin@ubiobio.cl

☯ These authors contributed equally to this work.

Introduction

Genetic analysis is a well-used tool for the conservation and management of animal populations and can help in determining genetic patterns, differentiating populations, resolving taxonomic uncertainties, and addressing evolutionary questions [1–5]. Small populations of endangered species have particularly benefited from genetic surveys and have contributed to the design of management strategies for breeding and reintroductions into the wild [6–8]. Genetic studies have also had an important management role in describing the effects of demographic changes (e.g. founder effects, bottlenecks) on the genetic variation of populations after capture, translocation, and release of wild individuals [9–15]. However, few studies have genetically surveyed wild or endangered species with well-documented population dynamics in order to test hypotheses about the impact of bottlenecks on genetic variation, and subsequent population persistence [16].

The guanaco (*Lama guanicoe*) is a native ungulate of South America. It is distributed widely from Perú (8° S) in the north, through Bolivia, Paraguay and Argentina until the southernmost part of Chile (55°S) [17,18]. Large guanaco populations occur from sea level to nearly 5000 m elevation [19], inhabiting deserts, some arid portions of the Andean mountains, and the shrublands and steppe plains of Patagonia [17,20]. A large population is also found on the large island of Tierra del Fuego [17,21–23], possibly arriving there at the time when sea levels were lower at the end of the Pleistocene 10,000-11,000 years ago and land bridges spanned over the Magellan Strait [24,25]. Besides the Tierra del Fuego population, the only other natural island population occurs on Navarino Island [26], but how guanacos spread from the continent to this island is still unclear [27].

The guanaco has successfully colonized new and marginal areas because of its capacity for adapting to an array of arid environments characterized by low annual rainfall, little-to-no winter snow-cover, low primary productivity, and high seasonality [17]. The species has been re-introduced to regions where natural populations were highly reduced or extirpated either by anthropogenic causes or natural events, such as, volcano eruptions and fires [28–32]. Various characteristics of guanaco biology (e.g. longevity, fertility, reproductive strategies, flexible social groupings, generalist-feeding habits, and a mating-birthing season encompassing the best weather and vegetation-growing season) have contributed to its ability to survive in small isolated populations [17,20,29,33].

However, the genetic impacts of these population dynamics have not been studied in this species. Recent genetic studies using mitochondrial DNA indicate that on a broad geographic scale, the guanaco is a monophyletic species with genetic lineages that support the previous classification of two subspecies: *L. g. cacsilensis* and *L. g. guanicoe* [34–36]. On a finer scale, microsatellite-based comparisons among populations on the South American mainland and Tierra del Fuego show that the guanaco is a diverse species with low to moderate population structure [25,36,37].

The guanaco population on Staats Island, in the Falkland Island archipelago (also known as Islas Malvinas) in the South Atlantic Ocean, offers a natural and on-going laboratory for testing the effects of a founder event on a guanaco population. This unique population has remained small and completely isolated from other populations for over 70 years. Together with several other Patagonian species, guanacos were introduced in the late-1930s by John Hamilton in an attempt to diversify the local economy [38]. Young animals were plausibly captured near Rio Gallegos (Argentina), or Pali-Aike (Chile) on the South American mainland, and then shipped to Sedge Island (11 animals) and Staats Island (15 animals) in the Falkland/Malvinas archipelago [38]. Only the guanacos on Staats Island survived despite several attempts 25 years later to eradicate them to reduce overgrazing and convert the island to sheep husbandry. This severe culling in the late-1950s drastically reduced the population from 300 animals to its second bottleneck and smallest size of 10 to 20 individuals in the early 1960s [38,39]. The population was then permitted to increase, with only sporadic and low levels of poaching and culling [38,39]. In 2004 the population numbered approximately 400 guanacos [40]. To best understand and interpret the relative genetic status and health of the island population, we compared it with mainland-guanaco populations of southern Patagonia.

Here we present the first genetic assessment of a small-island population of guanacos. We hypothesize that the population is characterized by low levels of genetic diversity as a result of two demographic bottlenecks in its short history as impacted by inbreeding, when each time it was reduced to less than 20 surviving animals. We also establish the phylogenetic relationship between this population and mainland populations in order to determine the most likely source of guanaco-genetic variation on Staats Island. Finally, based on these results we propose alternatives for the management of guanaco populations on the continent and Staats Island itself.

Materials and Methods

Ethics Statement

Guanaco samples did not come from any endangered guanaco populations in Chile or Argentina, where the guanaco is classified as "Least Concern" by the Red List, IUCN [41]. Liver samples were opportunistically taken collected from carcasses of adults animals hunted for meat production in Valle Chacabuco authorized by Chilean Government. Blood samples were obtained after chemical immobilization of adult guanacos in Torres del Paine National Park and Tierra del Fuego. The present study did not require the capture or handling of animals on Staats Island, Bosques Petrificados, San Julián, Monte León, Pali-Ayke and 8 animals from Tierra del Fuego, because our samples came from faeces and/or carcasses (muscle and skin). Guanaco samples from natural populations were obtained under permits and supervision of Servicio Agrícola y Ganadero (SAG) in Chile (permits Numbers 447, 263 and 1843) and samples from Staats Island and fecal samples obtained in Argentina were imported to Chile for analysis under CITES authorizations (Numbers 22920 and 22967). Samples taken within Chilean National Parks were authorized by Corporación Nacional Forestal (CONAF, permit number 6/02, 2002). DNA sequences were deposited in GenBank (accessing numbers JX678477 - JX678596). Individual-by-individual microsatellities data are available in the Dryad data repository at doi:10.5061/dryad.06g5v.

Study area

Staats Island is a small (500 ha) isle on the far western edge of the 750-island Falkland/Malvinas archipelago in the South Atlantic Ocean (51°53' S latitude and 61°11' W longitude) 600 km from the South American coast. Six research expeditions were conducted on Staats Island in December (early summer and beginning of guanaco birth season) from 1999 to 2008 and biological samples were collected for genetic analyses in 2004 and 2005. Staats Island is treeless and hilly. The south end is dominated by the Staats Plateau, a tableland surrounded by a coastline of formidable cliffs and monoliths. The island is characterized by large sink holes, cuts and gulches that are caused by connections to the sea along the coastline, especially on Staats Plateau. North of the plateau the balance of the island is dominated by domed peaks (max. 140 m) separated by steep slopes and four valleys of short grass and forb meadows. Dominant plant communities are Oceanic Heath, Grass, Cushion Plants, and Greens (meadows known as vegas and mallines on the mainland) [40,42].

Sample collection and DNA extraction

For comparison, material suitable for DNA analysis (170 samples) was collected from eight localities throughout southern Patagonia and Staats Island (Figure 1, Table 1). DNA was extracted: i) from muscle or skin samples from 36 dead animals from Staats Island and Tierra del Fuego (Chile), ii) from blood samples of 34 wild-caught adults following chemical immobilization [43,44] at Torres del Paine National Park and Tierra del Fuego (Chile), and iii) from 74 fresh fecal samples from different dung piles at Bosques Petrificado National Parck, San Julian and Monte Leon (Argentina) and Pali-Ayke National Park (Chile). DNA was also obtained opportunistically from liver samples of 26 adult males in Valle Chacabuco, Chile. Sample sites and the geographic position of individuals collected at each site are given in Figure 1 and Table 1. All samples were stored at −70°C in the Laboratorio de Genómica y Biodiversidad, Departamento de Ciencias Básicas, Facultad de Ciencias, Universidad del Bio-Bío, Chillán, Chile. We followed guidelines of the American Society of Mammalogists during the collection and handling of animals [45]. Total genomic DNA was extracted from blood using the Wizard Genomic DNA Purification Kit (Promega, Madison, Wisconsin). DNA from liver, skin and muscle samples was purified using proteinase-K digestion and a standard phenol-chloroform protocol [46]. DNA from feces was extracted using the QIAamp DNA

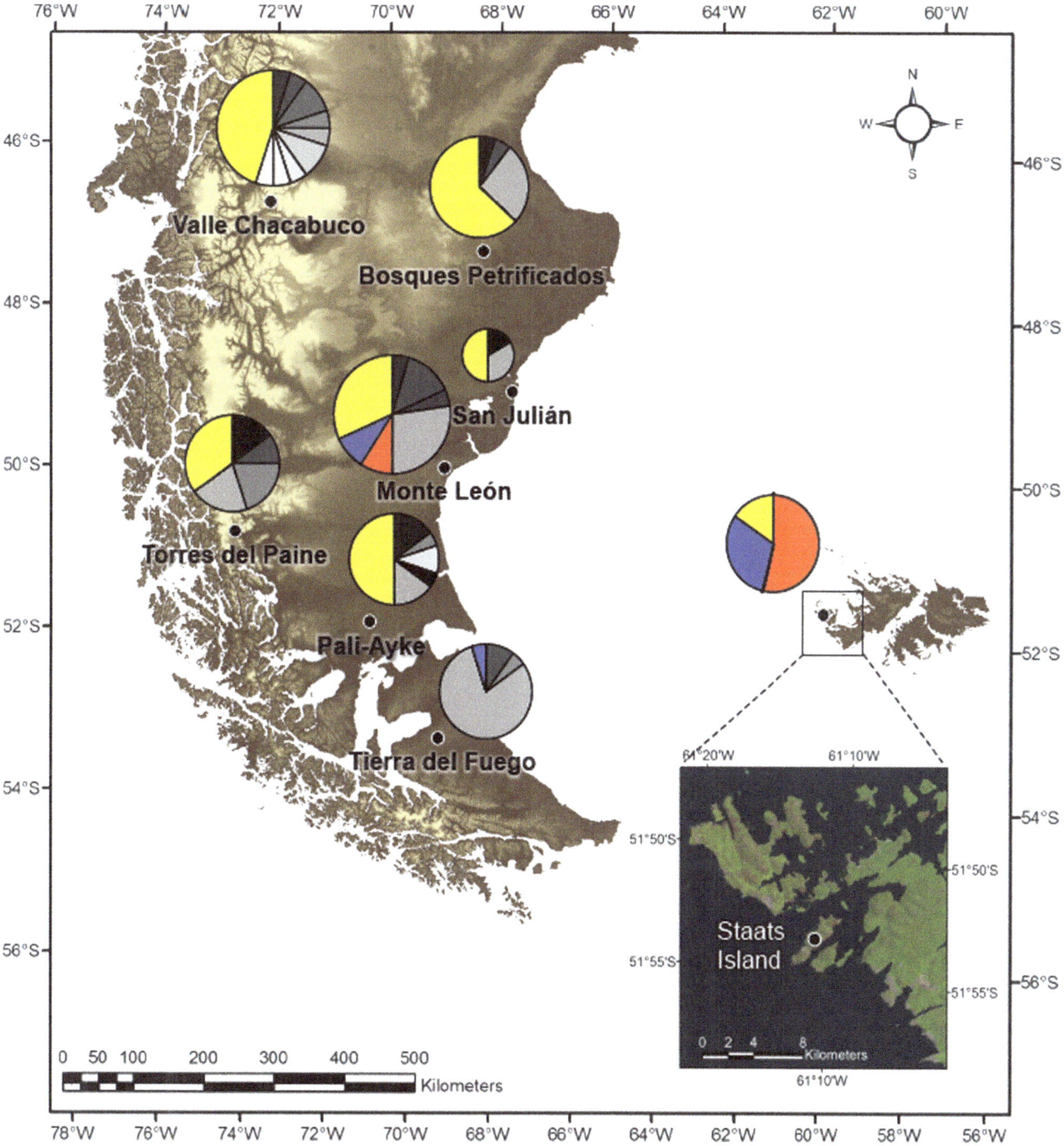

Figure 1. Map of southern South America and the Falkland/Malvinas Islands with sampled localities of *Lama guanicoe* analysed. Each location indicates the relative proportion of each haplotype in yellow, blue and red relative to Staats Island. Mainland haplotypes not found on Staats Island are shown in grey scale. The size of the sphere is proportional to the number of individuals sampled.

Stool Mini Kit (QIAGEN, Valencia, California) in a separate non-genetic-oriented laboratory.

Mitochondrial DNA analysis

A 513 bp long fragment of the 5' side of the mitochondrial Control Region was amplified from 156 samples using the primers LThr-ARTIO 5′ TCC TTT TTC GGC TTA CAA GAC C 3′, Hloop550G 5′ ATG GAC TGA ATA GCA CCT TAT G 3′, Lloop0007G 5′ GTA CTA AAA GAA AAT ATC ATG TC 3′ and H362 5′ GGT TTC ACG CGG CAT GGT GAT T 3' and H15998 5′ CCA GCT TCA ATT GAT TTG ACT GCG 3′ [34,35]. The polymerase chain reactions (PCRs) followed standard protocols by Marín *et al.* [34,35]. All PCR products were purified with the QIAquick PCR Purification Kit (QIAGEN) and sequenced in a ABI-3100 DNA sequencer (Perkin Elmer Applied Biosystems). PCR products were sequenced in both directions twice to test sequence fidelity. Sequences were aligned with Geneious Aligment implemented in Geneious Pro 5.3.4 (Biomat-

Table 1. Summary of the *Lama guanicoe* samples used in the genetic analyses, including localities, abbreviations, geographic positions, type of sample (B= blood, F= fecal, M= muscle, S= skin, and L =liver), and total number (N) of samples used from each locality for each genetic marker.

Locality, country	Abbreviation	Geographic positions	Sample type	Samples mtDNA (N=156)	Samples microsatellites (N=164)
Bosques Petrificados, Argentina	BP	47° 20′ S, 67° 30′ W	F	20	20
Valle Chacabuco, Chile	VC	47° 36′ S, 72° 27′ W	L	20	26
San Julian, Argentina	SJ	49° 22′ S, 67° 40′ W	F	8	8
Monte León, Argentina	ML	50° 16′ S, 68° 51′ W	F	22	26
Torres del Paine, Chile	TP	51° 03′ S, 72° 55′ W	B	20	23
Pali-Aike, Chile	PA	52° 04′ S, 69° 47′ W	F	20	20
Tierra del Fuego, Chile	TF	53° 18′ S, 70° 11′ W	B, S	20	21
Staats Island, United Kingdom	SI	51° 53′ S, 61° 11′ W	M, S	26	20

ters Ltd.) and the alignment was checked visually. Within population genetic variation was measured by the number of polymorphic sites (S), haplotype number (k), number of private haplotypes, nucleotide diversity (π), and haplotype diversity (H) using ARLEQUIN 3.5.1.2 [47]. Genetic differentiation among populations was expressed as pairwise fixation indices (Φ_{ST}) calculated with ARLEQUIN 3.5.1.2 [47] using 10,000 permutations to assess significance with K80+I, and the best fit model was tested with jModeltest.

Microsatellite DNA analysis

Fifteen autosomal dinucleotide microsatellite loci were analysed: YWLL08, YWLL29, YWLL36, YWLL38, YWLL40, YWLL43, YWLL46 [48], LCA5, LCA19, LCA22, LCA23 [49], LCA65 [50], LCA82 [51] and LGU49, LGU68 [52]. Amplification was carried out in a 10 μL reaction volume containing 50 – 100 ng of template DNA, 1.5 – 2.0 mm $MgCl_2$, 0.325 μm of each primer, 0.2 mm dNTP, 1X polymerase chain reaction (PCR) buffer (QIAGEN) and 0.4 U *Taq* polymerase (QIAGEN). All PCR amplifications were performed in a PE9700 (Perkin Elmer Applied Biosystems) thermal cycler with the following cycling conditions: initial denaturation at 95°C for 15 min, followed by 40 cycles of 95°C for 30 s, 52–57°C for 90 s and 72°C for 60 s, and a final extension of 72°C for 30 min [36]. Amplification and genotyping of DNA from samples was repeated 3 times. One primer of each pair was labelled with a fluorescent dye on the 5′-end, and fragments were analysed on an ABI-3100 sequencer (Perkin Elmer Applied Biosystems). Data collection, sizing of bands and analyses were carried out using GeneScan software (Applied Biosystems).

We identified multiple samples that came from the same individual by searching for matching microsatellite genotypes using the Excel Microsatellite Toolkit [53] and eliminated samples from the study if they showed more than 85% overlap. We excluded three loci (LCA19, LCA22 and LCA82) after checking for null alleles (i.e. non-amplifying alleles) using Micro-Checker [54]. Within population allele frequencies, Hardy-Weinberg equilibrium (HWE), linkage disequilibrium, observed heterozygosity (H_O), and expected heterozygosity (H_E) were estimated using FSTAT [55]. FSTAT software was also used to estimate population pairwise F_{ST} values with 10,000 permutations to assess significance. The difference between the expected heterozygosity in Staats Island and mainland populations were statistically tested using Welch's t-test to account for sample heteroskedasticity utilizing R software (http://www.R-project.org). The inbreeding coefficient F_{IS} was calculated using GENETIX version 4.05 [56], and its deviation from zero was assessed with 10,000 permutations across loci.

The program STRUCTURE 2.3.3 [57] was used to determine the number of clusters which best partitions the microsatellite data under different scenarios of population independence and admixture. We ran five independent runs assuming no admixture and independent allele frequencies for values of K from 1 to 8 with 200,000 burn-in steps for the MCMC and 200,000 data collection steps. The value of K best representing the division of the samples was identified with the method of Evanno [58].

Three methods were used to assign individuals to the clusters found by STRUCTURE. The first method is based on the probabilities of admixture inferred for each individual by STRUCTURE (q). An individual was considered assigned with high confidence if it presented a $q \geq 0.75$ for a single cluster. The second method consisted of assigning individuals to STRUCTURE's clusters using the likelihood-based method of [59] implemented in GENECLASS 2.0 [60]. From this analysis, the proportion of individuals assigned to the cluster that they were originally sampled from was reported.

Lastly, we also determined the continental population from which the Staats Island individuals came from using the trained clustering method in BAPS 5.2 [61]. In this analysis the individuals from Staats Island were allowed to cluster to any of the eight continental populations, or alternatively, if these populations did not reflect their population of origin, they were also allowed to form their own cluster. For that purpose we used a prior distribution of the number of clusters in the dataset between 1 and 10, so that the Staats Island individuals could also form new clusters beyond the eight potential clusters formed by the continental populations.

Due to the known changes in population size experienced by the Staats Island population, we also tested whether there was evidence of population bottlenecks in the microsatellite data. For this purpose we used the program BOTTLENECK [60] that aims to detect the excess of heterozygosity left after a bottleneck. These analyses were performed under the stepwise mutation model (SMM [62]) and the multiple step stepwise mutation model (TPM [63]). Under the TPM model the proportion of single-step mutation events was set at 90% with a mutation size-variance of 12%. Observed and expected heterozygosities were compared using a Wilcoxon sign-rank test as suggested by Piry *et al.* [60]. Complementary to the BOTTLENECK analysis, we also tested for deviations between the observed and expected number of alleles in each locus and their size range using the M-ratio method implemented in the M_P_VAL and Critical_M software [64]. During a bottleneck low frequency alleles become extinct, leaving gaps in the allele range distribution of a microsatellite. The ratio between the observed number of alleles and the allele size range is compared to the one expected under mutation-drift-equilibrium using simulations. If 95% of the simulations present a larger ratio (M_c) than the observed data it is considered that the analysed population passed through a bottleneck. The simulations are performed under the TPM model parameterised by the proportion of single step mutations (p_s), the size of the multistep mutations (Δ_g) and the neutral evolution rate ($\theta = 4N_e\mu$). As the M_c threshold is sensitive to these parameters and there is no species-specific information on N_e, we performed simulations under various biologically plausible θ values from 0.01 to 20. To ensure this range of values was relevant, we estimated θ from the expected heterozygosity (Eq. 3.15 from [65]) using a common microsatellite mutation rate (μ) recommended by Garza & Williamson [64]: 5.0×10^{-4} mutants/generation/locus [66].

Simulation of population variation size on genetic diversity

Current genetic diversity of the Staats Island population was compared with those of simulated populations using the software BOTTLESIM [67]. The model was run for 10,000 iterations with non-constant population size, random mating assuming that males have a chance to reproduce at any times of their lives [68]; reproductive maturity at 5-years of age [68], age of senescence ~15 years [21]; 70% generation overlap considering a ratio 4–5/15 (first years of life without reproduction/lifespan). An initial population of 10,000 individuals was constructed with the genetic characteristics of the continental Patagonian population, which was subsequently subjected to a bottleneck that reduced the population size to 15 animals. Temporal variation size of the Staats Island population (Figure 2) was constructed in excel and based on historical records obtained by Franklin and Grigione [38], and was used as a model in post-introduction population size on Staats Island. BOTTLESIM provided means for H_E, and alleles per locus (A) for the initial population and each of the 65 years of simulation.

Results

Mitochondrial DNA diversity

Among the 156 samples, there were 23 haplotypes and 15 polymorphic sites (2.9%) from the 514 bp Control-Region fragment (Table 2). A dominant haplotype (H_5) was observed in seven of eight localities, including Staats Island. Haplotype (h) and nucleotide diversity (π) are detailed in Table 3. As expected, the continental populations presented a higher diversity than the island populations (0.75±0.08, mean and standard deviation of h; Table 3). The Staats Island population had a similar diversity ($h = 0.61$) to the Bosques Petrificados population in southern Argentina and was higher than the island population of Tierra del Fuego ($h = 0.36$), which has a large population of more than 80,000 individuals [69,70]. Of the 23 haplotypes observed among our samples, the Staats Island population only contained three haplotypes, all of which were shared with the Monte Leon population (the only continental population that contained all three haplotypes observed on Staats Island) (Table 2, Figure 1).

Microsatellite diversity

Among the 166 replicated microsatellite genotypes, we found two pairs of samples with the same allelic profiles suggesting that two individuals had been sampled twice. For each of these individuals one of the samples was discarded. Combining the continental and Staats Island samples, 150 alleles were detected in the 12 polymorphic loci genotyped from 164 guanacos. The number of alleles per locus ranged from 3 to 23, and the average number of private alleles per population was 2.6. Within populations, we found no significant departures from HWE equilibrium (P>0.0011 after Bonferroni correction), and no significant evidence of linkage disequilibrium among loci (P>0.0005 after Bonferroni correction for each pair of loci across all populations). Consistent with the other measures of genetic variability, we found high levels of population expected heterozygosity (0.70±0.058**,** mean and standard deviation respectively) and high values of mean number of alleles per locus (6.2±1.17) (Table 3).

While there was variation in the expected heterozygosity per population, these differences were not significant between the continental sample set considered as a whole and the Staats Island population (Welch t-test p-value = 0.5424), or between the latter population and any of the continental populations (i.e. Monte

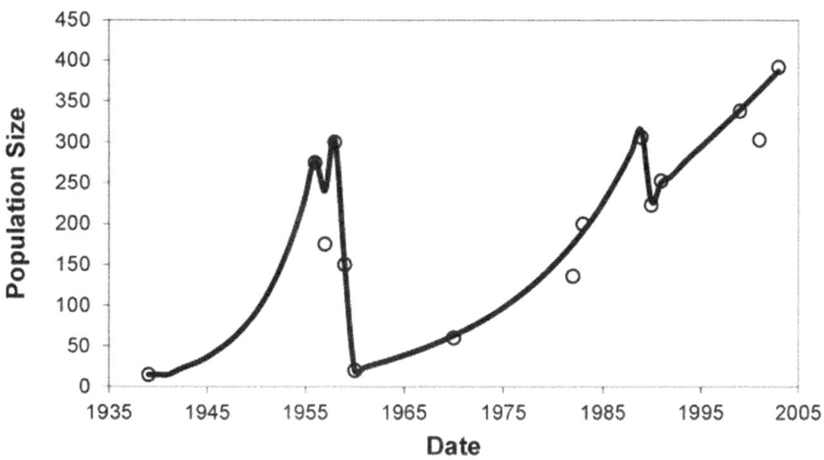

Figure 2. Reconstruction of population size (solid line) of guanacos on Staats Island since introduction based on historical records reported by [38] (circles).

Table 2. Distribution of the 23 Control Region haplotypes (frequency) observed in 156 guanacos from 8 localities.

Haplotypes																Localities								N
	1	**2**	**15**	**44**	**76**	**125**	**174**	**190**	**191**	**211**	**242**	**270**	**286**	**320**	**508**	**PB**	**VC**	**SJ**	**ML**	**TP**	**PA**	**TF**	**SI**	
1	C	C	G	A	A	A	T	C	G	C	A	T	C	C	C		1(0.05)							1
2	T	T	.	T	.	.	.	.	A	.	.	.	.	.	.		1(0.05)							1
3	A	T	A	.	.	.	.	.	A	.	.	.	.	.	.		2(0.10)							2
4	A	T	.	.	.	.	.	.	A	.	.	.	.	.	.		1(0.05)				2(0.10)			3
5	A	T	.	.	.	.	.	.	.	.	.	.	.	.	.	**12(0.60)**	**9(0.45)**	**3(0.50)**	**7(0.32)**	**7(0.35)**	**10(0.50)**		**4(0.15)**	52
6	A	T	.	T	.	.	.	.	.	.	.	.	.	.	.	5(0,25)	1(0.05)	2(0.30)	6(0.27)	4(0.20)	3(0.15)	16(0.80)		37
7	A	A	.	.	.	.	.	.	.	.	.	.	.	.	.		1(0.05)							1
8	A	T	.	.	.	G	.	.		.	.	.	.	.	.	1(0.05)	1(0.05)		3(0.14)					5
9	A	T	.	.	.	.	.	.	A	.	.	C	.	.	.		2(0.10)							2
10	A	T	.	.	.	.	C	.	.	.	.	.	.	.	.		1(0.05)							1
11	A	.	A	T	G	.	.	.	.	.	.	.	.	.	.	1(0.05)								1
12	A	T	.	.	.	.	.	.	.	.	.	.	G	.	.	1(0.05)								1
13	A	.	.	T	.	.	.	T	.	.	.	.	.	.	.				**2(0.09)**				**14(0.54)**	16
14	A	.	.	T	.	.	.	.	.	.	.	.	.	.	.			1(0.20)		4(0.20)	1(0.05)			6
15	A	.	.	T	.	G	.	.	A	.	.	.	.	.	.				1(0.05)					1
16	A	T	.	T	.	.	.	.	A	.	.	.	.	.	.				**2(0.09)**			**1(0.05)**	**8(0.31)**	11
17	A	.	.	T	G	.	.	.	.	.	.	.	.	.	.				1(0.05)					1
18	A	.	.	T	.	.	.	T	.	.	.	.	.	T	.					2(0.10)				2
19	A	T	.	.	.	.	.	.	.	T	G	.	.	.	.					3(0.15)				3
20	A	T	.	T	.	G	.	.	.	.	.	.	.	.	.						3(0.15)			3
21	A	T	.	T	.	.	.	T	.	.	.	.	.	.	.						1(0.05)			1
22	A	T	.	T	.	.	.	.	.	.	.	C	.	.	.							2(0.10)		2
23	A	T	.	T	.	.	.	.	.	.	.	.	.	.	G							1(0.05)		1
Total																20	20	6	22	20	20	20	26	154

The vertical numbers indicate the position of polymorphic sites relative to haplotype 1. Bold numbers emphasized the haplotypes presented in Staats Island.

Table 3. Genetic diversity indices from mtDNA Control Region sequences and 12 microsatellite loci by localities (defined in Table 1 and Figure 1).

Localities	mtDNA				Microsatellites					
	n	*np*	*h* ± SD	*π*± SD	*A* ± SD	*Ap*	H_O ± SD	H_E ± SD	F_{IS}	*Welch p-value*
BP	5	2	0.600±0.101	0.0018±0.0006	7.583±3.298	2	0.808±0.163	0.756±0.095	0.1170**	n.p.
VC	10	6	0.800±0.089	0.0031±0.0005	6.333±2.498	1	0.633±0.164	0.715±0.105	−0.0699	n.p.
SJ	3	0	0.800±0.164	0.0019±0.0005	4.500±1.624	2	0.677±0.268	0.686±0.110	0.0809*	n.p.
ML	7	2	0.810±0.053	0.0031±0.0005	7.167±2.918	5	0.699±0.141	0.741±0.115	0.0579*	0.294
TP	5	2	0.805±0.050	0.0036±0.0006	6.750±3.279	5	0.624±0.187	0.706±0.121	0.1177**	0.945
PK	6	1	0.726±0.090	0.0022±0.0003	6.750±2.768	3	0.694±0.130	0.715±0.109	0.0252	0.734
TF	4	2	0.363±0.131	0.0007±0.0003	4.417±1.975	1	0.570±0.200	0.567±0.180	−0.0054	n.p.
SI	3	0	0.615±0.063	0.0034±0.0002	6.167±2.443	2	0.660±0.159	0.715±0.095	0.0791**	

The right most column shows the result of Welsh's t-test between the populations of Staats Island and each its three potential continental source populations. *n*: number of haplotypes; *np*: number of private haplotypes; *h*: haplotype diversity; *π*: nucleotide diversity; *A*: mean number of alleles per locus; *Ap*: privates alleles; *He*: mean expected heterozygosity; *Ho*: mean observed heterozygosity; *Welch p-value*: significance value of Welch's t-test between the SI population and each of the other populations (e.g. Welch t-test between SI and ML has a *p-value* of 0.2948), *n.p.*: test not performed. Deviations from zero *$p<0.05$, **$p<0.001$.

Leon, Torres del Paine and Pali-Ayke) likely to be the source for the Staats Island population (Table 3). While other pairwise comparisons revealed no significant differences in H_e between populations, it is still interesting that, consistent with the mtDNA results, the Staats Island population seems to present a higher H_e and average number of alleles per locus than the Tierra del Fuego population (Table 3). However, a global test indicated a significant heterozygote deficiency in Bosques Petrificados, San Julian, Monte León, Torres del Paile and Staats Island as indicated by a positive F_{IS} (Table 3).

Genetic structure and population differentiation

The population comparisons show, with both types of markers, low levels of genetic differentiation among the continental samples (average $\varphi_{ST}=0.026\pm0.069$ and average $F_{ST}=0.06\pm0.029$; Table 4). Interestingly, the average divergence between the Staats Island population and the continental populations was not significantly different from the average divergence between the continental populations when measured with the microsatellites (Welch t-test p-value: 0.82). However, the same test based on the estimates from the mtDNA data showed a significantly higher divergence between Staats Island and the continental populations than between the continental populations (Welch t-test p-value: 0.0002). In contrast, the Tierra del Fuego population presented a significantly higher divergence with respect to the continent than did Staats Island for the microsatellites (Welch t-test p-value: 0.0002), while for the mtDNA data it was as divergent from the continental populations as the Staats Island population (Welch t-test p-value: 0.413). Consistent with the stronger effect of drift on island populations, the divergence between the Tierra del Fuego population and Staats Island was among the highest values of divergence observed with each marker type (Table 4).

The STRUCTURE analysis suggested that a partition of the microsatellite dataset into three clusters (K = 3) had the highest posterior probability, as measured with the ΔK Evanno method. The three clusters corresponded to the group of samples from i) northern Patagonia, ii) the samples from southern Patagonia including the guanacos from Staats Island, and iii) the samples from the island of Tierra del Fuego (Figure 3).

Genetic Origin of the Staats Island population

Haplotypic diversity indicated that Monte Leon is the most similar continental population to Staats Island. The island population had three haplotypes, which were observed together

Table 4. Pairwise population differentiations between guanaco populations.

Localities	BP	VC	SJ	ML	TP	PA	TF	SI
BP	-	0.0567*	−0.0424	0.0369	0.0702	−0.0219	0.4185**	0.3747**
VC	0.0542**	-	0.0842	0.1183**	0.1522**	0.0712*	0.4623**	0.3775**
SJ	0.1116**	0.0734**	-	−0.0679	−0.0629	−0.0590	0.3215**	0.2284*
ML	0.0678**	0.0490**	0.1035**	-	0.0225	−0.0166	0.1935**	0.1942**
TP	0.0464**	0.0422**	0.0949**	0.0297**	-	0.0536	0.2421**	0.1953**
PA	0.0489**	0.0425**	0.1007**	0.0291**	0.0213*	-	0.3190**	0.3112**
TF	0.1857**	0.1430**	0.1865**	0.1363**	0.1418**	0.1499**	-	0.3521**
SI	0.0835**	0.0697**	0.1138**	0.0334**	0.0408**	0.0463**	0.0937**	-

Genetic structure and statistical significance corrected by Bonferroni among Patagonia guanacos using mtDNA (pairwise Φ_{ST} over diagonal) and microsatellite (F_{ST} below diagonal) markers. Deviations from zero *$p<0.05$, **$p<0.001$.

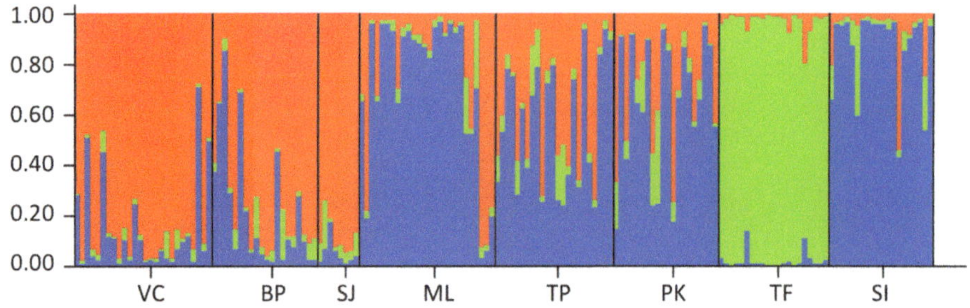

Figure 3. Clustering solution of the Patagonia guanaco populations. Plot of posterior probability of assignment for 164 guanacos (vertical lines) to three genetic clusters based on Bayesian analysis of variation at 12 microsatellite loci. Individuals are grouped by locality, and localities are indicated along the horizontal axis. Genetic Cluster 1: North Patagonia group (VC, BP, SJ); Genetic Cluster 2: Tierra del Fuego group (TF); Genetic Cluster 3: South Patagonia group (ML, TP, PK, SI). Population name abbreviations follow Table 1.

only in the Monte Leon population. In comparison to Monte Leon, no other population presented more than one haplotype shared with the Staats Island. Moreover, the dominant haplotype in Staats Island (frequency of Hap_20 = 0.54) was only recorded in the Monte Leon population, where it occurred at a moderately low frequency (0.09). This finding was supported by assignments tests, which showed that most of the Staats Island individuals could be assigned to the Monte Leon population.

Of the 164 individuals analysed in the STRUCTURE analyses, 125 were assigned to a single cluster with a q value greater than 75%. Twenty-two percent of these individuals were assigned to the Tierra del Fuego cluster, while 39% were assigned to either the North Patagonia cluster or the South Patagonia cluster. Out of the 20 Staats Island individuals, 19 were assigned to the South Patagonian cluster. However, of these samples only 16 presented a *q* value higher than 75% (Table 5). The only Staats Island individual not assigned to South Patagonia was assigned by STRUCTURE to North Patagonia but with a q value of 0.56. The BAPS and GENECLASS analyses placed all the Staats Island individuals in the South Patagonia cluster. However, while both methods assigned most of the Staats Island individuals to the population of Monte Leon (65%), BAPS and GENECLASS assigned between 5 and 7 individuals to the other two South Patagonia clusters/populations (Table 5).

Table 5. Results of population assignment algorithms STRUCTURE, BAPS and GENECLASS for guanacos of Staats Island.

Staats Island Individuals	STRUCTURE		BAPS		GENECLASS	
	Population assignment	*q*	Population assignment	Posterior probability	Population assignment	Likelihood Ratio
1	South Patagonia	0.665	TP	1.0	TP	12,384
2	South Patagonia	0.968	ML	1.0	ML	16,604
3	South Patagonia	0.965	ML	1.0	ML	15,898
4	South Patagonia	0.976	ML	1.0	ML	14,999
5	South Patagonia	0.895	ML	1.0	TP; ML	15,911; 15,977
6	South Patagonia	0.697	TP	1.0	TP; ML	11,614; 11,847
7	South Patagonia	0.981	ML	1.0	PK; ML	18,382; 18,751
8	South Patagonia	0.980	ML	1.0	ML	17,130
9	South Patagonia	0.973	ML	1.0	ML; PK	15,190; 15,269
10	South Patagonia	0.973	ML	1.0	TP	14,056
11	South Patagonia	0.972	ML	1.0	ML	16,778
12	South Patagonia	0.959	ML	1.0	PK	17,717
13	South Patagonia	0.979	ML	1.0	PK; ML	18,475; 18,710
14	North Patagonia	0.558	ML	1.0	ML	18,140
15	South Patagonia	0.896	PK	1.0	PK	13,829
16	South Patagonia	0.915	TP	1.0	TP	14,000
17	South Patagonia	0.966	ML	1.0	ML	17,048
18	South Patagonia	0.974	ML	1.0	ML	16,695
19	South Patagonia	0.577	TP	1.0	TP	11,602
20	South Patagonia	0.962	ML	1.0	ML	16,328

Table 6. Summary of parameters and results for the *M*-ratio and BOTTLENECK analyses used to detect significant reductions in effective population size.

Localities					Bottleneck	
	N_e	θ	*M*-ratio	M_c	Mutation model	Heterozygote excess
BP	5	0.01	0.8248	0.8666	TPM	$P=0.1696$
	50	0.1	0.8248	0.8527	SMM	$P=0.8493$
	500	1	0.8248	**0.7826**		
	1,556*	3.11	0.8248	**0.7184**		
	5,000	10	0.8248	**0.6572**		
VC	5	0.01	0.6539	0.8626	TPM	$P=0.1901$
	50	0.1	0.6539	0.8541	SMM	$P=0.5151$
	500	1	0.6539	0.7871		
	1,093*	2.18	0.6539	0.7450		
	5,000	10	0.6539	0.6779		
SJ	5	0.01	0.7227	0.8626	TPM	$P=0.1018$
	50	0.1	0.7227	0.8527	SMM	$P=0.6889$
	500	1	0.7227	0.7741		
	1,261*	2.52	0.7227	**0.7067**		
	5,000	10	0.7227	**0.5772**		
ML	5	0.01	0.8015	0.8626	TPM	$P=0.51514$
	50	0.1	0.8015	0.8527	SMM	$P=0.96802$
	500	1	0.8015	**0.7837**		
	1,202*	2.4	0.8015	**0.7374**		
	5,000	10	0.8015	**0.6683**		
TP	5	0.01	0.7964	0.8666	TPM	$P=0.2119$
	50	0.1	0.7964	0.8540	SMM	$P=0.6889$
	500	1	0.7964	0.7873		
	1,522*	3.04	0.7964	**0.7249**		
	5,000	10	0.7964	**0.6691**		
PK	5	0.01	0.7088	0.8611	TPM	$P=0.3667$
	50	0.1	0.7088	0.8527	SMM	$P=0.9959$
	500	1	0.7088	0.7782		
	1,276*	2.55	0.7088	0.7299		
	5,000	10	0.7088	**0.6549**		
TF	5	0.01	0.6732	0.8657	TPM	$P=0.3667$
	50	0.1	0.6732	0.8531	SMM	$P=0.9959$
	500	1	0.6732	0.7865		
	661*	1.32	0.6732	0.7690		
	5,000	10	0.6732	**0.6598**		
SI	5	0.01	0.7565	0.8626	TPM	$P=$ **0.0067**
	50	0.1	0.7565	0.8527	SMM	$P=0.1330$
	500	1	0.7565	0.7833		
	661*	1.32	0.6732	0.7690		
	5,000	10	0.7565	**0.6589**		

Bold indicates the *M*-ratios used the N_e from $N_e = H_E/4\mu\ (1 - H_E)$ indicated by *, and the bottleneck signature in bold indicates significance ($P<0.05$). $M_c = p_s = 0.9$ and $\Delta_g = 3.5$.

Bottleneck and N_e of Staats Island population

The Staats Island population presented evidence of having passed through a bottleneck under the TPM model (Table 6). In contrast, none of the continental population showed evidence of having passed through a bottleneck. While this result may indicate some evidence of a deviation from mutation drift equilibrium in the Staats Island population, it is unexpected that after a bottleneck the genetic diversity (e.g. number of alleles per locus) in this isolated island population would remain within the range of that in the continental populations (Table 3).

The BOTTLENECK analysis was complemented with the M-ratio method [64]. For this method we estimated a pre-bottleneck N_e, which, based on the expected heterozygosity of the Staats Island population, resulted in values ranging between 661 to 1556 individuals [65]. While these values may seem large at first glance, they are not extremely different than our own field observations, which show that up to almost 400 individuals existed in the island by December 2003 [71]. Thus, our biologically plausible pre-bottleneck Ne values for the M-ratio analysis were set to 5, 50, 500, and 5,000 (Table 6). The population size of five represents the most liberal test for significant reductions in population size by effectively reducing the pre-bottleneck to θ, thereby increasing M_c values. However, we also used N_e values considerably larger than our estimates of the current N_e. Among the various continental populations only the Valle Chacabuco (Chile) population showed no evidence of bottleneck for any of the N_e values tested. In contrast, all other populations show M-ratio values below M_c indicating they may have passed through a bottleneck. However, these analyses seem to become only significant for large values of simulated N_e. Although the observed N_e estimates of our populations based on the expected heterozygosity sometimes fell within the range of the simulated N_e values that were significant, it is unlikely that such a large N_e currently characterizes our populations. This is particularly true for the Staats Island population, since the demographic records since its establishment have never exceeded more than 400 animals.

Effect of Population variation on genetic diversity detected by simulation

A population of 10,000 individuals was simulated on the basis of the allele frequencies detected in Patagonia. This population was bottlenecked to 15 animals (assuming 8 females) in order to mimic the founding of the Staats Island population 65 years ago (Figure 2). As expected, after the bottleneck a drastic reduction by ~29% in the mean number of alleles was observed from 12.8 (SE = 0.29) to 9.1 (0.16) (Figure 4), while the H_E decreased by only 3.5% from 0.84 (SE = 0.01) to 0.81 (SE = 0.01) (Figure 4). A second bottleneck after ~25 years since the introduction of guanacos to the island further reduced the allelic diversity to approximately 51% of the original variation (i.e. 6.6 (SE = 0.61)), although it had no effect on H_E. After the initial bottleneck H_E steadily decreased throughout the simulation until reaching a value of 0.71 (SE = 0.01) at the end of the simulations.

Discussion

Genetic diversity and population structure

The guanacos of the Staats Island were introduced to Falkland/Malvinas archipelago ~75 years ago, with our samples representing animals ~65 years after the original introduction. This population was started from a founding group of animals no larger than 15 animals [38,40]. While the initial guanaco population prospered on the island, the number of animals was dramatically reduced ~25 years later as part of a plan to eradicate the

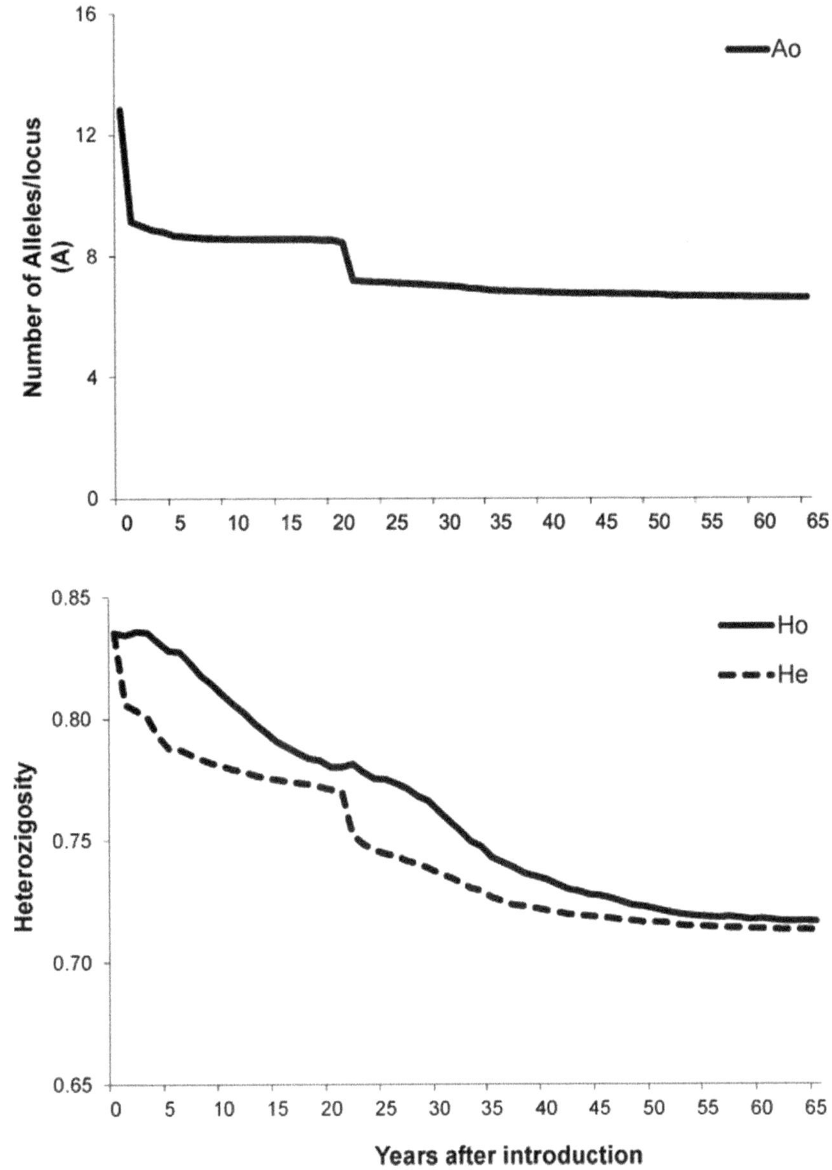

Figure 4. Temporal changes in observed alleles per locus (solid line, A_O), and observed (solid line, H_O) and expected (spotted line, H_E) heterozygosity as predicted by a simulation of the guanaco population after introduction in Staats Island. In the simulation, year 0 represents the pre-bottleneck (source in mainland) population and the bottleneck occurred in years 1 and 22. Population size changed as indicated in Figure 2.

population. This history of severe demographic reductions accompanied by isolation from other guanaco populations, suggests the Staats Island population should present a significant reduction in genetic variation due to the strong effect of genetic drift. Nevertheless, the analysis of mitochondrial and nuclear DNA variation in the southern South American guanaco populations that were examined showed unexpected and similar patterns of genetic variation and lack of divergence between the island and continental populations.

The Staats Island guanaco population is currently at similar levels of genetic diversity compared to continental natural populations in Patagonia and is higher than those measured in the Tierra del Fuego population despite its large size of ~80,000 animals. Although there are various mechanisms that could explain the retention of high genetic diversity in Staats Island (e.g. natural selection favouring diversity), the most likely candidate is the high genetic diversity due to a random sampling of individuals from the continental population, which were already diverse (Table 3, [36]).

The guanaco of southern South America were subdivided into three genetic populations that present various levels of divergence. The Staats Island was assigned to the clusters of South Patagonia, which presents the potential source populations from which its founder individuals were suspected have been taken [38]. Interestingly, the Tierra del Fuego island population appears to be a separate cluster that shows little admixture with continental populations. Although this may initially seem surprising, it is expected since this population was founded approximately 10,000–11,000 years ago and isolated 8,000 years ago when land bridges connected this island and the continent [25]. Consequently, this population has become genetically isolated from the continent [36] as no restocking of lost genetic variation has occurred via gene flow, and genetic drift has changed the population distribution of allelic/haplotypic variants. In contrast, the guanacos of Staats Island still remain genetically similar to the continental populations because of its short evolutionary history (~65 years).

Demographic history of the Staats Island population

The guanaco population on Staats Island surprisingly presented similar patterns of divergence as were observed in the various continental populations. Its nuclear DNA showed an average divergence from the continental populations similar to that between continental populations. This observation likely reflects the large genetic variation of Staats Island's founding population and a reduced effect of genetic drift on the distribution of allelic frequencies over the past ~65 years, as observed in other species [72].

Nevertheless, this finding was unexpected due to the recorded demographic history of Staats Island's guanaco population, which was founded by few animals [15] and later dramatically reduced in the early 1960s to approximately 10 – 20 individuals [38,39]. However, our analyses failed at detecting a consistent bottleneck signature in the Staats Island population. It is likely that if the demographic change had little effect on its genetic diversity (i.e. the N_e reduction was not very large and/or the reduction was not sufficiently long), the signature of a distant bottleneck cannot be picked up anymore [73]. Additionally, if after the demographic reduction the population grew exponentially, any weak signature left by the bottleneck was likely erased from the population's genetic make-up (Figure 2). As the bottlenecks experienced by the Staats Island guanaco population were only a few generations long and in both cases were followed by exponential demographic increases, it is expected that the bottleneck signatures were at least partially erased [74]. Although it is strongly suspected that young guanacos less than one year old were originally introduced to Staats Island [38], if older females were introduced and pregnant, then the effective number of individuals brought to the island would have been larger.

As revealed by the nuclear DNA analysis, genetic drift appears to have not played a major role in the microevolution of the guanaco population on Staats Island, yet it certainly had a role as observed in the mtDNA analyses. The mtDNA typically has a fourfold smaller N_e than the nuclear DNA in diploid species [75] and genetic drift is expected to have a larger effect on mtDNA. Consistent with this expectation, we found a significantly higher average divergence between Staats Island and the continental populations with the mtDNA compared with the microsatellite size variation (Welch t-test p-value: 0.00097). However, this estimate of divergence did not significantly differ between the continental populations and the island population of Tierra del Fuego. Both islands present an average divergence from continental populations approximately 12 times higher than the continental populations show with respect to each other ($F_{ST}=0.303$ and 0.026 respectively). Moreover, it is expected that the two island populations, with their smaller N_e (relative to the continent), have drifted from each other even more than they did from the continental population, and as expected, they presented the highest pairwise population differentiation ($\varphi_{ST}=0.53$, Table 4).

Origin of the introduced guanaco population

A continental-coastal origin of current guanacos in Staats Island is supported by shared haplotypes and results of assignment tests using microsatellites markers. The animals that were brought to Staats Island were shipped from the Rio Gallegos port in Argentina in two separate years, i.e. 5 animals in 1938 and 10 animals in 1939, but it is not clear from where they were collected [38]. It is known that at the time a guanaco population existed near Puerto Gallegos, but since then it has become locally extirpated or geographically displaced (WF pers. obs.). Today, the closest extant guanaco population to Rio Gallegos is Monte León, approximately ~150 km north. Consistent with an origin from South Patagonian populations, the Staats Island population groups within the cluster of South Patagonian populations (Monte León, Torres del Paine, and Pali-Ayke). Moreover, the Staats Island population shares all of their haplotypes with Monte León (MT), including its most frequent haplotype (H 20), which only occurs in Southern South America at Monte León. Nevertheless, Torres del Pine and Pali-Ayke also share one of its haplotypes with Staats Island, and as shown by the assignment tests, at least some animals can be placed in these two populations with high probability (Table 5). Thus, we conclude that while most of the genetic background of Staats Island appears to have originated in Monte León (or a Monte León-like population; Table 2 and 5), it is possible that some of the extant genetic variation originated from neighbouring South Patagonian populations, confirming historical records [38].

Conclusions and Conservation Implications

The Staats Island guanaco population presents a similar amount of genetic variation as continental guanaco populations. Interestingly and unexpectedly, this guanaco island population has been resilient to the effect of dramatic demographic changes as its genetic variation is not yet depauperated, i.e. become impoverished. Nevertheless, this by no means implies that potential severe cullings or hunts would not diminish the population's fitness. It is likely that the two dramatic population size reductions did not significantly affect the Staats Island population because the highly reduced numbers lasted only a few generations and surviving animals were a sufficiently large random sample of the pre-bottleneck genetic variation.

Similarly, continental guanaco populations also harbor high levels of genetic variation despite of the extensive pressure that was put on these populations in the recent past. Continental guanacos were hunted to increase the area intended for sheep flocks, because of a perceived forage competition between guanacos and sheep, and for the economic value of their pelts [17,76,77]. Chile exported some 35,000 pelts during the 20th century [78] and Argentina 223,000 units in only four years from 1976 to 1979 [77]. The vast majority of these pelts were obtained in Patagonia. Therefore, the current guanaco populations in Patagonia are survivors of massive hunting mainly on chulengos in the recent past, and the subdivision between the North and South Patagonia cluster probably reflects separate groups of animals that persisted throughout the last century, resulting in reconstituted extant populations. Currently, the biological and socio-economic worth of the Staats Island guanacos is high. The silk-like wool of the guanaco is worth US$60 to 80 per kg, an economic reality that has prompted Argentina's current management of Patagonian guanaco that is harvesting wool from annual, live-capture roundups [79]. The future of the privately owned guanacos on Staats Island is unknown, but its intrinsic and economic value is unmistakable.

Compared with the mainland, the relative population decrease on Staats due to killing is substantially higher when it was reduced to 15–20 animals. Therefore, it is surprising that this guanaco population has maintained genetic diversity much more than expected according its demographic history of two dramatic reductions in size. This has revealing implications and hope for surviving small remnant populations on the mainland that could be maintaining genetic diversity in spite of their low numbers. Nevertheless, it is important that although Staats Island and Patagonia guanacos harbor relatively high genetic variation, that their population numbers not be allowed to decline to severely low levels to avoid unforeseen, adverse-stochastic ecological events. Currently, guanacos in Staats Island are suffering neonatal malformations and mortality [38].

A genetic assessment, ideally with genomic-level data, of critically small and isolated guanaco populations from throughout their natural distribution [20], as well as populations established from translocation programmes of few individuals or frequency of reintroductions [31,32], is needed to document existing population sizes and biogeographic patterns. This would assist in ensuring a correct interpretation of guanaco genetic patterns and space in establishing and assessing informed and coordinated multidisciplinary management plans for the long-term recovery of guanaco populations.

Acknowledgments

We express our appreciation to Jérôme and Sally Poncet for their approval of the work on Staats Island; their conservation ethic and support for research are essential to this important work and the well-being of the island's guanacos. We thank Ricardo Baldi (CENPAT, Argentina) for storage and mailing the samples from Argentinean Patagonia, María Paz Acuña (Universidad de Chile) for developing maps and Angel Spotorno (Universidad de Chile) for useful information, discussions, advice and support. This article was prepared during the PhD studies of Benito A. González in the Programa de Doctorado en Ciencias Silvoagropecuarias y Veterinarias of the Universidad de Chile under the support of CONICYT scholarship as a part of a tutorial in Laboratorio de Ecología Molecular, Departamento de Ciencias Ecológicas, Facultad de Ciencias, Universidad de Chile, headed by Dr. Elie Poulin. Special thanks are due to the participants on research expeditions to Staats Island, for their dedicated assistance in the field, Homero Pezoa and Nicolás Aravena for support in the laboratory, and Patricio Cumsille (Universidad del Bio-Bio) for support in software analyses. Falkland Islands Department of Agriculture and Director Steve Pointing made it possible for storing and exporting biological samples from the islands. Thanks to Pamela Burger and two anonymous reviewers for comments.

Author Contributions

Conceived and designed the experiments: JCM BAG WLF. Performed the experiments: BAG RvB. Analyzed the data: JCM BAG PO-tW. Wrote the paper: BAG JCM PO-tW WEJ.

References

1. Fraser DJ, Bernatchez L (2001) Adaptive evolutionary conservation: towards a unified concept for defining conservation units. Mol Ecol 10: 2741–2752.
2. Hampton JO, Spencer PBS, Alpers DL, Twigg LE, Woolnough AP, et al. (2004) Molecular techniques, wildlife management and the importance of genetic population structure and dispersal: a case study with feral pigs. J Appl Ecol 41: 735–743.
3. Foll M, Gaggiotti O (2006) Identifying the environmental factors that determine the genetic structure of populations. Genetics 174: 875–891.
4. Johnson WE, Eizirik E, Pecon-Slattery J, Murphy WJ, Antunes A, et al. (2006) The Late Miocene Radiation of Modern Felidae: A Genetic Assessment. Science 311: 73–77.
5. Palsbøll PJ, Bérubé M, Allendorf FW (2006) Identification of management units using population genetic data. TREE 22: 11–16.
6. Manceau V, Crampe JP, Boursot P, Taberlet P (1999) Identification of evolutionary significant units wild goat, *Capra pyrenaica* (Mammalia, Artiodactyla). Anim Conserv 2: 33–39.
7. Marshall TC, Sunnucks P, Spalton JA, Greth A, Pemberton JM (1999) Use of genetic data for conservation management: the case of the Arabian oryx. Anim Conserv 2: 269–278.
8. Beheregaray LB, Ciofi C, Caccone A, Gibbs JP, Powell JR (2003) Genetic divergence, phylogeography and conservation units of giant tortoises from Santa Cruz and Pinzón, Galápagos Islands. Conserv Genet 4: 31–46.
9. Ralls K, Ballou J (1986) Captive breeding programs for populations with a small number of founders. TREE 1: 19–22.
10. Hedrick PW, Miller PS (1992) Conservation genetics – techniques and fundamentals. Ecol Appl 2: 30–46.
11. Caughley G (1994) Directions in conservation biology. J Anim Ecol 63: 215–244.
12. Ebenhard T (1995) Conservation breeding as a tool for saving animal species from extinction. TREE 10: 438–443.
13. Snyder NFR, Derrickson SR, Beissinger SR, Wiley JW, Smith TB, et al. (1996) Limitations of captive breeding in endangered species recovery. Conserv Biol 10: 338–348.
14. Wilson GA, Nishi JS, Elkin BT, Strobeck C (2005) Effects of a recent founding event and intrinsic population dynamics on genetic diversity in an ungulate population. Conserv Genet 6: 905–916.
15. Sigg DP (2006) Reduced genetic diversity and significant genetic differentiation alter translocation: Comparison of the remnant and translocated populations of bridled nailtail wallabies (*Onychogalea fraenata*). Conserv Genet 7: 577–589.
16. Le Page SL, Livermore RA, Cooper DW, Taylor AC (2000) Genetic analysis of a documented population bottleneck: introduced Bennett's wallabies (*Macropus rufogriseus rufogriseus*) in New Zealand. Mol Ecol 9: 753–763.
17. Franklin W (1982) Biology, ecology and relationship to man of the South American camelids. In: Mares MA, Genoways HH, editors. Mammalian biology in South America. Lab. of Ecol and Univ. of Pittsburgh, Pittsburgh: Vol. Pymatuning Symp. Ecol. Spec. Publ. Vol 6. pp. 457–489.
18. Wheeler JC (1995) Evolution and Present Situation of the South-American Camelidae. Biol J Linn Soc 52: 271–295.
19. Franklin WL (2011) Family Camelide (Camels). In: Wilson DE, Mittermeier RA, editors. Handbook of the Mammals of the World - Volume 2 - Hoofed Mammals. Barcelona, Spain: Lynx Edicions. pp. 206–246.
20. González BA, Palma RE, Zapata B, Marín JC (2006) Taxonomic and biogeographic status of guanaco *Lama guanicoe* (Artiodactyla, Camelidae). Mamm Rev 36: 157–178.
21. Raedeke K (1978) El guanaco de Magallanes, Chile, su distribución y biología. Corporación Nacional Forestal, Departamento de Conservación del Medio Ambiente, Santiago, Chile. 182 p.
22. Bas F, González B (2000) Current advances in research and management of the guanaco (*Lama guanicoe*) in Chile. Cienc Inv Agr 27: 51–65.
23. Montes C, De Lamo DA, Zavatti J (2000) Distribución de abundancias de guanacos (*Lama guanicoe*) en los distintos ambientes de Tierra del Fuego, Argentina. Mastozool Neotrop 7: 23–31.
24. McCulloch RD, Clapperton CM, Rabassa J, Current AP (1997) The natural setting, the glacial and post-glacial environmental history of Fuego-Patagonia. In: McEwan C, Borrero LA, Prieto A, editors. Patagonia natural history, prehistory and ethnography at the uttermost end of the earth. Princeton: Princeton University Press. pp. 12–31.
25. Sarno R, Franklin W, O'Brien S, Johnson W (2001) Patterns of mtDNA and microsatellite variation in an island and mainland population of guanacos in southern Chile. Anim Conserv 4: 93–101.
26. Massoia E, Chevez JC (1993) Mamíferos silvestres del Archipiélago Fueguino. Editorial L.O.L.A. Santa Fe, Argentina.
27. Estevez J, Piana E, Schiavini A, Juan-Muns N (2001) Archaeological Analysis of Shell Middens in the Beagle Channel, Tierra del Fuego Island. Int J Osteoarchaeol 11: 24–33.
28. Saba S, De Lamo DL (1994) Dynamic responses of mammals to the eruption of Volcan Hudson. Mastozool Neotrop 1: 113–122.
29. Cunazza C, Puig S, Villalba L (1995) Situación actual del guanaco y su ambiente. In: Puig S, editor. Técnicas para el manejo del guanaco. IUCN. pp 27–50.
30. Jaksic FM, Silva-Aranguiz E, Silva SI (2004) Fauna del Parque Nacional Bosque Fray Jorge: una revisión bibliográfica. In: Squeo F, Gutiérrez JR, Hernández IR, editors. Historia Natural del Parque Nacional Bosque Fray Jorge. Ediciones Universidad de La Serena, La Serena. pp. 93–114.
31. Barri FR, Fernández M (2011) Foraging and vigilance time allocation in a guanaco (*Lama guanicoe*) population reintroduced in Quebrada del Condorito National Park (Córdoba, Argentina). Acta Ethol 14: 103–107.
32. Flores CE, Cingolani AM, von Müller A, Barri FR (2012) Habitat selection by reintroduced guanacos (*Lama guanicoe*) in a heterogeneous mountain rangeland of central Argentina. Rangeland J 34: 439–445.
33. Franklin W (1983) Contrasting socioecologies of South America's wild camelids: the vicuña and the guanaco. In: Eisenberg JF, Kleiman D, editors. Advances in the Study of Animal Behaviour. American Society of Mammalogists, Special Publication 7. Allen Press. pp. 573–629
34. Marín JC, Zapata B, González BA, Bonacic C, Wheeler JC, et al. (2007) Sistemática, taxonomía y domesticación de alpacas y llamas: nueva evidencia cromosómica y molecular. Rev Chil Hist Nat 80: 120–141.

35. Marin JC, Spotorno AE, González BA, Bonacic C, Wheeler JC, et al. (2008) Mitochondrial DNA variation and systematics of the guanaco (*Lama guanicoe*, ARTIODACTYLA: CAMELIDAE). J Mammal 89:269–281.
36. Marín JC, González BA, Poulin E, Casey CS, Johnson WE (2012) The influence of the arid Andean high plateau on the phylogeography and population genetics of guanaco (*Lama guanicoe*) in South America. Mol Ecol 22: 463–482.
37. Maté ML, Bustamante A, Giovambattista G, De Lamo D, Von Thüngen J, et al. (2005) Genetic diversity and differentiation of guanaco populations from Argentina inferred from microsatellite data. Anim Genet 36: 316–321.
38. Franklin WL, Grigione MM (2005) The enigma of guanacos in the Falklands Islands: the legacy of John Hamilton. J Biogeogr 32: 661–675.
39. Franklin WL, Poncet S, Poncet J (2005) The history of Staats Island in the western Falklands: its guanacos, foxes, shanty and sojourners. The Falklands Islands Journal 8: 21–64.
40. Franklin W (2005) Penguins, dolphins and guanacos: an adventure to the Falkland Islands. International Camelid Quarterly 4: 7–20.
41. Baldi B, Lichtenstein G, González B, Funes M, Cuéllar E, et al. (2008) *Lama guanicoe*. In: IUCN 2010. IUCN Red List of Threatened Species. Version 2010.1. Available: http://www.iucnredlist.org/apps/redlist/details/11186/0. Accessed 25 August 2013.
42. Strange IJ (1992) A field guide to the wildlife of the Falkland Islands and South Georgia. Harper Collins, London.
43. Sarno RJ, Hunter RL, Franklin WL (1996) Immobilization of guanacos by use of tiletamine/zolazepam. JAVMA 208: 408–409.
44. Georoff TA, James SB, Kalk P, Calle PP, Martin-Flores M (2010) Evaluation of Medetomidine–Ketamine–Butorphanol anesthesia with Atipamezole–Naltrexone antagonism in captive male guanacos (*Lama guanicoe*). J Zoo Wildlife Med 41: 255–262.
45. Gannon WL, Sises RS, The Animal Care And Use Committee Of The American Society Of Mammalogists (2007). Guidelines of the American Society of Mammalogists for the use of wild mammals in research. J Mammal 88: 809–823.
46. Sambrook J, Fritsch EF, Maniatis T (1989) Molecular Cloning - A Laboratory Manual, Second Edition. Cold Spring Harbor Laboratory Press.
47. Excoffier L, Laval G, Schneider S (2005) Arlequin (version 3.0): An integrated software package for population genetics data analysis. Evol Bioinform Online 1: 47–50.
48. Lang KDM, Wang Y, Plante Y (1996) Fifteen polymorphic dinucleotide microsatellites in llamas and alpacas. Anim Genet 27: 285–94.
49. Penedo MCT, Casetano AR, Cordova K (1998a) Microsatellite markers for South American camelids. Anim Genet 29: 411–412.
50. Penedo MTC, Caetano AR, Cordova K (1998b). Eight microsatellite markers for South American camelids. Anim Genet 30: 161–168.
51. Penedo MCT, Caetano AR, Cordova K (1999) Eight microsatellite markers for South American camelids. Anim Genet 30: 166–167.
52. Sarno RJ, David VA, Franklin WL, O'Brien SJ, Johnson WE (2000) Development of microsatellite markers in the guanaco, *Lama guanicoe*: utility for South American Camelids. Mol Ecol 9:1922–1924.
53. Park SDE (2001) Trypanotolerance in West African cattle and the population genetic effects of selection. PhD thesis. University of Dublin.
54. Van Oosterhout C, Hutchinson WF, Wills DPM, Shipley P (2004) MICRO-CHECKER: software for identifying and correcting genotyping errors in microsatellite data. Mol Ecol Notes 4: 535–538.
55. Goudet J (2005) *FSTAT Version 2.9.4.: A Program to Estimate and Test Population Genetics Parameters*. Available from URL: http:// www2.unil.ch/popgen/softwares/fstat.htm.
56. Belkhir K, Borsa P, Goudet J, Bonhomme F (2005) *GENETIX 4.05: Logiciel sous Windows pour la genetique des populations*. Laboratoire Genome & Population, CNRS-UPR/Universite de Montpellier II, Montpellier, France.
57. Pritchard JK, Stephens M, Donnelly P (2000) Inference of population structure using multilocus genotype data. Genetics 155: 945–959.
58. Evanno G, Regnaut S, Goudet J (2005) Detecting the number of clusters of individuals using the software Structure: a simulation study. Mol Ecol 14: 2611–2620.
59. Rannala B, Mountain JL (1997) Detecting immigration by using multilocus genotypes. Proc Natl Acad Sci USA 94: 9197–9201.
60. Piry SG, Luikart G, Cornuet JM (1999) BOTTLENECK: a computer program for detecting recent reductions in the effective population size using allele frequency data. J Hered 90: 502–503.
61. Corander J, Sirén J, Arjas E (2008) Spatial modelling of genetic population structure. Computation Stat 23: 111–129.
62. Ohta T, Kimura M (1973) A model of mutation appropriate to estimate the number of electrophoretically detectable alleles in a finite population. Genet Res 22: 201–204.
63. Di Rienzo A, Peterson AC, Garza JC, Valdes AM, Slatkin M, et al. (1994) Mutation processes of simple-sequence repeat loci in human populations. Proc Natl Acad Sci USA 91: 3166–3170.
64. Garza JC, Williamson EG (2001) Detection of reduction in population size using data from microsatellite loci. Mol Ecol 10: 305–318.
65. Hartl DL, Clark AG (1989) Principles of population genetics. 2nd edition. Sinauer Associates, Sunderland, MA.
66. Weber JL, Wong C (1993) Mutation of human short tandem repeats. Hum Mol Genet 2: 1123–1128.
67. Kuo CH, Janzen FJ (2003) Bottlesim: a bottleneck simulation program for long-lived species with overlapping generations. Mol Ecol Notes 3: 669–673.
68. Young JK, Franklin WL (2004) Territorial fidelity of male guanacos in the Patagonia of southern Chile. J Mammal 85: 72–78.
69. Baldi R, Campagna C, Saba S (1997) Abundancia y distribución del guanaco (*Lama guanicoe*) en el NE del Chubut, Patagonia Argentina. Mastozool Neotrop 4: 5–15.
70. González BA (2010) ¿Que problemas de conservación tienen las poblaciones de guanaco en Chile? Ambiente Forestal 9: 26–36.
71. Franklin WL (2004) *Report on guanaco research expedition: Staats Island–Falkland Islands (December 2003)*, 11 pp. + 1 map. Department of Agriculture, Stanley, Falkland Islands.
72. Busch JD, Waser PM, DeWoody JA (2007) Recent demographic bottlenecks are not accompanied by a genetic signature in banner-tailed kangaroo rats (*Dipodomys spectabilis*). Mol Ecol 16: 2450–2462.
73. Cornuet JM, Luikart G (1996) Description and power analysis of two tests for detecting recent population bottlenecks from allele frequency data. Genetics. 144: 2001–2014.
74. Bonhomme M, Blancher A, Cuartero S, Chikhi L, Crouau-Roy B (2008) Origin and number of founders in an introduced insular primate: estimation from nuclear genetic data. Mol Ecol 17: 1009–1019.
75. Hare MP, Cipriano F, Palumbi SR (2002) Genetic evidence on the demography of speciation in allopatric dolphin species. Evolution 56: 804–816.
76. Dennler De La Tour G (1954) The guanaco. Oryx 2: 273–279.
77. Mares MA, Ojeda RA (1984) Faunal commercialization and conservation in South America. BioScience 34: 580–584.
78. Iriarte JA, Jaksic FM (1986) The Fur Trade in Chile: An overview of seventy-five years of export data (1910-1984). Biol Conserv 38: 243–253.
79. Lichtenstein G (2013) Guanaco management in Argentina: taking a commons perspective. Journal of Latin American Geography 12: 187–213.

15

Influence of Ethnolinguistic Diversity on the Sorghum Genetic Patterns in Subsistence Farming Systems in Eastern Kenya

Vanesse Labeyrie[1]*, Monique Deu[1], Adeline Barnaud[3], Caroline Calatayud[1], Marylène Buiron[1], Peterson Wambugu[2], Stéphanie Manel[4,5], Jean-Christophe Glaszmann[1], Christian Leclerc[1]

1 UMR AGAP, CIRAD, Montpellier, France, **2** National Genebank of Kenya, KARI, Nairobi, Kenya, **3** UMR DIADE, IRD, Montpellier, France, **4** UMR LPED, Université Aix-Marseille/IRD, Marseille, France, **5** UMR AMAP, CIRAD, Montpellier, France

Abstract

Understanding the effects of actions undertaken by human societies on crop evolution processes is a major challenge for the conservation of genetic resources. This study investigated the mechanisms whereby social boundaries associated with patterns of ethnolinguistic diversity have influenced the on-farm distribution of sorghum diversity. Social boundaries limit the diffusion of planting material, practices and knowledge, thus shaping crop diversity *in situ*. To assess the effect of social boundaries, this study was conducted in the contact zone between the Chuka, Mbeere and Tharaka ethnolinguistic groups in eastern Kenya. Sorghum varieties were inventoried and samples collected in 130 households. In all, 297 individual plants derived from seeds collected under sixteen variety names were characterized using a set of 18 SSR molecular markers and 15 morphological descriptors. The genetic structure was investigated using both a Bayesian assignment method and distance-based clustering. Principal Coordinates Analysis was used to describe the structure of the morphological diversity of the panicles. The distribution of the varieties and the main genetic clusters across ethnolinguistic groups was described using a non-parametric MANOVA and pairwise Fisher tests. The spatial distribution of landrace names and the overall genetic spatial patterns were significantly correlated with ethnolinguistic partition. However, the genetic structure inferred from molecular makers did not discriminate the short-cycle landraces despite their morphological distinctness. The cases of two improved varieties highlighted possible fates of improved materials. The most recent one was often given the name of local landraces. The second one, that was introduced a dozen years ago, displays traces of admixture with local landraces with differential intensity among ethnic groups. The patterns of congruence or discordance between the nomenclature of farmers' varieties and the structure of both genetic and morphological diversity highlight the effects of the social organization of communities on the diffusion of seed, practices, and variety nomenclature.

Editor: John P. Hart, New York State Museum, United States of America

Funding: This study was supported by Agropolis Fondation in the framework of the ARCAD project No.0900-001. V. Labeyrie was supported by Agropolis Fondation through a Ph.D. grant and received financial support from CIRAD for this study. The funders had no role in study design, data collection and analysis, decision to publish, or preparation of the manuscript.

Competing Interests: The authors have declared that no competing interests exist.

* E-mail: vanesse.labeyrie@gmail.com

Introduction

Identifying factors involved in crop evolution is of great importance for genetic resource conservation and crop improvement. Crop genetic diversity patterns result from selection, migration and genetic drift processes which are strongly influenced by human action. Recent studies combining linguistic, archeological and genetic data have unraveled the past domestication and diversification processes of crops such as banana [1] and sweet-potatoes [2], on a large time-space scale, by linking global diversity patterns to human migrations. However, the evolution of crops is still ongoing in smallholder farming systems under the pressure of agro-ecological conditions and farmers' management practices [3]. The study of these processes at the community scale is complementary to large time-space approaches and contributes to the general understanding of the *in situ* genesis of crop genetic patterns.

Social boundaries contribute to the evolution of crop populations both directly, by determining seed flows, and indirectly, by inducing the divergence of seed selection practices [4]. Previous studies notably showed that the ethnic organization of farming communities plays an important role in differentiating the domesticated populations of allogamous crops [5], vegetatively-propagated crops [6] and animals [7].

Sorghum (*Sorghum bicolor* L. Moench) is an annual cereal extensively cultivated in smallholder farming systems because of its ability to grow under harsh climatic conditions. De Wet and Huckabay [8] and Harlan et al. [9] suggested that the spatial distribution of sorghum botanical races in Africa was related to that of the ethnic groups, but this hypothesis was not further tested. In a study undertaken in Niger, Deu et al. [10] suggested that human ethnic diversity has probably a greater impact on sorghum diversity than recent environmental constraints. However, the authors were not able to assess this hypothesis as the spatial localization of the different ethnic groups in Niger corresponded to different agro-ecological regions. Thus, deciphering how the social organization of farmers affects the structure of sorghum diversity remains a challenge.

This article addresses the role of social boundaries in sorghum evolution and diversification processes. It set out to identify the mechanisms whereby social boundaries, associated with ethnolinguistic diversity patterns, shape sorghum genetic diversity on-farm. To study only the main effect of social boundaries, this study focused on an ethnolinguistic contact zone where both geographical distance between ethnic groups and agro-ecological variability were limited. If social boundaries do not limit seed-mediated gene flows and the diffusion of selection practices, then no relation should be observed between ethnic diversity patterns and both the genetic and morphological structure of sorghum diversity. Otherwise, it would reflect the impact of social boundaries on the evolutionary mechanisms that shape sorghum diversity *in situ.*

Farmers' varieties are relevant units for studying on-farm crop diversity as they are consciously defined and named by farmers for management, selection, seed exchanges and knowledge transmission purposes [11]. Farmer's nomenclature and taxonomy of crop varieties is a marker of knowledge diffusion and exchanges across communities [12], while the distribution of the genetic and morphological diversity of crop populations reflects gene flows and selection forces [5]. This study thus used molecular markers to estimate genetic diversity and compared the spatial distribution of varieties with genetic spatial patterns according to ethnic groups. These patterns were then discussed regarding the congruence between farmer's varieties and the structure of their genetic and morphological diversity. Combining these three approaches enabled us to investigate the influence of social boundaries on the evolutionary mechanisms that shape sorghum diversity *in situ.*

Clarifying the effect of social boundaries on crop evolutionary mechanisms has important applications for crop genetic resource collection, characterization and conservation. This study hence contributes to increasing the overall understanding of on-farm crop diversification processes. By highlighting the overall role of societies in shaping crop diversity, it stresses the relevance of multidisciplinary approaches for crop genetic diversity studies.

Materials and Methods

Ethics statement

This was a collaborative study between CIRAD and KARI-National Genebank of Kenya. KARI has the national mandate for the collection and conservation of all plant genetic resources and documentation of all accompanying information. Under this framework and mandate, the study was mounted and all laid down institutional and administrative procedures were carefully followed prior to undertaking the study. Based on the aforementioned mandate given to KARI, no specific permission was required to undertake the study. Though KARI does not have a body designated as ethical review board, it has equivalent committees and administrative organs which review proposed research activities before granting approval. Research clearance was therefore sought from these organs at all levels including the institutional legal office. Local government administrative as well as agricultural extension officers were informed of the study and kept updated of the activities.

During the survey, the mandate given to KARI as well as the importance of the study, both nationally and globally, was explained to the farmers and concurrence was sought before undertaking the study activities. According to KARI's procedures governing genetic resources collection and documentation, prior informed consent was obtained verbally and not recorded, all with the understanding that the process would only involve collection of genetic resources and no sensitive traditional knowledge. Where such consent was not granted, the germplasm collectors stopped any more activities in that particular household. In each household, we interacted with the female household head. Upon granting consent, they were interviewed mainly on their ethnicity and the sorghum varieties they grew. The survey was conducted by the authors among them V.L, A.B, P.W, and C.L with questions being translated by a local field assistant. We confirm that sorghum, the studied crop, is neither endangered nor protected.

Study site: Agro-ecological conditions and ethnic organization

This study was conducted on the eastern slope of Mount Kenya (0°24'27.88"S, 37°46'35.59"E), in an ethnolinguistic contact zone between Chuka, Tharaka and Mbeere groups (Figure 1). The three ethnolinguistic groups (hereafter ethnic) live within the same agro-ecological zone, as defined by Jaetzold et al. [13]. The study site was 15 km-square, and the elevation ranged from 810 to 946 m above sea level, so rainfall and temperature variability was limited. The mean temperature on the area ranges between 21.7°C and 23.9°C. The mean rainfall is about 700–800 mm per year, distributed across two rainy seasons with the Long Rains occurring from March to May and the Short Rains from October to December [14]. Soil characteristics are homogeneous in the area occupied by the three ethnic groups, corresponding to well drained Ferralsols, with a loamy-sand texture and moderate fertility [13].

The three ethnic groups, Chuka, Tharaka and Mbeere, migrated to the study area by the end of the 19th century, either because of a population increase or because of recurrent drought [15]. Social boundaries exist between Chuka, Tharaka and Mbeere groups as revealed by their distinct ethnic identity, and their current cultural and linguistic differences [16,17]. The Mbeere are closely related to the Embu group [18], while the Chuka and Tharaka are related to the Meru group. The Mbeere and Chuka had conflictual mutual relationships in the past [19], while the Chuka and Tharaka maintain strong social ties and consider they are kin [20]. Intermarriage is usual between the Chuka and Tharaka, while it is very uncommon between the Mbeere and Chuka or Tharaka (unpublished data). Men usually settle near their father's compound once they get married. The residence is thus patrilocal [16]. The three ethnic groups present a non-random spatial distribution. The Mbeere households are located in the southern part of the study area, the Tharaka mostly on the north-eastern side, and the Chuka on the north-western side (Figure 1). Consistently with the social relationships between groups depicted above, a clear spatial boundary was found between the Mbeere and both the Chuka and Tharaka, while the Chuka and Tharaka appeared to be spatially more mixed.

The three ethnic groups manage low-input cropping systems that harbor high specific and infra-specific crop diversity. Cropping systems are based on cereals and legumes that are usually intercropped. Sorghum (*Sorghum bicolor*), cowpeas (*Vigna unguiculata*), maize (*Zea mays*), mungo bean (*Vigna radiata*) and pearl millet (*Pennisetum glaucum*) are the main crop species grown in the area. Sowing is done either by hand-dibbling or by drilling, while plowing is done with animals. The different sorghum varieties are either grown in separate plots or mixed together within farmers' fields. Improved varieties, mainly disseminated by the extension services of the Kenyan Ministry of Agriculture, have also been adopted by the farmers. They are cultivated together in the same

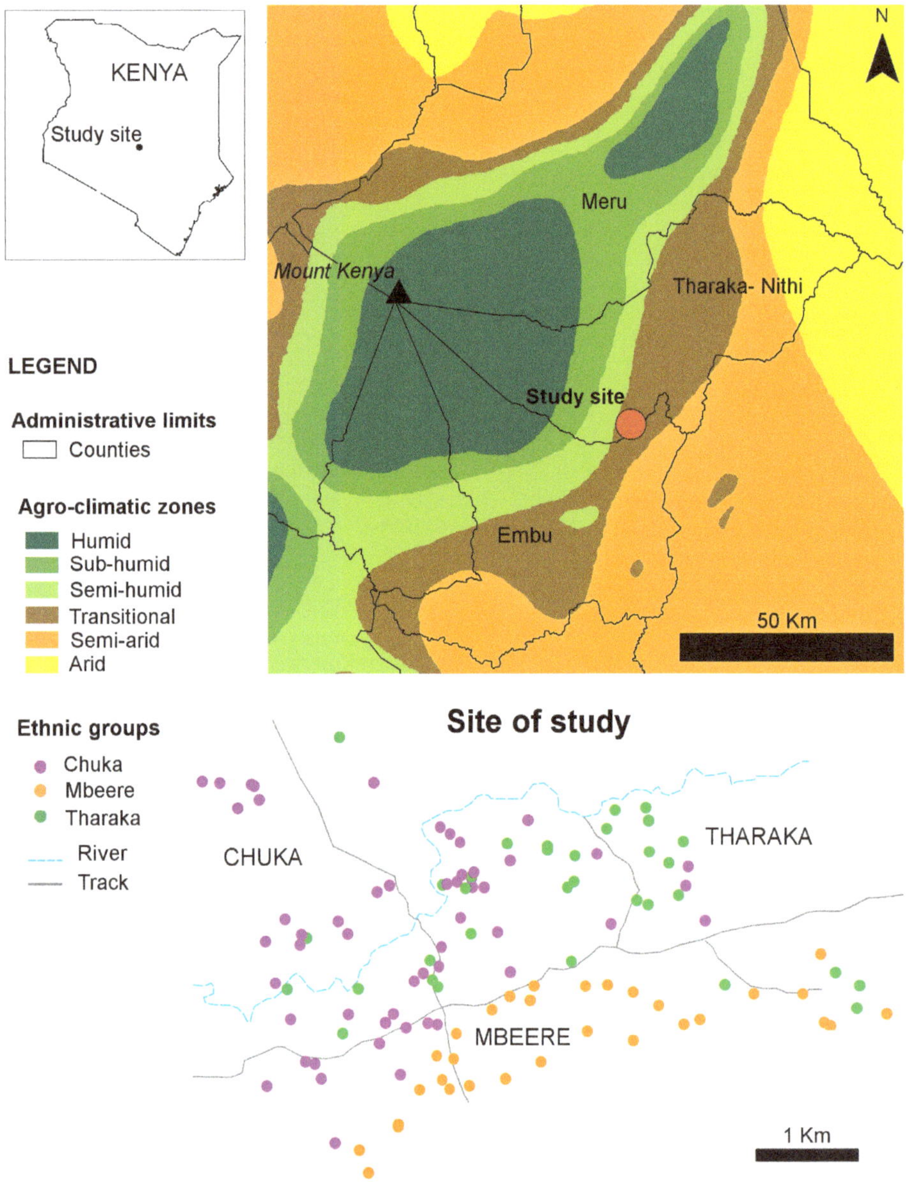

Figure 1. Study site location. Map of the eastern side of Mount Kenya and location of the farms where sorghum samples were collected (colors correspond to the ethnic identity of the male house-head).

field with the local varieties (or landraces). Farmers distinguish between short-cycle varieties that can be grown either from October to January or from March to June, and long-cycle varieties that are subjected to the ratooning practice [21] (Figure S1). These long-cycle varieties are sown in October, the vegetative part being cut before the grains are mature to stimulate regrowth from basal buds, and panicles are finally harvested in July.

Data collection

Sorghum inventory and germplasm collection. The field work consisted of two stages. A preliminary survey was carried out to estimate the frequency of varieties in the three ethnic groups. The strategy for on-farm germplasm collection was then based on that estimation of diversity, as it aimed at representing the diversity and frequencies of each variety in each ethnic group.

The preliminary inventory survey was conducted in both January (Short Rains cropping season) and June 2011 (Long Rains cropping season), just before harvesting and prior to germplasm collection. The inventory of sorghum varieties was based on the local names as reported by women farmers who were in charge of sorghum selection in each of the 124 households surveyed. Indeed, grain crop farming comprising seed sowing, harvesting, selection and trading is ensured by women ([16], personal observation). The ethnolinguistic identity of male house-heads or single women was also recorded, women becoming members of the family of their husband when they get married in this patrilinear society.

Sorghum panicles were then collected from a total of 130 households selected randomly. In 22 of these households, panicles were collected in both January and July, while the rest of the households were visited only in January (34 households) or in July (74 households). Half of these 130 households were visited during the preliminary survey described above. They represented about half of the total number of households in the area, hence insuring a good representativeness. 60 households belonged to the Chuka ethnic group, 35 to the Mbeere and 35 to the Tharaka. In order to be representative of the sorghum population of each ethnic group, all the varieties grown in each household were collected, except a highly dominant variety of improved origin (*Kaguru*). As this variety was much more abundant than the others, we limited the number of samples collected. We thus sampled *Kaguru* variety randomly in a maximum of 19 households per ethnic group. One or two individual panicles of each variety were collected in each household cultivating it. The mean number of varieties collected per household was 1.5 (min: 1, max: 6). It was similar across ethnic groups, as well as the mean number of panicles of each variety sampled per household (Table S1). The fraction of households where each variety was collected for the study of genetic diversity was correlated to the fraction of households where each variety was previously inventoried (Linear regression R^2: 0.77, Figure S2). In all, 290 samples were collected on-farm after harvest, each consisting of a single panicle. About 47% of the individual plants were sampled from the Chuka ethnic group, 30% from the Tharaka and 23% from the Mbeere. Information concerning the names, the origin (local or improved) and the cycle length of each sampled panicle was recorded from women house-heads, and we recorded the geographic coordinates of each household using a global positioning system (GPS).

DNA extraction and SSR genotyping. Seeds from the 290 panicles collected on-farm were sown in an experimental field *in-situ*, and the leaves of one sibling randomly chosen for each mother plant were collected and stored on silicagel. Leaves from seven individuals grown from certified seeds of the improved varieties *Serendo* and *Gadam* were also collected as controls. In total 297 individual plants were thus used for the genetic diversity study.

Twenty-two pairs of primers were selected for their high polymorphism in central Kenya (unpublished data) and West Africa [10]; twenty of them were part of a set of reference microsatellite markers proposed by Billot and colleagues ([22], http://sat.cirad.fr/sat/sorghum_SSR_kit/). Loci were distributed over the 10 chromosomes. DNA was extracted from dried leaves and the polymerase chain reaction amplifications were done following the procedure described previously [10,22]. The fluorescent dye–labeled PCR products from differentially labeled primers and with non-overlapping size were pooled and subjected together to capillary electrophoresis using a 24-capillary 3500xL System (Applied Biosystems). GeneMapper v 4.1 (Applied Biosystems) was used for genotype scoring. GeneScan 600 LIZ Size Standard v2.0 was added to each well, and three control samples were used to facilitate allele scoring [22]. Genotyping was done at the Montpellier Languedoc-Roussillon Genopole platform located on the CIRAD campus in Montpellier (France).

Four markers presenting either a high number of missing data, or low polymorphism (at a 99% threshold) were discarded from the analysis, so eighteen markers were kept, covering 9 chromosomes out of 10. The percentage of missing data for the 18 markers kept was 1%.Table S2 provides a list of these 18 markers and their description.

Panicle morphological characterization. Fifteen qualitative morphological traits were measured on the panicles of the 297 individuals that were genotyped (Table S3). Eight morphological descriptors were selected from the IPGRI descriptors [23] and were completed by seven additional descriptors for seeds and glumes characteristics that showed variability on the sorghum collected in our study area. Descriptors covered the characteristics of the whole panicle (panicle shape), seeds (color, presence of sub-coat, pericarp thickness, shape, endosperm texture and shattering) and glumes (color, adherence, covering, opening, texture, hairiness, awning and transversal wrinkle). Only qualitative traits were kept for these analyses because they are stable characteristics on which farmers base their nomenclature and classification [24]. Multiple characterizations of randomly sampled individuals enabled to check for morphological trait scoring consistency.

Data analysis

Comparing sorghum assemblages between ethnic groups. We characterized each household by its sorghum assemblage, which is the panel of co-occurring sorghum varieties that are cultivated by the household. The differentiation of sorghum assemblages across ethnic groups was tested using a non-parametric Multivariate Analysis of Variance (perMANOVA, [25]). The PerMANOVA was implemented under the *adonis* function in the R package *vegan* [26]. The presence/absence matrix for sorghum varieties in each household was transformed into a distance matrix using the Bray-Curtis index [27]. The *adonis* function partitions the distance matrix according to grouping factors (ethnic groups) and compares the sum of squared distances within groups (which is the sum of squared distances from individual replicates to their group centroid) and between groups (which is the sum of squared distances from group centroids to the overall centroid). A pseudo F-ratio is then computed and compared to its distribution under the null hypothesis simulated using 4000 random permutations of the raw data. Pairwise Fisher exact tests implemented in the R package *fmsb* [28] were then used to compare the occurrence frequencies of the most frequent varieties across the Chuka, Mbeere and Tharaka ethnic groups. The calculation of p-values was corrected for multiple comparisons using the False Discovery Rate (FDR) procedure [29] implemented in the *p.adjust* function.

Genetic diversity and genetic structure of sorghum populations. Genetic diversity within sampling populations. The genetic diversity of sorghum populations sampled in each ethnic group was assessed using several indexes. The observed number of alleles and the observed heterozygosity were calculated using GENETIX 4.05.2 software [30]. The allelic richness corrected for sample size [31], the unbiased gene diversity (expected heterozygosity) corrected for small sample size [32], and the F_{IS} [33] of multi-locus genotypes were estimated using the procedures implemented in FSTAT 2.9.3.2 software [34]. These indexes were compared among ethnic groups using paired Pairwise Wilcoxon tests with False Discovery Rate (FDR) correction implemented in R (package *stats*, *pairwise.wilcox.test* function).

Genetic structure assuming sampling populations. Pairwise F_{ST} [33] were computed among the sorghum populations collected in the three ethnic groups. The significance of the differences was assessed using a permutation test (3000 permutations) and corrected using a Bonferroni procedure [35]. A multilocus G-test of differentiation, known to be accurate for measuring the genetic differentiation between populations with unbalanced sizes [36], was used to test the genetic differentiation between the populations sampled in each ethnic group (10000 permutations). Calculations were carried out using FSTAT 2.9.3.2. Pairwise G-tests implemented in GENEPOP 4.2 [37] were used to estimate the genotypic differentiation among pairs of populations and p-values were corrected for multiple tests using FDR correction (*p.adjust* function in the R package *stats*).

Analysis at individual level. Two complementary approaches, Bayesian clustering and Neighbor-Joining tree, were used to assess the genetic structure without defining a-priori populations. First, the genetic structure of sorghum populations was characterized using the Bayesian clustering algorithm implemented in STRUCTURE 2.3.3 software [38] and run on the Bioportal server (http://www.bioportal.uio.no). The admixture model with correlated allele frequencies was used, assuming that the genome of each individual resulted from the mixture of K ancestral populations. The estimated proportions of each individual's genotype originating from each of the K ancestral populations (q) was calculated for K ranging from 2 to 10 ancestral populations (or clusters), with twenty runs for each K value. The burn-in period was set at 500 000 and 1 000 000 iterations were performed. The criterion suggested by Evanno et al. [39], based on the rate of change in the log probability of data between successive K values, was used to determine the most likely number of clusters (K). Second, a Neighbor-Joining tree [40] was built from a simple matching genetic dissimilarity index [41] using Darwin V5 software [42]. The results of both the Bayesian clustering and Neighbor-Joining methods were then compared to check for the consistency of the clusters. This led to what we refer to as an MMb (molecular-marker-based) classification scheme.

For further analysis, individuals whose estimated proportion of genome originating from one population (q, hereafter admixture coefficient) was below a 0.8 threshold were considered as resulting from admixture between the populations. Individuals whose q value was equal to or above 0.8 for a population were assigned to that population (hereafter cluster). To explain the MMb genetic structure, the assignment of individuals to clusters thus defined was crossed with information concerning their origin and cycle length as reported by farmers during the collection of samples *in situ*. The occurrence frequencies of each MMb genetic cluster were then compared across ethnic groups using Pearson's Chi-squared test, and pairwise Fisher exact tests with False Discovery Rate (FDR) correction for multiple comparisons.

To test a potential isolation-by-distance effect in cultivated sorghum, we applied Mantel test between pairwise genetic distances and geographical distances. The matrix of geographical distances among individuals was computed. The kinship coefficient of Loiselle et al. [43] was computed using SPAGeDI software [44] for each pair of cultivated sorghum individuals, producing a matrix of individual pairwise genetic distance. A stratified Mantel test implemented in the R package Vegan was used to test the significance of the correlation between the logarithm of the pairwise geographical distances [45] and sorghum individuals' pairwise genetic relatedness. 4000 permutations of the locations of samples were done within the genetic clusters previously identified using STRUCTURE software (stratified test), as recommended by Meirmans [46] for populations presenting a strong genetic structure.

Morphological structure of sorghum populations

To describe the structure of individual panicle morphological diversity, a dissimilarity matrix was computed on the basis of the 15 morphological traits coded through a total of 43 modalities using the simple matching index. The morphological similarity between individuals was then assessed using a Principal Coordinates Analysis (PCoA) using the R package *ade4*.

Results

Differences in variety assemblages across ethnic groups

On the basis of their local names, seventeen different varieties were inventoried among the 124 households visited during both the January and June surveys. 14 different varieties were respectively inventoried in the Chuka and Tharaka groups, and 10 in the Mbeere group, out of which 9 were shared by the three ethnic groups. The mean number of varieties inventoried in both cropping seasons per household was similar across ethnic groups (2.77, SE: 0.17 for the Chuka, 2.65, SE: 0.17 for the Mbeere, 3.02, SE: 0.21 for the Tharaka). The most frequent variety was *Kaguru*, (76% of the households), followed by *Gadam* (48% of the households), both of which are improved varieties. *Ngirigacha, Mugeta, Mbura imwe, Muruge mbura ciiri*, and *Muruge mbura imwe* were the most frequent local varieties (landraces) (Figure 2).

The non-parametric perMANOVA showed that sorghum variety assemblages differed significantly between ethnic groups (Table S4), even though the ethnic partition explained a limited part of variability (pseudo-$F_{2,121} = 4.971$, p-value $= 0.0002$, $R^2 = 0.076$). Pairwise Fisher exact tests confirmed that the frequency of three out of the five most frequent landraces differed significantly between ethnic groups, while the frequency of improved varieties (*Gadam, Kaguru* and *Serendo*) did not differ significantly between ethnic groups. *Muruge mbura imwe* and *Mugeta* were significantly less frequent in the Mbeere group than in the Chuka and Tharaka groups while *Ngirigacha* was significantly more frequent in the Mbeere group.

Genetic and morphological structure of cultivated sorghum

The most likely number of populations (K) identified by STRUCTURE was $K = 4$. Indeed, the log-probability of data increased up to $K = 4$, where it reached a plateau. This was congruent with Evanno's ΔK curve which presented a clear peak for $K = 4$. The populations (clusters) inferred by STRUCTURE for $K = 4$ (Figure 3.A) corresponded to distinct groups on the Neighbor-Joining tree (Figure 4.A). Cluster A and C were distinct and showed higher genetic uniformity than cluster B and D. Most of the individuals sampled (88%) showed an admixture coefficient (q) above or equal to $q = 0.8$, and they were thus assigned to the corresponding cluster. The remaining 12% of the individuals were considered to result from admixture between clusters.

The MMb genetic structure was found to be strongly related to the improvement status of the germplasm – improved varieties or local landraces, and by differences in growth-cycle length (Figure 3.B). Individuals assigned to the uniform clusters A and C were mostly improved varieties introduced by the extension services, while individuals assigned to the broader clusters B and D were mainly classified by farmers as local landraces. Moreover, almost all individuals assigned to cluster D were identified by farmers as long-cycle varieties (ratoon) while those individuals assigned to clusters A, B and C were mainly identified as short-cycle varieties.

Despite this global coherence, the characteristics of varieties reported by farmers showed some divergence from the MMb genetic classification. Twenty-two percent (22%) of the individual plants that were identified as long-cycle landraces by farmers during the collection were assigned to cluster B by STRUCTURE (Figure 3.B). A substantial proportion of individuals identified by farmers as short-cycle landraces were assigned to clusters A (13%) or C (10%). Indeed, young farmers may consider as local the varieties that were introduced a long time ago, perhaps before they began farming. Conversely, 14% of the individuals identified by farmers as improved varieties were assigned to cluster B.

The morphological diversity was summarized by the PCoA (Figure 4.B). The two first axes accounted for 29 and 13% of the variation, respectively. Axis 1 isolated a clear group on its positive side (II), corresponding to the major share of individuals assigned to MMb cluster D, while the rest of the individuals were broadly distributed along axes 1 and 2. Individuals assigned to MMb clusters A and C displayed narrow distributions indicating uniform morphological types, which is consistent with their improved

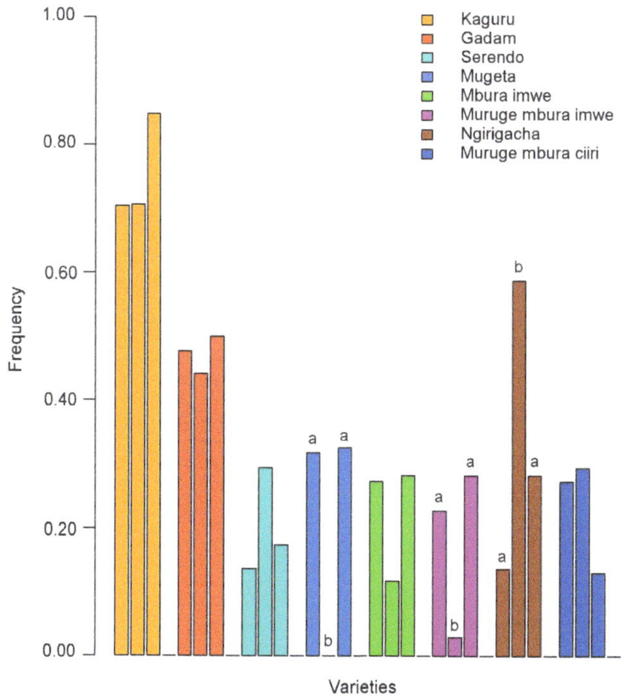

Figure 2. Frequency of the eight major varieties in each ethnic group. The vertical axis displays the percentage of farms where each variety was cultivated. Ethnic groups are present in the following order for each variety: Chuka, Mbeere, Tharaka. The letters (a, b) on top of the bars indicate the statistical significance of differences (Fisher test) at a 5% level after correction for multiple testing (FDR). For a given variety, ethnic groups with the same letter did not present significantly different frequencies.

origin and recent introduction. Individuals assigned to cluster C formed a distinct morphological group (I), discriminated on the third axis of the PCoA (expressing 10.7% of the total variation, data not shown). Individuals assigned to MMb cluster B displayed a broad distribution, reflecting high variability and continuous distribution across diverse morphotypes. It is noteworthy that some of the individuals assigned to MMb genetic clusters A and B displayed morphological similarity (Figure 4.B.), which may induce possible confusion in naming the recent improved variety and the local landraces (homonymy). Nevertheless, part of the individuals assigned to the MMb cluster B clustered in a separate morphological groups (III).

Genetic differentiation of sorghum populations across ethnic groups

Various indexes were used to characterize the diversity displayed within each ethnic group (Table 1). The unbiased gene diversity estimates (H_e) of Chuka and Tharaka sorghum populations were significantly higher than that of the Mbeere (Wilcoxon test: p-value < 0.01). Similar results were found for the unbiased allelic richness. F_{IS} was very high in the three groups, yet it was significantly lower in the Chuka population as compared to those of both the Tharaka and the Mbeere (Wilcoxon test: p-value $<$ 0.05 for both pairwise comparisons), in relation with the higher heterozygosity found within the Chuka sorghum population (0.033) compared to the other two populations (0.022 for the Mbeere and 0.023 for the Tharaka).

An exact G-test of genetic differentiation of sorghum across ethnic groups was significant (p-value $= 0.0205$). The differentiation was clearer (G-test p-value $= 0.0026$) when removing from the analysis the individuals assigned to cluster A, derived from the recent introduction of the *Gadam* improved variety. The Pairwise G-tests showed that genetic differentiation was highly significant between the sorghum populations of the three groups, being highest between the Chuka and both the Tharaka (p-value $<$ 0.0001) and Mbeere (p-value $=$ 0.0002) populations and lowest between the Tharaka and Mbeere populations (p-value $=$ 0.0083). The F_{ST} values between the sorghum populations of the three ethnic groups were low: 0.027 between the Chuka and Mbeere sorghum populations and 0.019 between the Chuka and Tharaka populations, both significant; and non significant between the Mbeere and Tharaka populations ($F_{ST} = 0.010$).

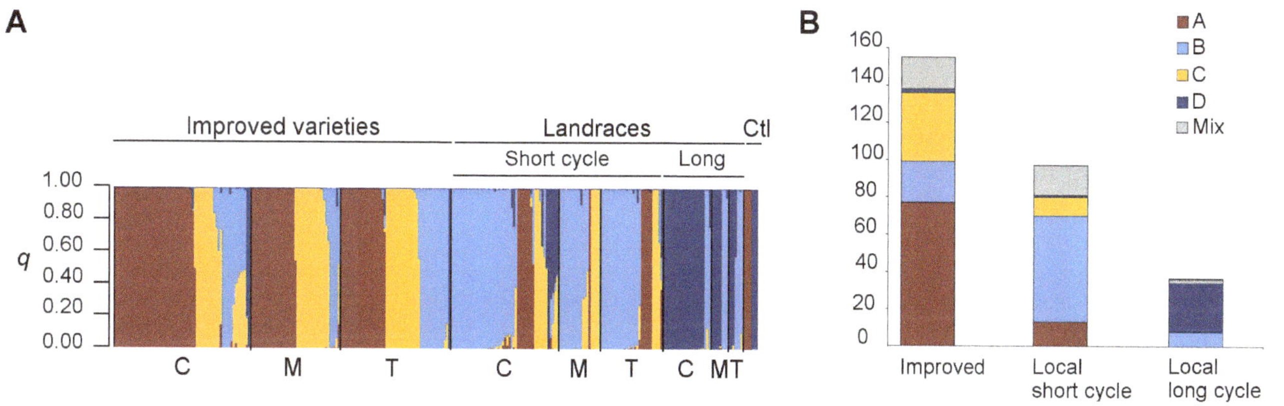

Figure 3. Genetic structure of the sorghum cultivated on the area of study. (A) Cluster assignment of 297 sorghum individuals estimated using STRUCTURE for $K=4$. The genome of each individual is represented by a vertical line, which is partitioned into K colored segments that represent the admixture coefficient (q), i.e the estimated proportion of membership of its genome in each of the K clusters (Red: cluster A, light blue: cluster B, yellow: cluster C, dark blue: cluster D). Thick black lines separate the individuals identified by farmers as improved varieties, short-cycle landraces or long-cycle landraces, and control individuals (Ctrl), as labeled above the figure. Thin black lines separate individuals sampled in the different ethnic groups (Chuka: C, Mbeere: M, Tharaka: T, as labeled below the figure. The figure shown is based on the highest probability run at $K=4$. (B) Number of individuals classified according to their origin and cycle length (farmers' information) assigned to each MMb genetic cluster. The vertical axis indicates the number of individuals assigned to each cluster. Individuals were assigned to a cluster when their estimated admixture coefficient (q) for this cluster was equal to or over 0.8. Admixed individuals are represented in gray.

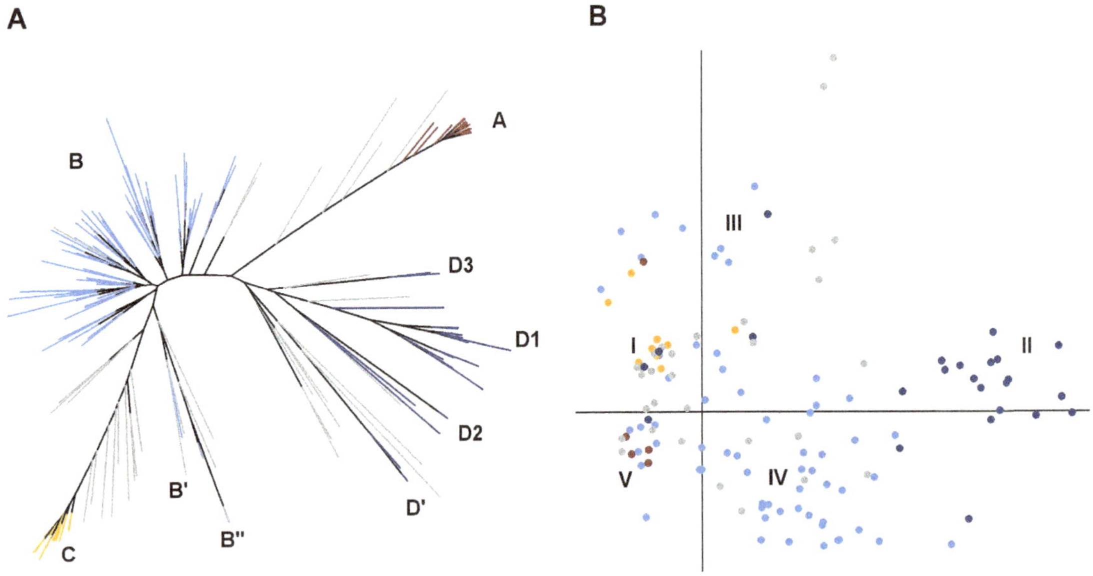

Figure 4. Genetic and morphological structure of the sorghum cultivated on the study area. (A) Neighbor-Joining tree based on 18 SSRs among sorghum plants using the simple matching index. Genetic clusters inferred by STRUCTURE are displayed using different colors (Cluster A: red, B: light blue; C: yellow, D: dark blue). Sub-clusters are identified by letters followed by a number. (B) Plot of the two first axes of the Principal Coordinates Analysis (PCoA) based on 15 panicle morphological traits using the simple matching index. The first axis (x) expresses 29.3% of the total variation and the second axis 13.2%. The main morphological groups are indicated by roman numerals and the MMb genetic assignment of individuals for $K=4$ is displayed with the same colors as in figure 4.A.

No significant relationship was found between the genetic relatedness of individuals and their geographical distance. The partial Mantel test was not significant ($r=-0.38$, p-value $=0.130$), and the correlogram did not display any significant spatial structure. Nevertheless, the spatial distribution of the four MMb genetic clusters was not uniform (Figure 5.C) and they were not evenly distributed across the three ethnic groups (Table 2). Pearson's Chi-squared test led to rejecting independence between the genetic clusters and the ethnic groups (p-value $=0.003$).

Correspondence between the genetic structure and farmers' variety names

The MMb cluster A was clearly separated from the others, as illustrated by the Neighbor-Joining tree. It included the four control individuals stemming from certified seeds of the *Gadam* improved variety, which has been disseminated in the area since 2009. Most of the other individuals assigned to cluster A were identified by farmers as *Gadam* (Chuka: 50%, Mbeere: 71%, Tharaka: 48%), confirming the cluster A – *Gadam* correspondence. Yet cluster A also included 46% of varieties collected under other names, mainly *Ngirigacha* (Chuka: 29%, Mbeere: 24%, Tharaka: 28%) and *Mbura-imwe* (Chuka: 11%, Tharaka: 16%). As a result, cluster A was distributed throughout the study area and its spatial distribution appeared more uniform than that of the individuals designated by farmers as *Gadam* (Figure 5.A & C).

The MMb cluster C was also clearly separated from the others, yet with an array of individuals that appeared as intermediates (along the branch of the Neighbor-Joining tree). The major share of the individuals assigned to cluster C was identified by farmers as an improved variety called *Kaguru*, which was introduced in the area about ten years ago (Chuka: 83%, Mbeere: 94%, Tharaka: 100%). *Kaguru* individuals originated uniformly from the study area (Figure 5.A) and in the three ethnic groups (Figure S1), but were less frequent in the Chuka area (Figure 5.C). The proportion of the Chuka sorghum individuals assigned to cluster C was significantly smaller (9%) as compared to the Tharaka (21%) population (Fisher test: p-value $=0.023$), and the Mbeere (25%) population (Fisher test: p-value $=$ 0.009). Half of the individuals (52%) collected under the name *Kaguru* in the Chuka farms were admixed, while this proportion was significantly lower in the Mbeere farms (15%, Pairwise Fisher test p-value: 0.0300) and in the Tharaka farms (10%, p-value $=$ 0.0190). Accordingly, the genetic diversity parameter estimates calculated for the *Kaguru* individuals collected in the Chuka farms were significantly higher than those for the Mbeere and Tharaka farms (Table 3). Altogether, these observations suggest that more admixture occurred between the *Kaguru* population and local landraces within the Chuka cropping systems than within the Tharaka and Mbeere systems.

The MMb cluster D appears clearly separated but rather heterogeneous on the Neighbor-Joining tree. On a morphological basis, these varieties mostly fall in a clearly distinct group. Most individuals assigned to cluster D were identified as long-cycle landraces by farmers (*Muruge mbura ciiri*, *Mugana*, *Muthigo*, *Mucuri*, *Kathirigwa)* and a few as short-cycle improved varieties (*Serendo* and *Musalama)*. The latter individuals identified as improved varieties, both collected on-farm and stemming from certified seeds, formed a distinct genetic sub-group D' on the Neighbor-Joining tree and STRUCTURE confirmed these results for $K=5$. The rest of the individuals assigned to cluster D were distributed across three major sub-clusters (Figure 4.A). Most *Muruge mbura ciiri* individuals clustered together in a separate branch on the Neighbor-Joining tree (D1). *Mugana* and *Kathirigwa* formed another branch (D2), and *Mucuri* a third one (D3). Hence, there was a clear correspondence between the farmers' nomenclature and the genetic structure of individuals assigned to MMb cluster D, as well as with the structure of panicle morphological diversity (Figure S4). Cluster D was mainly observed in the Chuka area, as seen on Figure 5.C and confirmed with pairwise Fisher tests (p-value <0.05). Interestingly, the few Tharaka households where we collected individuals assigned to cluster D were located in the Chuka area. Moreover,

Table 1. Summary of the genetic polymorphism indexes of sorghum individuals sampled in each ethnic group.

Ethnic group	N_i	N_{hh}	N_{Al}	R_{Al}	H_e	H_o	F_{IS}
Chuka	135	60	6.8	6.1[a]	0.590[a]	0.033	0.943[a]
Mbeere	68	35	4.7	4.6[b]	0.544[b]	0.022	0.959[b]
Tharaka	87	35	6.1	5.9[a]	0.569[a]	0.023	0.961[b]
Total	290	130	7.7	7.7	0.574	0.028	0.952

N_i: number of samples, N_{hh}: number of households, N_{Al}: Mean number of observed alleles over the 18 loci, R_{Al}: unbiased allelic richness corrected for sample size, H_e: unbiased gene diversity, H_o: observed heterozygosity, F_{IS}: fixation index. The letters (a, b) next to the R_{Al}, H_e and F_{IS} values indicate the statistical significance of their differences between ethnic groups (Wilcoxon test) at a 5% level after correction for multiple testing (FDR). For a given index, ethnic groups with the same letter did not present significant differences.

one household located on the eastern side presented several individuals assigned to cluster D, but it was a Chuka household settled in the Tharaka area (Figure 5).

The MMb cluster B is both central and diverse on the basis of molecular markers as well as morphological traits. The individuals assigned to cluster B were mainly identified as local landraces bearing various local names, whose occurrence frequency differed across ethnic groups (Table 4). Most of those collected in the Chuka and Tharaka farms were named *Muruge mbura imwe*, *Mugeta* and *Mbura imwe* while no or very few individuals collected in the Mbeere farms were named as such. Moreover, most of those collected in the Mbeere farms were named *Ngirigacha* (61%), while fewer individuals bore that name in the Chuka (8%) and Tharaka (18%) populations. Cluster B accounted for a uniformly large share among the farmers of the Tharaka (39%), the Chuka (27%) and the Mbeere (27%) ethnic groups. It showed little internal substructure with no clear correspondence to farmers' varieties, and a morphological differentiation between *Muruge mbura imwe* and *Mugeta* (Figure S4). As the only individuals with peculiar features, four individuals assigned to MMb cluster B for $K=4$ formed a separate branch on the genetic Neighbor-Joining tree (B') and their difference was confirmed by STRUCTURE for $K=5$. It could be explained by their foreign origin, as farmers reported purchasing these seeds at a lowland market. A fifth individual assigned to cluster B for $K=4$ formed a long branch (B'') indicative of a marked genetic differentiation. It was identified as *Muthigo wa mwimbi* which means that it was introduced from another ethnic group (*Mwimbi*).

Discussion

Our study showed that in a uniform agro-ecological environment, social boundaries associated with ethnolinguistic diversity patterns have impacted the distribution of sorghum varieties and their genetic spatial patterns. If seeds, knowledge and practices were freely exchanged across the three ethnic groups, we would expect their sorghum varieties to be similar and, because of their geographical proximity and similar environmental conditions, to display no genetic differentiation. Quite the contrary, we showed that ethnic groups maintained different sorghum landraces, whereas improved varieties were uniformly distributed across groups.

Factors structuring the distribution of sorghum genetic diversity

The genetic diversity of sorghum in the area of study, as assessed with molecular markers, is organized in four major groups. These groups reflected the influence of improved variety dissemination and a differentiation in terms of cycle duration and phenology. The improved varieties (groups A and C) and short-cycle landraces (group B) collected on the area of study clustered with the Caudatum accessions from eastern Africa and central Africa of a reference set representing the worldwide sorghum genetic diversity ([47]; Figure S3). The long-cycle landraces (group D) clustered with accessions from various origins (eastern & central Africa, India, Middle-East) and races (Durra, Caudatum, Bicolor and intermediates). Some new alleles, absent from the global reference set, were found in the local pool and notably among the long-cycle landraces, which could hence complement the reference set.

The overall distribution patterns of sorghum diversity on our study site were clearly associated with the farmers' ethnic partition. This genetic differentiation did not appear to result from isolation-by-distance as no significant relationship between the geographic distance and the genetic relatedness of individuals was detected.

Table 2. Number of individuals sampled in each ethnic group and assigned to each MMb genetic cluster.

MMb cluster	Chuka	Mbeere	Tharaka	Total	Chi²	P-value
A	44 (33%) [a]	21 (31%) [a]	25 (29%) [a]	90 (31%)	0.37	0.832
B	36 (27%) [a]	18 (27%) [a]	34 (39%) [a]	88 (30%)	4.49	0.106
C	12 (9%) [a]	17 (25%) [b]	18 (21%) [b]	47 (16%)	10.5	0.005
D	21 (15%) [a]	5 (7%) [ab]	4 (4%) [b]	30 (11%)	7.7	0.021
Mix	22 (16%) [a]	7 (10%) [a]	6 (7%) [a]	35 (12%)	4.7	0.097
Total	135 (100%)	68 (100%)	87 (100%)	290 (100%)		

Individuals with a *q* value equal to or above the threshold of 0.8 for a cluster were assigned to that cluster. The Chi-Square statistics and p-value compare, for each MMb cluster, the observed and the expected frequencies under the null hypothesis of independence. For each cluster, the letters indicate the statistical significance of the differences in its frequency between ethnic groups (Fisher test) at a 5% level after correction for multiple testing (FDR). For a given cluster, ethnic groups with the same letter did not present significant differences.

Long-cycle landraces formed a genetically distinct cluster which was more frequently encountered in the Chuka sorghum population than in the Tharaka sorghum population. The improved *Kaguru* variety showed more admixture with the local landraces in the Chuka sorghum population than in the Mbeere and Tharaka ones. As a result of the unbalanced frequency of the different genetic clusters across ethnic groups, the genetic differentiation of their sorghum populations was significant. The uneven distribution of named landraces across the Chuka, Tharaka and Mbeere ethnic groups is consistent with the results of Baco et al. [48], who reported that different ethnic groups in Benin cultivated different varieties of yam. A similar relationship between the structure of the genetic diversity of domesticated populations and farmers' social organization was found in taro populations across linguistic groups in Vanuatu [6], and in goat populations across ethnic groups in Vietnam [7]. However, a common caveat to such crop diversity studies conducted on large spatial scales is the difficulty involved in assessing whether the spatial patterns of crop diversity are related to variations in agro-ecological conditions, geographical distances, or to socio-cultural differences between human societies [10].

The field setting adopted in our study enabled us to limit the interference between socio-cultural factors and other environmental factors. Notably, climate and soil variations can influence the distribution of crop diversity. The climatic variation was neglectable on our study site regarding the limited gradient of altitude. In addition, we conducted a survey which did not highlight significant differences of soils' physical properties among the areas inhabited by the three ethnic groups (data not shown). Furthermore, farmers did not report that some varieties were better adapted to particular types of soils. Hence, the interference between socio-cultural factors and other uncontrolled environmental factors remains much unlikely, even though it cannot be totally left out.

The community-scale approach we used in this study revealed that social boundaries have contributed to the differentiation of sorghum populations across spatially-close ethnic groups living in the same agro-ecological environment. Such an approach is thus complementary to country or regional-scale studies. In addition, such an approach makes it possible to investigate the mechanisms behind the relationship by jointly analyzing the distribution of varieties and the structure of genetic and morphological diversity in relation to the social organization of the communities concerned.

The ethnic identity of human groups is maintained by social boundaries that impede their cultural homogenization [49]. Our results suggest that these social boundaries also maintain differences between crop populations across ethnic groups. Indeed, gene flows in crop populations greatly depend on the exchange of seed, which is facilitated by social relationships and limited by social boundaries [4]. In addition, farmers' seed selection practices have a strong impact on crop populations and can differ considerably across communities [5,50]. The comparison of the structure of the genetic and morphological diversity of sorghum populations provides information concerning gene flows and selection forces, while the study of the nomenclature given to farmers' varieties tracks the diffusion of knowledge across farming communities. Thus, by combining the two approaches it is possible to investigate the respective influence of seed exchanges and the diffusion of selection practices across ethnic groups on sorghum genetic diversity patterns.

Limited diffusion of long-cycle landraces across ethnic groups

Long-cycle landraces formed a distinct MMb genetic cluster, whose frequency differed across ethnic groups. It was more frequent among the Chuka than among the Tharaka, and, interestingly, these results confirmed farmers' reports stating that long-cycle landraces were "*Muvia wa Chuka*", the sorghum of Chuka people. Moreover, certain sub-types within this cluster (sub-clusters D2 and D3) were not present in the Mbeere population and corresponding landraces were not inventoried in that ethnic group. The relation between the spatial distribution of the MMb genetic clusters and that of ethnic groups suggests that social boundaries limit the diffusion of planting material. Indeed, in most rural societies, seed exchanges depend on social networks as trust is required for seed transactions [51]. On the one hand, social relationships directly shape the seed exchanges because they facilitate access to seed [52,53,54]. On the other hand, the social network is the major pathway for information exchange [55] and indirectly helps shape seed exchanges, as farmers tend to imitate relatives [56]. The joint action of these two mechanisms can thus explain the uneven distribution of long-cycle landraces across ethnic groups. In addition, the small grains and the bitter taste could explain the low economic value of these landraces, which probably helps limit their diffusion.

Management practices of improved varieties differ across ethnic groups

In contrast to the case of some landraces, improved varieties were uniformly distributed and their frequencies did not differ between ethnic groups. The recently introduced *Gadam* variety was

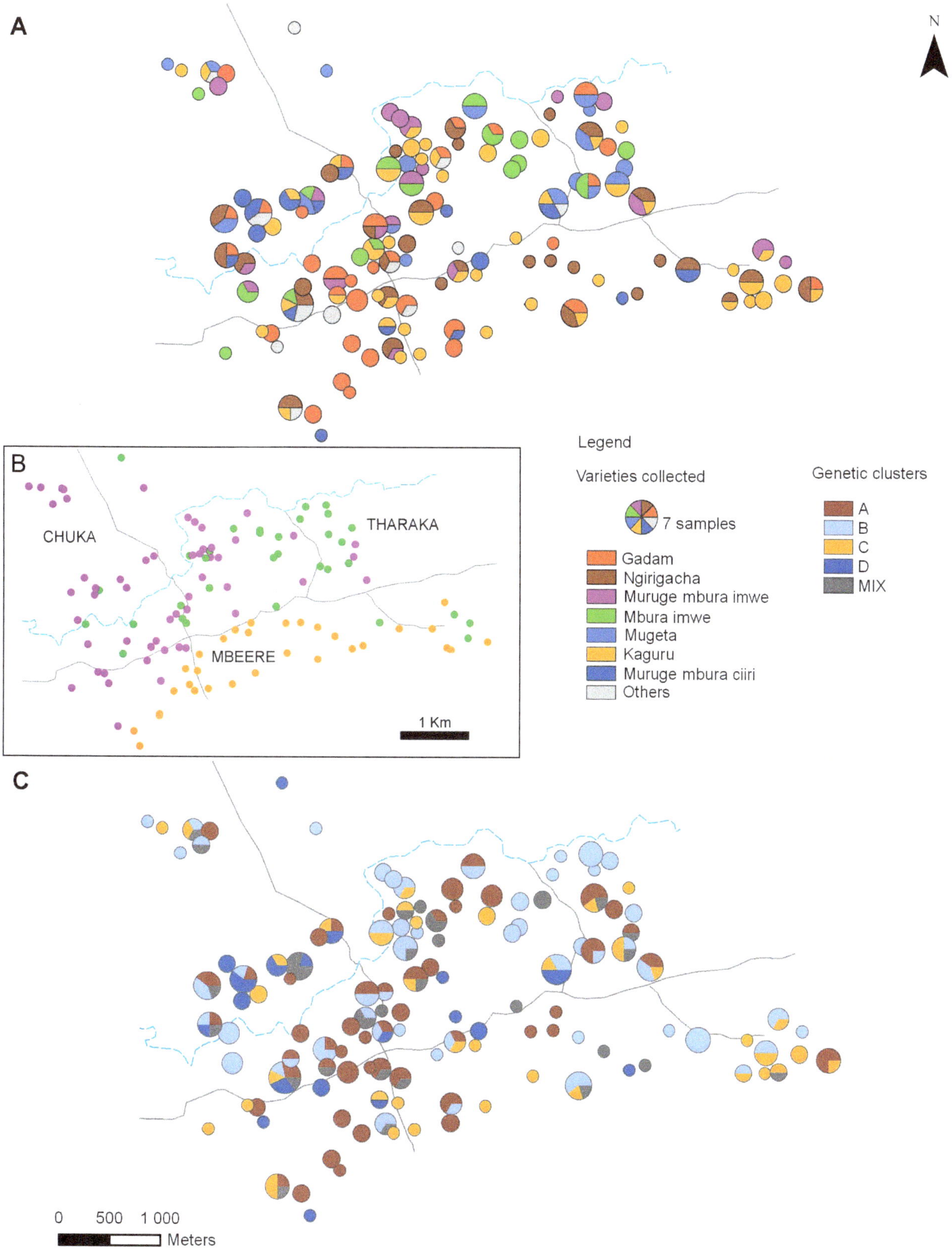

Figure 5. Spatial distribution of the sorghum varieties, ethnic groups, and sorghum genetic clusters. (A) Map of the named varieties collected in each ethnic group. Pie charts represent the number of samples of each variety collected in each household. The size of each circle is proportional to the number of individuals sampled. (B) Location of the ethnic groups (Purple: Chuka, Green: Tharaka, Orange: Mbeere). (C) Map of the number of sorghum individuals in each household assigned to each of the four MMb genetic clusters. Individuals were assigned to a cluster if their estimated genome fraction to that cluster, i.e. admixture coefficient (*q*), was higher than 0.8.

Table 3. Summary of the genetic polymorphism indexes of the *Kaguru* variety sampled in the three ethnic groups.

Ethnic group	N_{hh}	N_i	R_{AL}	*He*	*Ho*	F_{IS}
Chuka	19	22	3.29 [a]	0.339 [a]	0.049	0.857
Mbeere	17	20	2.33 [b]	0.184 [b]	0.003	0.985
Tharaka	13	20	1.83 [b]	0.091 [c]	0.006	0.939

N_i: number of samples, N_{hh}: number of households, R_{Al}: unbiased allelic richness. H_e: unbiased gene diversity, H_o: observed heterozygosity, F_{IS}: fixation index. The letters (a, b, c) next to the R_{Al} and H_e values indicate the statistical significance of their differences between ethnic groups (Wilcoxon test) at a 5% level after correction for multiple testing (FDR). For a given index, ethnic groups with the same letter did not present significant differences.

genetically distinct from the landraces and showed limited introgression from the other genetic clusters. It was genetically uniform and complied with certified control. However, farmers also gave the names of local and already known variety to individuals that have the same genetic profile as *Gadam*, an improved variety. This can be explained by a morphological similarity. Yet it raises the question of the consequences this will have for the on-farm evolution of the improved variety. *Kaguru*, for instance, which was introduced in the area 10–15 years ago, seems to have evolved differently across ethnic groups. High admixture was detected between *Kaguru* and the local landraces in the Chuka population, resulting in a range of genetically diverse materials still called *Kaguru*, while this variety was found to be genetically more uniform in the Mbeere and Tharaka populations. As a result, the genetic diversity of *Kaguru*, as identified by farmers, was greater in the Chuka population than in the Mbeere and Tharaka populations. According to the farmers, the variety was introduced simultaneously in the three ethnic groups but little information is available concerning the origin of the seed lots. The divergence of the *Kaguru* variety across ethnic groups within a few decades could thus be the result of differences in their management practices, be it in planting (spatial arrangement of the varieties) or in seed selection. The higher admixture rate between *Kaguru* and the local landraces among the Chuka could be due to more intense gene flows within fields or to less stringent selection practices. As our observations suggest that the cropping systems used by the three ethnic groups were similar, the hypothesis of different selection practices is more likely. Cases of divergent selection practices between geographically close communities were observed by Pressoir and Berthaud [50], and by Perales et al. [5], who hypothesized that social boundaries impede the homogenization of selection practices. However, the hypothesis of the introduction of seed lots with different genetic characteristics in the three ethnic groups cannot be excluded.

Divergence in the nomenclature of the landraces between ethnic groups

Comparing the genetic structure of short-cycle landraces, their morphological characteristics and the farmers' nomenclature raises interesting questions concerning the relation between farmers' nomenclature and the diffusion of planting material. Indeed, the frequencies of the majority of named short-cycle landraces differed significantly between ethnic groups even though they were assigned to the same genetic pool and no clear correlation was detected between named landraces and the MMb genetic sub-structure. The molecular markers used did not discriminate the three major short-cycle landraces whose frequency varied markedly across ethnic groups and which display different morphological characteristics. The short-cycle landraces grown by the different ethnic groups thus appeared to belong to the same genetic pool. Yet, the analysis of the morphological characteristics of the panicles suggests that the landraces presented morphological differences that were not detected with neutral genetic markers. *Mugeta* and *Muruge mbura imwe*, mainly grown by the Chuka and Tharaka, corresponded to two distinct morphological groups while *Ngirigacha*, which is mainly grown by the Mbeere, was distributed over the entire PCoA plot (Figure S4.A). These results suggest that ethnic groups use different names for landraces with similar morphotypes: the Chuka and Tharaka appear to identify and name two main short-cycle landraces corresponding to distinct morphotypes while the Mbeere mainly

Table 4. Proportion of individuals of each variety assigned to MMb cluster B regarding their collection ethnic group.

Variety	Chuka	Mbeere	Tharaka
Muruge mbura imwe	14 (39%)	1 (5%)	10 (29%)
Mugeta	10 (28%)	-	8 (24%)
Ngirigacha	3 (8%)	11 (61%)	6 (18%)
Mbura imwe	7 (19%)	-	8 (23%)
Muthigo wa mwimbi	1 (3%)	-	-
Others :			
Gadam	1 (3%)	2 (12%)	2 (6%)
Kaguru	-	1 (5%)	-
Muruge mbura ciiri	-	3 (17%)	-
Total	36 (100%)	18 (100%)	34 (100%)

Percentages in brackets.

use the name *Ngirigacha* for all the morphotypes corresponding to the short-cycle landraces group.

This difference in folk-nomenclature and classification between the Mbeere and both the Chuka and Tharaka groups may result from limited knowledge diffusion. This is consistent with a number of observations concerning the conflictual relationship between the Chuka and Mbeere groups [19]. The impact of social relationships on the diffusion of folk-taxonomy and nomenclature among farming communities was demonstrated by Boster [57], who showed that the cassava nomenclature used by kin-related women was more similar than that used by non-kin in the Aguaruna community in Peru. Nuijten and Almekinders [12] also reported that the naming of rice varieties was more consistent within villages than between villages in Gambia. They pointed out that information concerning varieties, such as names, is not necessarily passed on with the seed lots. Hence, seed exchanges between communities can be more intense than knowledge diffusion, leading to the use of different names for similar morphotypes and genotypes. Further comparison of farmers' nomenclature and taxonomy between ethnic groups is required to confirm this hypothesis.

Previous studies showed that different sorghum varieties may display no genetic differentiation despite being morphologically distinct. Notably, in Cameroon, Barnaud et al. [3] showed that considerable gene flows existed between Guinea sorghum landraces while farmers kept on selecting them for their morphological distinctiveness. Rabbi et al. [58] reported similar results in western Kenya, while varieties collected in eastern Sudan were clearly genetically distinct. He explained these results by the varietal isolation practiced in Sudan, while Kenyan farmers mixed varieties within their fields. Soler et al. [59] found that landraces were distinct genetic units, but in that study each landrace was sampled in a single field belonging to one farmer, which considerably limits the variability. As farmers' taxonomy and nomenclature is based on morphological traits with a simple genetic determinism, morphological differences can be maintained even though gene flows occur within farmers' fields. The 18 SSRs used in our study were selected because they revealed high polymorphism in previous diversity studies, and they proved to be adequate for characterizing the genetic sub-structure of long-cycle landraces. However, their resolution power may not be sufficient to reveal a finer-scale genetic sub-structure in short-cycle landraces. The use of high-density markers may help to evidence finer-scale genetic structure and could hence contribute to decipher the evolutionary mechanisms that molded the landraces.

Effect of community social organization on the diffusion of seeds, knowledge and farmers' practices

According to local elders, the three ethnic groups migrated to the study area about a century ago. Our results suggest that even though they have lived in proximity since then, the way knowledge, practices and seeds are diffused has maintained differences between sorghum populations across ethnic groups. Ethnographic observations of community social organization provide explanations for such limited exchanges across geographically close communities. Indeed, information transmission and diffusion appear to be confined within the residential groups (parents and married sons) first, which is common in patrilineal and patrilocal societies [60], and next within the neighborhood group, which is a major social institution among eastern Kenyan Bantu communities [16,61], (Linsig pers.com). The way knowledge is transmitted and diffused is very conservative and favors cultural differentiation between communities [62,63]. It thus probably plays a major role in maintaining differences in nomenclature and practices between ethnic groups, and maybe also in limiting seed exchanges.

Conclusion

Our study highlights the importance of local short-scale studies to investigate farm crop evolution processes. To date, emphasis has been placed on the effect of agro-ecological conditions on crop evolution processes, as in the study of the evolution of wild plants. The influence of the cultural diversity and social organization of farming communities has consequently been neglected, although the major role of smallholders in the management of crop diversity has been acknowledged [64]. Crop evolution is still ongoing in smallholder farming systems and such systems occupy a substantial proportion of croplands in developing countries, especially in Africa [65]. Most of these rural communities have retained pre-colonial social institutions that continue to shape the relationships between people. Sixty-eight living language groups were inventoried in Kenya and about 2146 linguistic groups in Africa [66], so the situation of ethnic co-existence described in this paper is not an isolated case. This study confirms the influence of the ethnolinguistic patterns of rural communities on gene flows and on farmers' selection practices that shape crop diversity *in situ*. Crop diversity patterns, thus result not only from an interaction between genetic and environmental factors, G × E, but from a three-way interaction G × E × S, where "S" stands for effects of the social boundaries [4]. Investigating this relation in other communities, with different social organizations and rules for the transmission of knowledge, would thus help gain a clearer picture of crop evolution dynamics in subsistence farming systems.

A further study is now needed to probe the mechanisms involved. Notably, the link between seed exchange networks and social organization deserves more investigation to confirm whether seed exchanges are confined within ethnic groups. This would explain why the diffusion of long-cycle landraces is more limited than that of short-cycle landraces. Moreover, further comparison of the local sorghum nomenclature and classification systems (folk taxonomy) across ethnic groups would make it possible to test whether their definition of landraces differs, and whether it influences their seed selection practices.

The uneven distribution of the genetic clusters across ethnic groups within a restricted geographic area highlights the need to take the social relationship and exchanges into account in the characterization, collection, and conservation of crop diversity. Accounting for the impact of human practices on crop populations would help capture their diversity more efficiently and, to this end, ethnic contact zones are of major interest for their potentially high genetic diversity. This study paves the way for participatory plant breeding as it shows that farmers' individual choices concerning planting material are not only determined by agro-ecological conditions or economic interest, but also by their cultural background.

Supporting Information

Figure S1 Diagram displaying the rain seasons and the growth-cycle of sorghum on our study site. Inventories' dates are symbolized by the letter I (orange points) and collections' dates by the letter C (red points).

Figure S2 Comparison between the inventory of varieties and their sampling. (A) Percentage of households where each variety was sampled for the genetic diversity study on a total of 130 households. (B) Linear correlation between the proportions of households where each variety was inventoried (vertical axis, 124 households) and where it was collected (horizontal axis, 130 households) in each ethnic group.

Figure S3 Neighbor-Joining tree based on the genetic dissimilarity among the individuals sampled on our study site (in Black) and the accessions of a global reference set (Billot et al. 2013). The genetic dissimilarities were calculated on 16 SSRs using the simple matching index. The sorghum individuals sampled on our study site are displayed in black. The genetic assignment (A, B, C, D - $q>0.8$) or unassignment (Unassigned - $q\leq0.8$) of our individuals is indicated on the figure. Colors represent the ten genetic groups identified in Billot et al. 2013, and described as following by the authors: "Group 1 [Dark orange] included Caudatum, Caudatum-Bicolor and Durra from Eastern Asia; Group 2 [Light orange] encompassed Durra and Bicolor from the Indian subcontinent, while Group 3 [Light green] exhibited Durra from Eastern Africa. Bicolor and Durra-Bicolor from Eastern Africa were assigned in Group 4 [Light blue]. Group 5 [Dark blue] included Guinea and Guinea margaritiferum from Western Africa and Bicolor from North America. Group 6 [Red] appeared as a well-separated group made predominantly of Guinea accessions from western Africa, accompanied by intermediate race Durra-Caudatum materials from western Africa while Group 7 [Magenta] was made essentially of materials collected from eastern Africa and central Africa generally classified as race Caudatum (visible along FA axis 3). Group 8 [Dark green] was a small and heterogeneous group made of Durra and Caudatum race accessions from central Africa. Group 9 [Pink] was made essentially of Guinea race accessions from the Indian subcontinent and southern/eastern Africa with Guinea-Caudatum (GC) intermediate race accessions from various parts of Africa. Group 10 [Purple] was made almost exclusively of accessions from southern Africa of race Kafir or intermediate race Kafir-Caudatum (KC)." Unassigned individuals in the global reference set are displayed in grey.

Figure S4 Structure of the morphological and genetic diversity within the MMb clusters B (top) and D (bottom). (A) Plot of the two first axes of the Principal Coordinates Analysis (PCoA) done on the sorghum plants assigned to the MMb cluster B and based on 15 panicle morphological traits. The first axis (x) expresses 35.1% of the total variation and the second axis 13.1%. Varieties are displayed using the following color code: Blue: *Mugeta*, purple: *Muruge mbura imwe*, green: *Mbura imwe*, brown: *Ngirigacha*, Red: *Gadam*, yellow: *Kaguru*, salmon: *Muthigo wa mwimbi*. (B) Neighbor-Joining tree based on the genetic dissimilarity among individuals assigned to the MMb cluster B calculated on 18 SSRs using the simple matching index. (C) Plot of the two first axes of the Principal Coordinates Analysis (PCoA) done on the sorghum plants assigned to the MMb cluster D and based on 15 panicle morphological traits. The first axis (x) expresses 51.5% of the total variation and the second axis 18.0%. Varieties are displayed using the following color code: Yellow: *Serendo*, orange: *Musalama*, light-pink: *Kathirigwa*, Fushia: *Mugana*, Greenish blue: *Mucuri*, dark-blue: *Muruge mbura ciiri*, black: *Muthigo*, blue: *Mugeta*. (D) Neighbor-Joining tree based on the genetic dissimilarity among individuals assigned to the MMb cluster D calculated on 18 SSRs using the simple matching index.

Table S1 Summary of the sampling of planting material. Mean number of varieties collected per household (Mean no. varieties/household) and mean number of samples of each variety collected per household (Mean no. samples/variety/household) in each ethnic group, followed by their standard error (SE).

Table S2 Summary of information and genetic diversity estimates per locus. Minimum and maximum size of alleles (Size), chromosome where the locus is located (Ch), percentage of missing data per locus (Miss), number of sampled alleles (N_{Al}), *He*: unbiased gene diversity, F_{IS}: Fixation index.

Table S3 Morphological descriptors used for panicle description.

Table S4 Results from the perMANOVA comparing the effect of ethnic groups on sorghum variety assemblages. Df: degrees of freedom, Ssq: sequential sum of squared distance between individuals and their group's centroïd, Mean Ssq = Ssq/Df, F.Model: pseudo F ratio, R^2: coefficient of determination [Ssq Etnic group/Ssq Total].

Acknowledgments

We wish to thank Joseph I. Kamau and Catherine Wanjira for their contribution to field work, the staff of the National Genebank of Kenya and its director Zachary Muthamia, as well as Philippe Letourmy and Xavier Perrier for their help with statistical analysis. Thanks are also due to the participating farmers for their collaboration in our study. We feel indebted to the headmaster, teachers and students of Kabururu primary school for hosting our experimental trials on their school farm. We are also deeply grateful to the extension and local administrative staff in the study area. The support of the chiefs and assistant chiefs is particularly acknowledged. We thank three anonymous reviewers and the editor for their helpful comments.

Author Contributions

Conceived and designed the experiments: CL VL AB MD. Performed the experiments: CL VL PW AB MB CC. Analyzed the data: VL MD SM AB CL. Contributed reagents/materials/analysis tools: SM AB PW. Wrote the paper: VL CL MD JCG AB SM PW.

References

1. Perrier X, De Langhe E, Donohue M, Lentfer C, Vrydaghs L, et al. (2011) Multidisciplinary perspectives on banana (*Musa* spp.) domestication. Proc Nat Acad Sci USA 108: 11311–11318.
2. Roullier C, Benoit L, McKey DB, Lebot V (2013) Historical collections reveal patterns of diffusion of sweet potato in Oceania obscured by modern plant movements and recombination. Proc Nat Acad Sci USA 110: 2205–2210.
3. Barnaud A, Deu M, Garine E, McKey D, Joly H (2007) Local genetic diversity of sorghum in a village in northern Cameroon: structure and dynamics of landraces. Theor Appl Genet 114: 237–248.
4. Leclerc C, Coppens d'Eeckenbrugge G (2012) Social organization of crop genetic diversity. The G × E × S interaction model. Diversity 4: 1–32.
5. Perales H, Benz B, Brush S (2005) Maize diversity and ethnolinguistic diversity in Chiapas, Mexico. Proc Nat Acad Sci USA 102: 949–954.
6. Sardos J, Noyer JL, Malapa R, Bouchet S, Lebot V (2012) Genetic diversity of taro (Colocasia esculenta (L.) Schott) in Vanuatu (Oceania): an appraisal of the distribution of allelic diversity (DAD) with SSR markers. Genet Resour Crop Evol 59: 805–820.
7. Berthouly C, Do Ngoc D, Thévenon S, Bouchel D, Van T, et al. (2009) How does farmer connectivity influence livestock genetic structure? A case-study in a Vietnamese goat population. Mol Ecol 18: 3980–3991.
8. De Wet JMJ, Huckabay JP (1967) The origin of Sorghum bicolor. II. Distribution and domestication. Evolution 21: 787–802.

9. Harlan JR, De Wet JMJ, Stemler ABL (1976) Origins of African plant domestication. The Hague: De Gruyter Mouton. 498 p.
10. Deu M, Sagnard F, Chantereau J, Calatayud C, Hérault D, et al. (2008) Niger-wide assessment of *in situ* sorghum genetic diversity with microsatellite markers. Theor Appl Genet 116: 903–913.
11. Bellon MR, Brush SB (1994) Keepers of maize in Chiapas, Mexico. Econ Bot 48: 196–209.
12. Nuijten E, Almekinders CJ (2008) Mechanisms explaining variety naming by farmers and name consistency of rice varieties in the Gambia. Econ Bot 62: 148–160.
13. Jaetzold R, Schmidt H, Hornetz B, Shisanya C (2007) Farm management handbook. Vol II, Part C, East Kenya. Subpart C1, Eastern Province. Nairobi Ministry of Agriculture/GTZ.
14. Camberlin P, Boyard-Micheau J, Philippon N, Baron C, Leclerc C, et al. (2012) Climatic gradients along the windward slopes of Mount Kenya and their implication for crop risks. Part 1: climate variability. Int J Climatol. Available: http://onlinelibrary.wiley.com/doi/10.1002/joc.3427/full. Accessed 2 July 2013.
15. Ambler C (1988) The Great Famine, 1897–1901. In: Ambler C, editor. Kenyan Communities in the Age of Imperialism: The central region in the late Nineteenth Century. New Haven, London: Yale University Press. pp.123–149.
16. Middleton J (1953) The central tribes of the north-eastern Bantu; Forde D, editor. London: International African Institute. 105 p.
17. Moehlig WJG, Guarisma G, Platiel S (1980) La dialectométrie: une méthode de classification synchronique en Afrique. In: Guarisma G, Platiel S editors. Dialectologie et comparatisme en Afrique Noire. Paris: SELAF. pp. 27–45.
18. Chesaina C (1997) Chapter One: Historical and cultural background. In: Chesaina C editor. Oral literature of the Embu and Mbeere. Nairobi: East African Educational Publishers. pp. 3–10.
19. Glazier J (1970) Ritual and social conflict: circumcision and oath-taking in Mbeere. Nairobi: Institute of Development Studies, Nairobi University. 17 p.
20. Fadiman JA (1993) When we began there were witchmen: an oral history from Mount Kenya. Los Angeles: University of California Press. 395 p.
21. Plucknett DL, Evenson JP, Sanford WG (1970) Ratoon cropping. Adv Agron 22: 285–330.
22. Billot C, Rivallan R, Sall M, Fonceka D, Deu M, et al. (2012) A reference microsatellite kit to assess for genetic diversity of Sorghum bicolor (Poaceae). Am J Bot 99: e245–e250.
23. International Plant Genetic Resources Institute (IPGRI) (1993) Descriptors for Sorghum [*Sorghum bicolor* (L.) Moench]. Rome: IBPGR/ICRISAT. 38 p.
24. Gibson RW (2009) A review of perceptual distinctiveness in landraces including an analysis of how its roles have been overlooked in plant breeding for low-input farming systems. Econ Bot 63: 242–255.
25. Anderson MJ (2001) A new method for non-parametric multivariate analysis of variance. Austral Ecol 26: 32–46.
26. Oksanen J, Blanchet FG, Kindt R, Legendre P, Minchin PR, et al. (2012) Vegan: Community Ecology Package. 2.0-5 ed.
27. Bray JR, Curtis JT (1957) An ordination of the upland forest communities of southern Wisconsin. Ecol Monogr 27: 325–349.
28. Nakazawa M (2013) fmsb: Functions for medical statistics book with some demographic data. 0.3.8 ed.
29. Benjamini Y, Hochberg Y (1995) Controlling the false discovery rate: a practical and powerful approach to multiple testing. J R Stat Soc Series B Stat Methodol 1: 289–300.
30. Belkhir K, Borsa P, Chikhi L, Raufaste N, Catch F (2004) GENETIX, software under Windows for the genetics of populations. 4.05 ed. Montpellier: University of Montpellier.
31. Petit RJ, El Mousadik A, Pons O (1998) Identifying populations for conservation on the basis of genetic markers. Conserv Biol 12: 844–855.
32. Nei M (1987) Molecular Evolutionary Genetics. New York: Columbia University Press.
33. Weir BS, Cockerham CC (1984) Estimating F-statistics for the analysis of population structure. Evolution 38: 1358–1370.
34. Goudet J (2001) FSTAT, a program to estimate and test gene diversities and fixation indices 2.9.3 ed. Lausanne: Institute of Ecology.
35. Rice WR (1989) Analyzing tables of statistical tests. Evolution 43: 223–225.
36. Goudet J, Raymond M, de Meeüs T, Rousset F (1996) Testing differentiation in diploid populations. Genetics 144: 1933–1940.
37. Raymond M, Rousset F (1995) GENEPOP version 1.2: population genetics software for exact tests and ecumenicism. J Hered 86: 248–249.
38. Pritchard JK, Stephens M, Donnelly P (2000) Inference of population structure using multilocus genotype data. Genetics 155: 945–959.
39. Evanno G, Regnaut S, Goudet J (2005) Detecting the number of clusters of individuals using the software STRUCTURE: a simulation study. Mol Ecol 14: 2611–2620.
40. Saitou N, Nei M (1987) The Neighbor-Joining method: a new method for reconstructing phylogenetic trees. Mol Biol Evol 4: 406–425.
41. Sokal R, Michener CD (1958) A statistical method for evaluating systematic relationships. University of Kansas Science Bulletin 38: 1409–1438.
42. Perrier X, Jacquemoud-Collet JP (2006) DARwin software 5.0.156 ed. Montpellier: CIRAD.
43. Loiselle BA, Sork VL, Nason J, Graham C (1995) Spatial genetic structure of a tropical understory shrub, *Psychotria officinalis* (Rubiaceae). Am J Bot 82: 1420–1425.
44. Hardy OJ, Vekemans X (2002) SPAGeDi: a versatile computer program to analyse spatial genetic structure at the individual or population levels. Mol Ecol notes, 2: 618–620.
45. Rousset F (2000) Genetic differentiation between individuals. J Evol Biol 13: 58–62.
46. Meirmans PG (2012). The trouble with isolation by distance. Mol Ecol 21: 2839–2846.
47. Billot C, Ramu P, Bouchet S, Chantereau J, Deu M, et al. (2013). Massive sorghum collection genotyped with SSR markers to enhance use of global genetic resources. PLOS ONE 8: e59714.
48. Baco MN, Biaou G, Pham JL, Lescure JP (2008) Geographical and social factors of cultivated yam diversity in northern Benin. Cah Agric 17: 172–177.
49. Barth F (1969) Introduction In: Barth F, editor. Ethnic groups and boundaries: The social organization of culture difference. Boston: Little, Brown. pp. 9–38
50. Pressoir G, Berthaud J (2004) Population structure and strong divergent selection shape phenotypic diversification in maize landraces. Heredity 92: 95–101.
51. Badstue LB, Bellon MR, Berthaud J, Ramírez A, Flores D, et al. (2007) The dynamics of farmers' maize seed supply practices in the central valleys of Oaxaca, Mexico. World Dev 35: 1579–1593.
52. McGuire S (2008) Securing access to seed: Social relations and sorghum seed exchange in eastern Ethiopia. Hum Ecol 36 217–229.
53. David S, Sperling L (1999) Improving technology delivery mechanisms: lessons from bean seed systems research in Eastern and Central Africa. Agric Human Values 16: 381–388.
54. Bellon MR (2004) Conceptualizing interventions to support on-farm genetic resource conservation. World Dev 32: 159–172.
55. Van den Broeck K, Dercon S (2011) Information flows and social externalities in a Tanzanian banana growing village. J Dev Stud 47: 231–252.
56. Bandiera O, Rasul I (2006) Social networks and technology adoption in northern Mozambique. Econ J 116: 869–902.
57. Boster JS (1986) Exchange of varieties and information between Aguaruna manioc cultivators. Am Anthropol 88: 428–436.
58. Rabbi IY, Geiger HH, Haussmann BI, Kiambi D, Folkertsma R, et al. (2010) Impact of farmers' practices and seed systems on the genetic structure of common sorghum varieties in Kenya and Sudan. Plant Genet Resour 8: 116–126.
59. Soler C, Saidou AA, Hamadou TVC, Pautasso M, Wencelius J, et al. (2013) Correspondence between genetic structure and farmers' taxonomy - a case study from dry-season sorghum landraces in northern Cameroon. Plant Genet Resour 11: 36–49.
60. Herbich I, Dietler M (2008) The long arm of the mother-in-law: Learning, postmarital resocialization of women, and material culture style. In: Stark MT, Bowser BJ, Horne L, editors. Breaking down boundaries: Anthropological approaches to cultural transmission and material culture. Tucson: University of Arizona Press. pp. 223–244.
61. Labeyrie V, Rono B, Leclerc C (2013) How social organization shapes crop diversity: an ecological anthropology approach among Tharaka farmers of Mount Kenya. Agric Human Values. In press.
62. Reyes-García V, Broesch J, Calvet-Mir L, Fuentes-Peláez N, McDade TW, et al. (2009) Cultural transmission of ethnobotanical knowledge and skills: an empirical analysis from an Amerindian society. Evol Hum Behav 30: 274–285.
63. Cavalli-Sforza LL, Feldman MW (1981) Cultural transmission and evolution: a quantitative approach. Princeton: Princeton University Press. 388 p.
64. Brush SB (2000) Genes in the field: on-farm conservation of crop diversity; Brush SB, editor. Rome: Copublished by International Plant Genetic Resources Institute. 288 p.
65. International Fund for Agricultural Development (IFAD) (2001) Rural Poverty Report 2001: The Challenge of Ending Rural Poverty. Oxford: Oxford University Press for IFAD.
66. Lewis MP, Gary FS, Fennig CD (2013) Ethnologue: Languages of the World, Seventeenth edition. Dallas: SIL International. Available: http://www.ethnologue.com. Accessed 25 August 2013.

Splitting or Lumping? A Conservation Dilemma Exemplified by the Critically Endangered Dama Gazelle (*Nanger dama*)

Helen Senn[1]*, Lisa Banfield[2], Tim Wacher[3], John Newby[4], Thomas Rabeil[5], Jennifer Kaden[1], Andrew C. Kitchener[6,7], Teresa Abaigar[9], Teresa Luísa Silva[8,9,10], Mike Maunder[11], Rob Ogden[1]

1 WildGenes Laboratory, Royal Zoological Society of Scotland, Edinburgh, United Kingdom, **2** Conservation Department, Al Ain Zoo, Al Ain, Abu Dhabi, United Arab Emirates, **3** Conservation Programmes, Zoologicial Society of London, Regents Park, London, United Kingdom, **4** Sahara Conservation Fund, L'Isle, Switzerland, **5** Sahara Conservation Fund, Niamey, Niger, **6** Department of Natural Sciences, National Museums Scotland, Chambers Street, Edinburgh, United Kingdom, **7** Institute of Geography, School of Geosciences, University of Edinburgh, Drummond Street, Edinburgh, United Kingdom, **8** CIBIO/InBIO, Centro de Investigção em Biodiversidade e Recursos Genéticos da Universidade do Porto, Vairão, Portugal, **9** Estación Experimental de Zonas Áridas, Consejo Superior de Investigaciones Científicas (CSIC), Almería, Spain, **10** Departamento de Biologia da, Faculdade de Ciências da Universidade do Porto, Porto, Portugal, **11** College of Arts and Sciences, Florida International University, Miami, Florida, United States of America

Abstract

Managers of threatened species often face the dilemma of whether to keep populations separate to conserve local adaptations and minimize the risk of outbreeding, or whether to manage populations jointly to reduce loss of genetic diversity and minimise inbreeding. In this study we examine genetic relatedness and diversity in three of the five last remaining wild populations of dama gazelle and a number of captive populations, using mtDNA control region and cytochrome b data. Despite the sampled populations belonging to the three putative subspecies, which are delineated according to phenotypes and geographical location, we find limited evidence for phylogeographical structure within the data and no genetic support for the putative subspecies. In the light of these data we discuss the relevance of inbreeding depression, outbreeding depression, adaptive variation, genetic drift, and phenotypic variation to the conservation of the dama gazelle and make some recommendations for its future conservation management. The genetic data suggest that the best conservation approach is to view the dama gazelle as a single species without subspecific divisions.

Editor: Ulrich Joger, State Natural History Museum, Germany

Funding: Funding for genetic analysis was provided by Al Ain Zoo and the Royal Zoological Society of Scotland. The Sahara Conservation Fund, the Zoological Society of London, Saint Louis Zoo, Al Ain Zoo, the Addax and Oryx Foundation, and HH Sheikh Mohammed bin Zayed funded the collection of the dama gazelle samples from the wild. The authors' time during the project was funded by their own institutions and TLS's work was additionally partially supported by FCT-Fundação para a Ciência e a Tecnologia (SFRH/BD/73680/2010). Some funders to the project (Al Ain Zoo, Royal Zoological Society of Scotland, Sahara Conservation Fund, Zoological Society of London) were directly involved in the study design, data collection and analysis, decision to publish, and preparation of the manuscript via the participation of their employees in the project. These individuals are authors on this manuscript. The other funders (Saint Louis Zoo, the Addax and Oryx Foundation, FCT-Fundação para a Ciência e a Tecnologia and HH Sheikh Mohammed bin Zayed) had no role in study design, data collection and analysis, decision to publish, or preparation of the manuscript.

Competing Interests: The authors have declared that no competing interests exist.

* Email: hsenn@rzss.org.uk

Introduction

Fragmented populations of endangered species present a conservation dilemma [1]. If numbers dwindle to the extent that conservation intervention is necessary, should they be managed separately to conserve locally adapted genetic diversity and minimise the risk of outbreeding? Or should they be managed jointly to preserve evolutionary potential and to reduce the risk of inbreeding? This dilemma has been discussed recently both from taxonomic [2–3] and management [4] perspectives and is present in many wild and captive management scenarios [5–10]. The dilemma is typified in regard to the Critically Endangered dama gazelle (*Nanger dama*) [11].

There are around 300 dama gazelles left in the wild [11–12] and their range has shrunk by 99% during the period 1960–90 [13–14]. They now only exist in confirmed isolated populations in Chad, Niger, and Mali (Figure 1) and possibly in Sudan. Approximately 550 individuals are thought to exist within zoos and breeding centres across Europe, the USA and the Middle East, with further populations in private collections in the Middle East and particularly the USA accounting for an estimated 1000 individuals.

The gazelle's range formerly spanned the entire Sahel and Atlantic Saharan region (Figure 1) from south of the Anti-Atlas Mountains to the west coast of Morocco, south to Senegal and Mauritania, and east to the River Nile in Sudan. It is likely this range was once continuous (Figure 1), although a possible break in southern Mauritania has been debated [15–16]. The current wild populations are situated in the central part of the former range. Captive populations are known to have originated from two distant points of the original geographical distribution and are currently maintained as separate breeding populations (Figure 1, Table 1).

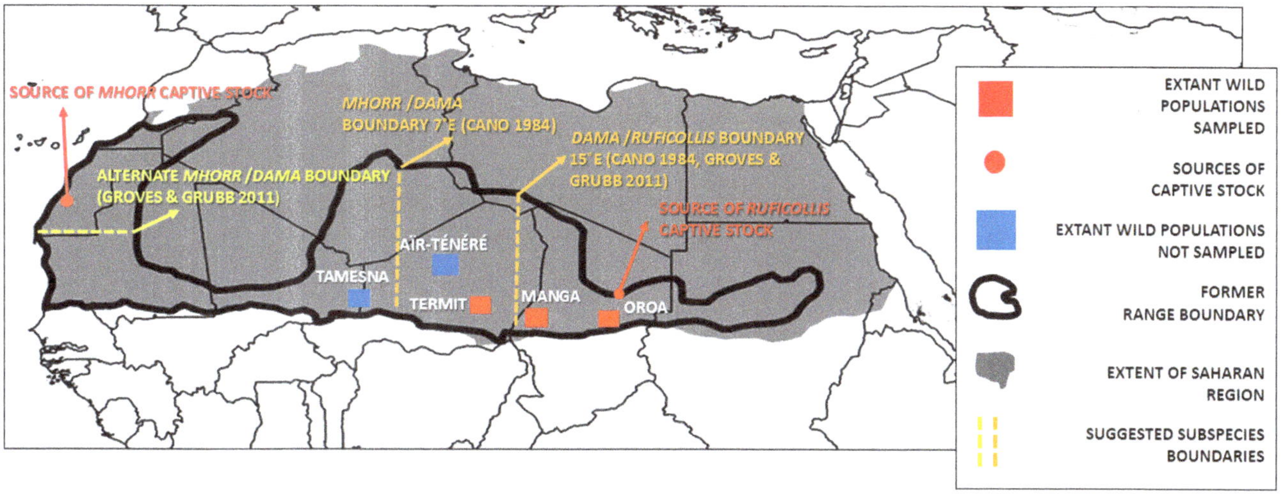

Figure 1. A map depicting the former and current range of the dama gazelle (adapted from [14]**).** Wild sampling sites and places of origin for the captive populations are depicted. The suggested subspecies boundaries according to different authors are listed (see in text for detail).

The dama gazelle is the largest of the gazelle species [17]. It shares the genus *Nanger* with Soemmering's gazelle (*N. soemmerringii*) and the Grant's gazelle complex (*N. granti* and related species) [18]. Once previously included in the genus *Gazella*, molecular phylogenetic analysis revealed inclusion of these larger gazelles would make the genus *Gazella* paraphyletic and hence they have been separated into the genus *Nanger* [19]. The dama gazelle has been subdivided into a variable number of subspecies (see below). The coloration of the dama gazelle's pelage is variable and, according to taxonomic tradition, was described by Cano [15] as following a northwest to south and east cline across the species' range: Animals in the east are predominantly white with reddish-brown coloration confined to the neck and shoulders, whereas animals in the west have extensive reddish-brown coloration, which extends further down the back, flanks and haunches. In the far north-west the animals are predominantly chestnut with a white underbelly and rump (Figure 2). However, this is an over-simplification of the dama gazelle's geographical variation, especially in the center of its range. Among former and remaining wild populations in Chad, considerable individual variation in the

Table 1. Details of the 124 samples in this study and the populations that they originated from.

Population	Site Code (suffix indicates putative subspecies)	Details	Putative sub-species	Number of Samples collected
WILD				
Chad (Ouadi Rimé-Ouadi Achim)	OROA_R	Wild Population in Ouadi Rimé-Ouadi Achim Game Reserve in Central Chad(~N14.9027, E19.8318)	*N.d.ruficollis*	18
Chad (Manga)	MANGA_R	Wild Population in Manga region of Western Chad (~N15.33087, E15.1277)	*N.d.ruficollis*	6
Niger (Termit)	TERMIT_D	Wild population(s) in the Central (~N16.1047, E11.4171) & Northern (~N16.3706, E11.4581) massif of the Termit mountains	*N.d.dama*	12
ZOO/CAPTIVE				
Al Ain Zoo *'mhorr'*	AIN_M	Origin unrecorded, highly likely to be descended from animals in the EEP (originally from EEZA).	*N.d.mhorr*	42
Al Ain Zoo *'ruficollis'*	AIN_R	Origin unrecorded, likely to stem from the North American Regional Studbook for addra (*ruficollis*) gazelle as it records the transfer of two females and a male to Al Ain Zoo in 1982.	*N.d.ruficollis*	20
Dama gazelle EEP	EEP_M	Animals sampled from City of Belfast Zoo, EEZA and Montpellier all ultimately originating from EEZA	*N.d.mhorr*	12
Marwell Zoo *'ruficollis'*	MAR_R	Origin is the North American Regional Studbook for addra (*ruficollis*) gazelle.	*N.d.ruficollis*	5
Katané, Ferlo North Game Reserve, Senegal	SEN_M	Ultimately originating from EEZA via Réserve Spéciale de Faune de Guembeul, Senegal	*N.d.mhorr*	3
Safia Reserve, Morocco	SAF_M	Ultimately originating from EEZA via R'Mila Royal Reserve, Morocco.	*N.d.mhorr*	6

Figure 2. Variation in pelage coloration in different populations listed from northwest to southeast. a&b) Captive *mhorr* at Al Ain (a) and Frankfurt Zoo (b) which are descended from four founders caught in the Dora-Hagunia and Tichla-Bir Ganduz area of Western Sahara in 1958 (in the case of Al Ain zoo the origin is unrecorded, but it is highly likely that this is where they come from). c) animals from the population in Termit, Niger. d) animals from the population in Manga, Chad. e) animals from the population in Ouadi Rimé-Ouadi Achim, Chad. f) animals from the captive population at Al Ain Zoo, most likely descended from 20 founders taken from the wild in around Ouadi Haouach close to Ouadi Rimé-Ouadi Achim, Chad. Animals from the most north-westerly populations have the most extensive dark coloration, which descends down the legs. Moving to the south and east, this dark coloration fades upwards and forwards. Note that there is also phenotypic variation within populations, for example the width of the thigh marking differs in Manga and the animals in OROA exhibit presence or absence of the ham-shaped mark on the thigh [15].

extent of the dark dorsal coloration was and is obvious even within the same herd [20–21].

Geographical variation and putative subspecies

This apparent cline in dorsal coloration led to the designation of a number of subspecies, based mostly on single specimens, which are often poorly described [22–25]. Cano [15] examined 50 museum specimens and captive populations to determine only three subspecies, a classification that is followed generally today: *N. dama ruficollis* occurs east of c.15°E, *N. dama dama* occurs from c.7°E to c.15°E, and *N. dama mhorr* occurs west of c.7°E. This tri-part classification was also followed by Groves and Grubb [18], although they presented a difference in the location of the boundaries between *dama* and *mhorr* locating it further northwest, stating that *mhorr* did not occur further south than the Western Sahara (Figure 1). This difference is due to historical confusion over the type locality of the species, then "*Antilope dama*" described by Pallas in 1766 [22] based on a specimen described and figured by Buffon in 1764 [23] which was collected from Senegal. It was later assumed that this specimen had in fact been collected in the Lake Chad region and not Senegal based on the pelage coloration [15]. Groves & Grubb [18] argue that it is more likely however that it did indeed come from Senegal because Lake Chad was not discovered by Europeans until 1823, more than 50 years after the species was described. The figure by Buffon [23] is, however, not a very biologically accurate representation of the species, illustrating the difficulty of making reliable inferences from historical references. Owing to the confusing number of subspecies names described for this species, we reproduce here the synonyms that Groves and Grubb [18] provide for each putative subspecies:

N. d. dama (Pallas, 1766), including *Antilope nanguer*, *Antilope dama* var. *occidentalis*, *Gazella dama permista*, *Gazella mhorr reducta*, *Gazella dama damergouensis*, *Gazella dama weidholtzi*

N. d. ruficollis (Hamilton Smith, 1827), including *Antilope addra*, *Antilope dama* var. *orientalis*

N. d. mhorr (Bennett, 1833), including *Gazella dama lozanoi*

N. d. mhorr is often referred to by the common names of mohor or mhorr gazelle and *N.d. ruficollis* is often called the addra gazelle. Here we use the term "dama gazelle" to cover all putative subspecies and populations of the species *Nanger dama* and for the time being we follow Cano's [15] definition of subspecies and Groves and Grubb's [18] geographical ranges for each subspecies. We highlight here again however that considerable variation within the subspecies has been recoded recently [12] [21] and that [20] observed phenotypic variation, claiming to have seen representatives of all three subspecies phenotypes within Eastern Chad.

History of captive populations

The early history of the captive stock is detailed by Cano [26]. In 1971, a captive breeding programme was established for *N. d. mhorr* in Almeria (Spain) at the Parque de Rescate de Fauna Sahariana (now known as "La Hoya" Experimental Field Station, Estación Experimental de Zonas Aridas), with animals that had originated from the Dora-Hagunia and Tichla-Bir Ganduz of Western Sahara in 1958 [26–27]. The last remaining *N. d. mhorr* were seen there in 1968 [26]. The captive population was based on one male and three females, which are the founders for all animals within the mhorr gazelle international studbook. There are currently 293 animals in 20 institutions across Europe, USA, Africa and the Middle East, of which the largest collection, ~100 individuals, is at the original location of the "La Hoya" Experimental Field Station. Additionally, this captive population has been either the direct or indirect source (via descendants in other institutions) of all animals for a number of reintroductions to fenced protected areas in North and West Africa; Bou Hedma National Park, Tunisia (22 animals in the early 1990s) [28–29]; Souss-Massa National Park, Morocco (12 animals in 2006); Domaine Royal de R'Mila, Morocco, Safia Reserve, Morocco; Guembeul Faunal Reserve, Senegal (seven gazelles in 1984) [30], of which five descendants were the founders for the population at Katané, Ferlo North Game Reserve, Senegal, in 2003 [31]. There are no other known sources of mhorr gazelles, so that in all likelihood the entire captive population today is descended from just four founders.

The decision was taken in 1995 by the European Association of Zoos and Aquaria (EAZA) (whose animals had originally come from USA populations) to phase out the captive management of *N. d. ruficollis* "addra" in favour of *N. d. mhorr* [32]. Similarly, AZA took the decision in 2008 to phase out management of *N. d. mhorr in* favour of *N. d. ruficollis*. Therefore, the current situation is that *N.d.ruficollis* is subject to coordinated breeding efforts in the USA and breeding of *N.d.mhorr* is coordinated in Europe. A number of institutions in the Middle East hold both subspecies

There has been historical debate over whether the *ruficollis* population is indeed *ruficollis* or in fact *dama* or a mixed subspecies population, but here we refer to it as *ruficollis* throughout, following the current consensus. Twenty-eight institutions across the USA hold a total of 168 animals [33]. The studbook was founded by five females and three males from the Ouadi Haouach region of north-east Chad [34], which crosses the north-east boundary of the Ouadi Rimé-Ouadi Achim Game Reserve sample site in this study (see OROA Table 1). Therefore, this captive population (which shows variation in coat coloration) is defined as *ruficollis*, because the original collecting locality is within this subspecies' range according to the Cano [15] classification and comes from a location close to where an extant wild dama population shows marked inter-individual variation in detail of basic pelage pattern. There are also a similar number of *N. d. ruficollis* in captivity in private collections, for which no additional founders are known.

Given the small number of animals left in the wild, the studbook-managed populations represent a considerable proportion of the species' global population (~300 in the wild, ~550 in captivity). At least as many animals again exist in captivity outside the two studbooks, notably on private ranches in Texas [35]. So if there can be any positive news for this species at all, it is the captive insurance populations are reasonably large in size. However, it cannot be assumed that all animals alive are currently of equal genetic value to the long-term conservation of this species. Both wild and captive populations are likely to be experiencing higher rates of inbreeding relative to the past (wild populations due to reduction in numbers and fragmented populations, and captive populations due to the small number of founders). However, levels of inbreeding (and inbreeding depression) may differ considerably between populations. Additionally remaining (wild and captive) populations are unlikely to be equally closely related to each other.

Study Aims

The purpose of this study was to assess the relatedness and relative genetic diversity in extant populations and to use this information to assess the conservation value of different populations and the validity of the current subspecies classification system in order to inform future conservation management strategies for the dama gazelle.

Methods

Sample collection

In total 124 samples were collected. This included 36 samples from the wild that were collected as a result of six collecting trips on public land between 2009 and 2013 with the authority and participation of the Termit & Tin Toumma National Nature Reserve, Niger and the Ouadi Rimé-Ouadi Achim Game Reserve and Government of Chad. Samples were collected opportunistically as pellets from distinctly separated fresh piles of faeces, in most cases following direct observation of wild dama gazelles. Samples from captive and reintroduced populations represent a mixture of blood and tissue samples taken from routine veterinary treatment or during autopsy (following natural death) or via faecal sampling and were provided by the relevant institutions or from the collections of National Museums Scotland (see Supplementary Table S1 for details of specific samples), and as a result of two collection trips to Safia and Katané reserves (public land) with the permission of the National Parks in Senegal and the Haut Commissariat des Eaux et Fôrets et de la Lutte Contre la Desertification in Morocco (Table 1: summary details of samples used in this study). CITES permits were in place for the samples that required them. No animals were harmed or killed for the purpose of this study.

DNA extraction

Tissue and blood samples were extracted using DNeasy blood and tissue kits (Qiagen) and faecal samples with the QIAamp DNA Stool Mini Kit (Qiagen), according to standard protocols. In the case of faecal samples, a single pellet was used per extraction.

Control region mtDNA sequencing

Primers were designed from existing sequences on Genbank to amplify a 560 basepair fragment from the centre of the control region: nang-fwd (5′ to 3′): CTA TGT CCT GTT ACC ATT GAC, nang-rev (5′ to 3′): GAT TGT CCA CAT GCA TAT AAG C. PCR amplification of the fragment was conducted with 1 µl of template DNA (10–50 ng ml^{-1})1 µl each of forward and reverse primer (10 µM) and 7 µl of Maxima Hot Start PCR Master Mix (Thermo-Fisher). Amplification was performed with an initial denaturation step of 5 mins. at 95°C, followed by 35 cycles of 1 min. (denaturation) at 95°C, 1 min. (primer annealing) at 58°C, 1 min. (elongation) at 72°C and ending with a 72°C extension for 10 mins. Negative controls were run as standard.

The fragments were examined by running them out on a 1% agarose gel and successfully amplified products were cleaned up by addition of 0.5 µl of the enzymes EXO1 and FastAP (Fisher) with an incubation step of 37°C×45 min and a denaturation step of 80°C×10 min. Fragments were sequenced in the forward direction using the BigDye Terminator Kit® (Applied Biosystems), using 3 µl of PCR product and conditions according to manufacturer's instructions. Sequences were run on a capillary ABI 3730 DNA Analyzer sequencer® (Applied Biosystems). In the case of faecal samples, sequencing was additionally conducted in the reverse direction and for a subset of samples with unique haplotypes the PCR was repeated and sequenced.

Cytochrome b sequencing

Since cytochrome b evolves at a slower rate than the control region, a subset of 16 samples exhibiting unique control region DNA haplotypes were sequenced for a 421 base pair region of the mtDNA cytochrome b gene using the primers MCB 398 and MCB 869 [36]. PCR reaction volumes were identical to those for the control region: thermocycling conditions were: were 95°C×5 -

min./[95°C×30 s, 50°C×90 s, 72°C×30 s]×35 cycles/ 60°C×30 min. PCR clean-up and sequencing were conducted as for the control region (see above). Sequencing was conducted in a single direction, using the primer MCB 398 and verified in the reverse direction for a subset of unique haplotypes.

Alignment

Chromatograms were analysed with Geneious Pro v6.1.4 [37]. Primer sequences were trimmed from the alignments. The cytochrome b sequences were translated to verify the absence of stop codons in the sequences. Sequences aligned with the ClustalW algorithm implemented in the Geneious Pro software and final corrections done by eye.

Tree and network building

Tree building was performed separately for control region and cytochrome b fragments, using two different methods for each fragment. Firstly, a simple distance-matrix method was used to build the trees. This was done using neighbour joining with the Tamura-Nei model of genetic distance. The trees were built in Geneious Pro v6.1.4 and one-hundred bootstrap iterations were performed. Secondly, Bayesian inference was carried out with MrBayes v3.1.2 [38]. Posterior probabilities were calculated using four heated Markov chains run (chain temp. 0.2) for 2,000,000 Metropolis-coupled MCMC generations. Tree sampling was conducted every 500 generations and a burn-in of 200,000 trees was used. The model used for nucleotide substitution was the General Time Reversible model with among-site substitution-rate heterogeneity described by a gamma distribution and a fraction of sites constrained to be invariable (GTR+Γ+I). This was chosen as a conservative measure, because it has been previously been shown that Bayesian inference of phylogenetic trees performs better, if the chosen substitution model tends toward being over- rather than under- parameterised [39]. The Bayesian analysis was replicated twice to examine its stability.

Control region trees were built using the Genbank haplotype JN632666 (*Nanger granti*) as an outgroup and incorporating sequence JN632665 from a dama gazelle, originating from San Diego Zoo. Here it is referred to as control region haplotype S.

Haplotype networks of the control region data were created using TCS v1.21, treating gaps as a 5th state (there is in fact only a single in-del in the alignment). The network was examined at a number of different probabilities of parsimony (Templeton *et al.* 1992).

Cytochrome b trees were built, including Genbank sequences KC188777, JN632667, JF728776 (*Nanger soemmerringii*) and JN632666 (*Nanger granti*) as outgroups. Sequence JN632665 from the same dama gazelle, originating from San Diego Zoo, was also incorporated into the tree [40]. Here it is referred to as cytochrome b haplotype 4.

Nucleotide diversity

Haplotype and nucleotide diversity and matrices of mutational distance between haplotypes were calculated using Arlequin 3 [41].

Results

Control region

In total, 122 control region sequences were obtained from the 124 samples, with only two samples failing to give any result (they were faecal samples collected from the sites EEP_M and OROA). 16 haplotypes were discovered and these sequences have been named A to O & R and have been placed on Genbank under accession KJ848615–30. The majority of the haplotype diversity was spread between the wild populations (13 haplotypes). Captive and captive-derived populations (SEN_M, SAF_M) only contained three haplotypes, two within *mhorr* and one within *ruficollis*. No haplotypes were shared between the different wild populations, or between wild and captive populations (Figure 3). Haplotype diversity [42], defined as the probability that two randomly chosen haplotypes are different in the sample, was highest in the populations OROA and Manga (0.840+/−0.059 s.d., 0.870+/−0.129 s.d.) and lower, but roughly equivalent between the Niger and EEP_M populations (0.510+/−0.100 s.d., 0.485+/−0.106 s.d.) (Figure 3). Nucleotide diversity, defined as the probability that two randomly chosen homologous nucleotides are different, was highest in OROA and Manga (0.031+/−0.016 s.d., 0.031+/−0.018 s.d.). Nucleotide diversity was somewhat higher in the EEP_M (0.013+/−0.007 s.d.) than in Niger (0.006+/−0.004 s.d.) (Figure 3).

For the Bayesian inference of phylogeny the average standard deviation of split frequencies was 0.0069 after 2,000,000 Metropolis-coupled MCMC generations for both of the two replicates.

Phylogenetic analysis of the mtDNA control region data revealed tree topologies that were roughly concordant with Neighbour-Joining and Bayesian methods. However, support was generally low apart from for clusters (S, L), (O,J,K), (H,I) and (R, A,B) (Figure 4a).

Network analysis revealed a similar certainty surrounding resolution, with four clusters being resolved at a 95% connectivity limit (S,L), (H,I,K,J,O), (R,A,B) and (E,F). Relaxation of the connectivity criteria to below 90% was required to draw a complete network, with 18 mutational steps required to link the entire tree (Figure 5). Pair-wise differences between control region haplotypes can be found in Table 2.

Cytochrome b

In total seven haplotypes of cytochrome b were discovered, which were named MCB1–3, 5–8 and placed on Genbank under accession KJ848631–7. Individuals carrying control region haplotypes H and I corresponded to MCB Haplotype 1. MCB2 corresponded to control region B,O,A,D,K,J and R; MCB3 to C; MCB5 to L; MCB6 to N, MCB7 to G,M,F and MCB 8 to E. As previously mentioned, MCB4 corresponds to S (JN632665 whole mitochondrial DNA sequence from animals from San Diego Zoo). These cytochrome b haplotype groupings have been overlaid on the control region network (Figure 5).

For the Bayesian inference of phylogeny the average standard deviation of split frequencies was 0.007 and 0.0042 after 2,000,000 Metropolis-coupled MCMC generations for the two replicates. Graphical output of the change in Log Likelihood during the analysis can be found in supplementary material 1.

Resolution of the phylogeny was not much improved at the cytochrome b gene (Figure 4b), with only a clustering of haplotypes 4 and 5 showing strong support (Figure 4b).

Pair-wise differences between cytochrome b haplotypes can be found in Table 3. There was a 2.1% divergence at cytochrome b across the whole sample set and a 0.475% divergence between haplotypes M and R representing the captive mhorr and ruficollis populations.

Discussion

The purpose of this study was to assess the relatedness and relative genetic diversity in extant populations of dama gazelle and to use this information to begin an assessment of the conservation value of different populations.

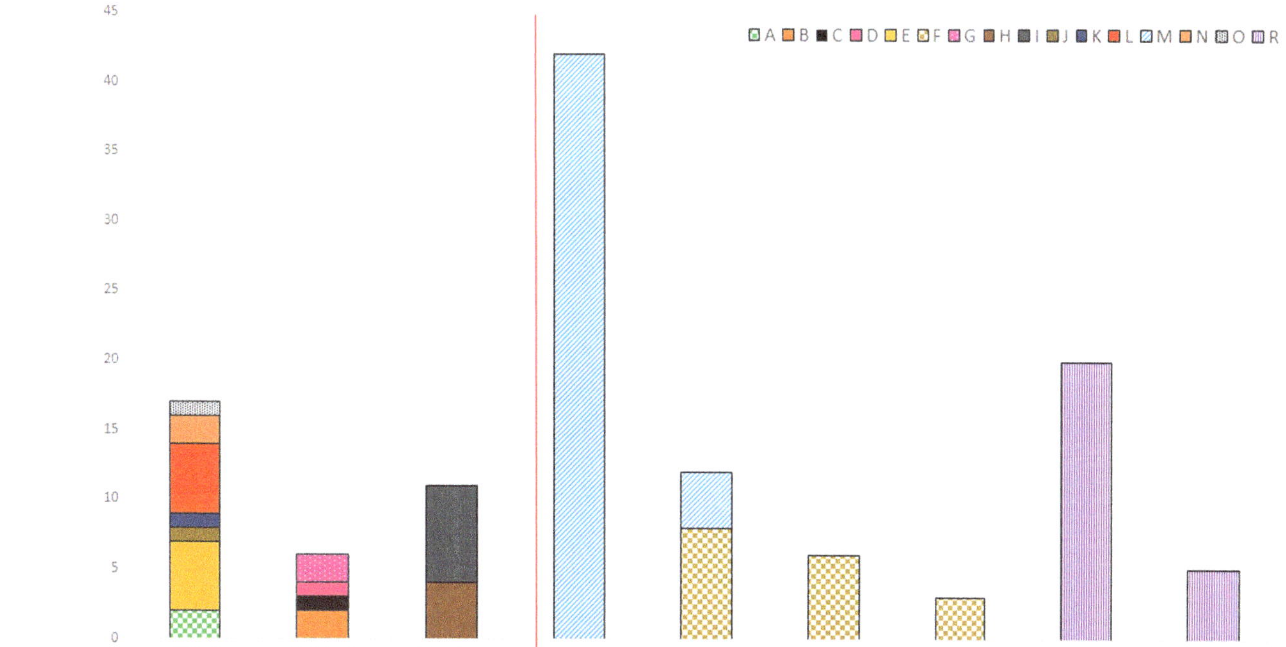

	OROA_R	MANGA_R	TERMIT_D	AIN_M	EEP_M	SAF_M	SEN_M	AIN_R	MAR_R
Haplotype diversity	0.84 +/- 0.059	0.87 +/- 0.129	0.51 +/- 0.100	0	0.485 +/- 0.106	0	0	0	0
Nucleotide diversity	0.031 +/- 0.016	0.031 +/- 0.018	0.006 +/- 0.004	0	0.013 +/- 0.007	0	0	0	0

Figure 3. Bar chart of control region haplotypes (A–O & R) found at different sampling sites in this study. Sites are enumerated by collecting locality (see Table 1) and putative subspecies; R (*N. d. ruficollis*), D (*N. d.* dama), M (*N. d. mhorr*). Wild and captive populations are separated by a red line. Particularly striking, but not unexpected, is the higher haplotype diversity in the samples from wild populations (OROA_R, MANGA_R,TERMIT_D) than in samples from captive and captive-derived *mhorr* populations (Ain_M, EEP_M, SEN_M, SAF_M) and captive *ruficollis* populations (Ain_R, Mar_R). No haplotypes are shared between wild populations, or between wild and captive populations.

Genetic diversity of dama gazelle populations

The mtDNA control region data show genetic diversity generally appears to be lower in captivity than in the wild populations (note that for samples exhibiting the same control region haplotype in the wild, we do not know to what extent resampling has occurred, but unintentional resampling would deflate not elevate diversity). Particularly striking is the much higher haplotype diversity in the wild populations as a whole (OROA_R, MANGA_R, TERMIT_D) compared with the captive and captive-derived *mhorr* populations (AIN_M, EEP_M, SEN_ M, SAF_M) and the captive *ruficollis* populations samples (AIN_R, MAR_R) (Figure 2). Low genetic diversity in captivity is unsurprising given what we know of the captive population's history (see Introduction). This survey represents a fairly representative sample of the captive *mhorr* population derived from the founding stock at La Hoya (represented by populations AIN_M, EEP_M, SEN_M & SAF_M). However, we have not been able to survey genetic diversity extensively within the captive *ruficollis* population, having only sampled from derived European (MAR_R) and Arabian (AIN_R) captive populations, and not the original population brought from Chad to the USA. We can expect USA populations to retain some additional genetic diversity not found in the derived populations sampled as part of this study. Understanding how the captive *ruficollis* population in the USA, both within zoos and private collections, relates to dama gazelle populations globally, represents an important future task. It is interesting to note that the animals sampled from the *ruficollis* captive population exhibit haplotypes that are closely related to those found at OROA (Figure 5), as indeed might be expected given the location that their ancestors were taken from (Figure 1), although see discussion of general lack of phylogeographic structure later. Genetic diversity in one of the wild populations (TERMIT_D) is equivalent to that of the captive *mhorr* populations. This may represent a worrying sign for this wild population, however mtDNA sequencing and the level of sampling available lack the resolution required to compare genetic diversity fully.

Captive populations of threatened species should be managed to maximise genetic diversity and to minimise inbreeding for two reasons: to reduce the likelihood of inbreeding depression, and to retain the greatest adaptive potential of the population.

Inbreeding has been shown on numerous occasions to have a detrimental effect on fitness in naturally outbreeding species [43–45] and review by [46]. This includes studies of captive populations [47–48] and populations released into the wild [49–51]. Pedigree estimates of inbreeding do not correlate with either body size or juvenile mortality in captive populations of *mhorr* at La Hoya ([52–53], although see earlier study [54] for contradictory results). It had been suggested that either captive conditions are artificially inflating fitness or that purging of deleterious alleles are responsible for this lack of effect. On the other hand, in the same population, genetic diversity, measured across 17 microsatellite loci (but not pedigree estimates of inbreeding), is correlated with semen quality, a trait that is directly related to fitness [55]. Therefore, Ruiz-Lopez *et al.* [55] suggest (based on results both from *mhorr* and a captive *Lynx pardinus*, population) that in highly inbred captive populations, pedigree estimates of inbreeding are often not reliable indicators of accumulated inbreeding, because base levels of inbreeding may already be high and assumptions of

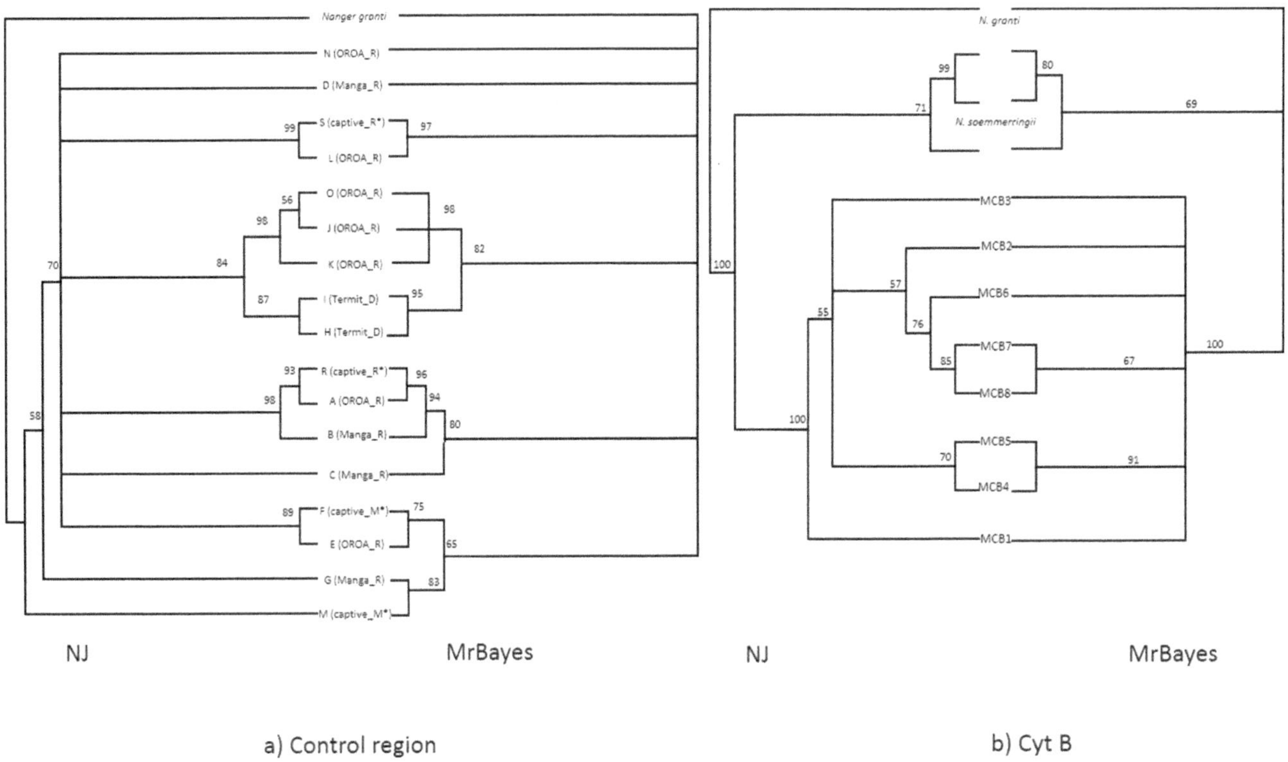

Figure 4. Evolutionary relationships at different genes a) Tree based on 560 bp of control region and b) 421 bp of cytochrome B. In each case Neighbourhood joining (left) and Bayesian Inference of Phylogeny via MrBayes (right) were conducted according to the conditions listed in the methods section. The putative *ruficollis* subspecies exhibits polyphyly at the control region (and cytochrome b, compare with Figure 5). * Captive populations have been combined (see Table 1).

non-relatedness of founders may not be met. It seems likely that captive *mhorr* populations are suffering from other effects of inbreeding, but it is hard to measure these effects due to lack of statistical power. Similar studies have not been conducted for *ruficollis*.

Inbreeding depression is not the only reason to be concerned about inbreeding in captive dama gazelles. Concurrent loss of genetic diversity may result in loss of adaptive potential, limiting the ability of the population to evolve [56–58], to be resilient to environmental change or disease [59–60], and to retain the ability of readapting to wild environments from captivity [61–63].

Although genetic diversity in the wild appears to be considerably higher than in captivity, diversity is undoubtedly being lost in wild populations and this is of concern for the same reasons as mentioned above. Populations are small and isolated, and in the long term may become vulnerable to inbreeding, if the situation remains the same. Interestingly no haplotypes were shared between different populations despite lack of genetic structure (see below). This may indicate a lack of recent gene flow between sites and genetic drift in small fragmented populations, or may simply be an artefact of the small sample sizes available to this study. Additionally, the comparatively small number of haplotypes found in Termit versus the other wild populations (Figure 3) may indicate elevated levels of inbreeding there. For example twice as many samples were collected from Termit as from Manga but half as many haplotypes were found. The small sample sizes involved make it hard to draw any strong conclusions (Figure 3). A critical task for dama gazelle conservation is to secure genetic diversity and population connectivity in the wild. However, hunting, habitat loss and competition for grazing due to pastoral development and inherent vulnerability to demographic fluctuations caused by small population size are likely to be more immediate threats to the species' survival than loss of genetic diversity [12].

Relatedness of dama gazelle populations

A crucial task for this study was to make an assessment of the relatedness of dama gazelle populations. Owing to its desperate conservation status, opportunities for sampling in the wild are limited. Despite this issue, a pattern emerged that suggests that phylogeographical structure of mtDNA across the range of the dama gazelle is weak or even absent altogether (Figure 5). There is a low level of support for basal nodes of the control region and cytochrome b trees (Figure 4), which means that it is not possible to resolve the entire structure of interrelatedness of the mitochondrial genome with confidence, but there is higher confidence in substructures within the trees, which cluster together haplotypes from disparate geographical locations. For example, control region haplotypes A and B, which come from OROA and Manga, are supported with 98% bootstrap (NJ) and 94% posterior probability (Mr Bayes), and differ by only 3 bp mutations. Haplotypes E and F, which come from the captive *mhorr* population (originating from Western Sahara, see introduction) and Manga, are supported by 89% bootstrap/75% posterior probability and differ by 8 bp mutation steps. Haplotypes G and M from Manga, and the captive *mhorr* population, cluster with 83% posterior probability (no NJ support) and differ by 12 base pair mutations (Figure 4 & 5, Table 2).

Critically, the clustering of these groups is also supported by the pattern of relatedness at the more slowly evolving cytochrome b

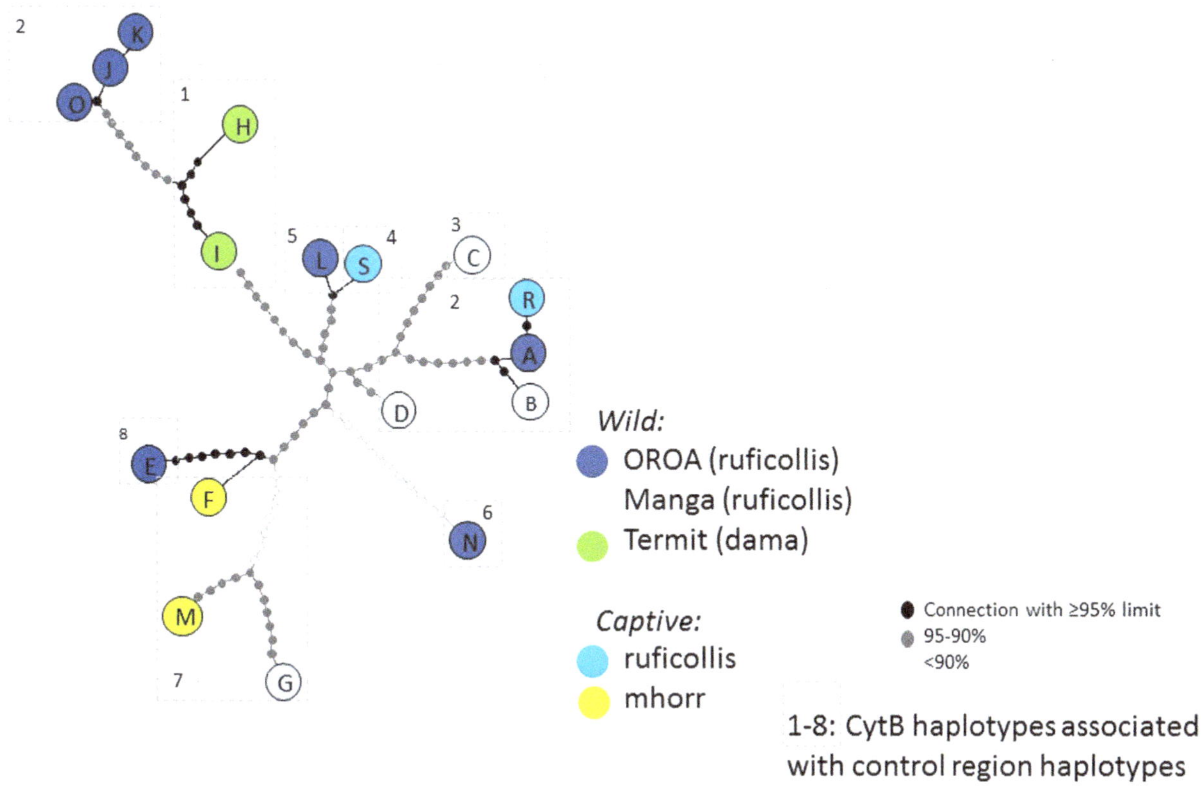

Figure 5. Haplotype network of the control region haplotypes present in this study. Each haplotype is colour coded according to population of origin, and single base-pair step-wise mutations between haplotypes are colour coded according to their connection limit. Relatedness of haplotypes does not correspond to subspecies divisions or geographical structure.

gene, because closely related control region haplotypes are associated with identical or closely related cytochrome b haplotypes (Figure 4, 5). Notably the closely related control region haplotypes M (*mhorr*), F (*mhorr*) and G (*ruficollis*/Manga) are associated with an identical cytochrome b haplotype.

The result of examination of both control region and cytochrome b data is that the putative *ruficollis* subspecies grouping is likely to be polyphyletic with respect to both *dama* and *mhorr* (Figure 4), although support for the arrangement is not conclusive (75% posterior probability/89% bootstrap). Sequencing of a larger section of the mtDNA or addition of more samples to the tree might result in greater resolution in the future.

Lack of phylogeographical structure at mtDNA genes has been found in two other large Saharan mammals across their range: the Dorcas gazelle, *Gazella dorcas* [8,64] and the African wolf *Canis lupus lupaster* [65]. Other genetic studies of large mammals native to the Sahara are absent, perhaps because there are few or no animals left to study [14]. Evidence for phylogeographical structuring does exist in smaller non-flying vertebrates native to this region and is particularly associated with mountains ranges and micro-scale water features [66].

Phenotypic variation

The lack of phylogeographical structure across the dama gazelle range does present a contradiction because the cline in pelage coloration (Figure 2) points to the fact that some level of genetic divergence must be present, assuming that coat-colour is a trait under genetic control. The existence of clines in phenotypic traits is widespread and well documented (for review see [67–68]). Clines in coat coloration in the oldfield mouse (*Peromyscus polionotus*) and rock pocket mouse (*Chaetodipus intermedius*) have been linked to variation in substrate colour exerting differential selection pressure on specific genes [69–70] and micro-habitat variations in enivormantal luminosity have been linked to divergence in the sympatiric mtDNA clades of the Saharan jerboa (*Jaculau jaculus*) [71]. The case of the oldfield mouse has interesting parallels to that of the dama gazelle: The mouse inhabits a sandy habitat and the coat colour polymorphism in question is a gradation in the extent of a dark cape over a pale body, with similarities to that of dama gazelle. This trait has been linked to selection on the regulatory region of the *Agouti* gene, a gene that has also been implicated in coat colour variation in a large number of mammals, including Soay sheep, *Ovis aries* [72], dogs, *Canis familiaris*, [73] and other domesticated animals [74].

There are distinct climatic differences between the Sahel and the Atlantic Sahara [66], which may be responsible for variation in coat colour as the darker coat coloration is roughly concordant with the Atlantic Sahara [15]. However it is not known how closely correlated trends in climate are to trends in phenotype (and note for example the relatively minor change in extent of coloration between animals originating from Western Sahara and those from Termit in the Sahel approx. 2800 km inland from the coast, Figure 2). It is also not known whether dama gazelles display other phenotypic traits (e.g. skeletal measures, behavioural traits) that are coincident with the change in pelage colour.

If the phenotype is not under selection, one possible explanation for the observed variation in pelage coloration is that genetic drift occurred in separate refugia, perhaps either side of palaeolake Mega-Chad, which was larger than the Caspian Sea 5,000–6,000 years ago and has undergone changes in size associated with

Table 2. Matrix of pair-wise mutational differences between control region haplotypes.

	A	E	K	J	O	L	N	C	B	D	G	I	H	M	F
A															
E	23														
K	17	20													
J	18	21	1												
O	20	21	3	2											
L	19	22	18	19	19										
N	22	25	19	20	22	21									
C	16	20	19	20	22	19	21								
B	3	22	18	19	21	18	21	15							
D	14	17	15	16	18	11	18	14	15						
G	26	21	21	22	22	17	24	25	23	22					
I	22	23	15	14	14	15	22	18	21	14	22				
H	25	22	14	13	13	22	23	23	22	21	23	7			
M	26	21	25	26	26	25	28	29	25	24	12	26	25		
F	19	8	18	19	19	16	19	20	18	13	17	19	20	15	
R	2	23	19	20	22	21	22	17	5	16	26	24	27	26	19

Table 3. Matrix of pair-wise mutational differences between cytochrome b haplotypes.

	1	2	3	4	5	6	7	8
1								
2	1							
3	2	1						
4	4	3	4					
5	3	2	3	1				
6	3	2	3	5	4			
7	3	2	3	5	4	2		
8	4	3	4	6	5	3	1	

climatic cycles for at least the last 23,000 years [75]. It is possible that divergence resulted in evolution of separate phenotypes, which came together again after a period of isolation, forming a broad zone of secondary contact. The apparent absence of structure at mtDNA loci does not support this theory. However, there may be a cline at nuclear markers, but not at mtDNA markers. Species with male-biased dispersal (which is often the case in mammals, although nature of dispersal is unknown in the case of the dama gazelle) have paradoxically higher rates of mtDNA intogression than at nuclear loci [76], and so it is possible that differentiation at mtDNA markers might disappear first and that we are viewing the results of extensive introgression over a very large area.

It seems unlikely however that pelage coloration is neutral to selection given the (historical) presence of predators [14] and perhaps more importantly, the high level of solar radiation present in the Sahara/Sahel region. The desert-dwelling antelopes, addax (*Addax nasomaculatus*) and Arabian oryx (*Oryx leucoryx*), are thought to have evolved a predominantly white pelage to reflect solar radiation for thermoregulation and advertisement when in the open and to aid with camouflage when in the shade [77–78]. Does the varying degree in white present on the dama pelage point to a varying need to reflect solar energy? Or perhaps a tension between this and camouflage either as adults or juveniles? Whether secondary or primary in origin, it is likely, given the mtDNA results that the cline in pelage is caused by selection on a small number of genes. Importantly considerable variation is apparent in pelage within populations (localities) [20] and to our best knowledge transition in phenotype gradually occurs across a very large area [15], suggesting a concurrent gradual change in selection pressure if one exists.

Until evidence to the contrary is produced, phenotype should be assumed to have some selective relevance as this is the most conservative scenario for conservation. The key here is that it is the preservation of the diversity of genes controlling phenotype that is important as this is the raw material on which selection can act (in different environments), not the phenotype *per se*. Genetic management of dama gazelles should not seek to deliberately breed for phenotype as this will certainly result in further inbreeding of captive populations and may lead to selection for unforeseen traits, because genes controlling coat colour are known to have pleiotropic effects on other (often fitness-negative) traits [74].

Conservation Dilemma

The issues behind the apparent cline in dama gazelle phenotype come to the fore when dealing with the conservation of its populations, creating a classic dilemma for modern-day conservation: are the now fragmented (wild and captive) populations distinct enough to justify being managed separately?

By mixing populations that are distinct, the risk is that the population will suffer from hybrid sterility or outbreeding depression. The risk of maintaining them separately is inbreeding depression and loss of genetic diversity (see above). Frankham *et al.* [4] argued that when considering the option of mixing different populations for conservation purposes, the risk of outbreeding is generally much lower than the risk of inbreeding (see above), but that conservation practitioners tend to overplay the former risk. Predicting the probability of outbreeding depression in advance is not an easy task and Frankham *et al.* [4] propose the use of a flow chart for evaluating risk, where the risk of outbreeding depression is smaller with the absence of chromosomal differences, absence of gene flow for <500 years, and lack of substantial environmental differences between the populations (Figure 1 of [4]).

In the case of the dama gazelle it seems likely the range has been contiguous within the last 500 years and dama gazelles are known to migrate seasonally in search of water [26,79] suggesting high connectivity (also suggested by mtDNA data). As environmental differences have not been determined, it is tempting to suggest that there is little variation, but given that phenotypic variation is present, we cannot be sure. There is some evidence of karyotypic differences between male animals from Almeria (*mhorr*) and San Diego (*ruficollis*) in the form of a single chromosomal rearrangement (Robertsonian translocation, centric fusion) between the short acrocentric Y_1 male-specific sex chromosome and an autosome of pair 14 in the San Diego animals. However a karyotypic polymorphism is also present within animals from Almeria (*mhorr*); in this case a Robertsonian polymorphism of chromosome 1 [80–81]. A complex of four Robertsonian translocations was also found in a female *mhorr* in a private collection at Taif in Saudi Arabia, although the origin of this animal is not known [82]. Chromosomal translocations are commonly responsible for reproductive isolation, but polymorphism is known to be common in gazelles and has been observed in *Gazella subgutturosa*, *G. gazella* (from northern Israel), *Eudorcas thomsonii* and *Antilope cervicapra* [80,83]. In summary, karyotypic evidence is inconclusive, but the presence of karyotypic differences suggest that we must treat seriously the possibility of genetic incompatibility between the captive populations.

The dangers of the sampling effect

It must also be noted that the captive animals studied here are descended from few individuals (see introduction) and thus represent a very narrow sampling of the former range. This "sampling effect" must always be borne in mind when comparing captive populations, not only for karyotype, but also for phenotype (which may have become more constrained owing to intentional or unintentional inbreeding) and genetic variation (at both nuclear and genetic markers), which will have been subject to extreme levels of drift. The only way to understand patterns of karyotypic, genetic and phenotypic variation is to examine them in the wild, a task which is all but impossible both for logistical reasons and due to widespread population extirpation and fragmentation. Even when examining nuclear structure in wild populations, care must be taken not to conflate recent drift events caused by population fragmentation with deeper historical substructure. The best examination would involve the use of large numbers of linked markers [84] in both current and historic (museum) samples, but this may not be feasible.

Dama gazelle taxonomy and conservation

The problem with the historical preference for splitting into subspecies is that the burden of proof rests with the contemporary researcher. The null hypothesis becomes: "there are *n* subspecies" even if subspecies were not classified on a rigorous scientific basis. A large number of putative dama gazelle subspecies were recognised historically, based on rather limited evidence, which the paper by Cano [15] rationalised to only three based on the best available phenotypic and geographical data at the time. We would argue that in retrospect perhaps even this revision did not go far enough, owing to the apparent clinal and non-discrete nature of the variation. Today, although the genetic data do not prove polyphyly of the three subspecies conclusively (thereby disproving a null hypothesis of three putative subspecies), the joint evidence at the mtDNA control region and cytochrome b are suggestive of polyphyly and it is crucial to note that the data would not support a split into three subspecies from an *a priori* assumption of one. The levels of genetic divergence at cytochrome b (0.5% between captive *mhorr* and *ruficollis* haplotypes and 2.1% as a whole) are at within-species levels both for the best studied Saharan gazelle species, the dorcas gazelle [8,64], and for other a mammalian taxa [85].

In the light of this study and the evidence presented here, we conclude with the following statements relevant to the conservation of the dama gazelle:

1. To the best available knowledge, phenotypic variation was originally more or less clinal across the species range. Today, we see phenotypic variation between populations, but also within captive and wild populations. Phenotypic differences between subspecies in captivity may be exaggerated by (in)breeding of the current populations from small numbers of founder individuals. Uneven and incomplete sampling of historical wild populations, and absence and fragmentation of contemporary populations, has in addition possibly presented a distorted view of the original phenotypic variation throughout the species' range. There is no *a priori* reason to divide the cline into three discrete units and lack of coincident mtDNA genetic structure and the possible polyphyly of the nominate *ruficollis* subspecies supports this view.
2. The conservation of the dama gazelle will be greatly advanced if it is considered a single species without subspecies division, even though it exhibits phenotypic variation. Under the "three subspecies view" artificial impermeable boundaries are erected and only individuals which conform to a particular phenotype from within the same subspecies should be bred together, and used for reintroduction and population augmentation. Under the "monotypic species view" there is a continuum of suitability of donors to a population, where all else being equal, the geographically most proximate population is the most suitable, but there is no *a priori* barrier to exchange between any populations provided the risks of exchange have been evaluated properly.
3. Inbreeding depression and loss of genetic diversity must be taken seriously during ongoing management efforts. Evidence for inbreeding depression in a captive population has already been found [55]. Captive populations of dama gazelle should be managed to maximise genetic diversity and minimise inbreeding. To achieve this aim, the continuation and improvement of the coordination and monitoring of captive breeding efforts across the world is vital.
4. Unless there is evidence to the contrary, phenotype should be assumed to be under some degree of selection in the wild as this is the most conservative scenario to conservation. However, no attempt should be made to breed or select for "a true phenotype" in captivity as this will result in further loss of genetic diversity and possibly unintended selection for traits linked to phenotype.
5. The possible risks associated with interbreeding of animals from phenotypically distinct and geographically distant populations should be taken seriously, however the potential conservation benefits of such interbreeding should be evaluated. Small-scale scientifically monitored experimental crosses of captive *mhorr* and *ruficollis* individuals should be conducted as part of a risk-benefit analysis for future actions [57]. This analysis will provide information to evaluate the possibility of more extensive mixing of captive populations and will contribute useful information if translocation of animals between wild populations or between the wild and captivity are to occur in the future.

Acknowledgments

From Al Ain Zoo, we thank the Director General Mr Ghanim al Hajeri for his support for the project and Dr Arshad Toosy, Myyas al Qarqaz and their team for collecting the blood samples from Al Ain Zoo's population. We thank the Director of the Estación Experimental de Zonas Aridas (EEZA, CSIC, Almeria, Spain) for the access to animals in the "La Hoya" Field Station, as well as to the National Parks in Senegal and the Haut Commissariat des Eaux et Fôrets et de la Lutte Contre la Desertification (Morocco) for access to the Katané and Safia Reserves respectively; Marwell Zoo (Peter Small and Peter Bircher) and the City of Belfast Zoo (Michael Griffith and Mark Challis) for donations of dama gazelles to National Museums Scotland; Lauri Bingaman Lackey for help with the European *ruficollis* studbook; Eulalia Moreno Mañas and the EEZA-BRB (Biologic Resource Bank of the Estación Experimental de Zonas Áridas (CSIC)) for samples and discussion; Gerardo Espeso Pajares the EAZA *mhorr* EEP coordinator for advice and discussion; Montpellier Zoo for samples; Ross McEwing for d-loop primer development; Caroline Whitson for assistance with figure design; Hannes Lerp for comments on the manuscript. Later revisions of this manuscript benefited from discussion during the creation of the Conservation Review of the Dama Gazelle [12], the authors of this report are acknowledged and thanked for indirect input that was obtained.

Author Contributions

Conceived and designed the experiments: HS MM RO JN TW TR. Performed the experiments: HS JK. Analyzed the data: HS. Contributed reagents/materials/analysis tools: ACK TR TW JN TA TLS LB. Wrote the paper: HS.

References

1. Edmands S (2007) Between a rock and a hard place: evaluating the relative risks of inbreeding and outbreeding for conservation and management. Molecular Ecology, 16: 463–75.
2. Frankham R, Ballou JD, Dudash MR, Eldridge MDB, Fenster CB, et al. (2012). Implications of different species concepts for conserving biodiversity. Biological Conservation, 153: 25–31.
3. Zachos FE, Apollonio M, Bärmann EV, Festa-Bianchet M & Göhlich U, et al. (2013) Species inflation and taxonomic artefacts - A critical comment on recent trends in mammalian classification. Mammalian Biology - Zeitschrift für Säugetierkunde 78 (1): 1–6.
4. Frankham R, Ballou J, Eldridge M, Lacy RC, Ralls K, et al. (2011) Predicting the probability of outbreeding depression. Conservation Biology, 25: 465–475.
5. Iyengar A, Gilbert T, Woodfine T, Knowles JM, Diniz FM, et al. (2007) Remnants of ancient genetic diversity preserved within captive groups of scimitar-horned oryx (*Oryx dammah*). Molecular Ecology, 16: 2436–49.
6. Nsubuga A, Holzman J (2010) The cryptic genetic structure of the North American captive gorilla population. Conservation Genetics, 161–172.
7. Witzenberger KA, Hochkirch A (2013) Evaluating ex situ conservation projects: Genetic structure of the captive population of the Arabian sand cat. Mammalian Biology - Zeitschrift Für Säugetierkunde, 1–4.
8. Godinho R, Abáigar T, Lopes S, Essalhi A, Ouragh L, et al. (2012) Conservation genetics of the endangered Dorcas gazelle (*Gazella dorcas* spp.) in Northwestern Africa. Conservation Genetics, 13: 1003–1015.
9. Simons ND, Wagner RS, Lorenz JG (2013) Genetic diversity of North American captive-born gorillas (*Gorilla gorilla gorilla*). Ecology and Evolution, 3: 80–88.
10. Hvilsom C, Frandsen P, Børsting C, Carlsen F, Sallé B, et al. (2013) Understanding geographic origins and history of admixture among chimpanzees in European zoos, with implications for future breeding programmes. Heredity, 110: 586–93.
11. Newby J, Wacher T, Lamarque F, Cuzin F, De Smet K (2008) Nanger dama. In: IUCN 2013. IUCN Red List of Threatened Species. Version 2013.2. www.iucnredlist.org
12. RZSS & IUCN Antelope Specialist Group (2014) Dama gazelle, Nanger dama, Conservation Review. Royal Zoological Society of Scotland, Edinburgh, UK.
13. Devillers P, Devillers-Tershuren J, Beudels-Jamarr RC (2005) Gazella dama. In: Beudels-Jamar RC, Devillers P, Lafontaine RM, Devillers-Tershuren J & Beudels MO (2006) Sahelo-Saharan antelopes. Status and perspectives. Report on the conservation status of the six Sahelo-Saharan Antelopes. CMS Technical Series Publication No. 11. UNEP/CMS, Bonn. pp. 57–70.
14. Durant SM, Wacher T, Bashir S, Woodroffe R, De Ornellas P, et al. (2014) Fiddling in biodiversity hotspots while deserts burn? Collapse of the Sahara's megafauna. Diversity and Distributions, 20: 114–122.
15. Cano Perez M (1984) Revision Systematik von Gazella (Nanger) dama. Zeitschrift der Köelner Zoo, 27: 103–107.
16. Jebali A (2008) Déclin de la faune Sahélo-Saharienne et tentative de réintroduction d'antilopes dans des habitats restaurés. PhD thesis, Museum of Natural History Paris.
17. Kingdon J, Hoffmann M (eds) (2013) Mammals of Africa:Vol. VI : Pigs, Hippopotamuses, Chevrotain, Giraffe, Deer and Bovids. Bloomsbury Publishing, London.
18. Groves C, Grubb P (2011) Ungulate Taxonomy. Baltimore, The John Hopkins University Press.
19. Hernández Fernández M, Vrba ES (2005) A complete estimate of the phylogenetic relationships in Ruminantia: a dated species-level supertree of the extant ruminants. Biological Reviews of the Cambridge Philosophical Society, 80: 269–302.
20. Malbrant R (1952) Faune du Centre africain français (mammiferes et oiseaux). Paris, Lechevalier.
21. SCF reports (2010–2013) www.saharaconservation.org
22. Buffon G-LL (1764) Histoire naturelle générale et particulière avec la description du cabinet du roy. Tome XII.
23. Pallas PS (1766) Miscellanea zoologica, quibis novae imprimis atque obscurae animalum apecies describuntur et observationsibus iconibusque illustrantur. The Hague: Kessinger.
24. Hamilton-Smith CH (1827) Supplement to the order Ruminantia. In: Griffith E. et al. The animal kingdon arranged in conformity with its organisations by the Baron Cuvier. London, Whittaker. pp. 33–428.
25. Bennet ET (1833) Chracter of a new species of antelope (Antilope mhorr) presented by E.W.A Drummond Hay in. Proc. Zool. Soc. London 1–2, London.
26. Cano M (1991) El antílope mohor (*Gazella (Nanger) dama mhorr* Bennett 1832) en cautividad. Ph. D. Unviersidad de granada, Spain.
27. Valverde JA (2004) Sáhara,Guinea,Marruecos.Expediciones africanas. Memorias de un biólogo heterodoxo.Editorial Quercus
28. Abaigar T, Cano M, Espeso G, Ortiz J (1997) Introduction of Mhorr gazelle in Bou-Hedma National Park, Tunisia, International Zoo Yearbook, 35: 311–316.
29. Wiesner H, Müller P (1998) On the Reintroduction of the Mhorr Gazelle in Tunisia and Morocco. Naturwissenschaften, 85: 553–555.
30. Cano M, Abaigar T, Vericad JR (1993) Establishmnet of a group of Dama gazelles (Gazella (Nanger) dama) for reintroduction in Senegal. Internaltional Zoo Yearbook 32: 98–107.
31. Abáigar T, Youm B, Niaga M, Ensenyat C, Cano M (2005) The role of Senegal in the recovery of the Sahelo-Saharan antelope species: the case of the reintroduction of Dorcas Gazelle. Gnusletter, 28: 6–8.
32. Rietkerk F (2008) Regional collection planning for antelope and giraffes. EAZA News 64: 12–13. www.eaza.net/News/EAZA_Magazine
33. Petric A (2012) AZA North American Regional Studbook for Addra Gazelle (Nanger dama ruficollis). Sponsored by Saint Louis Zoo.
34. Scholte P (2013) Nanger dama Dama Gazelle; pp 382–387 In Kingdon, J. & Hoffmann, M. (eds) Mammals of Africa: Volume I: Pigs, Hippotamuses, Chevrotain, Giraffes, Deer and Bovids. London, Bloomsbury Publishing,
35. Mungall C (2007) Exotic Animal Field Guide, Texas A&M University Press.
36. Verma SK, Singh L (2003) Novel universal primers establish identity of an enormous number of animal species for forensic application. Molecualr Ecology 3: 28–31.
37. Drummond AJ, Ashton B, Buxton S, Cheung M, Cooper A, Heled J, et al. (2010) Geneious v6.1.4 http://www.geneious.com.
38. Huelsenbeck JP, Ronquist F (2001) MR.BAYES: Bayesian inference of phylogeny. Bioinformatics 17: 754–755.
39. Ronquist F, Deans AR (2010) Bayesian phylogenetics and its influence on insect systematics. Annual Review of Entomology, 55: 189–206.
40. Hassanin A, Delsuc F, Ropiquet A, Hammer C, van Vuuren BJ, et al. (2012) Pattern and timing of diversification of Cetartiodactyla (Mammalia, Laurasiatheria), as revealed by a comprehensive analysis of mitochondrial genomes. Comptes Rendus Biologies, 335: 32–50.
41. Excoffier L, Laval G, and Schneider S (2005) Arlequin ver. 3.0: An integrated software package for population genetics data analysis. Evolutionary Bioinformatics Online 1:47–50.
42. Nei M (1987) Molecular Evolutionary Genetics. Columbia University Press, New York.
43. Darwin C (1876) The Effects of Cross and Self Fertilisation in the Vegetable Kingdom. John Murray, London.
44. Crnokrak P, Roff D (1999) Inbreeding depression in the wild. Heredity, 83 (Pt 3): 260–70.
45. Spottiswoode C, Møller AP (2004) Genetic similarity and hatching success in birds. Proceedings. Biological sciences/The Royal Society, 271: 267–72.
46. Frankham R (2010) Inbreeding in the wild really does matter. Heredity, 104: 124.

47. Ralls K, Ballou J (1983) Extinction: lessons from zoos. In:Schonewald-Cox CM, Chambers SM, MacBryde B, Thomas L(eds) Genetics and conservation: a reference for managing wild animal and plant populations. Benjamin Cummings, Menlo Park, CA
48. Alados CL, Escós J (1991) Phenotipic and genetic characters affecting lifetime reproductive success in female Cuvier's, Dama and Dorcas gazelles. Journal of Zoology, 223:307–321.
49. Frankham R (1995) Conservation genetics. Annual Review of Genetics, 29, 305–27.
50. Kephart SR (2004) Inbreeding and reintroduction: Progeny success in rare Silene populations of varied density. Conservation Genetics, 5: 49–61.
51. Vilas C, Miguel ESAN, Amaro R, Garcia C (2006) Relative Contribution of Inbreeding Depression and Eroded Adaptive Diversity to Extinction Risk in Small Populations of Shore Campion. Conservation Biology, 20: 229–238.
52. Ibáñez B, Moreno E, Barbosa A (2011) No inbreeding effects on body size in two captive endangered gazelles. Mammalian Biology - Zeitschrift für Säugetierkunde, 76: 748–754.
53. Ibáñez B, Moreno E, Barbosa A (2013) Parity, but not inbreeding, affects juvenile mortality in two captive endangered gazelles (I Gordon, H Fritz, Eds,). Animal Conservation, 16: 108–117.
54. Alados CL, Escós J (1991) Phenotipic and genetic characters affecting lifetime reproductive success in female Cuvier's, Dama and Dorcas gazelles. Journal of Zoology, 223:307–321.
55. Ruiz-López MJ, Gañan N, Godoy JA, Del Olmo A, Garde J, et al. (2012) Heterozygosity-fitness correlations and inbreeding depression in two critically endangered mammals. Conservation Biology: the journal of the Society for Conservation Biology, 26: 1121–9.
56. Swindell WR, Bouzat JL (2005) Modeling the adaptive potential of isolated populations: experimental simulations using Drosophila. Evolution - International Journal of Organic Evolution, 59: 2159–69.
57. Latta RG (2008) Conservation genetics as applied evolution: from genetic pattern to evolutionary process. Evolutionary Applications, 1: 84–94.
58. Frankham R, Ballou JD, Briscoe DA (2002) Introduction to conservation genetics. Cambridge University Press, Cambridge.
59. Sgrò CM, Lowe AJ, Hoffmann AA (2011) Building evolutionary resilience for conserving biodiversity under climate change. Evolutionary Applications, 4: 326–337.
60. Weeks AR, Sgro CM, Young AG, Frankham R, Mitchell NJ, et al. (2011) Assessing the benefits and risks of translocations in changing environments: a genetic perspective. Evolutionary Applications, 4: 709–725.
61. Araki H, Cooper B, Blouin MS (2007) Genetic effects of captive breeding cause a rapid, cumulative fitness decline in the wild. Science, 318: 100–3.
62. Frankham R (2008) Genetic adaptation to captivity in species conservation programs. Molecular Ecology, 17: 325–33.
63. Christie MR, Marine ML, French RA, Blouin MS (2012) Genetic adaptation to captivity can occur in a single generation. Proceedings of the National Academy of Sciences of the United States of America, 109: 238–42.
64. Lerp H, Wronski T, Pfenninger M, Plath M (2011) A phylogeographic framework for the conservation of Saharan and Arabian Dorcas gazelles (Artiodactyla: Bovidae). Organisms Diversity & Evolution, 11: 317–329.
65. Gaubert P, Bloch C, Benyacoub S, Abdelhamid A, Pagani P, et al. (2012) Reviving the African wolf Canis lupus lupaster in North and West Africa: a mitochondrial lineage ranging more than 6,000 km wide. PloS One, 7: e42740.
66. Brito JC, Godinho R, Martínez-Freiría F, Pleguezuelos JM, Rebelo H, et al. (2013) Unravelling biodiversity, evolution and threats to conservation in the Sahara-Sahel. Biological reviews of the Cambridge Philosophical Society, 1, doi: 10.1111/brv.12049
67. Endler JA (1977) Geographic variation, speciation, and clines. Princeton Univ. Press, Princeton, NJ.
68. Hedrick PW (2006) Genetic Polymorphism in Heterogeneous Environments: The Age of Genomics. Annual Review of Ecology, Evolution, and Systematics, 37: 67–93.
69. Hoekstra HE, Drumm KE, Nachman MW (2004) Ecolgiocal Genetics of Adaptive Color polymorphism in pocket mice. Evolution, 58: 1329–1341.
70. Mullen LM, Hoekstra HE (2008) Natural selection along an environmental gradient: a classic cline in mouse pigmentation. Evolution; international journal of organic evolution, 62: 1555–70.
71. Boratyński Z, Brito JC, Campos JC, Karala M, Mappes T (2014) Large spatial scale of the phenotype-enviromental color matching in two cryptic species of African desert Jerboas. PloS One, 9 (4): e94342.
72. Gratten J, Pilkington JG, Brown EA, Beraldi D, Pemberton JM, et al. (2010) The genetic basis of recessive self-colour pattern in a wild sheep population. Heredity, 104: 206–14.
73. Schmutz SM, Berryere TG (2007) Genes affecting coat colour and pattern in domestic dogs: a review. Animal Genetics, 38: 539–49.
74. Cieslak M, Reissmann M, Hofreiter M, Ludwig A (2011) Colours of domestication. Biological reviews of the Cambridge Philosophical Society, 86: 885–99.
75. Gasse F (2000) Hydrological changes in the African tropics since the Last Glacial Maximum. Quaternary Science Reviews, 19: 189–211.
76. Petit RJ, Excoffier L (2009) Gene flow and species delimitation. Trends in Ecology & evolution, 24: 386–93.
77. Newby JE (1984) Larger mammals of the Sahara. In Cloudsley-Thompson, J. L. (ed.). Sahara Desert. Key Environments Series, Pergamon Press, Oxford. 348 pp.
78. Stanley-Price M (1989) The Arabian oryx in Oman, Cambridge University Press, New York
79. Newby JE (1978) The Ecological Resources of the Ouadi Rimé Ouadi Achim Faunal Reserve, Chad. Direction des Parcs Nationaux, N'Djaména.
80. Effron M, Bogart MH, Kumamoto AT, Benirschke K, Box PO (1976) Chromosome studies in the mammalian subfamily antilopinae, Genetica, 46: 419–444.
81. Arroyo Nombela J, Rodriguez Murcia C, Abaigar T & Vericad J (1990) GTC-banded karyotype of Gazella dama mhorr. Z Säugetierkunde, 55: 194–201.
82. Vassart M, Greth A, Durand V, Cribiu E (1993) An unusual Gazella dama karyotype. Annales de Génétique, 36: 117–111.
83. Vassart M, Séguéla A, Hayes H (1994) Chromosomal evolution in gazelles. The Journal of Heredity, 86: 216–27.
84. Lawson DJ, Hellenthal G, Myers S, Falush D (2012) Inference of population structure using dense haplotype data. PLoS genetics, 8: e1002453.
85. Bradley RD, Baker RJ (2001) A test of the genetic species concept: cytochrome-b sequence and mammals. Journal of Mammalogy, 82(4): 960–973

Genetic Variability and Population Divergence in the Rare *Fritillaria tubiformis* subsp. *moggridgei* Rix (Liliaceae) as Revealed by RAPD Analysis

Marco Mucciarelli[1]*, Diana Ferrazzini[2], Piero Belletti[2]

1 Department of Life Sciences and Systems Biology, University of Torino, Torino, Italy, **2** Department of Agricultural, Forest and Food Science, University of Torino, Grugliasco (Torino), Italy

Abstract

Fritillaria tubiformis subsp. *moggridgei* Rix. is a rare alpine geophyte with shiny yellow flowers. This plant is sporadically distributed across the southwestern Alps where it is biogeographically close to *F. tubiformis* var. *burnatii* Planch. The latter has dark purple flowers and ranges in the majority of the Western and Central Alps. In order to develop appropriate strategies of conservation, a RAPD based analysis was conducted to study the genetic status of these taxa and the distribution of genetic variability of the subspecies by sampling seven populations distributed across the subspecies' range. Four populations of var. *burnatii* were chosen within this range and included in the genetic analysis. Some 264 individuals were analysed and 201 polymorphic loci were scored. Genetic diversity scored in the subspecies was in line with expectations for endemic species (H_e = 0.194). *F. tubiformis* var. *burnatii* showed lower intraspecific diversity (H_e = 0.173), notwithstanding a wider range than the subspecies. Most of the total phenotypic variation (about 83%) was allocated within populations, and significant lower proportions between taxa (6.45%) and between populations of the same taxon (10.64%). Moreover, PCoA analysis and Bayesian clustering separated populations into two genetically differentiated groups corresponding with the subspecific taxa. However, three populations ascribed to the subsp. *moggridgei* repeatedly showed genetic admixture with var. *burnatii* populations. Our findings suggest that: i) although the different flower colour, the two taxa are genetically very similar and share a consistent part of their gene pool, ii) the majority of genetic variability is allocated within populations rather than among them, iii) a representative amount of genetic diversity can be preserved by sampling from a restricted number of populations. The efficacy of RAPD markers in analysing genetic variation, and the contribution of the results to the preservation of biodiversity of the species, are discussed.

Editor: Massimo Labra, University of Milano Bicocca, Italy

Funding: The authors have no support or funding to report.

Competing Interests: The authors have declared that no competing interests exist.

* Email: marco.mucciarelli@unito.it

Introduction

Genetic erosion is one of the most important threats to the survival of many wild species. Nowadays these species are subjected to many stress factors, most of which are due to human activities: namely environment alteration, habitat fragmentation, urbanisation, pollution, irrational forest management and the introduction of allochthonous species. Current climate changes will probably determine a decline in growth and survival, and an alteration of genetic structure and diversity for many wild species. This pressure may be strengthened in marginal and peripheral populations of the species distribution, mainly in European southern populations of plants [1]. In order to survive these threats, and to persist over time, a high adaptive potential is fundamental: this is mainly determined by the within-species genetic diversity [2]. Genetic differentiation derives from several evolutionary forces, among which selection pressure as a response to different ecological conditions and genetic drift are the most important. Gene flow, in contrast to this pattern, leads to homogenisation of allelic frequencies [3].

The characterisation of patterns of genetic diversity within species, and among populations, is a fundamental necessity to establish any program aimed at preservation of biodiversity. Genetic markers are basic for supplying information on the genetic structure of populations, to analyse the pattern of within-species variability distribution [4] and to support the management of seed supply [5]. In particular, knowledge of genetic variation should be the basis for actions of ecological restoration [6]. This is even more important where the preservation of local, marginal and probably well adapted and/or differentiated gene pools must be achieved, through the management of small, marginal and inbreed populations [7,8]. As a consequence, the goal is to focus not only on scattered and rare species, but also on those characterised by wide distribution that could appear to be less sensitive to genetic erosion [9].

Fritillaria tubiformis subsp. *moggridgei* (Boiss. & Reut. ex Planchon) Rix (Liliaceae) is a rare taxon sporadically distributed along the Ligurian and Maritime Alps. The species is limited to a few main sites within the Parc National du Mercantour in France

(Muséum National d'Histoire Naturelle. National inventory of natural heritage. Website: http://inpn.mnhn.fr [accessed 17 04 2014]), and to seven populations of no more than 1000–10000 individuals found across Piedmont and Liguria, in Italy.

Conversely, *Fritillaria tubiformis* Gren. var. *burnatii* Planch. has a wider range, including the majority of the Western and Central Alps [10]. In the past, the two taxa have been confused and often fully synonymised by Italian botanists with *F. tubiformis* Gren. & Godr. [11], from which they differ by leaf morphology and the colour of tepals, which in *moggridgei* are yellow instead of dark purple. Results of a phylogenetic analysis carried on combined DNA sequences of the partial *matK* gene and the *rpl16* intron supported this view, and showed that subsp. *moggridgei* is genetically more closely related to var. *burnatii* than to any other species in the genus. However, in the same study, using plastid length-variable microsatellites a number of plastid haplotypes were found to be differently fixed in the two subspecies [12].

Owing to this taxonomic uncertainty, our genetic analysis has been focused on the seven Italian populations of subsp. *moggridgei*. In addition, we have used a data set based on DNA from four populations of *F. tubiformis* var. *burnatii*, sampled at the periphery of the subspecies distribution. A biosystematic study is also underway aimed at clarifying the exact phylogenetic relationships between the two taxa and their distribution range.

Random Amplified Polymorphic DNA (RAPD) technology is based on the amplification, using the Polymerase Chain Reaction (PCR), of short segments of DNA randomly chosen by the use of arbitrary, usually short (8–12 base pairs) primers. Despite having some drawbacks, the most important of which are the dominant expression and the low level of reproducibility [13,14], the RAPD technique analyses a larger number of loci, and thus provides a more general evaluation of the genome with a high potential for detecting polymorphism. Furthermore, the technique has a low cost, and can give a large amount of data in a short time; it requires small amounts of DNA and there is no requirement of prior knowledge of the genome being studied.

The aims of the present research were to estimate the amount and distribution of genetic diversity within and among *Fritillaria tubiformis* subsp. *moggridgei* populations from north-western Italy using RAPDs, and to evaluate possible effects of habitat fragmentation on genetic variation. The genetic relationship between the subspecies and *Fritillaria tubiformis* var. *burnatii* has also been investigated. Spatial structure and the expected association between genetic and geographical distances among populations were assessed, in order to provide recommendations for in situ preservation of the species.

Materials and Methods

Plant material

Seven populations of *F. tubiformis* subsp. *moggridgei* were considered in the study, distributed across the Maritime and Ligurian Alps, which represent the natural range of the subspecies in Italy. In general, populations were of small size and individuals therein limited to several hundreds. In addition, four populations of *F. t.* var. *burnatii* were also considered and were sampled within an area that partly overlaps the distribution range of the former subspecies. Figure 1 and Table 1 summarise names and locations of the populations analysed. Ente di gestione del Parco naturale del Marguareis (Chiusa di Pesio, province of CN) issued the permit for leaf sampling within PLU, MAR and SER populations (see Table 1 for their location). For the remaining locations (Table 1) specific permission was not required since these locations are

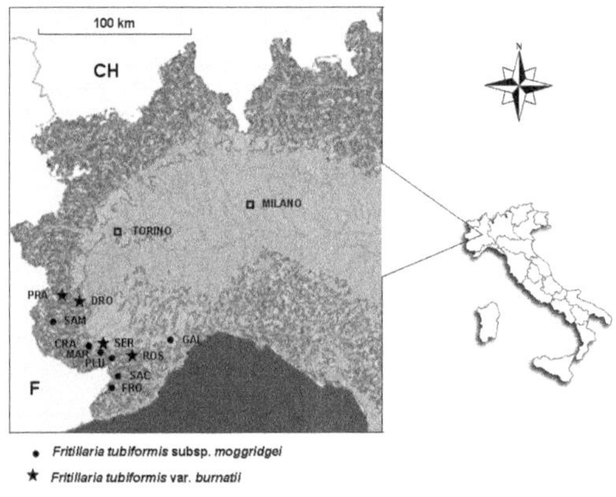

Figure 1. Geographical distribution of the Italian populations of *Fritillaria tubiformis* subsp. *moggridgei* (circles) and *Fritillaria tubiformis* var. *burnatii* (stars) analysed in this study.

outside protected areas and leaf collection does not represent a threat to the sampled individuals.

Leaves were sampled from about 30 non-adjacent blooming individuals in each population, in order to avoid selection of clonal individuals. In total, 264 individuals were analysed. After collection, the vegetative material was frozen at −20°C until DNA extraction.

Molecular analysis

Total DNA was extracted from frozen tissue using E.Z.N.A. SP Plant DNA Kit (Omega Bio-Tek, Norcross, GA), and purified on columns following the manufacturer's protocol. Amplifications were carried out in a Perkin Elmer DNA thermal cycler (PE 9600). The optimal reaction for RAPD analysis was set using the following conditions: 25 µl volume containing 1× reaction buffer, 1 unit of Taq DNA polymerase (Promega), 2.5 mM magnesium chloride, 0.2 µM decamer primers, dNTPs mix 0.2 mM and 20 ng DNA. The amplification conditions were as follows: the first step at 95°C for 5 min., followed by 40 cycles of 1 min. at 94°C, 1 min. at 36°C and 2 min. at 72°C. The reaction was completed with a final run at 72°C for 8 min. We screened a total of 16 primers from Life Technologies. The amplification products were separated in 1.5% agarose gel with ethidium bromide in 0.5× TBE buffer. The banding patterns were visualised under UV light and acquired using a fluorimeter linked with a gel documentation system (GelDoc 2000). The bands were scored visually, and those of similar molecular size, scored for the same primer, were assumed to be homologous. Since RAPD markers are dominant it was assumed that each band was an independent locus with two alleles: presence or absence of the band. The reproducibility of the amplification products was tested twice for each sample and each primer.

Data processing

Amplified fragments were scored as presence or absence of bands, and a binary matrix of RAPD phenotypes was assembled. Only polymorphic loci were used. The data analysis were further restricted to bands whose observed frequencies in each population were less than 1 – (3/N), where N is the number of plants analysed [14] to avoid significant bias in estimates of population genetic parameters. Non-random associations between pairs of loci were investigated using the Spearman rank correlation, with the SPSS 12.0 computer package. Intrapopulational genetic diversity was assessed as the proportion of polymorphic loci (P, using the 95%

Table 1. Details of the site characteristics of *Fritillaria tubiformis* populations from Italy which were sampled for the study.

Taxon*	Code	Locality	Location	Elevation (m a.s.l.)
M	SAM	Sambuco (Stura Valley)	44° 20′ N–7° 05′ E	1430
M	CRA	Cravina (Pesio Valley)	44° 14′ N–7° 39′ E	1420–1540
M	MAR	Marguareis (Pesio Valley)	44° 11′ N–7° 40′ E	1430
M	PLU	Pian del Lupo (Pesio Valley)	44° 11′ N–7° 41′ E	1990
M	SAC	Saccarello (Tanaro Valley)	44° 04′ N–7° 43′ E	1900
M	FRO	Frontè (Tanaro Valley)	44° 03′ N–7° 45′ E	1890–1980
M	GAL	Galero (Tanaro Valley)	44° 09′ N–8° 00′ E	1690
B	PRA	Prazzo (Maira Valley)	44° 31′ N–7° 02′ E	1960–2020
B	DRO	Droneretto (Maira Valley)	44° 31′ N–7° 12′ E	1700–1800
B	SER	Serpentera (Pesio Valley)	44° 13′ N–7° 41′ E	1890–2000
B	ROS	Pian Rosso (Tanaro Valley)	44° 09′ N–7° 46′ E	1800–1910

*M = Fritillaria tubiformis subsp. moggridgei, B = Fritillaria tubiformis var. burnatii.

criterion), mean number (N) and effective number (N_e) of alleles per locus, Nei's gene diversity [15] (H_e, that was adopted assuming the populations to be in Hardy-Weinberg equilibrium, although we were not able to investigate this since dominant markers were used), and Shannon's information index. The latter can be considered a measure of phenotypic diversity ($I = -\Sigma p_i \log_2 p_i$, where p_i is the frequency of presence or absence of a given RAPD fragment; [16]), assuming that populations are not in Hardy-Weinberg equilibrium. This index is frequently used in RAPD analysis because it is insensitive to bias that may be introduced into data owing to undetectable heterozygosity [17]. Calculations were performed using the GENALEX 6.5 software package [18].

The distribution of genetic diversity within and among populations was assessed using Nei's genetic differentiation degree (G_{ST}) [19]. Analysis of Molecular Variance (AMOVA), based on squared Euclidean distances between all pairs of RAPD phenotypes [20], was employed using the Arlequin software [21]. The AMOVA procedure was performed in order to further partition the total genetic variation among taxa, among population within subspecies and within populations, and to compute a pairwise population F_{ST} value matrix according to Weir and Cockerham [22]. The statistical significance of the covariance components was estimated by nonparametric randomisation tests using 10000 permutations. The null distribution of pairwise F_{ST} values under the hypothesis of no differences between the populations was also tested by using a permutational approach (10000 replicates).

Cluster analysis was performed on pairwise F_{ST} distances using the Unweighted Pair Group Method with Arithmetic Averages (UPGMA; [23]) with the SAHN program in NTSYS-pc 2.10j [24]. A cophenetic value matrix (COPH in NTSYS) was produced from the dendrogram and compared with the genetic distance matrix by using the MXCOMP program in NTSYS to estimate the goodness of fit of the cluster analysis. Principal coordinate analysis (PCoA) was also performed, based on pairwise F_{ST} distances matrix (DCENTER and EIGEN procedures in NTSYS), to better understand genetic relationships among populations.

The genetic structure of the populations was also explored using Bayesian clustering, performed using the software Structure [25]. The program uses a Markov chain Monte Carlo (MCMC) algorithm to cluster individuals into populations on the basis of multilocus genotypic data. Individual multilocus genotypes are first assigned probabilistically to genetic clusters (K) without considering sampling origins. Admixed or hybrid individuals can be identified as they will have a fraction of their alleles derived from each genetic cluster. Posterior probabilities of K were calculated from the means of 20 runs for each value of K (from 1 to 6), and the optimum K determined using the method of Evanno et al. [26].

Results

Variation of RAPD loci

Among the primers tested, OPA01, OPA02, OPA03, OPA4, OPA6, OPAF04, OPAF12 and OPAS18 gave non-reproducible amplification products and were therefore excluded from the analysis. The remaining eight selected primers showed reliable banding patterns and generated a total of 204 consistent and differential amplification products, ranging in size from 170 to 990 bp. Each primer amplified between 15 (OPA10) and 30 (OPA19) fragments, with a mean number of scored bands per primer of 25.5 (Table 2). Almost all these bands were polymorphic, being monomorphic only three, OPA10-4, OPA12-1 and OPA19-1. Many of the bands were fixed at the population level, and several even at the subspecific level (16 within subsp. *moggridgei* and 4 within var. *burnatii*). Five bands were unique to a single population: OPA05-19, OPA05-30, OPA05-21 and OPA09-27 in GAL, OPA05-22 in ROS. All bands fulfilled the Lynch and Milligan criterion [14], and were therefore included in further analyses. No band pair showed any correlation with any other, as indicated by Spearman's rho correlation coefficients (data not showed). Therefore, each of the 201 polymorphic bands was considered as an independent locus.

Genetic variation within populations

Genetic variation within populations was estimated for the 11 populations using various methods (Table 3). The mean number of alleles per locus ranged from 1.48 (PRA) to 1.72 (GAL), with a mean value of 1.59 (standard error, SE = 0.017). The mean effective number of alleles per locus was 1.29 (SE = 0.006), and the range was from 1.25 (ROS) to 1.32 (MAR). The percentage of polymorphic loci ranged from 72 (PRA) to 85 (GAL), with a mean value of 77.5 (SE = 1.19), and the expected heterozygosity was on average 0.187 (SE = 0.003), ranging from 0.165 (ROS) to 0.206 (CRA). Lastly, the Shannon diversity index, based on phenotype frequency, ranged from 0.268 (ROS) to 0.327 (GAL), with a general mean of 0.299 (SE = 0.005). In general, populations

Table 2. Characteristics of fragments generated by the 8 primers selected for the genetic analysis.

Primer	Nucleotide sequence (5′ → 3′)	Molecular weight range (bp)	Total number of bands	Number of polymorphic bands
OPA05	AGGGGTCTTG	190–930	26	26
OPA07	GAAACGGGTG	170–940	28	28
OPA09	GGGTAACGCC	180–700	27	27
OPA10	GTGATCGCAG	340–920	15	14
OPA12	TCGGCGATAG	280–990	26	25
OPA15	TTCCGAACCC	210–920	29	29
OPA18	AGGTGACCGT	200–850	23	23
OPA19	CAAACGTCGG	240–870	30	29
Total	-	-	**204**	**201**

belonging to *Fritillaria tubiformis* subsp. *moggridgei* showed an amount of internal genetic variability slightly higher than those of *Fritillaria tubiformis* var. *burnatii*.

Genetic variation among populations and groups

Partitioning of the total gene diversity (H_T = 0.293), following Nei's statistics [19], revealed that about 12.9% of the total variation was allocated among taxa and 4.5% among populations belonging to the same taxon. The remaining genetic variation was due to differences between individuals within populations. Within taxa, *F. tubiformis* subsp. *moggridgei* showed a genetic differentiation slightly higher than *F. tubiformis* var. *burnatii* (values of G_{ST} respectively of 0.135 and 0.117). The hierarchical analysis of molecular variance (AMOVA) confirmed these findings, demonstrating that about 83% of the variation was within populations and the remaining variance was attributable partly to differences between groups (6.45%) and partly among populations (10.64%), providing corroborating evidence for the genetic structure predicted by the Nei's indices analysis (Table 4).

Cluster analysis, based on pairwise F_{ST} distances between all pairwise combinations of individuals, was used to generate an UPGMA dendrogram in order to examine relationships between populations (Figure 2). Populations clustered thoroughly in two distinct main groups, showing low geographical correspondence. The first cluster contained all the four populations of var. *burnatii* but with intermixed populations of subsp. *moggridgei*. The DRO and PRA populations (both sampled in Maira Valley) clustered together as a sister group to ROS (Tanaro Valley) but also to SAC and SAM belonging to subsp. *moggridgei*. The second cluster grouped together the remaining populations belonging to subsp. *moggridgei* and precisely CRA and MAR (from Pesio Valley) as a sister group to the populations FRO and PLU. No precise assignation was possible for the remaining two populations, SER and GAL, belonging respectively to var. *burnatii* and subsp. *moggridgei*.

The PCoA analysis was performed to visualise the grouping of populations more clearly, providing a graphical representation of the relationships between them (Figure 3). Calculated pairwise Euclidean distances were used as an input for PCoA. The first two principal coordinate axes accounted for 49.76% and 21.35% of the total variation (cumulative value 71.11%). The first principal coordinate clearly distinguished subspecific taxa, but when referred to the second coordinate, populations were broadly scattered and clustered independently of each other (Figure 3).

Following the method of Evanno et al. [26], the Bayesian clustering results obtained with STRUCTURE indicate that K = 2 clusters represents the most informative representation of the overall genetic structure that we analysed (Figure 4). We found that most individuals from subsp. *moggridgei* (CRA, FRO, GAL, MAR, and PLU) clearly belong to cluster 1, whereas var. *burnatii* populations (DRO, PRA, ROS and SER) are primarily composed by individuals from cluster 2. Exceptions are represented by two admixed populations belonging to subsp. *moggridgei* (SAC and SAM), that already in the UPGMA analysis were shown to be more similar to var. *burnatii* than to other populations of the same taxon.

Discussion

This study investigated genetic variation of *Fritillaria tubiformis* subsp. *moggridgei* endemic of the South-Western Alps. Genetic structure in this subspecies was compared with four populations of *Fritillaria tubiformis* var. *burnatii* sampled within the range of the subspecies.

RAPD markers proved to be an efficient tool to analyse genetic variation both within and among populations. However, it must be underlined that RAPD variation cannot be linked with variation in adaptive genes [27], so that further research is needed to better understand the relationship between the observed variation and the adaptive potential of these taxa. The use of RAPD markers to estimate genetic parameters in outcrossing populations has been questioned: since RAPDs are dominant markers, null alleles present only in heterozygous genotypes cannot be detected, thus underestimating gene diversity [14].

In species with complex life histories, the choice of appropriate techniques of genetic analysis is not obvious. In a previous analysis run on chloroplast DNA with different type of sequence repeats, we showed that only insertion/deletion events were effective in discriminating genetically between yellow and dark purple flowered-morphotypes of *F. tubiformis* [12]. In the present study, RAPDs amplified mostly non-coding nuclear DNA sequences, which are subject to weaker selection pressures, thereby allowing the scoring of a higher amount of genetic variation.

F. tubiformis subsp. *moggridgei* displayed a level of genetic diversity in line with expectations. Expected heterozygosity (H_e = 0.194) (Table 3) was similar to the average value of H_{pop} reported by Nybom [28] for endemic species (0.20), and for species with seeds dispersed by gravity (H_{pop} = 0.19).

The average percentage of polymorphic loci (75–85%) (Table 3) and total gene diversity (0.293) were in the range previously reported with RAPDs for other endemic or endangered species [17,29,30,31]; in contrast, percent polymorphism was higher in

Table 3. Genetic variation within populations of *Fritillaria tubiformis* from Italy, based on 201 RAPD markers.

Population	Taxon*	S	N	N_e	P	H_e	I
SAM	M	24	1.63	1.26	79	0.173	0.284
CRA	M	24	1.64	1.33	80	0.206	0.327
MAR	M	24	1.55	1.32	75	0.203	0.318
PLU	M	24	1.62	1.31	79	0.196	0.312
SAC	M	24	1.58	1.28	77	0.183	0.295
FRO	M	24	1.66	1.31	81	0.198	0.318
GAL	M	24	1.72	1.30	85	0.199	0.322
PRA	B	24	1.48	1.28	72	0.179	0.284
DRO	B	24	1.55	1.28	75	0.178	0.285
SER	B	24	1.56	1.26	76	0.171	0.278
ROS	B	24	1.49	1.25	73	0.165	0.268
General mean		**264**	**1.59 (0.017)**	**1.29 (0.006)**	**77.5 (1.19)**	**0.187 (0.003)**	**0.299 (0.005)**
F. t. moggridgei	M	168	1.63 (0.016)	1.30 (0.006)	79.4 (1.22)	0.194 (0.004)	0.311 (0.006)
F. t. burnatii	B	96	1.52 (0.019)	1.27 (0.006)	74.0 (1.20)	0.173 (0.003)	0.279 (0.006)

*M = Fritillaria tubiformis subsp. moggridgei, B = Fritillaria tubiformis var. burnatii.
S = sample size, N = mean number of alleles per locus, N_e = effective number of alleles per locus, P = percentage of polymorphic loci (95% criterion), H_e = Nei's (1978) unbiased expected gene diversity, I = Shannon's index over loci. Standard errors are given in parenthesis.

Table 4. Analysis of molecular variance (AMOVA) within and among the populations studied.

Source of variation	df	Sum of squares	Variance components	Percentage of variation	P
Among groups	1	357.245	2.046	6.45	<0.001
Among population within groups	9	965.217	3.373	10.64	<0.001
Within populations	253	6.651.667	26.291	82.91	<0.001

subsp. *moggridgei* than in the Iranian endemic *F. imperialis* L. a much more widespread congeneric [32].

Despite the relatively restricted geographical range covered by this investigation, *F. tubiformis* populations exhibited genetic divergence. Using the AMOVA procedure, we showed that although most of the phenotypic variation was allocated within populations (82.91%), a significant proportion was attributable to genetic differences between intraspecific taxa (6.45%), as well as between populations (10.64%) (Table 4).

The overall F_{ST} value (0.171) was significantly different from zero ($P<0.001$) (Table 4), providing evidence for a strong genetic structuring among populations. This pattern of differentiation is expected for species pollinated by insects, and with scattered distribution. In this case genetic differentiation is generally higher than that occurring in wind-pollinated species with a more continuous distribution [33,34]. Significant levels of variation among populations were also observed by partitioning of variability based on Nei's gene diversity (G_{ST} = 17.4%), confirming the existence of a genetic structuring among populations.

The results also indicate that most of the total genetic variation was present within populations. Patterns of high genetic diversity at both the species' and local population levels have already been documented in other mountain plants [35] and critically endangered species [36,37], especially when relatively large populations have been considered [38]. Our findings, however, were in contrast with two studies on *F. imperialis* and *F. camschatcensis* using RAPD markers, where variation within populations accounted for only 54% and 35.7% of the total variation, respectively [32,39]. As stressed in the study with chloroplast DNA, when molecular variation is based on allele frequencies, a slight over-estimation of the within population component of variation is possible [12]. This is likely to have occurred in the present study. Moreover, other ecological and demographical characteristics are likely to have promoted the maintenance of high levels of genetic variation in subsp. *moggridgei* populations. This is supported by previous RAPD-based analyses which showed that long-lived, outcrossing or late successional species retain most of their genetic variability within populations [35,40]; and by contrast, annual, selfing and/or early successional plants allocate most of the genetic variability among populations [28]. The results stress that *F. tubiformis* subsp. *moggridgei* has not experienced population bottlenecks, and the general low levels of inbreeding likewise suggest a predominantly outcrossing mating system for this taxon [40], in accordance with field surveys and breeding features of related species [41].

Population size and geographical separation may be strong determinants of the spatial distribution of genetic variation in populations of neighboring species with fragmented distributions [42–43]. These conditions, in combination with life-history features such as growth form, mating system and seed dispersal mechanisms [44], may have conditioned the genetic history of *F. tubiformis* subsp. *moggridgei* populations. SAM, for example, has little chance of gene flow. This is in fact a very small population at a relatively low altitude, which has suffered the impact of agricultural activities and human settlements. GAL and SAC populations, on the contrary, are isolated populations, which nevertheless might receive frequent visits of pollinators thanks to their position on the very summit of two mountains.

Analyses of population genetic structure (UPGMA, PCoA and Structure) showed that the studied populations belonged to two distinct genetic pools corresponding to the subspecific taxa. The varied geological, topographical and climatic conditions of the Alpine chain can boost isolation and divergence of populations even over short geographical distances, thus explaining the fragmented distribution and genetic complexity of some endemic species of the Maritime Alps [45]. This genetic separation, however, was never rigorous, and three populations of subsp. *moggridgei* showed repeated admixture with var. *burnatii*, leaving open the problem of their exact taxonomic attribution. The profile

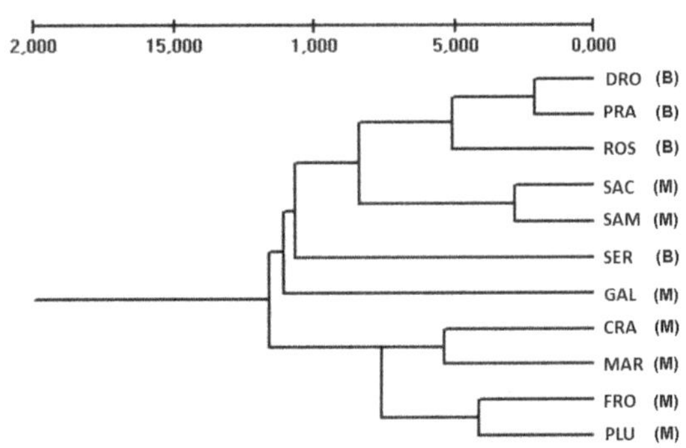

Figure 2. Dendrogram based on UPGMA of RAPDs profiles, showing relationships among the 11 populations of *Fritillaria tubiformis* considered in the study. M = *Fritillaria tubiformis* subsp. *moggridgei*, B = *Fritillaria tubiformis* var. *burnatii*.

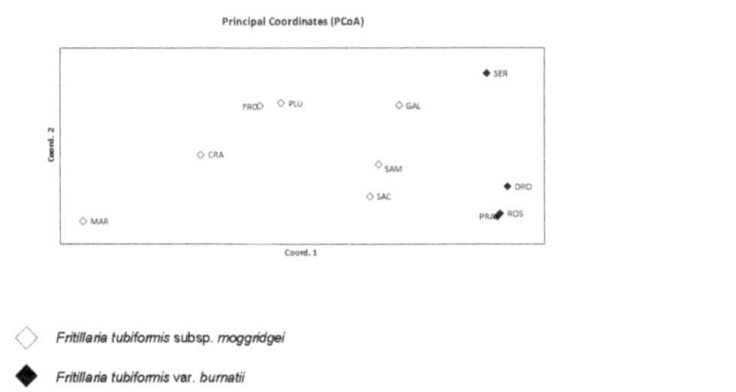

Figure 3. PCoA plot based on RAPDs variation, with the first two principal components showing relationships among the 11 populations of *Fritillaria tubiformis* studied.

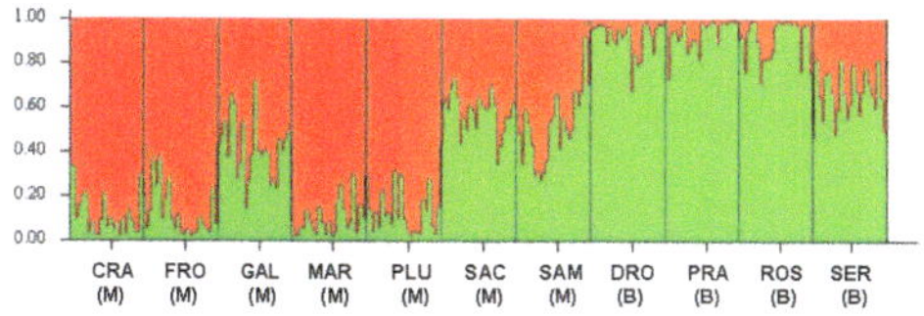

Figure 4. Probability of assignment of 264 plants of *Fritillaria tubiformis* from Italy to the two genetic clusters identified by hierarchical STRUCTURE analysis, and corresponding with subspecific taxa. Each vertical bar corresponds with a distinct genotype and different colours indicate the part of its genome assigned to each cluster. M = *Fritillaria tubiformis* subsp. *moggridgei*, B = *Fritillaria tubiformis* var. *burnatii.*

of the UPGMA tree supported a certain degree of genetic admixture across SAC and SAM populations, which positioned in the tree within populations of var. *burnatii* (Figure 2). When analyzed by Structure, genetic admixture was present to some extent also in GAL population (Figure 4). These findings are in accordance with a previous phylogenetic analysis, where we showed that subsp. *moggridgei* was more phylogenetically related to var. *burnatii* (*matK* and *rpl16* genes) than to any other *Fritillaria* species considered in the study [12].

If the ambiguous positioning of GAL population in these analysis can be explained by its geographical position and low demographic size, no explanations seems possible for the other two populations. Moreover, clustering of apparently distant populations as SAC (Val Tanaro, Ligury) and SAM (Val Stura, Piedmont) demonstrated that population positioning in the UPGMA tree was almost geographically unrelated. In support of this finding, it has been observed that when analysed by RAPD markers, population diversity is little affected by geographical range [28]. The UPGMA results were supported by PCoA that showed that populations were distributed along the first coordinate according to the intraspecific taxa (almost 50% variation), but showed admixture according to the second coordinate (SAC and SAM in subspecies *moggridgei* and SER in the variety *burnatii*) (Figure 3).

The position of the three populations of subsp. *moggridgei* is ambiguous and it remains unclear whether they are a hybrid of recent origin between subsp. *moggridgei* and var. *burnatii* [46] or whether they represent a separate evolutionary lineage. The low levels of genetic divergence among populations encountered in this study, seem to support the first hypothesis. However, broader taxon sampling in the territory of France might shed more light on this interesting problem.

F. tubiformis sensu lato is increasingly rare in the wild, due to habitat loss, and some populations are already under threat because of human activities. In conclusion, this study has shown that *F. tubiformis* subsp. *moggridgei* has a substantial degree of genetic diversity, which may counteract genetic erosion and maintain high adaptive potential in the subspecies. Indeed, this potential is mainly determined by the species genetic diversity [2]. The results show that despite being a narrow endemic with a relatively small distribution and fragmented, isolated populations, *F. tubiformis* subsp. *moggridgei* exhibits high level of genetic diversity and, in the perspective of a long-term conservation, a restricted number of its populations would be variable enough to represent the genetic variability of the subspecies.

A literature and herbarium-based survey of subsp. *moggridgei* and var. *burnatii* has shown that the taxonomic status of the Italian populations is uncertain [11]. A thorough phylogenetic analysis aiming at assessing their genetic relationships with respect to other widespread congeneric taxa of the Alpine chain (for example *F. tubiformis* Gren. & Godr.), will provide information essential for the formulation of appropriate management strategies for the conservation of all these populations.

Supporting Information

Dataset S1 *Fritillaria tubiformis* subsp. *moggridgei* RAPD profile and Analysis of Molecular Variance (AMOVA). The pairwise F_{ST} matrix of all population comparisons and step by step pairwise distances along with population labels are also given and can be used to manually calculate AMOVA results.

Acknowledgments

Thanks are due to colleagues from "Translations Group" for the language revision of the manuscript (www.translation-group.com/).

Author Contributions

Conceived and designed the experiments: MM PB. Performed the experiments: DF MM. Analyzed the data: PB DF. Contributed reagents/materials/analysis tools: PB MM. Contributed to the writing of the manuscript: PB MM.

References

1. Reich PB, Oleksyn J (2008) Climate warming will reduce growth and survival of Scots pine except in the far north. Ecol Lett 11: 588–597.
2. Kramer AT, Havens K (2009) Plant conservation genetics in a changing world. Trends Plant Sci 14: 599–607.
3. Jump AS, Peñuelas J (2006) Genetic effects of chronic habitat fragmentation in a wind-pollinated tree. Proc Natl Acad USA 103: 8096–8100.
4. Pautasso M (2009) Geographical genetics and the conservation of forest trees. Perspect Plant Ecol 11: 157–189.
5. Escudero A, Iriondo JM, Torres ME (2003) Spatial analysis of genetic diversity as a tool for plant conservation. Biol Conserv 113: 351–365.
6. Väli Ü, Einarsson A, Waits L, Ellegren H (2008) To what extent do microsatellite markers reflect genome-wide genetic diversity in natural populations? Mol Ecol 17: 3808–3817.
7. Rice KJ, Emery NC (2003) Managing microevolution: restoration in the face of global change. Front Ecol Environ 1: 469–478.
8. Frankham R (2005) Genetics and extinction. Biol Conserv 126: 131–140.
9. Eckert CG, Samis K, Lougheed SC (2008) Genetic variation across species' ranges: the central-marginal hypothesis and beyond. Mol Ecol 17: 1170–1188.
10. Conti F, Abbate G, Alessandrini A, Blasi C (2005) An annotated checklist of the Italian vascular flora. Roma: Palombi.
11. Bartolucci F, Peruzzi L (2012) Typification of *Fritillaria tubiformis* Gren. & Godr., *Fritillaria delphinensis* f. *moggridgei* Planch. and *Fritillaria delphinensis* var. *burnatii* Planch. (Liliaceae) from SW Europe. Candollea 67: 23–29.
12. Mucciarelli M, Fay M (2013) Plastid DNA fingerprinting of the rare *Fritillaria moggridgei* (Liliaceae) reveals population differentiation and genetic isolation within the *Fritillaria tubiformis* Complex. Phytotaxa 91: 1–23.
13. MacPherson J, Eckstein P, Scoles G, Gajadhar A (1993) Variability of the random amplified polymorphic DNA assay among thermal cyclers, and effects of primer and DNA concentration. Mol Cell Probes 7: 293–299.
14. Lynch M, Milligan BG (1994) Analysis of population genetic structure with RAPD markers. Mol Ecol 3: 91–99.
15. Nei M (1973) Analysis of gene diversity in subdivided populations. Proc Nat Acad Sci USA 70: 3321–3323.
16. Lewontin RC (1972) The apportionment of human diversity. Evol Biol 6: 381–398.
17. del Hoyo A, López-Pujol J, Chung MY, Lasso de la Vega B (2012) Population genetics and conservation of the extremely narrow Pyrenean palaeoendemic *Glandora oleifolia* (Boraginaceae). Plant Ecol Divers 4: 501–511.
18. Peakall R, Smouse PE (2006) GENALEX 6: genetic analysis in Excel. Population genetic software for teaching and research. Mol Ecol Notes 6: 288–295.

19. Nei M (1987) Molecular evolutionary genetics. New York, NY: Columbia Univ Press.
20. Excoffier L, Smouse PE, Quattro JM (1992) Analysis of molecular variance inferred from metric distances among DNA haplotypes: application to human mitochondrial DNA restriction data. Genetics 131: 479–491.
21. Schneider S, Roessli D, Excoffier L (2000) Arlequin ver. 2.000: a software for population genetics data analysis. University of Geneva, Switzerland: Genetics and Biometry Laboratory.
22. Weir BS, Cockerham CC (1984) Estimating F-statistics for the analysis of populations structure. Evolution 38: 1358–1370.
23. Sneath PHA, Sokal RR (1973) Numerical Taxonomy: The principles and practice of numerical classification. San Francisco, CA: Freeman, WH and Co.
24. Rohlf FJ (2001) NTSYS-pc Numeral Taxonomy and Multivariate Analysis System, version 2.10j. New York, NY: Exeter Publications.
25. Pritchard JK, Stephens M, Donnelly P (2000) Inference of population structure using multilocus genotype data. Genetics 155: 945–959.
26. Evanno G, Regnaut S, Goudet J (2005) Detecting the number of clusters of individuals using the software structure: a simulation study. Mol Ecol 14: 2611–2620.
27. Bekessy SA, Ennos RA, Burgman MA, Newton AC, Ades PK (2003) Neutral DNA markers fail to detect genetic divergence in an ecologically important trait. Biol Conserv 110: 267–275.
28. Nybom H (2004) Comparison of different nuclear DNA markers for estimating intraspecific genetic diversity in plants. Mol Ecol 13: 1143–1155.
29. Li XJ, Yang HL, Liu JQ (2008) Genetic variation within and between populations of the Qinghai-Tibetan Plateau endemic *Gentiana straminea* (Gentianaceae) revealed by RAPD markers. Belg J Bot 141: 95–102.
30. Rustaiee AR, Hassani ME, Tabatabaei SMF, Omidbaigi R (2010) Evaluation of genetic diversity among some populations of *Thymus daenensis* Celak., a vulnerable medicinal plant from Iran. Hort Environ Biotechnol 51: 335–342.
31. Sözen E, Özaydın B (2010) A study of genetic variation in endemic plant *Centaurea wiedemanniana* by using RAPD markers. Ekoloji 19: 1–8.
32. Badfar-Chaleshtori S, Shiran B, Kohgard M, Mommeni H, Hafizi A, et al. (2012) Assessment of genetic diversity and structure of Imperial Crown (*Fritillaria imperialis* L.) populations in the Zagros region of Iran using AFLP, ISSR and RAPD markers and implications for its conservation. Biochem Syst Ecol 42: 35–48.
33. Hamrick JL, Godt MJW, Sherman-Broyles SL (1992) Factors influencing levels of genetic diversity in woody plant species. New Forest 5: 95–124.
34. Nybom H, Bartish IV (2000) Effects of life history traits and sampling strategies on genetic diversity estimates obtained with RAPD markers in plants. Perspect Plant Ecol Evol Syst 3: 93–114.
35. Medrano M, Herrera CM (2008) Geographical structuring of genetic diversity across the whole distribution range of *Narcissus longispathus*, a habitat-specialist, mediterranean narrow endemic. Ann Bot 102: 183–194.
36. Wang BY, Shi L, Ruan ZY, Deng J (2011) Genetic diversity and differentiation in *Dalbergia sissoo* (Fabaceae) as revealed by RAPD. Genet Mol Res 10: 114–120.
37. Liu J, Shi S, Chang E, Yang W, Jiang Z (2013) Genetic diversity of the critically endangered *Thuja sutchuenensis* revealed by ISSR markers and the implications for conservation. Int J Mol Sci 14: 14860–14871.
38. Ellstrand NC, Elam DR (1996) Population genetic consequences of small population size: implications for plant conservation. Annu Rev Ecol Syst 24: 217–242.
39. Yamagishi M, Nishioka M, Kondo T (2010) Phenetic diversity in the *Fritillaria camschatcensis* population grown on the Sapporo campus of Hokkaido. Landscape Ecol Eng 6: 75–79.
40. Barrett SCH, Cole WW, Herrera CM (2004) Mating patterns and genetic diversity in the wild daffodil *Narcissus longispathus* (Amaryllidaceae). Heredity 92: 459–465.
41. Zynch M, Stpiczyńska M (2012) Neither protogynous nor obligatory out-crossed: pollination biology and breeding system of the European Red List *Fritillaria meleagris* L (Liliaceae). Plant Biol 14: 285–294.
42. Young A, Boyle T, Brown T (1996) The population genetic consequences of habitat fragmentation. Trends Ecol Evol 11: 413–418.
43. Leimu R, Mutikainen P, Koricheva J, Fischer M (2006) How general are positive relationships between plant population size, fitness and genetic variation? J Ecol 94: 942–952.
44. Hamrick JL, Godt MJW (1996) Effects of life history traits on genetic diversity in plant species. Philos Trans R Soc Lond B Biol Sci 351: 1291–129.
45. Casazza G, Grassi F, Zecca G, Mariotti MG, Guerrina M, et al. (2013) Phylogeography of *Primula allionii* (Primulaceae), a narrow endemic of the Maritime Alps. Bot J Linn Soc 173: 637–653.
46. Wolf PG, Soltis PS, Soltis DE (1993) Phylogenetic significance of chloroplast DNA restriction site variation in the *Ipomopsis aggregata* complex and related species (Polemoniaceae). Syst Bot 18: 652–655.

18

Individual Identification and Genetic Variation of Lions (*Panthera leo*) from Two Protected Areas in Nigeria

Talatu Tende[1,2]*, Bengt Hansson[2], Ulf Ottosson[1], Mikael Åkesson[3], Staffan Bensch[2]

1 A.P. Leventis Ornithological Research Institute, Department of Zoology, University of Jos, Plateau State, Nigeria, **2** Department of Biology, Lund University, Lund, Sweden, **3** Grimsö Wildlife Research Station, Department of Ecology, Swedish University of Agricultural Sciences, Riddarhyttan, Sweden

Abstract

This survey was conducted in two protected areas in Nigeria to genetically identify individual lions and to determine the genetic variation within and between the populations. We used faecal sample DNA, a non-invasive alternative to the risky and laborious task of taking samples directly from the animals, often preceded by catching and immobilization. Data collection in Yankari Game Reserve (YGR) spanned through a period of five years (2008 –2012), whereas data in Kainji Lake National Park (KLNP) was gathered for a period of three years (2009, 2010 and 2012). We identified a minimum of eight individuals (2 males, 3 females, 3 unknown) from YGR and a minimum of ten individuals (7 males, 3 females) from KLNP. The two populations were found to be genetically distinct as shown by the relatively high fixation index (F_{ST} = 0.17) with each population exhibiting signs of inbreeding (YGR F_{IS} = 0.49, KLNP F_{IS} = 0.38). The genetic differentiation between the Yankari and Kainji lions is assumed to result from large spatial geographic distance and physical barriers reducing gene flow between these two remaining wild lion populations in Nigeria. To mitigate the probable inbreeding depression in the lion populations within Nigeria it might be important to transfer lions between parks or reserves or to reintroduce lions from the zoos back to the wild.

Editor: Claire Wade, University of Sydney, Australia

Funding: Kolmården zoo conservation fund Sweden and the Leventis foundation provided the funds for the field and laboratory work. The study was supported by grants from the Swedish Research Council (621-2009-4945 to BH; 621-2010-5277 to SB). The funders had no role in study design, data collection and analysis, decision to publish, or preparation of the manuscript.

Competing Interests: The authors have declared that no competing interests exist.

* E-mail: talatu.tende@biol.lu.se

Introduction

Human pressure, agricultural developments and industrialization are leading increasingly to the destruction, fragmentation and isolation of natural populations [1]. A consequence of these changes is loss of genetic variability [2] and increasing risk of extinction [3–5]. Their requirements for large home ranges, low fecundity and low numbers have made mammalian carnivores vulnerable to local extinction in fragmented habitats [6]. Mammals are often the dominant carnivores where they occupy the top position in the food chain, thereby serving as ecologically fundamental species for the stability of the ecosystem. Their decline or extinction may disrupt the food chains and alter the structure of ecological communities [7]. Sadly over the years, the activities of man have put on verge the future prospects of existence of many wild mammal species, especially large carnivores [8].

Lions (*Panthera leo*) once roamed most parts of Africa, Southern Europe, the Middle East and Asia [9]. Today they are only found in sub-Saharan Africa and at one locality in India where they are being increasingly restricted to protected areas and often in declining numbers [10–11]. In West Africa, lions are found only in protected areas such as national parks, game reserves and zoos. In Nigeria, the only protected areas known to still have wild lions are Yankari Game Reserve and Kainji-Lake National Park. The number of lions in these two isolates in West Africa has been poorly investigated. Population size estimate is an important biological parameter necessary for proper implementation of conservation measures [12]. Thus it is important to get adequate information on population size and connectivity between fragments of populations for proper conservation and management of a species [13]. Lions just like other large terrestrial carnivores are usually very difficult to count due to their elusive behavior and ability to cover large home ranges. An effort to conduct complete counts of a lion population is thus likely to be both organizationally difficult and time consuming [14].The alternative is to interpolate population sizes using different sampling strategies.

The use of DNA from non-invasive samples such as faeces, saliva, hairs or feathers for individual genetic tags can provide useful information for population monitoring as well as contributing with important genetic parameters [15]. Non-invasive sampling is widely used in genetic studies of elusive animal populations [16–18].This method is of prime importance in conservation genetics and behavioral ecology because it allows for genetic studies of wild animals without having to catch or even directly observe the animals under study [19], thereby reducing the possible amount of stress and harm inflicted on the animal. Various studies have employed the use of non-invasive sampling to identify individuals in a population, estimate population size [20,12,21,16,22], to monitor population sizes over time [23–24] and to also estimate the home range of individuals [20]. By studying the appropriate nuclear markers (most often microsatellites) analysis of non-invasive genetic samples (e.g. faeces) collected opportunistically from the field can provide individual identification, adequate information on population size, sex identification as well as genetic polymorphism within and between populations [25]. Knowledge of past events in a population and genetic structure is very important to assess the risk of extinction and chances of regional persistence of lion populations in Nigeria and elsewhere. Conservation of the genetic diversity of a species is important for preserving endangered wildlife [26]. This is because genetic diversity is the raw material for evolutionary change; that will allow the population to evolve in response to catastrophic changes such as new disease outbreak, pests, competition and predators.

In a pilot study carried out on the lions in Yankari Game Reserve (YGR from here on), Central North-Eastern Nigeria, between 2007–2008, we employed a population survey using genetic analysis of faecal DNA and showed the feasibility and reliability of the method [27]. In the present study, we employed the same method of faecal sample DNA analyses increasing the number of loci from two to nine to improve the precision of individual assignment. We also extended the survey to Kainji-Lake National Park (KLNP from here on), western Nigeria, which is a protected area that contains the most closely located and presumably only other Nigerian lion population to Yankari. We aimed to identify individual lions in YGR and KLNP and to determine the degree of genetic variability within and between these populations. Because the lions in the two parks have been isolated for several years, we expect them to be genetically differentiated. Also, the small population sizes suggest that there is a substantial level of inbreeding, which should be reflected by a positive inbreeding coefficient.

Materials and Methods

Ethics

The study was carried out with permission from the National Park service in the case of KLNP. In YGR, the A.P. Leventis Ornithological Research Institute had a Memorandum of Understanding (MoU) with the Bauchi State government to conduct any type of ecological research. Faeces are not part of the animal and as such faecal samples are not banned by CITES from being transported between countries.

Study sites

Faecal samples were collected from two protected areas: Yankari Game Reserve (YGR) and Kainji-Lake National Park (KLNP) (Fig. 1).

The YGR is located in Central North-Eastern Nigeria with a landmass of 2,244 km^2 (9° 50′N and 10° 30′E). Detail site description for YGR is contained in Tende et al. [27].

KLNP is located in the western part of Nigeria (10° 22′N 04° 33′E) and occupies a landmass of 5,340 km^2. The vegetation is made up primarily of Guinea savanna woodland; common woodland species include *Terminalia macroptera* found along the Oli River which flows in the centre of the Park, *Detarium microcarpum* and *Borkea africana* woodland occupy about 70% of the Park area. *Isoberlinia tomentosa* woodland play vital role in providing shelter and cover for game. The mean annual rainfall is between 1000 and 1200 mm per year and occurs between April and October, with the highest peak of rain in September [28].

Sample collection and extraction

A total of 3724 hours were spent sampling in YGR during data collection between 2008 and 2012, and 294 hours were spent sampling in KLNP in 2009, 2010 and 2012. The Global Positioning System (GPS) was used to record the position of each sample collected. Field methods are described in detail in Tende *et al.* [27]. All samples were preserved in 95% ethanol at room temperature, thereafter taken to Lund University, Sweden and kept in a freezer at −40°C before DNA extraction. DNA was extracted from all samples collected from YGR (n = 836) and KLNP (n = 93). DNA extraction was carried out using the QIAamp® DNA Stool Mini Kit (Qiagen).

Contamination of DNA during extraction or the PCR process can be a major problem when using non-invasive DNA; this was carefully taken care of by conforming to guidelines to avoid this through the use of a blank to control for contamination during the extraction and PCR processes.

Mitochondrial DNA

A short (206 bp) portion of the mitochondrial cytochrome *b* gene was amplified and sequenced using PCR-based methods with locus specific primers to confirm samples that were from lions and to exclude a few cases of spotted hyena (*Crocuta crocuta*) or striped hyena (*Hyaena hyaena*). The primers LIHYF (5′- ATGACCAA-CATTCGAAAATCWC-3′) and LIHYR (5′-ATGTGGGT-SACTGATGAG-3′) were designed to avoid amplification of human and ungulate DNA in general, in order to promote detection of the target species [27]. All amplifications were done using 2X Qiagen multiplex PCR kit in 10 µl reaction volume containing 5 µl Qiagen multiplex PCR buffer mix; 0.2 µM forward primer (Applied Biosystems), 0.2 µM reverse primer, 2.6 µl water and 2 µl of DNA extract with a hot start at 95°C for 15 minutes. PCR profiles consisted of 35 cycles as follows: 90°C for 30 seconds; annealing temperature of 52°C for 30 seconds with elongation period of 72°C for 30 seconds. A blank control (reagents only) from DNA extraction process was included in all PCRs to monitor for contamination. The results of the PCR were evaluated by electrophoresis using 2% agarose gels and GelRed™ (Biotium) staining. Positive samples were sequenced using LIHY forward primer (BigDye sequencing kit; Applied Biosystems) in an ABI Prism® 3100 capillary sequencer (Applied Biosystems).

The sequences were aligned against species reference sequences (Ascension numbers; EF437586.1, AJ809332.1 and EF107524.1) obtained from Genbank to determine species identity.

Microsatellite amplification and genotyping

All lion samples were scored for allelic variability at nine polymorphic microsatellite primers (FCA001, FCA008, FCA026, FCA031, FCA045, FCA077, FCA126, FCA506 and FCA567 [29]. PCR amplifications were performed in 6 µl reactions containing 0.12 µl (concentration: 10 µM) dye-labelled (6-Fam, Hex or Ned) F-primer, 0.12 µl unlabelled R-primer (concentration: 10 µM), 3 µl of 2X Qiagen Master mix, 0.76 µl double distilled water and 2 µl DNA extract. PCRs were done in a GeneAmp 9700 thermocycler (Applied Biosystems) with the following profiles: 95°C for 15 min; 40 cycles at 94°C 30 s, 52°C 90 s, and 72°C 90 s; followed by an elongation period at 72°C for 10 min. Primers were multiplexed together in batches based on differences in fragment length and dye. The primer combinations are as follows: FCA001-FCA026-FCA031, FCA008-FCA045-FCA126, and FCA077-FCA506-FCA567. After amplification, alleles of the PCR products of the multiplex 3 loci, labelled with different dyes and of different lengths were separated using capillary electrophoresis in an ABI PRISM 3730 Genetic Analyzer (Applied Biosystems). Alleles were sized relative to GS500 ROX size standard and proof read and scored in Geneious vs. 5.6.6 (Biomatters).

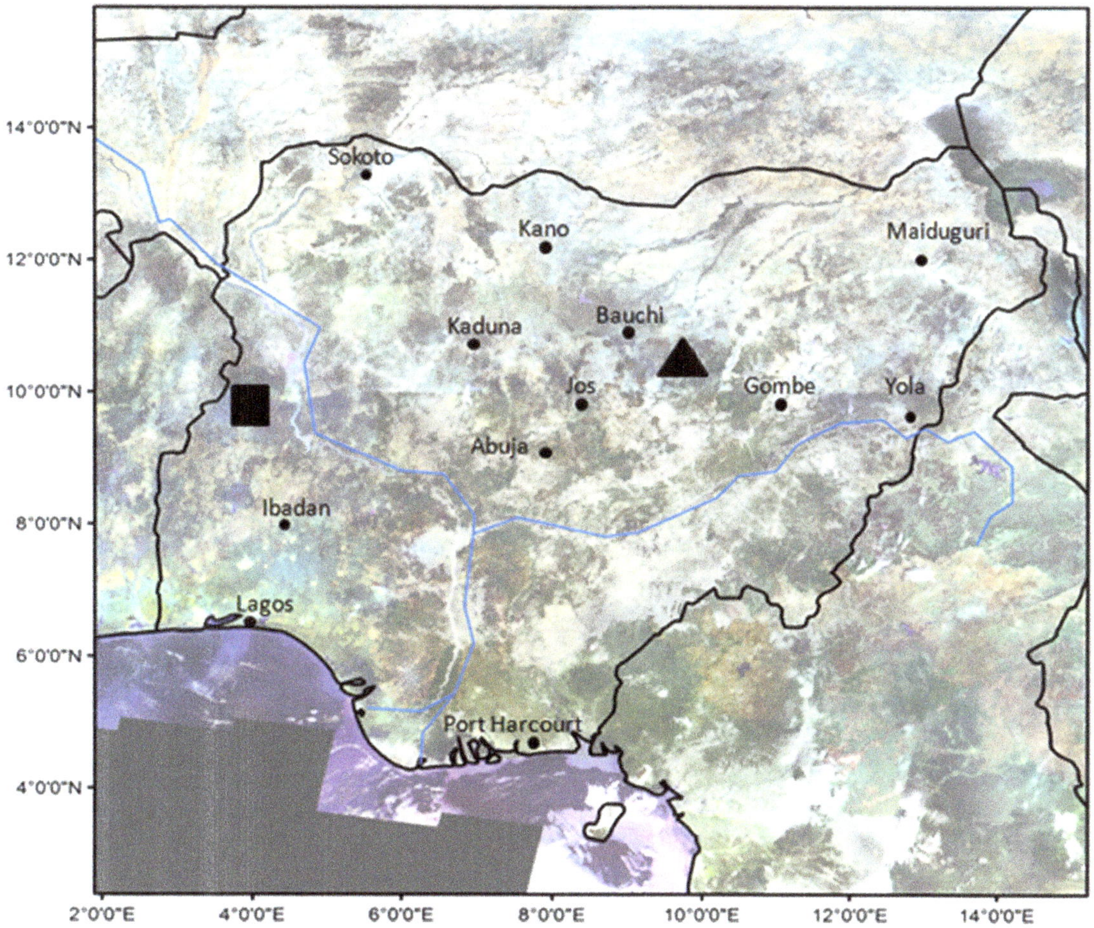

Figure 1. Map of Nigeria with some major cities and position of the two survey sites. Kainji-Lake National Park (KLNP) in black rectangle, Yankari Game Reserve (YGR) in black triangle.

Molecular sexing

Sex of identified individuals were determined using X and Y specific primers (SMCX17 and DBY7); [30]. The primers have been designed to avoid non-target amplification [31]. PCR amplifications were performed in 6 µl reactions containing 0.12 µl F-primer (concentration: 10 µM), 0.12 µl R-primer (concentration: 10 µM), 3 µl of 2X Qiagen Master mixes, 0.76 µl double distilled water and 2 µl DNA extract). PCR profile conditions for the multiplex amplification of the SMCX17 and DBY7 fragments are as follows: An initial denaturation of 95°C for 15 min, 20 cycles of 94°C for 30 sec, 60°C for 40 sec (decreasing 0.5 per cycle) and 72°C for 90 sec, followed by 20 cycles of 94°C for 30 sec, 50°C for 40 sec and 72°C for 90 sec and a final elongation period of 72°C for 15 min. After amplification, 2.5 µl of each PCR product was evaluated using 2% agarose gel with GelRed^{TM} (Biotium) staining and samples with one band were scored as females (XX) and samples with two bands as males (XY).

Allelic drop-out

Most non-invasive sampling studies are often confronted with low quantity and quality DNA (i.e. degraded DNA) making it ideal to use PCR primers that amplify short DNA fragments [32,33]. Highly degraded DNA may cause allelic drop-out which results in heterozygotes being typed as homozygotes due to failure of amplification of one of the alleles. To minimize or avoid genotype errors due to allelic drop out and false alleles three independent PCRs were performed for each locus and samples as suggested by Taberlet *et al.* [34]. No alleles were retained in further analysis unless they had been detected at least twice.

Partial genotypes are assigned in some individuals at some loci where only one allele could be observed more than once. Although there is a possibility that the partial genotypes might belong to a new individual, this method of assigning them with matching samples ensure conservative population estimation [35] by minimizing individuals created through erroneous, multi-locus genotypes (non-existent individuals). Only samples that amplified at between 4 and 9 loci were included in further analysis. According to Murphy *et al.* [36] a minimum of four loci are sufficient for accurate individual identification.

We scored samples as being from the same individual if they had been scored as the same sex, identical genotypes at ≥ four loci, and if the mismatching locus could be explained by allelic dropout. Such analyses allowed us to discern both the number of unique individuals as well as the number of "re-captured individuals". Faeces with the same multilocus genotypes are considered as recaptures.

Genetic Analysis

We used the identity analysis module in the program CERVUS [37] to identify individuals with unique genotypes within the data set. We also calculated number of alleles (K), allelic richness (A), observed (H_{OBS}) and expected (H_{EXP}) levels of heterozygosity, probability of observing identical genotypes by chance among unrelated samples ($P_{(ID)}$) and probability of observing identical genotypes by chance among siblings ($P_{(ID)sibs}$)from the microsatellite genotype data. The probability of identity, $P_{(ID)}$, describes the probability that two individuals which are drawn at random from a population will have the same genotypes at multiple loci [25]. The software program CREATE [38] was used to create input files for use in the software program FSTAT v2.93 [39]. Inbreeding coefficient (F_{IS}), population fixation index (F_{ST}) and Jost's estimate of genetic differentiation (D_{est}) were calculated using FSTAT and GenAlEx 6.5 [40]. Test for deviations from Hardy-Weinberg equilibrium exact test within populations was calculated based on 1000 randomisations, bootstrapping over loci at 95% CI. The nominal statistical significance value of 5/1000 was adjusted for multiple comparisons using the Bonferroni correction to minimize possible type I error. F_{ST} is used instead of R_{ST} [41] because it is considered to be a more reliable estimate of genetic differentiation when using small data set with less than 20 loci.

Waypoints of sampled genotypes were laid out on site maps to provide an overview of areas they were sampled during the survey in the two study sites (Fig. 2a and b). The software package Wild1 [42] in R was used to obtain Minimum Convex Polygon of all individuals and afterwards the extension Xtools (version 9.2) in ArcView 3.3 was used to calculate the area covered by each individual in order to have an overview of their movement patterns. For individuals that were encountered more than three times, their home ranges were estimated by using the minimum convex polygon method. Since at least a minimum of three points are needed to make a polygon, this was computed only for individuals that were encountered at least three times during the survey.

Results

Out of 929 samples collected from the two protected areas (YGR n = 836 and KLNP n = 93), 713 were successfully amplified for the partial cytochrome *b* (YGR, n = 625 and KLNP n = 88). The 216 samples that did not amplify at all were excluded from further analyses. Hence, the overall amplification success rate was about 77%. This is higher than the success rate of 40% (108 samples) that was obtained in the 2008 data from YGR [27]. There was no observed difference in the number of samples that successfully amplified in YGR and KLNP.

We determined species identity by aligning our sequences to Genbank reference sequences of the mitochondrial cytochrome *b* gene of lions, spotted hyena and striped hyena. Based on five informative nucleotide positions, 300 (42%) sequences out of the 713 samples matched with spotted hyena, while the remaining samples (n = 413) were found to match with lion.

To get individual identification, the 413 samples found to contain lion DNA from the two areas (YGR n = 352 and KLNP n = 61) were genotyped at the nine polymorphic microsatellites loci.

For the nine loci we obtained complete genotypes (at all 9 loci) for 115 samples from YGR and for 39 samples from KLNP. Partial genotypes (<9 loci) were obtained from additional 185 and 22 samples from Yankari and Kainji, respectively. Fifty two samples that amplified at less than four markers (i.e. 0–3 loci) were discarded from further analyses. Hence, the total number of samples with informative genotypes (≥4 loci) was 361.

In YGR the nine loci had a mean of 3.33 alleles with an expected heterozygosity (H_{EXP}) of 0.63 (assuming Hardy-Weinberg equilibrium) and an observed heterozygosity (H_{OBS}) of 0.32, whereas KLNP had a mean of 6.44 alleles with an expected heterozygosity (H_{EXP}) of 0.82 and observed heterozygosity (H_{OBS}) of 0.51 (Table 1). The lion populations in Kainji and Yankari showed significant signs of inbreeding with $F_{IS} = 0.49$ in YGR and $F_{IS} = 0.38$ in KLNP ($p<0.001$, Table 1). The two populations were found to be genetically differentiated with $F_{ST} = 0.17$ and $D_{est} = 0.65$ (bootstrapping over loci the ±95% CI of F_{ST} was between 0.10–0.23, $p = 0.004$).

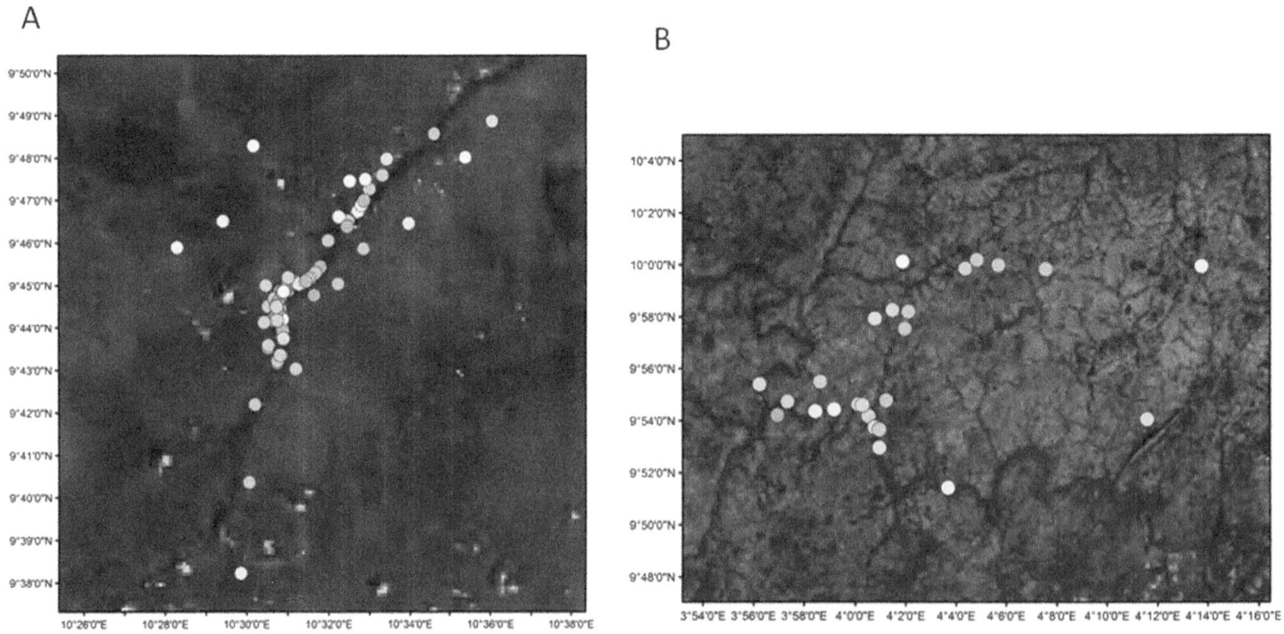

Figure 2. a: All genotypes sampled along the core area of Yankari Game Reserve. b: All genotypes sampled along the core area of Kainji-Lake National Park.

Table 1. Summary of genetic diversity; number of alleles (K), allelic richness (A), sample size (N), Observed and Expected heterozygosity (H_{OBS} & H_{EXP}) and inbreeding coefficient (F_{IS}) in the two populations over the years.

Locus	N	K	A	H_{OBS}	H_{EXP}	F_{IS}	N	K	A	H_{OBS}	H_{EXP}	F_{IS}
FCA001	8	4	3.61	0.50	0.60	0.29	5	4	4.00	0.25	0.82	0.72
FCA008	5	3	2.77	0.20	0.51	0.63	10	6	4.06	0.40	0.77	0.50
FCA026	8	3	2.88	0.37	0.62	0.41	10	9	5.66	0.70	0.90	0.23
FCA031	4	2	2.00	0.50	0.65	0.14	10	6	4.71	0.20	0.83	0.76
FCA045	8	2	2.00	0.00	0.53	1.00	8	6	4.49	0.57	0.79	0.29
FCA077	8	6	4.37	0.70	0.80	0.07	8	6	4.61	0.83	0.87	−0.05
FCA126	8	2	2.00	0.00	0.53	1.00	8	5	3.80	0.37	0.71	0.49
FCA506	7	3	2.44	0.14	0.69	0.86	8	7	5.05	0.50	0.85	0.43
FCA567	8	5	3.93	0.50	0.72	0.32	10	9	5.43	0.80	0.87	0.08
Mean	-	3.33	2.88	0.32	0.63	0.49	-	6.44	4.64	0.51	0.82	0.38

Yankari Game Reserve (N = 8) Kainji-Lake National Park (N = 10)

A total of eight individuals (2 males, 3 females, 3 unknown) were identified in YGR if we assume that allelic drop-outs had affected the scored genotypes (Table 2), while ten individuals were identified in KLNP (7males, 3 females, Table 3). The 3 individuals in YGR whose sex could not be determined molecularly were due to shortage of DNA template. Based on the observed allele frequencies there was a low probability of observing identical genotypes $P_{(ID)}$ from two randomly sampled individuals from the same population in both YGR ($P_{(ID)}$ = 0.00000213, $P_{(ID)sibs}$ = 0.00259) and KLNP($P_{(ID)}$ = 0.00000000150, $P_{(ID)sibs}$ = 0.000188). The mean polymorphic information content (PIC) values in both YGR (0.50) and KLNP (0.73) were high.

Within both survey areas, some of the identified individual genotypes were encountered in more than one year (YGR; n = 6, KLNP; n = 3), while some were encountered several times within a year and others were encountered only once or twice. In YGR individual Y#2 (female) and Y#5 (male) were observed to be present in the population all through the five year period. Individual Y#6 (female) was sampled from the beginning of the survey until 2011 (Fig. 3a). The combined genotypes of Y#2 and Y#5 are compatible with the hypothesis that the genotypes of Y#6 and Y#8 are their offspring (Table 2).

In both YGR and KLNP all individuals were sampled within the core area of the reserve (Fig. 2a and b). The home range analysis using the minimum convex polygon method gave an average home range of 11.91 km^2, ±2.1 SD (1.71– 47.62 km^2) in YGR (Appendix S1) and 26.75 km^2 ± 1 SD (14.63– 39.37 km^2) in KLNP (Appendix S2).

Discussion

Number of individuals

Based on the encounter of unique genotypes during the different survey years, a minimum of eight individuals was identified in YGR between 2008 and 2012, while ten individuals were identified in KLNP between 2009, 2010 and 2012. During the course of the survey there has been at least one observation of new cubs in the field in 2010 in YGR. Recently, there have also been sightings (between 8th and 12th March 2013) of 2 adult females with 3 cubs, 2 adult females without any cubs, 1 adult male with a mane together with 3 cubs in YGR. The sightings of these adults within YGR (1 male and 4 females, or 1 male and 2 females, (if we assume that it was the same females that were seen at different times) contradict the findings of Wildlife Conservation Society (WCS) who has reported, through call-up stations, that only two adult lions exist within the reserve [43]. These sightings are also confirmed by our molecular findings (Table 2), where females Y#2 and Y#6 and male Y#5 have been present most parts of the

Table 2. Identified individuals in Yankari Game Reserve.

SampleID	IndvID	Sex	#times sampled	FCA001	FCA026	FCA031	FCA567	FCA007	FCA506	FCA008	FCA045	FCA126
YGR321	Y#1	-	4	129/155	128/128	***/***	85/85	135/144	244/244	***/***	125/125	124/124
YGR242	Y#2	F	107	127/127	128/137	242/244	94/96	146/153	214/214	117/117	127/127	124/124
YGR211	Y#3	-	5	129/155	128/128	***/***	85/85	130/135	244/244	***/***	125/125	124/124
YGRB1	Y#4	M	6	127/129	128/130	***/***	85/85	135/141	244/244	124/124	125/125	124/124
YGRN1	Y#5	M	21	127/127	128/128	242/242	94/105	141/153	191/191	117/117	127/127	127/127
YGR13	Y#6	F	14	127/127	128/137	244/244	96/105	141/153	191/214	117/117	127/127	127/127
YGR56	Y#7	F	1	137/155	130/130	***/***	85/85	135/135	***/***	***/***	125/125	124/124
YGR7	Y#8	-	1	127/127	137/137	242/244	94/107	153/153	214/214	117/117	127/127	***/***

Shown are; Sample identity (SampleID), individual identity (IndvID), sex and allelic length at the nine scored loci, *** (indicates missing data).
Microsatellite Loci

Table 3. Identified individuals in Kainji-Lake National Park.

Microsatellite loci

SampleID	IndvID	Sex	#times sampled	FCA001	FCA026	FCA031	FCA567	FCA077	FCA506	FCA008	FCA045	FCA126
KLNP13	K#1	M	4	127/129	130/137	234/246	98/103	148/153	193/***	129/133	127/127	***/***
KLNP18	K#2	M	1	***/***	122/130	224/224	78/78	***/***	229/229	129/129	***/***	129/191
KLNP19	K#3	M	1	***/***	139/141	238/238	81/85	137/141	203/214	125/125	139/153	133/151
KLNP4	K#4	M	3	***/***	134/139	238/252	83/85	137/139	195/203	119/133	139/139	139/139
KL26	K#5	M	1	145/145	141/141	238/238	83/85	137/153	195/195	125/125	139/139	133/133
KL18	K#6	F	7	***/***	137/137	246/246	103/105	148/151	191/191	133/133	145/149	131/139
KL11	K#7	F	7	153/153	134/141	238/238	83/85	137/139	195/195	125/125	139/139	133/133
KL9	K#8	M	1	129/129	128/130	234/234	98/111	148/153	85/191	133/133	***/***	139/139
KL33	K#9	F	1	153/153	134/139	***/***	85/85	153/153	193/208	117/133	***/***	133/133
K22	K#10	M	1	145/153	144/148	244/244	83/94	137/***	197/203	127/129	149/153	133/***

Shown are: Sample identity (SampleID, individual identity (IndvID), sex and allelic length at the nine scored loci, *** (indicates missing data).

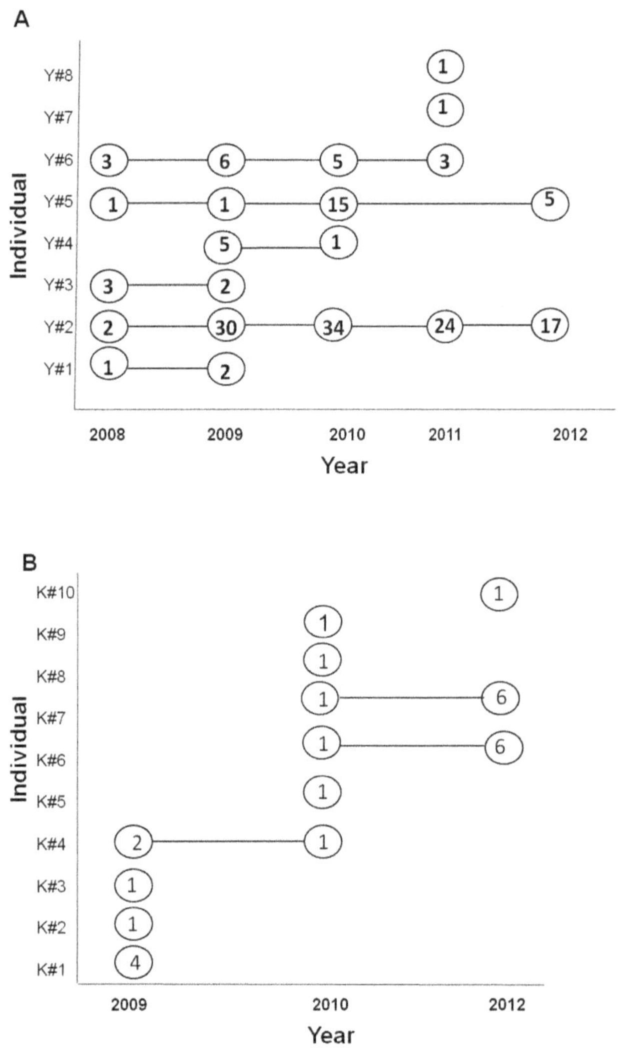

Figure 3. a: Sampling frequency per individual and year in Yankari Game Reserve. b: Sampling frequency per individual and year in Kainji-Lake National Park.

survey periods. Six of the eight identified individuals have been sampled multiple times (3-107 times) during the survey (Fig 3a). Our estimate gives the minimum number of lions that exist in both YGR and KLNP. Since several individuals have been encountered only once, especially in KLNP, the true number is potentially higher. The low estimates of probability of identity ($P_{(ID)}$ and $P_{(ID)sibs}$) obtained in our study both in YGR and KLNP gives confidence to the identity of individuals identified in these two survey areas and support the uniqueness of the identified individuals. Waits *et al.* [44] suggested that these values should be between $P_{(ID)} = 0.01 - 0.0001$ in studies estimating population size.

Despite the fact that there were fewer visits and also few samples collected from KLNP, we observed more individuals as well as a higher genetic diversity than in YGR. This suggests that there are and have been more individuals in the recent history in KLNP, which have contributed to a larger effective population size and a higher genetic diversity. Alternatively, the high diversity in KLNP may have resulted from connecting with gene flow with lions in the W-Arli-Pendjari complex, a vast protected area in Benin, Burkina Faso and Niger.

Comparison with previous study

A number of studies conducted on the natural populations of African lions have been carried out mostly in Eastern, Central and Southern Africa and in India. These studies have employed opportunistic field observation and/or genetic data to understand behavioral, social and breeding structure of the lions [45–53]. Researchers have incorporated molecular methods to understand the genetic status of the lions in their natural populations (c.f. [50,54]), characterize their evolutionary history, as well as to study disease outbreaks [55–58]. Our study has employed non-invasive samples with aid of molecular techniques to make available information on the population size as well as the genetic status of the remaining relict wild lion populations in Nigeria. The method does not have any negative impact on the study species.

The number of individuals estimated in YGR in the present study is lower than the number recorded during the pilot survey conducted by Tende *et al.* [27] where eleven individuals were identified from two microsatellite loci. This difference could be due to either a disappearance from the population of some individuals due to natural deaths or to the activities of poachers within the reserve, or it could be an overestimation in our previous study due to the low number of loci used. The use of polyacrylamide gel to genotype individuals during our pilot study [27] could possibly also have biased our estimation due to bad gels where stutter bands might have been typed as true bands.

During the course of the laboratory analysis and efforts to optimize different primers for the study, we ran short of DNA template from most of the extracted samples from the first pilot study in YGR. Hence, not all samples analyzed in our 2008/2009 pilot survey could be rerun on the new microsatellite primers used. However, in some samples where we still had DNA template it was observed that certain individuals have persisted in the population throughout the survey years.

The higher amplification success rate attained in this study as compared to 2007/2008 could be attributed to the new PCR kit/ technique (Qiagen multiplex PCR kit) that was employed and also we concentrated our sample efforts on fairly fresh faeces. The shorter time the faeces are left out in the environment, the less degradation of DNA and this is especially true in tropical environments where high temperature and UV radiation, can cause fast degradation of DNA [59].

Home range estimates

Many of the identified individual genotypes were encountered several times during the study. Lions are known to use large home ranges to satisfy energetic demands, but this can be limited where required resources have clumped distribution [50,60]. This is the case during the dry season, for both study sites, when game concentrates close to rivers - Gaji River in YGR and Oli River in KLNP. Both rivers run through the core areas of the respective reserve/park and offer lush vegetation for prey species, as well as cover for the lions to rest and also stalk. It is also the main source of drinking water both for the predators and the prey. A similar survey by Spong *et al.* [50] to determine space use by lions in Selous Game Reserve, Tanzania, showed the most intensively used area to be small within the reserve and averaged only 11.7 km^2, which is consistent with our finding in YGR where individuals were sampled mostly within the core areas which are intensively used and this averaged at 11.9 km^2 (SD: $\pm$2.1). They found out that prides often had close relatives in neighbouring prides but in areas with high prey abundance the home ranges between individuals, irrespective of relatedness tended to overlap more [50]. A survey by Lehmann *et al.* [60] on home range utilization of lions in Karongwe Game Reserve, South Africa, showed the home range used by a pride ranged between (10.3 – 64.4 km^2) and for a single male ranged between (5.0 – 56.3 km^2). Their findings about home range size are slightly larger than what we have obtained in our study (YGR: 1.71– 47.62 km^2; KLNP: 14.63–39.37 km^2). In our study a female (Y#2) was observed to have the largest home range (47.52 km^2), this might probably be due to the need to hunt and feed the cubs and the male. This female has also been observed to persist in the population since the onset of this survey in 2007/2008. Both of the above studies [50,60] employed the use of telemetry. Our study has shown that genotype data also can be used to determine the home range of lions, and possibly other mammals, enhancing sample size and reducing disturbance to the animal under study.

Inbreeding

The populations of lions in both KLNP and YGR exhibit significant signs of inbreeding. This is not surprising given their small population sizes. The inbreeding level found in YGR in the present study is in line with our pilot study conducted in 2008 [27] when the inbreeding coefficient was estimated to be 0.21, whereas in this present survey the value was found to be 0.49 (Table 1). The inbreeding levels in both YGR (0.49) and KLNP (0.38) are high and comparable to what has been recorded in some other carnivore species. For instance the estimated inbreeding coefficient in the Scandinavian wolf (*Canis lupus*) population was up to 0.41 [61,62,17] before the population was genetically rescued by one immigrant from Finland [63]. The arrival of this immigrant into the Scandinavian wolf population provided the possibility to avoid inbreeding, decrease the risk of inbreeding depression and cause population growth. High inbreeding coefficient reaching up to 0.37 has also been recorded in the brown bear (*Ursus arctos*) [64]. Inbreeding and subsequent negative effects of inbreeding have been reported in the lions in Ngorongoro Crater in Tanzania [65,48]. The number of alleles at a set of microsatellite loci for the Etosha lion population (A = 4.6) reported by Lyke et al. [53], and that reported by Antunes et al. [58] (A = 4.4) is similar to that detected in KLNP (A = 4.6), but higher than in YGR (A = 2.8), where Lyke et al. [53] detected no sign of inbreeding in the Etosha population (F_{IS} = 0.03). The inbreeding coefficient recorded in our study is higher than what has been recorded in the lion population in Etosha National Park in Namibia. Some studies (e.g., [4,66–68]) have shown that if populations remain small and isolated for many generations they are bound to face increased inbreeding and gradual erosion of genetic variability. The lion populations in Nigeria are small, isolated and restricted to two protected areas (YGR and KLNP) where their populations may be declining. This is of course a threat to the long-term survival of these populations and to the lion population throughout West Africa [69].

Low genetic variability has been reported in lions in Ngorongoro Crater in Tanzania [48,55,70,71]. Although there are recent observations of cubs within one of our study systems (YGR) the genetic status of this population as shown by our study is poor. This might affect individual survival and reproductive success in the long term if proper measures are not implemented. O'Brien [72] has observed that there is a strong correlation between genetic variation and reproductive parameters in lions. This means that with time the fitness of individuals within the lion populations in Kainji and Yankari will decline due to accumulation and expression of recessive and detrimental alleles with subsequent inbreeding depression, which might consequently drive the population to extinction.

Population structure

The YGR and KLNP populations were found to be genetically differentiated (F_{ST} = 0.17, D_{est} = 0.65). This is not surprising due to the fact that these populations are small and isolated from each other; about 1000 km apart and they are separated by dispersal barriers including highways, agricultural landscapes and cities. Without any corridor for dispersal, isolation is expected to build up the observed pattern of allelic differentiation between the two populations. Moreover, high human and livestock densities characterize most of the surroundings of these protected areas, which can increase mortality risk of cubs and "possible dispersers" because of overlap with human habitations and livestock. All these factors may act as barriers to gene flow between the two populations and possible populations in neighboring countries. Studies have shown that geographical and environmental features can influence gene flow and genetic variation within and between populations [73,74]. Studies carried out in California on mountain lion (*Puma concolor*), and coyote (*Canis latrans*) and bobcat (*Lynx rufus*) to assess the level of gene flow and differentiation in these carnivores showed how anthropogenic obstacles such as roads, agriculture and urbanization were great barriers to dispersal and gene flow [75,76]. It was shown how these barriers imposed artificial home range boundaries on territorial carnivores thereby decreasing genetically effective migration, which leads to population differentiation.

A study carried out by Antunes *et al.* [58] to assess genetic variation from 357 lions from most of its current range in Africa and Asia using microsatellite data revealed significant population structure (F_{ST} = 0.03–0.79) within East-African lion populations despite the low geographic distance within them. Their finding is in accordance with our finding about the two lion populations in Nigeria (F_{st} = 0.17).

The existence and maintenance of genetic diversity and good connectivity between subpopulations are essential factors for long-term viability of a population [77], which should serve as the primary target of any acceptable conservation management program. Genetic diversity is the raw material needed in order to evolve the ability to cope with environmental challenges such as disease outbreak and parasites [78].

Conservation genetics and management

The two lion populations are small and genetically different with signs of inbreeding within each population. This might elevate their risk of extinction in the face of sudden environmental catastrophes in the future. It has been shown that small populations are often faced with a higher risk of extinction from environmental catastrophes [3,4,5,79–83] than large interconnected populations [84,77,85].

The main aim of a conservation genetic approach is to maintain diversity within and between populations so as to enhance the evolutionary potential of the population to cope with environmental changes. Laboratory and translocation experiments have indicated that small and inbred populations can be rescued by contribution of minimal number of immigrants [86–88,62,63]. This can help to decrease inbreeding and inbreeding depression [86,89], and bring about profound changes in genetic structures [90,91,63]. Thus to mitigate the inbreeding condition and subsequent probable inbreeding depression in the lion population within Nigeria it might be important to transfer lions between parks or reserves or to reintroduce lions from the zoo back to the wild. This is necessary in order to enhance their genetic diversity for long term survival and reproduction. The bringing together of genetically dissimilar mates, hybrid vigor [89–91] can be advantageous because it will enhance reproductive success and also fitness [92–93,68]. However, it would be recommended that lions that are genetically similar to the receiving population should be preferably used in any translocation program in order to avoid introduction of potentially non-locally adapted genotypes.

Acknowledgments

We will like to thank the following people for their enormous contributions to the success of this survey; the director and staff of A.P. Leventis Ornithological Research Institute (APLORI) Jos for all the logistic supports. A big thank you to all the staff of APLORI research centre Yankari Game Reserve and Taiwo Crossby Omotoriogun who was our faithful, hardworking and dedicated field assistant in 2009. Thanks to all friends and colleagues who helped to collect samples in Yankari: Onoja Joseph, Adetutu Ojomo, Himma Bakam, Shomboro Karau Dauda and Haladu Idi. We also thank Dr. Yahkat Barshep, Chima Nwogu, Martin Stervander and Onoja Joseph for helping with maps and analysis of the home range. Finally, we thank the management of Yankari Game Reserve and Kainji Lake National Park for allowing field work. This is contribution no. 68 from A.P. Leventis Ornithological Research Institute.

Author Contributions

Conceived and designed the experiments: TT UO. Performed the experiments: TT UO. Analyzed the data: TT MA BH. Contributed reagents/materials/analysis tools: BH SB. Wrote the paper: TT BH MA UO SB.

References

1. Ceballos G, Ehrlich PR (2002) Mammalian population losses and the extinction crisis. Science 296: (5569)904–907.
2. Frankham R (1995) Inbreeding and Extinction- a Threshold Effect. Conserv Biol 9 (4): 792–799.
3. Saccheri I, Kuussaari M, Kankare M, Vikman P, Fortelius W, et al. (1998) Inbreeding and extinction in a butterfly metapopulation. Nature 392:491–494.
4. Bijlsma R, Bundgaard J, Boerema AC (2000) Does inbreeding affect the extinction risk of small populations? predictions from Drosophila. J Evol Biol 13: 502–514.
5. Higgins K, Lynch M (2001) Metapopulation extinction caused by mutation accumulation. Proc Natl Acad Sci USA 98: 2928–2933.
6. Crooks KR (2002) Relative sensitivities of mammalian carnivores to habitat fragmentation. Conserv Biol 16: 488–502.
7. Crooks KR, Soule ME (1999) Mesopredator release and avifaunal extinctions in a fragmented system. Nature 400: 563–566
8. Morris DW, Kotler BP, Brown JS, Sundararaj V, Ale SB (2009) Behavioral indicators for conserving mammal diversity. Ann New York Acad Sci 1162: 334–356
9. Bauer H, Vander Merwe S (2004) Inventory of free ranging lions *Panthera leo* in Africa. Oryx 38: 26–31.
10. Nowell K, Jackson P, Compilers and editors (1996) Wild cats: Status survey and Conservation Action Plan. IUCN/SSC Cats Specialist Group. IUCN, Gland, Switzerland.
11. Wilson DE, Mittermeier RA, editors (2009) Handbook of the Mammals of the World.Vol.1. Carnivores. Lynx Ediciones. Barcelona.
12. Kohn MH, York EC, Kamradt DA, Haught G,Sauvajot RM, et al. (1999) Estimating population size by genotyping faeces. Proc R Soc Lond B 266:1–7.
13. Creel S, Spong G, Sands JL, Rotella J, Zeigle J, et al. (2003) Population size estimation in Yellowstone wolves with error-prone noninvasive microsatellite genotypes. Mol Ecol 12:2003–2009.
14. Balme GA, Hunter LTB, Slotow R (2007) Evaluating Methods for Counting Cryptic Carnivores. J Wildl Mgt 73 (3): 433–441.
15. Bellemain E, Taberlet P (2004) Improved non-invasive genotyping method: application to brown bear (*Ursus arctos*) faeces. Mol Ecol Notes 4:519–522.
16. Ernest HB, Penedo MCT, May BP, Syvaren M, Boyce WM (2000) Molecular tracking of mountain lions in the Yosemite Valley region in California: genetic analysis using microsatellites and faecal DNA. Mol Ecol 9:433–441.
17. Bensch S, Andren H, Hansson B, Pedersen HC, Sand H, et al. (2006) Selection for heterozygosity gives hope to a wild population of inbred wolves. *PLoS ONE* 1(1): e72.doi:10.1371/journal.pone.0000072.
18. Zhang HM, Guo Y, Li DS, Wang PY, Fang SG (2009) Sixteen novel microsatellite loci developed for the giant panda (*Ailuropoda melanoleuca*). Conserv Genet 10:589–592.
19. Taberlet P, Luikart G (1999) Non-invasive genetic sampling and Individual Identification. Biol J Lin Soc 68: 41–55.
20. Taberlet P, Camarra JJ, Griffin S, Uhres E, Hanotte O, et al. (1997) Nonivasive genetic tracking of the endangered Pyrenean brown bear population. Mol Ecol 6: 869–876.
21. Woods JG, Paetkau D, Lewis D, McLellan BN, Proctor M, et al. (1999) Genetic tagging free-ranging black and brown bears. Wildl Soc Bull 27: 616–627.
22. Waits LP, Luikart G, Taberlet P (2001) Estimating the probability of identity among genotypes in natural populations: cautions and guidelines. Mol Ecol 10: 249–256.
23. Kendall KC, Metzgar LG, Patterson DA, Steele BM (1992) Power of sign surveys to monitor population trends. Ecol Appl 2: 422–430.

24. Schwartz M, Tallmon D, Luikart G (1998) Review of DNA-based census and effective population size estimators. Anil Conserv 1:293–299.
25. Kohn MH, York EC, Karmradt DA, Haught G, Sauvajot RM, et al. (1999) Estimating population size by genotyping faeces. Proc R Soc Lond B 266: 657–663.
26. Uphyrkina O, O'Brien SJ (2003) Applying molecular genetics tools to the conservation and action plan for the critically endangered Far Eastern leopard (*Panthera pardus orientalis*).Comp R Biol 326: 93–97.
27. Tende T, Ottosson U, Hansson B, Åkesson M, Bensch S (2010) Population size of lions in Yankari Game Reserve as revealed by faecal DNA sampling. J Afr Ecol 59: 7–12.
28. Afolayan TA (1978) The effect of fires on the vegetation in Kainji lake National Park, Nigeria. Oikos 33: 376–382.
29. Menotti RM, David VA, Lyons LA, Schaffer AA, Tomlin JF, et al. (1999) A genetic linkage map of microsatellites in the domestic cat (Felis catus). Genomics 57: 9–23.
30. Hellborg L, Ellegren H (2004) Low levels of nucleotide diversity in mammalian Y chromosomes. Mol Biol Evol 21:158–163.
31. Seddon JM (2005) Canid-specific primers for molecular sexing using tissue or non-invasive samples. Conserv Genet 6:147–149.
32. Taberlet P, Waits LP, Luikart G (1999) Noninvasive genetic sampling: look before you leap. Trends Ecol Evol 14(8): 323–327.
33. Pompanon F, Bonin A, Taberlet P (2005) Genotyping errors: causes, consequences and solutions. Nature 6: 847–859.
34. Taberlet P, Griffin S, Goossens B, Questiau S, Manceau V, et al. (1996) Reliable Genotyping of Samples With Very Low DNA Quantities Using PCR: Nucleic Acids Res 24: 3189–3194.
35. Frantz AC, Pope LC, Carpenter PJ, Roper TJ, Wilson GJ et al. (2003) Reliable microsatellite genotyping of the Eurasian badger (*Meles meles*) using faecal DNA. Mol Ecol 12: 1649–1661.
36. Murphy MA, Waits LP, Kendall KC, Wasser SK, Higbee JA, et al. (2002) An evaluation of long-term preservation methods for brown bear (*Ursus arctos*) faecal DNA samples. Conserv Genet 3: 435–440.
37. Marshall TC, Slate J, Kruuk LEB, Pemberton JM (1998) Statistical confidence for likelihood-based paternity inference in natural populations. Mol Ecol 7: 639–655.
38. Coombs JA, Letcher BH, Nislov KH (2008) CREATE: a software to create input files from diploid genotypic data for 52 genetic software programs. Mol Ecol Res 8: 578–580.Version.1.31.
39. Goudet J (2002) FSTAT, a computer program to estimate and test gene diversities and fixation indices Version 2.9.3.2.
40. Peakall R, Smouse PE (2006) GenAlEx 6: genetic analysis in Excel. Population genetic software for teaching and research. Mol Ecol Notes 6: 288–295.
41. Slatkin M (1995) A measure of population subdivision based on microsatellite allele frequencies. Gen Mol Biol 139: 457–462.
42. Sargeant GA (2011) wild1: R tools for wildlife research and management. R package version 1.09. Geological Survey Northern Prairie Wildlife Research Center, Jamestown, ND, USA.
43. Nyanganji G, Saidu Y, Henschel P, Dunn A (2012) 2011 survey of lion (Panthera leo) in Yankari Game Reserve and Kainji Lake National Park, Nigeria (A submitted report National Park service).
44. Waits LP, Luikart G, Taberlet P (2001) Estimating the probability of identity among genotypes in natural populations: cautions and guidelines. Mol Ecol 10: 249–256.
45. Packer C, Pusey AE (1982) Cooperation and competition within coalitions of male lions: kin selection or game theory? Nature 296:740–74.
46. Pusey AE, Packer C (1987) The evolution of sex-biased dispersal in lions. Behaviour 101: 275–310.
47. Gilbert DA, Packer C, Pusey AE, Stephens JC, O'Brien SJ (1991) Analytical DNA fingerprinting in lions: parentage, genetic diversity, and kinship. J Heredity 82: 378–386.
48. Packer C, Pusey AE, Rowley H, Gilbert DA, Martenson J, et al. (1991) Case study of a population bottleneck: Lions of the Ngorongoro Crater. Conserv Biol 5 (2): 219–230.
49. Spong G, Creel S (2001) Deriving dispersal distances from genetic data. Proc R Soc Biol Sci Ser B 268: 2571–2574.
50. Spong G, Stone J, Creel S, Björklund M (2002) Genetic structure of lions (*Panthera leo L.*) in the Selous Game Reserve: implications for the evolution of sociality. J Evol Biol 15: 945–953.
51. Kays RW, Patterson BD (2002) Mane variation in African lions and its social correlates. Can J Zool 80: 471–478.
52. Dubach J, Patterson BD, Briggs MB, Venzke K, Flamand J, et al. (2005) Molecular genetic variation across the southern and eastern geographic ranges of the African lion, Panthera leo. Conserv Genet 6:15–24.
53. Lyke MM, Dubach J, Briggs MB (2013) A molecular analysis of the African lion (*Panthera leo*) mating structure and extra-group paternity in Etosha National Park. Mol Ecol DOI: 10. 1111/mec.12279.
54. Shankaranarayanan P, Banerjee M, Kacker RK, Aggarwal RK, Singh L (1997) Genetic variation in Asiatic lions and Indian tigers. Electrophoresis 18:1693–1700.
55. Brown EW, Yuhki N, Packer C, O'Brien SJ (1994) A lion lentivirus related to feline immunodeficiency virus: epidemiologic and phylogenetic aspects. J Virol 68:5953–5968.
56. Roelke ME, Munson L, Packer C, Kock R, Cleaveland S, et al.(1996) A canine distemper virus epidemic in Serengeti lions (Panthera leo).Nature 379:441–445.
57. Packer C, Altizer S, Appel M, Brown E, Martenson J, et al. (1999) Viruses of the Serengeti: patterns of infection and mortality in African lions. J Anil Ecol 68:1161–1178.
58. Antunes A, Troyer JL, Roelke ME, Pecon-Slattery J, Packer C, et al. (2008) The Evolutionary Dynamics of the Lion *Panthera leo* Revealed by Host and Viral Population Genomics. Plos Gen 4 (11) e1000251.
59. Piggott MP, Bellemain E, Taberlet P, Taylor AC (2004) A multiplex pre-amplification method that significantly improves microsatellite amplification and error rates for faecal DNA in limiting conditions. Conserv Genet 5: 417–420.
60. Lehmann MB, Funston PJ, Owen CR, Slotow R (2008) Home Range Utilization and Territorial Behaviour of Lions (*Panthera leo*) on Karongwe Game Reserve, South Africa. PLOS ONE3(12):e3998.doi:10.1371/journal.pone.0003998.
61. Laikre L, Ryman N (1991) Inbreeding depression in a captive Wolf (Canis lupus) population. Conserv Biol 5:33–40
62. Liberg O, Andren H, Pedersen CH, Sand H, Sejberg D, et al. (2005) Severe inbreeding depression in a wolf *Cani lupus* population. Biol Lett 1:17–20.
63. Vila C, Sundqvist AK, Flagstad Ø, Seddon J, Blörnerfeldt S, et al. (2003) Rescue of a severely bottlenecked wolf (*Canis lupus*) population by a single immigrant. Proc R Soc Lond B 270: 91–97.
64. Laikre L, Andren R, Lars H, Ryman N (1996) Inbreeding depression in Brown Bear (Ursus arctos). Biol Conserv 76:69–72.
65. O'Brien SJ, Martenson JS, Packer C, Herbst L, de Vos V, et al. (1987) Biochemical genetic variation in a geographic isolates of African and Asiatic lions. Natl Geo Res 3:114–124.
66. Lande R (1995) Mutation and conservation. Conserv Biol 9: 782–791.
67. Laikre L (1999) Conservation genetics of Nordic carnivores: lessons from zoos. Hereditas 130 (3): 203–216.
68. Keller LF, Waller DM (2002) Inbreeding effects in wild populations. Trends Ecol Evol 17: 230–241.
69. Frankham R, Ballou JD, Briscoe DA (2002) Introduction to Conservation Genetics. Cambridge University Press, Cambridge, UK.
70. Munso L, Brown JL, Packer C, Janssen D, Reiziss SM, et al. (1996) Genetic diversity affects morphology in free-ranging lions (Panthera leo) of the Serengeti Plains and Ngorongoro Crater. J Rep Fer 108:11–15.
71. Wildt DE, Bush M, Goodrowe KL, Packer C, Pusey AE, et al. (1987) Reproductive and genetic consequences of founding isolated lion populations. Nature 329:328–331.
72. O'Brien S (1994) A role for molecular genetics in biological conservation. Proc Natl Acad Sci USA 91:5748–5755.
73. Manel S, Schwartz MK, Luikart G, Taberlet P (2003) Landscape genetics: combining landscape ecology and population genetics. Trends Ecol Evol 18:189–197.
74. McRae BH, Beier P, Dewald LE, Huynh LY, Keim P (2005) Habitat barriers limit gene flow and illuminate historical events in a wide-ranging carnivore, the American puma. Mol Ecol 14:1967–1977.
75. Ernest HB, Boyce WM, Bleich VC, May B, Stiver SJ, et al. (2003) Genetic structure of mountain lion (*Puma concolor*) populations in California. Conserv Genet 4:353–366.
76. Riley SPD, Pollinger JP, Sauvajot RM, York EC, Bromley C, et al. (2006) A southern California freeway is a physical and social barrier to gene flow in carnivores. Mol Ecol 15:1733–1741.
77. Frankham R (1996) Relationship of genetic variation to population size in wildlife. Conserv Biol 10:1500–1508.
78. Frankham R, Kingslover JG (2004) Response to environmental change: Adaptation or extinction. In: Ferrier R., Dieckman U., Couvet D (Eds.) Evolutionary Conservation Biology. Cambridge University Press, Cambridge pp.85–100.
79. Frankel OH (1970) Variation, the essence of life. Proc Linn Soc New South Wales 95:158–169.
80. Frankel OH (1974) Genetic conservation: our evolutionary responsibility. Genetics 78:53–65.
81. Lande R (1988) Genetics and demography in biological conservation. Science 214:1455–1460.
82. Woodroffe R., Ginsberg JR (1998) Edge effects and extinction of populations inside protected areas. Science 280:2126–2128.
83. Dunham J, Peacock M, Tracy CR, Nielsen J, Vinyard G (1999) Assessing extinction risk: integrating genetic information. Conserv Ecol [online] 3(1): 2. Available: http://www.consecol.org/vol3/iss1/art2/.
84. Lande R (1993) Risks of population extinction from demographic and environmental stochasticity and random catastrophes. Am Nat 142:911–927.
85. Lacy RC (2000) Considering threats to the viability of small populations using individual-based models. Ecol Bull 48: 39–52.
86. Spielman D, Frankham R (1992) Modelling problems in conservation genetics using captive Drosophila populations: improvement of reproductive fitness due to immigration of one individual into small partially inbred populations. Zoo Biol 11:343–351.
87. Westemeier RL, Brawn JD, Simpson SA, Esker TL, Jansen RW, et al. (1998) Tracking the long-term decline and recovery of an isolated population. Science 282:1695–1698.
88. Madsen T, Shine R, Olsson M, Wittzell H (1999) Conservation biology: restoration of an inbred adder population. Nature 402:34–35.

89. Ebert D, Haag C, Kirkpatrick M, Riek M, Hottinger JW, et al. (2002) A selective advantage to immigrant genes in a Daphnia meta population. Science 295:485–488.
90. Ball SJ, Adams M, Possingham HP, Keller MA (2000) The genetic contribution of single male immigrants to small, inbred populations: a laboratory study using Drosophila melanogaster. Heredity 84: 677–684.
91. Saccheri IJ, Brakefield PM (2002) Rapid spread of immigrant genomes into inbred populations. Proc R Soc Lond B 269:1073–1078.
92. Amos W, WorthingtonWilmer J, Fullard K, Burg TM, Croxall JP, et al. (2001) The influence of parental relatedness on reproductive success. Proc R Soc Lond B 268:2021–2027.
93. Hedrick PW, Kalinowski ST (2000) Inbreeding depression in conservation biology. Ann Rev Ecol Syst 31:139–162.

19

Comparative Transcriptome of Wild Type and Selected Strains of the Microalgae *Tisochrysis lutea* Provides Insights into the Genetic Basis, Lipid Metabolism and the Life Cycle

Gregory Carrier[1], Matthieu Garnier[1], Loïc Le Cunff[3], Gaël Bougaran[1], Ian Probert[2], Colomban De Vargas[2], Erwan Corre[4], Jean-Paul Cadoret[1], Bruno Saint-Jean[1]*

1 IFREMER-PBA, Nantes, France, **2** CNRS-UPMC, UMR 7144, Station Biologique de Roscoff, Roscoff, France, **3** UMT Geno-Vigne®, Montpellier, France, **4** CNRS-UPMC, ABiMS, Station Biologique de Roscoff, Roscoff, France

Abstract

The applied exploitation of microalgae cultures has to date almost exclusively involved the use of wild type strains, deposited over decades in dedicated culture collections. Concomitantly, the concept of improving algae with selection programs for particular specific purposes is slowly emerging. Studying since a decade an economically and ecologically important haptophyte *Tisochrysis lutea* (Tiso), we took advantage of the availability of wild type (Tiso-Wt) and selected (Tiso-S2M2) strains to conduct a molecular variations study. This endeavour presented substantial challenges: the genome assembly was not yet available, the life cycle unknown and genetic diversity of Tiso-Wt poorly documented. This study brings the first molecular data in order to set up a selection strategy for that microalgae. Following high-throughput Illumina sequencing, transcriptomes of Tiso-Wt and Tiso-S2M2 were *de novo* assembled and annotated. Genetic diversity between both strains was analyzed and revealed a clear conservation, while a comparison of transcriptomes allowed identification of polymorphisms resulting from the selection program. Of 34,374 transcripts, 291 were differentially expressed and 165 contained positional polymorphisms (SNP, Indel). We focused on lipid over-accumulation of the Tiso-S2M2 strain and 8 candidate genes were identified by combining analysis of positional polymorphism, differential expression levels, selection signature and by study of putative gene function. Moreover, genetic analysis also suggests the existence of a sexual cycle and genetic recombination in *Tisochrysis lutea.*

Editor: María Mar Abad-Grau, University of Granada - Q1818002F, Spain

Funding: This work was partly funded by region of Pays de Loire "Nouvelle équipes, nouvelles thématique" program and by the Agence National de la Recherche, Facteur 4 project. There are no non-financial competing interests. The funders had no role in study design, data collection and analysis, decision to publish, or preparation of the manuscript.

Competing Interests: The authors have declared that no competing interests exist.

* E-mail: Bruno.Saintjean@ifremer.fr

Introduction

Interest in microalgae as a potential source of economic benefit is booming [1]. Applications are envisaged in very different domains such as human or animal alimentation due to their high nutritional value [2], bio-remediation such as water purification [3], and pigment production for the cosmetic industry, for food-processing or for human health [4]. In the last decade new biotechnological applications have emerged [5]. The use of microalgae as cell factories would offer numerous advantages for the production of safe and complex recombinant pharmaceutical proteins [6–8]. Recently, several research projects have focused on the use of microalgae for biofuel production [9], citing very attractive biomass and oil productivities compared to oleaginous land plants. Mass production of microalgae would not enter into competition with agricultural food production as they can be cultivated on non-arable land, saline or wastewater sources [10]. In the context of the energy crisis, certain concepts of production of biofuels from algal oils are currently at the demonstration stage. However, the production of microalgae is currently not sufficiently economically efficient compared to the use of fossil fuels to be envisioned at a large scale [7].

Among the very large diversity of microalgae, a culture strain originally isolated from Tahiti and designated as *Isochrysis* affinis *galbana* but recently renamed *Tisochrysis lutea* (Tiso) [11] has been historically extensively studied due to it's widespread use in aquaculture as a feedstock for shellfish and shrimps that reflects an attractive fatty acid content [12]. However, the lack of molecular data for this strain has limited investigation of aspects such as the metabolism and life cycle of this microalga. Rapid developments in Next Generation Sequencing (NGS) technologies now make it relatively easy to conduct large-scale genotyping of species for which full genome sequences are not yet available [13].

Beyond the obvious economic interest of Tiso, it is also of significant fundamental interest as a member of a microalgal lineage (the division Haptophyta) which is diverse and often ecologically dominant in the planktonic photic realm [14]. Tiso is a member of the haptophyte order Isochrysidales that comprises two families, the

Isochrysidaceae and the Noëlaerhadaceae. Members of the Noëlaerhabdaceae exhibit a heteromorphic haplo-diploid life cycle in which the diploid phase produces calcified plates (coccoliths) and the haploid phase is non-calcifying [14]. The best-known member of this family is *Emiliania huxleyi*, which is by far the most abundant coccolithophore in modern oceans and consequently an extremely important actor in global carbon cycling. By contrast, isochrysidaceaens such as Tiso have only one known non-calcifying morphological form. No information about sexual reproduction or ploidy levels are available for this family [16,17].

Sophisticated selection programs like those implemented in terrestrial agriculture will probably play an important role in the future exploitation of microalgal resources [18]. Domestication of plants and animals by long-term selection programs led to the rise of modern agriculture. For example, the yield of wheat culture has increased 16-fold in 1200 years through directed selection of highly productive strains [11]. By comparison, domestication of microalgae is in its infancy. Recent studies in this domain have focused, for example, on production of the pigment prodigiosin in the rhodophyte *Hahella chejuensis* [19], carotenoid production in *Dunaliella salina* [20], and lipid production in *Nannochloropsis* sp. [21] and in Tiso [22]. Until recently, all these selection programs of microalgae consisted to identify and select the best individuals among a population. To date, microalgal selection programs generally start without prior knowledge of the natural diversity of the taxon in question. A number of studies have reported high levels of intra-specific genetic or metabolic diversity in different microalgal species [23–25]. In this study, we estimated the genetic diversity in a Tiso culture strain, this being a requirement for conception of improved selection programs [18]. Life cycles of microalgae are generally poorly known and consequently breeding programs have yet to be initiated, despite several real advances in the knowledge of certain microalgal groups such as diatoms [18].

In our laboratory, a selection program was performed starting from a wild type strain of *Tisochrysis lutea* (Tiso-Wt). This wild type strain consists in one ecotype (Tahiti) and is presumed to be composed of several genotypes characterizing a population of Tiso with an unknown diversity. A sequential mutation-selection procedure was performed from this wild type strain, involving: i) UVc treatment to induce mutations and thus increase the genetic (and metabolic) diversity of the strain, and ii) selection of the 10% of cells with the highest lipid content. This resulted in selection of a new certificate strain [26] (Tiso-S2M2) that accumulates twice the amount of neutral lipids in nitrogen limited culture conditions compared to the Tiso-Wt strain [22].

In context of algae selection program, this study using RNAseq approach, brings the first molecular data for an economically and ecologically important microalgae *Tisochrysis lutea*. Here, we describe the impact of a selection program on the genetic repertoire and gene expression of a selected strain. Comparison of transcriptomes of wild type and selected strains allowed identification of positional polymorphism and differentially expression levels. Analysis of polymorphisms generated as a result of the selection program provides detailed information on the genomic-level impact of the program, and notably identification of candidate genes that could account for lipid over-accumulation, as well as providing insights into the putative life cycle of this species.

Results

De novo sequencing and annotation of Tiso transcriptomes

Transcriptomes of two Tiso strains, the reference strain (Tiso-Wt) and a selected strain (Tiso-S2M2), were sequenced with Illumina HiSeq 2000 technology. Read pairs obtained for each strain were filtered and assembled into 44,983 and 44,564 transcripts for Tiso-Wt and Tiso-S2M2, respectively (Table 1). These two datasets were clustered to obtain a final consensus transcriptome of 46,687 transcripts. Of these, 34,374 were unique, representing a total length of 44.4 Mb. At least two isoforms were detected for the remaining (12,313) transcripts. The origin of these non-unique transcripts could be the sequencing of mature and non-mature RNA, alternatively spliced or the sequencing of duplicated genes [27]. The consensus transcriptome was annotated and 14,790 transcripts were associated with at least one Gene Ontology (GO) term. These transcripts with putative function were sorted according to cellular functions (Figure 1). Globally, enzymes of all major metabolic pathways such as nitrogen, photosynthesis or sugar metabolisms were represented and among those, 2,010 transcripts were assigned at least one known universal KEGG pathway (Kyoto Encyclopedia of Genes and Genomes, http://www.genome.jp/kegg/). Specifically, all the major enzymes of lipid metabolism were identified in the Tiso transcriptome and match in the universal KEGG lipid pathway (Figure S1). However, no putative function was identified for 57% of transcripts (19,584).

Comparison of Tiso transcriptomes with reference microalgal transcriptomes

The consensus Tiso transcriptome was compared with transcriptomes from eight reference microalgal taxa chosen for their diverging phylogenetic position (2 chlorophytes, 2 diatoms, 1 rhodophyte, 1 glaucophyte, 1 dinoflagellate and 1 haptohyte). For each algae transcriptome, we considered only the coding transcripts known to produce a putative protein in order to avoid biases due to the existence of non-coding transcripts. The number of transcripts in each transcriptome and the number of homologous genes between Tiso and these reference microalgae are shown in Figure 2. Tiso has many more putative unique transcripts (more than 30,000) with the most closely related species in the list, *Emiliania huxleyi*, than with the other reference microalgae. Unsurprisingly, *E. huxleyi* possesses the highest proportion (25%) of genes that are homologous with those of Tiso, the proportion being less than 11% with the reference microalgae from other lineages. Finally, only 750 genes that were homologous between all of these algae were identified. The high diversity and very ancient divergence of these reference algae [28] explains this low number of homologous genes. Indeed, 66% (22,664) of Tiso genes have not homologous from these microalgae.

Codon usage bias and transcriptomic G+C content were also compared between the reference algae (Figures S2 and S3). The codon usage bias refers to differences in the frequency of occurrence of synonymous codons. Generally, the codon preference reflects a balance between mutations and natural selection for translational optimization [29]. The G+C content in coding DNA is variable between species and is known to play a role in the codon usage bias [30]. Our results indicate that the codon usage bias of Tiso is similar to that of the other haptophyte *E. huxleyi*, but also to that of the dinoflagellate *Alexandrium ostenfeldii* (Figure S3). Surprisingly, Tiso has a transcriptomic G+C content (60.2%) more similar to that of the dinoflagellate A. *ostenfeldii* (58.3%) than that of *E. huxleyi* (Figure S2).

Impact of the selection program on the transcriptome of Tiso-S2M2. The Tiso wild type strain is monospecific, but is presumed not to be a clonal strain but composed of several genotypes characterizing a population. During the selection program [22], diversity of Tiso-Wt was not explored. The

Table 1. Tiso transcriptome characteristics.

Libraries	Read pairs, raw data	Read pairs produced with good quality	Assembly transcripts	Consensus transcripts between the libraries	Read pairs aligned on unique *loci*	Read pairs aligned on multi *loci*	Read pairs not aligned
Wt	177 M	140 M (79.1%)	44,983	46,687 among them	100 M (71.6%)	30 M (21.7%)	9 M (6.7%)
S2M2	189 M	156 (82.4%)	44,564	34,374 unique	115 M (73.7%)	30 (19.3%)	6 M (4.0%)

The table summarizes the different steps in building the Tiso transcriptome using RNAseq data obtained from Illumina HiSeq technology.

identification of transcriptome polymorphisms gives some light concerning the molecular diversity of this strain. Moreover, the selection program applied to Tiso was accomplished in closed conditions [22]. Therefore, the polymorphism observed between Tiso-Wt and Tiso-S2M2 was generated only by mutation and/or within-strain selection effects and not by allele flux from the environment. In these conditions, the impact of the selection program on the genetic diversity of Tiso-S2M2 could be studied.

We first investigated the molecular diversity in each strain at the transcriptome level and its evolution during the selection program. Polyallelic *loci* were identified and analyzed in the transcriptome of each strain (Figure 3) to estimate the molecular diversity. For SNPs, 925 and 883 biallelic *loci* were detected for Tiso-Wt and Tiso-S2M2, respectively. As for the indels, 782 and 784 biallelic *loci* were detected, respectively. These results confirm that both strains are composed of several genotypes. Allelic frequency of polymorph *loci* was compared between both strains to look for theevolution of the genetic diversity during selection program,. Only transcripts without differentially expressional level were considered, in each population. A Jost's D genetic index [31] was used and a score of 0.161 was estimated between both strains showing a high conservation of genetic diversity between the populations. Indeed, the majority of polyallelic *loci* detected in Tiso-Wt, i.e. 798 (86%) SNPs and 762 (97%) indels, were conserved in Tiso-S2M2. For each conserved polyallelic *locus*, the differential allelic frequency between the two strains was measured (Figure 4). On average, the differential allelic frequency per *locus* was 15.6% for SNPs and 8.5% for indels confirm a global conservation of genetic diversity at the end of the selection program. 92 SNPs and 20 indels were *loci* with a differential allele frequency greater than 35% (Figure 4), reflecting selection events during the selection program. Five transcripts containing more than five polyallelic *loci* with a differential allele frequency are of particular interest, these being considered as transcripts with a clear selection signature.

Thereafter, we focused on the positional polymorphisms and differentially expression levels generated during the selection program. For positional polymorphisms, 241 SNP and 22 indel polymorphs were detected between the strains. SNP and indel polymorphisms were distributed among 146 and 19 transcripts, respectively (Figure S4). The large majority of these transcripts (124/165) contained only one polymorphism. However, seven hot spots of polymorphism with more than five polymorphisms in a specific region of one transcript were identified (frequency superior to 1 polymorphism per 100 bases). Among these seven hot spots, four were localized in UTR regions and three in coding regions.

Concerning differential gene expression levels, 84 transcripts were over-expressed and 207 under-expressed in Tiso-S2M2 compared to Tiso-Wt. Of these, 32 had a high expression difference (>100 fold) and 18 transcripts showed a specific expression in only one strain (Table S1). More precisely, eleven transcripts were specifically expressed in Tiso-S2M2 and seven transcripts were specific to Tiso-Wt. These results show that expression of some genes was clearly affected as a result of the selection program.

Origin of polymorphisms. The origin of polymorphisms (mutation or selection events) was identified to evaluate the evolution strength producing from the selection program. The selection program was conducted in closed conditions and consequently the 241 SNP and 22 indel polymorphisms between the strains could be sorted into different classes (Table 2). Class 1 comprises the polymorphisms for which there was one allele in Tiso-Wt and two alleles in Tiso-S2M2 (Figure 3). This class refers to polymorphisms generated by mutation events occurring during

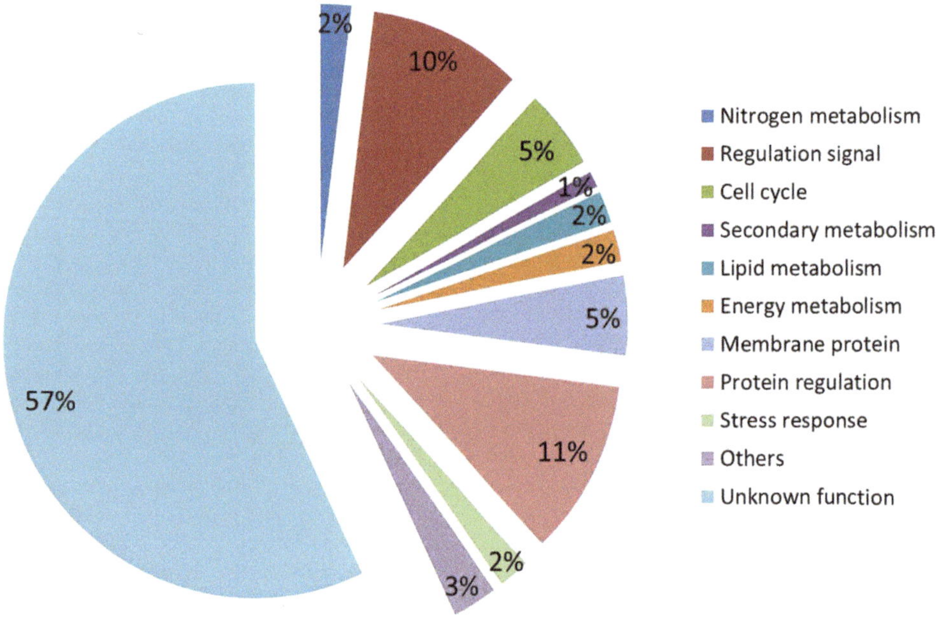

Figure 1. Annotation of Tiso transcriptome. Transcripts were annotated by BLAST in the NCBI nr database. Transcripts were sorted among identified major cell functions from Gene Ontology.

the selection program (85 SNPs and 12 indels; Table 2). Class 2 comprises polymorphisms for which there were two alleles in Tiso-Wt, only one of which was conserved in Tiso-S2M2 (Figure 3). This implies a loss of one allele by selection pressure (126 SNPs and 10 indels; Table 2). Finally, class 3 comprises polymorphisms for which there was one allele in Tiso-Wt and a different allele in Tiso-S2M2 (Figure 3). This refers to polymorphisms appearing as a result of mutation followed by selection (30 SNPs and 0 indels; Table 2).

Annotation and selection of candidate genes for lipid over-accumulation. Transcripts with position polymorphism or differentially expressed between Tiso-Wt and Tiso-S2M2 were manually annotated and sorted according to metabolic pathways (Figure 5). A large number of differentially expressed transcripts involved in stress responses, protein signal regulation and membrane proteins (transporters and structural proteins) were identified. Few genes were classed as being involved in energy, lipid or nitrogen metabolism.

The selected strain Tiso-S2M2 is characterized by an over-accumulation of lipids in nitrogen-limited culture conditions and we focused on this aspect. Candidate genes were identified by a combined analysis of polymorphism between the strains. Firstly, transcripts with a selection signature generated by the impact of the selection program could be interesting candidates. Among positional polymorphisms, three hot spots located in coding regions of *TisoTranscripts-291*, *TisoTranscripts-227* and *TisoTranscripts-46* were detected (21, 7 and 6 SNPs were identified for each transcripts respectively). These hot spots contain polymorphisms only produced by selection effects and consequently are considered as transcripts with a selection signature. Moreover, some polymorphisms (12, 3 and 1 for *TisoTranscripts-291*, *TisoTranscripts-227* and *TisoTranscripts-46* respectively) were non synonymous and could induce a change of protein activities. Annotation of these candidate genes revealed a putative Na^+ solute transporter (*TisoTranscripts-291*) and two proteins with zinc finger domains (*TisoTranscripts-227* and *TisoTranscripts-46*). One homologous gene of *TisoTranscripts-291* was exclusively identified in *E. huxleyi* (*Emihu-233786*) suggesting a haptophyte specific gene. Five transcripts with hot spots of *loci* with differential allelic frequency were also detected (*TisoTranscripts-440*, *TisoTranscripts-441*, *TisoTranscripts-442*, *TisoTranscripts-443* and *TisoTranscripts-444*), but no function was found for these candidate genes bearing a selection signature.

In the same ecophysiological conditions, a total of 298 transcripts were differentially expressed between the two strains. Gene expression was therefore clearly affected by the selection program and this could explain lipid over-accumulation in Tiso-S2M2. A total of 18 transcripts were expressed specifically in only one strain (Table S1) and these were considered as promising candidate genes for further studies. The putative annotation of only three of these genes was possible. *TisoTranscripts-9* was positively assigned as a putative transcription factor, *TisoTranscripts-57* as an ankyrin protein-like and *TisoTranscripts-59* a Glycosyl-PhosphatIdylinositol protein-like (GPI protein). No homologous genes of these candidates were identified in reference microalgae suggesting that these candidates are specific to Tiso.

Because the selected Tiso-S2M2 strain is characterized by lipid over-accumulation, the transcripts with a putative function linked to lipid metabolism and with positional polymorphism or differential gene expression were considered as good candidates. *TisoTranscripts-288* and *TisoTranscripts-160*, annotated as a long chain fatty acid ligase and a GDSL lipase respectively, showed differential expression (under expressed 53- and 24-fold in TisoS2M2, respectively). Their respective expression was confirmed by a RT-qPCR experiment (Data S1) and we consider them as excellent candidates for future studies. Moreover, homologous gene of *TisoTranscripts-288* was identified in *E. huxleyi* (*Emihu-456684*), *P. tricornutum* (*Phatr-45510*) and *C. merolae* (*CMG-147C*). However, *TisoTranscripts-160* was only one homologous identified in *E. huxleyi* (*Emihu-213608*).

Discussion

The haptophyte *Isochrysis* affinis *galbana* is a species of major economic importance in aquaculture with good potential for biofuel

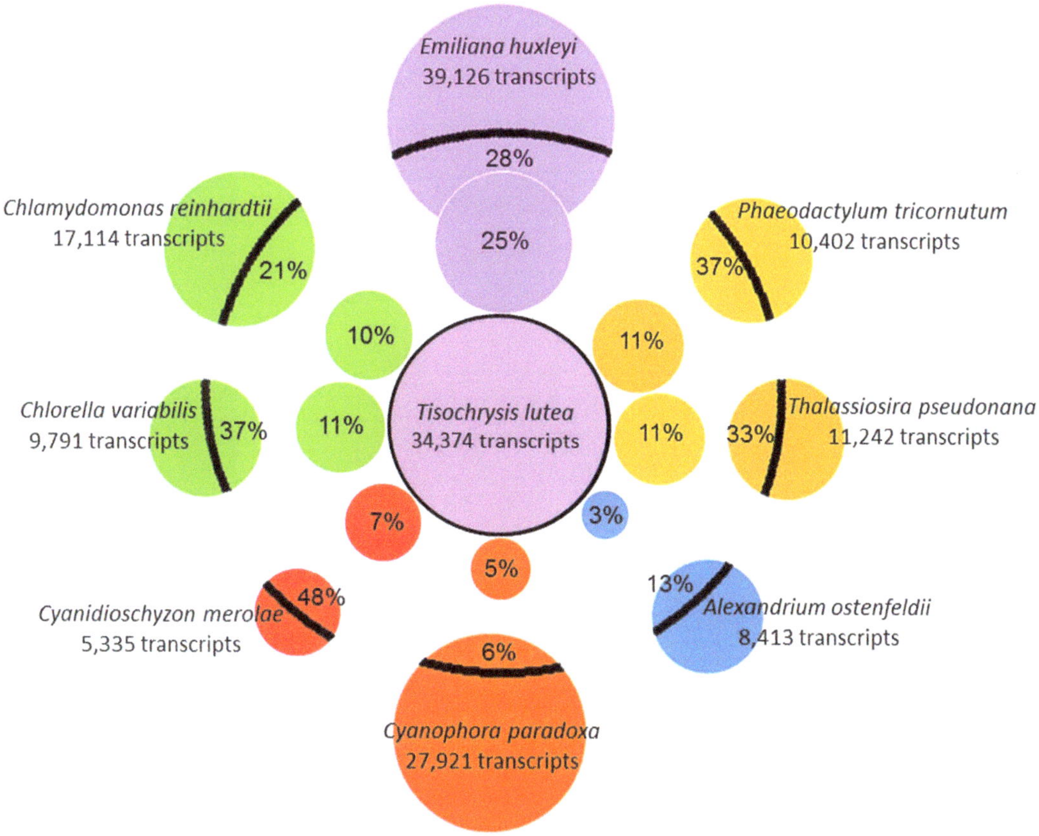

Figure 2. Comparative analysis of consensus Tiso transcriptome with those of reference microalgae. The size of circles represents transcript number. Color coding: green circle for chlorophytes, red for rhodophyte, orange for glaucophyte, blue for dinoflagellate, yellow for diatoms and purple for haptophyte. The number in each peripheral circles show proportion (%) homologous genes with Tiso. The number in each inner circles show proportion (%) homologous genes with reference microalgae.

production [12]. A laboratory selection program was developed and a strain obtained (Tiso-S2M2) that accumulates more lipid in nitrogen-limiting conditions than the wild type [22]. Molecular evolution of Tiso-Wt resulting from the selection program was studied and this allowed us to establish a list of candidate genes that could play a role in lipid over-accumulation. Moreover, fine-scale analysis of the transcriptomes of the wild type and selected strains provides new insights concerning the life cycle of Tiso.

First molecular knowledge of Tiso

In this study, transcriptomic data was produced for Tiso for the first time using high-throughput sequencing. The transcriptomes were annotated and all major metabolic pathways were identified. Annotation of these major pathways is usually relatively easy because they contain highly conserved protein domains [32]. However, no putative function was found for 57% of transcripts. To date, the number of genes with known function is very low for

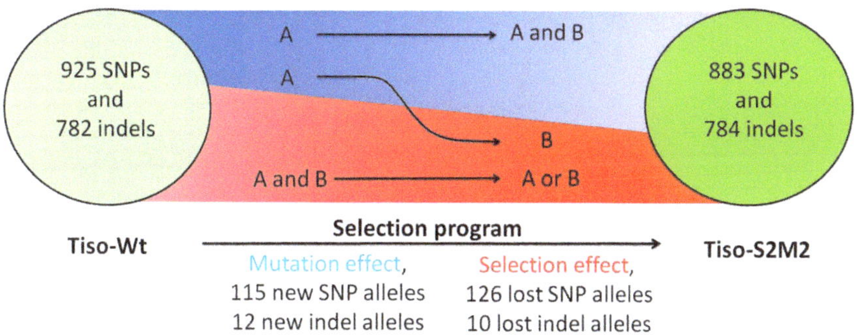

Figure 3. Impact of the selection program on the Tiso transcriptome. This figure is a schematic representation of the selection program. The number in each circle represents genetic diversity for each strain. Between these circles, evolutionary origins of polymorphisms between strains are illustrated: in red, polymorphism generated by loss of alleles (selection events), the narrower of area represent a reduction of diversity; in blue, new alleles (mutation events) the top of the area represents an increase of diversity; the arrow from blue to red, new and selected alleles (mutation and selection events).

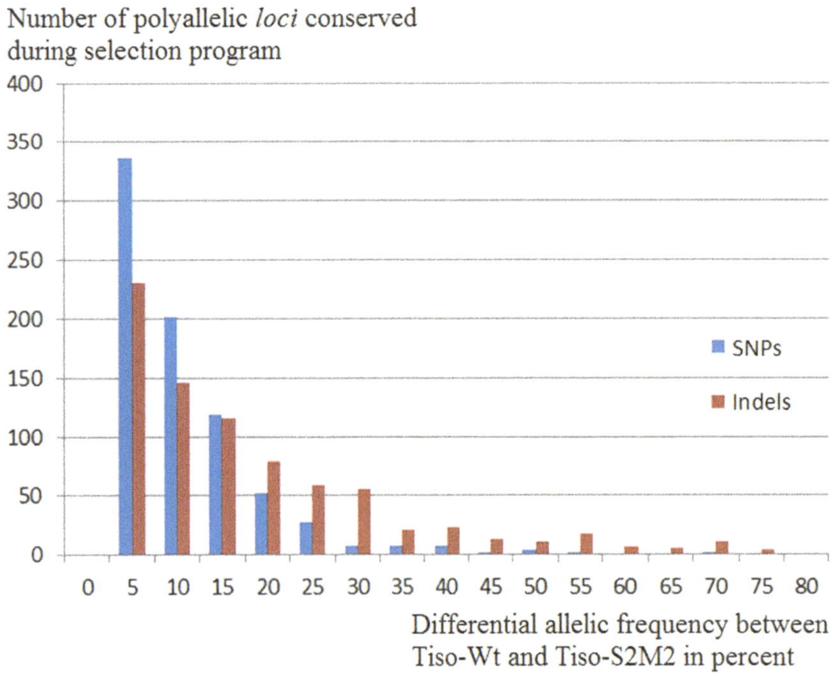

Figure 4. Distribution of differential allelic frequency between Tiso-Wt and Tiso-S2M2. This histogram represents the number of polymorph loci (SNP and indel) following their conservation rate through the selection program. Conservation rate defined as percent represents the differential allelic frequency between Tiso-Wt and Tiso-S2M2.

microalgae and the high diversity of microalgae makes annotation difficult [14].

The consensus transcriptome of Tiso was compared with eight other transcriptomes from reference microalgae. These microalgae were chosen because of their phylogenetic position and the availability of transcriptomic data. We compared the total number of transcripts and homologous genes, the transcriptomic G+C content and the codon usage bias. In general and as expected, these results confirmed that Tiso is genetically very closely related to *Emiliania huxleyi*, which is a member of the same haptophyte order, the Isochrysidales. Surprisingly, the codon usage bias and transcriptomic G+C content of Tiso were more similar to those of the dinoflagellate *Alexandrium ostenfeldii* than to those of *E. huxleyi*. Although these characteristics are not significant in an evolutionary sense, they reflect a high level of genomic speciation since the divergence between Tiso and *E. huxleyi*, estimated around 120 Mya [33,34].

Table 2. Genotype class of polymorphic *loci* between wild type Tiso-Wt and selected Tiso-S2M2.

Class	Tiso-Wt		Tiso-S2M2	Polymorphism number
1	*Locus* A	=>	*Locus* A/B	85 SNPs
2	*Locus* A/B	=>	*Locus* A or B	126 SNPs
3	*Locus* A	=>	*Locus* B	30 SNPs

Class 1 shows polymorphisms generated by a mutation event. Wild allele (A) and mutated allele (B) were observed in Tiso-S2M2. Class 2 shows polymorphisms generated by selection events only. Only one allele, (A) or (B) was observed in Tiso-S2M2 strain, whereas the two alleles were found in Tiso-Wt strain. Class 3 shows polymorphisms generated by both mutation and selection events. Only the allele (B) was detected in Tiso-S2M2.

Transcriptome evolution as a result of the selection program

The lipid over-accumulating selected strain Tiso-S2M2 was obtained from the wild type strain by a selection program [22]. This selection program consisted of increasing the genetic diversity by UVc exposition followed by selection of lipid-rich cells by flow cytometry.

In this study, we measured and analyzed the molecular modifications produced by the selection program on the population of Tiso-Wt at transcriptome level. These results could help to refine future selection programs. To measure the impact of this selection program, the genetic diversity of Tiso-Wt and Tiso-S2M2 was evaluated and compared at the transcriptome level. Polyallelic *loci* identified for both strains confirm the non-clonal character of Tiso-Wt and Tiso-S2M2. Interestingly, polyallelic *loci* detected in Tiso-Wt were mostly conserved in Tiso-S2M2 and their allele frequency did not change significantly after the selection program.

RNAseq strategy allowed us to analyze the gene expression and the positional polymorphism. Differential gene expression level observed for almost any cell functions reveals important modifications in all metabolism pathways. Studies of the inter-genic regions and the large polymorph insertions (requiring genome sequencing approaches) are underway. In turn, positional polymorphisms on coding region were observed between both strains. Among these polymorphisms, SNP mutations were more frequently detected than indels (i.e: 242 SNPs and 22 indels). This could be explained by a stronger selection pressure on indels than SNPs. Indeed, it is known that indel polymorphism has a stronger impact on phenotype because indel mutations are more harmful [35].

The selection program was undertaken in non-clonal strain in closed conditions and consequently the polymorphism observed between the strains was generated only by mutation or selection

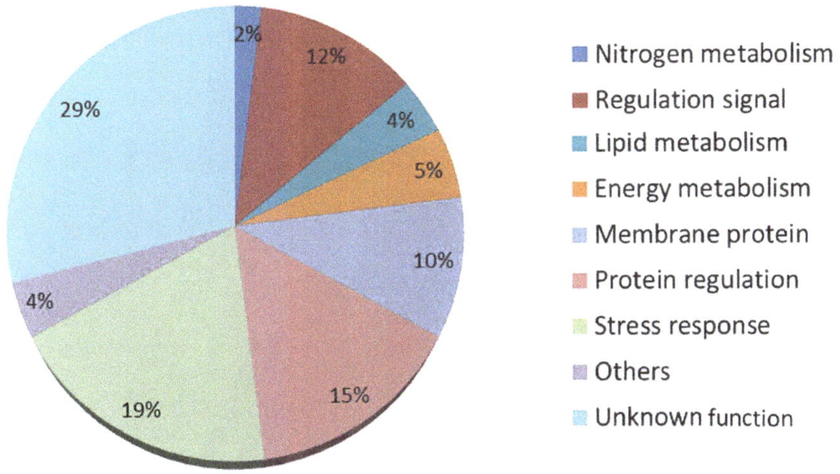

Figure 5. Expert annotations of transcripts containing polymorphisms or differential gene expression between Tiso-Wt and Tiso-S2M2. Transcripts were sorted according to major cell function.

[22]. We sorted polymorphisms (Figure 3) according to whether they were: (i) new alleles generated by mutation events (class 1 and class 3, Table 2); (ii) loss of alleles in Tiso-S2M2 by a selection effect (class 2, Table 2). It is noteworthy that the numbers of new and lost alleles were approximately equal. This suggests that mutation and selection strengths were balanced [36,37]. The selection program was thus characterized by maintenance of genetic diversity [38], which has the theoretical advantage of producing a strain with higher capacity to adapt to environmental variability compared to a clonal strain [39]. The lipid over-accumulating phenotype in Tiso-S2M2 has been conserved in the absence of selective pressure about lipid contents for more than two years [22]. This could be explained by the selected trait being more easily maintained in a population with high genetic diversity [40,41].

The clonal diversity of Tiso-S2M2 will be used in a next program to further select lipid accumulating Tiso strains. Two selection strategies will be investigated: One will be to select the best clone for lipid content from the Tiso-S2M2 population. However, conservation of clonal strains is usually risky on a long term basis and cryopreservation could be the solution [42]. Another selection strategy could call for a different mutagenesis methods and higher selection pressure.

List of candidate genes for lipid over-accumulation in the selected Tiso strain (Tiso-S2M2)

The selection program modified globally all cell functions (Figure 5). The selection of lipid trait generated large modifications in cells. These large modifications could be explained by selection program not affecting only lipid trait but, for example, selection of cells with UVc tolerance. Furthermore, the lipid metabolism is in relationship with all other metabolisms and a weak modification can affect entire cells. We focused on the lipid metabolism as biofuel production by microalgae is considered as one of the most promising sources for future energy production [9]. Identification of genes or alleles that play roles in lipid over-accumulation could be of great interest for future selection and metabolic engineering programs. A combined approach was used to produce a list of candidate genes that included: (i) transcripts with a selection signature; (ii) transcripts with specific expression in only one of the strains; (iii) transcripts with a putative function in lipid metabolism and with a positional polymorphism or differential gene expression.

In this study, eight transcripts with a clear selection signature were detected. Of these, only three could be annotated. Two of these genes (*TisoTranscripts-227* and *TisoTranscripts-46*) encode proteins with zinc finger domains which are known to play an important role in the regulation of gene expression [43]. The target genes of these two candidates are not currently known, but their selection signatures imply a role in increasing the fitness of Tiso-S2M2 in terms of lipid accumulation or UVc tolerance. No homologous genes were identified in reference microalgae suggesting a specific regulation of Tiso. According to the annotation, the third transcript (*TisoTranscripts-291*) is a Na^+ solute transporter-like. The most closely related protein domain that we could identify is the SLC5 family [44]. This family of co-transporters is known to exchange Na^+ solutes with different substrates such as glucose, urea or amino acids [45]. Allele selection of this gene could allow Tiso-S2M2 to optimize the transport of a molecule involved in the resistance to the selection program. A functional study of these genes would be of great interest to understand their role in the Tiso-S2M2 phenotype and specifically in lipid metabolism.

Gene expression was affected by the selection program. Of the 34,374 transcripts of Tiso, 291 exhibited differential expression between the strains. This differential gene expression could have several origins, such as mutations in regulation regions or epigenetic variations. Of these transcripts, 18 were specifically expressed in Tiso-Wt or Tiso-S2M2. With the highest differential expression, these are considered as serious candidates to explain the selection result. Other genes with differential expression could also be good candidates, but their differential expression need to be confirmed by a qRT-PCR approach [46]. Among the 18 main differentially expressed candidates, a putative function was found for only three genes. These candidates as well have no homologous genes in reference microalgae. *TisoTranscripts-9* encodes a putative transcription factor. Differential expression of this type of gene is known to modify expression of target genes and thus metabolic pathways. *TisoTranscripts-57* codes for an ankyrin protein-like. The ankyrin domain is considered to be the most common location for protein–protein interactions [47] and can play a role in several cellular functions. *TisoTranscripts-59* was identified as a GPI protein-like, a category of proteins known for their roles in communication between cells [48]. Given their putative functions, a potential implication of these latter two genes in the lipid over-accumulation observed in Tiso-S2M2 is not obvious, but they should be kept in mind for further studies.

The targeted phenotypic character of the selected strain is lipid over-accumulation and we therefore focused on the annotated genes of lipid pathways. Transcripts with putative annotation linked to lipid metabolism and containing a positional polymorphism or differential expression level were selected. Two candidate genes were identified (*TisoTranscripts-288* and *TisoTranscripts-160*) with differential expression between the strains. These published results rely on RNAseq dataset and because the implementation of this technique is however rather new, we confirmed the results using the well-established RT-qPCR technique (data S1) in a biological replicates. *TisoTranscripts-288* was identified as encoding a putative long chain fatty acid ligase (ACLS, EC: 6.2.1.3). This enzyme family is known to esterify free fatty acids containing C14/C20 carbon chains into fatty acyl-coenzyme A (acyl-CoA) [49,50]. Esterification into fatty acyl-CoA is a key step in numerous lipid metabolism pathways, in particular those involved in lipid catabolism. Analysis of the Tiso transcriptome revealed that *TisoTranscripts-288* was the only transcript annotated as coding for ACLS proteins. Consequently, the under-expression of this enzyme in Tiso-S2M2 may suggest a defect in lipid catabolism causing the lipid over-accumulation phenotype. Functional study of this enzyme is under way and will allow testing this hypothesis. The other candidate gene (*TisoTranscripts-160*) is assigned as a putative GDLS lipase. This family is composed of hydrolytic enzymes with multifunctional properties such as broad substrate and region specificities [51]. Two other GDSL lipase transcripts were identified in the Tiso transcriptome, but these were not differentially expressed. Functional study of the under-expressed GDLS lipase is also required to determine whether it is implicated in lipid catabolism. Surprisingly, the expression of known and identified genes involved in either fatty acid or triacyglycerol biosynthesis was similar between the two strains. In contrast, these candidate genes (*TisoTranscripts-288* and *TisoTranscripts-160*) showed the catabolic pathway related to fatty acid oxidation and hydrolytic lipase activity appears to have been affected.

Insights into the life cycle of Tiso

Many haptophytes, including *Emiliania huxleyi*, are known to undergo dimorphic haplo-diplontic life cycles, in which both haploid and diploid phases are capable of independent asexual division [15]. Ploidy level and sexual stages have never been reported for Tiso, or for any other member of the Isochrysidaceae. In culture, isochrysidaceaens have a single non-calcifying morphotype that resembles the haploid phase of noelhaerhabdaceaens (i.e. non-calcifying, usually flagellate). Diploid phase calcification is thought to have evolved only once at the origin of coccolithophores, and was thus apparently lost early in the evolutionary history of the Isochrysidaceae [52]. It is not clear whether this reflects a complete loss, or reduction, of the diploid phase in these species (i.e. clonal or haplontic life cycle, respectively), or whether they undergo a haplo-diplontic life cycle like other haptophytes, but with isomorphic non-calcifying haploid and diploid stages.

Knowledge of ploidy level is the basis for the understanding of genetics of species [53] and knowledge of the life cycle of this microalga could allow addition of breeding steps during the selection program [18]. This type of information would enable, for example, the study of heritability, gene functions or regulation, and allele interactions. Study of the evolution of genetic diversity resulting from the selection program and observation of evolutionary origin for positional polymorphisms provides insights into the life cycle of Tiso in laboratory conditions.

Selection events typically generate a loss of genetic diversity in populations [54]. Numerous (136) alleles present in Tiso-Wt were selected in Tiso-S2M2 (class 2, Table 2) and several (30) new alleles that appeared as a result of mutation were also selected in Tiso-S2M2 (class 3, Table 2). The majority of genetic diversity observed in Tiso-Wt was conserved in Tiso-S2M2. The conservation of the observed diversity despite the selection strength necessarily suggests allelic recombination during the selection program. We suggest that this allelic recombination implies a sexual step during our selection program. In other microalgae, such as diatoms [55], sexual reproduction has also been observed in response to particular environmental conditions [56,57]. Observation and physiological conditions of the fertilization step in Tiso will be confirmed in future studies.

Evolutionary origins of positional polymorphisms were analyzed to determine the ploidy of Tiso. Polymorphisms were sorted into three classes (Table 2). Because we observed multiallelic *loci* in both strains, the first proposed hypothesis is that Tiso is diploid. In this case, class 1 and class 2 polymorphisms (Table 2) could be explained by a mutation event on one allele (class 1) or selection events which had selected one of the two alleles in Tiso-Wt (class 2). However, for the third class, the origin of polymorphisms is, in this case, difficult to explain. The first possibility is a mutation event on each allele at the same *locus* and selection of both mutations in TisoS2M2. The probability of a mutation impacting the two alleles at the same *locus* is near null ($P = 1/(\text{number of nucleotides in genome})^2$). Consequently, this hypothesis is not conceivable. A second possibility is that a mutation appears on one allele and this individual undergoes self-fertilization. A part of this progeny is homozygote for the new allele and can be selected. However, in this case a high loss of genetic diversity in Tiso-S2M2 would be expected, which was not the case. These considerations lead us to suggest that Tiso is not diploid during the major stage of its life cycle, but rather that the Tiso-Wt and Tiso-S2M2 strains are composed of haploid cells. This latter hypothesis is attractive because it can explain the presence of polymorphisms in the third class by one mutation and one selection event.

The Tiso strains appear thus to be haploid in our culture conditions and capable of sexual reproduction, which means they either have a haplontic or a haplo-diplontic life cycle. Isolation and crossing of clonal haploid cultures could therefore be feasible in the context of development of a breeding program.

Conclusion

This study brings molecular basis of *Tisochrysis lutea* in a perspective of intensive selection program but complementary studies such as genome sequencing project or large study of genetic diversity will be conducting. Here, we added value to a previous program by analyzing at the genetic level the events provoked by the mutation/selection procedure. This analysis shows that balanced selection led to production of a strain that over-accumulates lipids. A comparative analysis of polymorphisms in the strains allowed identification of 8 genes that are candidates for involvement in provoking this phenotypic difference. These are promising targets for functional studies in the perspective of developing marker-assisted selection and genetic engineering protocols.

Materials and Methods

Microalgae strains and culture conditions

Isochrysis affinis *galbana* recently renamed *Tisochrysis lutea* clone Tahiti (Tiso) [11] was provided by the Culture Centre of Algae and Protozoa (CCAP 926/14). This Tiso wild type strain is monospecific and was isolated by Haines in the late 70s and kept in algae bank until today. Tiso wild type strain has been used in the selection program previously realized in our laboratory [22] and is considered as the reference strain in this study (Tiso-Wt). The selection program allowed to acquire a new strain (Tiso-S2M2) with higher lipid content than Tiso-Wt in nitrogen starved conditions [22]. This selected strain was certificate (IFR 32B85, [26]). Strains were grown in 2 L flasks containing modified Conway medium [57] with a modified nitrate concentration of 0.12 mM and bubbled with 0.22 µm filtered-air. Cultures were maintained at a constant temperature of 21°C and under constant irradiance of 100 µmol m^{-2} s^{-1}. The harvesting of microalgae was undertaken at the same time of day for both strains at the onset of nitrogen starvation when Tiso-S2M2 over-accumulated more lipids than Tiso-Wt (Figure S5).

RNA extraction, cDNA library construction and sequencing

Total RNA was extracted from each strain (Tiso-Wt and Tiso-S2M2) using the TRIZOL reagent (Invitrogen, USA) according to the manufacturer's instructions. DNase treatment (DNase RQ1, Promega) was used to remove residual genomic DNA. The quality and quantity of purified total RNA were determined by measurement of absorbance (260 nm/280 nm) using a Nanodrop ND-1000 spectrophotometer (LabTech, USA). Poly(A) mRNA was isolated from total RNA using oligo(dT) magnetic beads (MicroPoly(A)PuristTM Kit, Ambion) according to the manufacturer's instructions. The first and second-strand cDNA synthesis was performed on purified mRNA using the SuperScript Double-Stranded cDNA Synthesis Kit (Invitrogen, USA) according to the manufacturer's protocol. The two cDNA libraries were constructed and sequenced with an Illumina HiSeq 2000 sequencer (Illumina Corporation Inc.). Approximately 4–5 µg of cDNA were used for library construction, undertaken by the Genoscope platform (http://www.genoscope.cns.fr). Sequencing was performed using the paired-ends method with a read length of 100 bases, producing an average of 183 Million of read pairs per transcriptome. Read pairs obtained were analyzed with FastQC software developed by S. Andrews in the Babraham Institute (www.bioinformatics.bbsrc.ac.uk) in order to validate run qualities (read number, quality score of nucleotide sequenced in Q-phred scale [58], composition of reads). We sequenced 35 and 37 Gb for Tiso-Wt and Tiso-S2M2 cDNA libraries respectively (Table 1) with a mean sequencing quality score per read at Q35. Sequence data for this article have been deposited in the National Center for Biotechnology Information and are accessible in: SRR823264 for Tiso-S2M2 and SRR824147 for Tiso-Wt.

Construction of the reference transcriptome for Tiso

Read pairs obtained for each sample were filtered to select only read pairs with the correct quality to be assembled. First, reads containing Illumina sequencing adapter and Illumina control sequences were eliminated with CutAdapt software [59]. In a second step, reads were filtered on a quality score of the last nucleotides for each read because quality of sequencing decreases proportionally with read length [60,61]. The last nucleotides were eliminated until detection of a nucleotide with a quality score of Q25. In a third step, reads were filtered on their length. Reads inferior to 75 bp were eliminated. In a final step, reads were filtered on a mean quality score of all nucleotides. Reads with a quality score inferior to Q25 were eliminated. After these different screens, 79.1% and 82.4% total read pairs for Tiso-Wt and Tiso-S2M2 respectively were conserved (Table 1).

Read pairs of each library were assembled with the Trinity pipeline [62] with the parameters advised by Zhao Q-Y *et al.*, [63]. The putative transcripts assembled for each library were clustered with CD-hit-EST software with 90% identity [64] to obtain the consensus transcripts considered as the reference transcriptome of Tiso (Table 1). Transcriptome datasets of each strain (Tiso-Wt and Tiso-S2M2) were aligned using the reference transcriptome with MosaikAssembler software (Wan-Ping Lee and Michael Strömberg, Marth lab). On average, 72.7% ($\sigma = 1.4$) read pairs were aligned on unique *locus*, 20.5% ($\sigma = 1.6$) aligned on multiple *loci* and 5.4% ($\sigma = 1.9$) were not aligned (Table 1). Transcripts composed at 90% of reads aligned in multiple *loci* were considered as genes with 2 or more isoforms.

Research of homologous genes from several reference microalgae and comparison of codon bias

Transcriptome data from Tiso were compared with data from 8 other reference microalgae. Data for 5 of these (the haptohyte *Emiliana huxleyi* [65], the diatoms *Phaeodactylum tricornutum* [66] and *Thalassiosira pseudonana* [67], the chlorophytes *Chlorella variabilis* [68] and *Chlamydomonas reinhardii* [69]) were produced by the US Department of Energy Joint Genome Institute (JGI http://www.jgi.doe.gov/) and are publically available in the web site. The other reference microalga used were the dinoflagellate *Alexandrium ostenfeldii* [70], the glaucophyte *Cyanophora paradoxa* [71] and the rhodophyte *Cyanidioschyzon merolae* [72]. Homologous transcripts were detected by BLAST analysis [73] between the transcriptome data of reference microalgae and Tiso (tblastx with alignment length greater than 100 amino acids and identity score greater than 30%). Codon usage bias was calculated with the Sequence Manipulation Suite software [74] on coding regions of each reference taxon. For Tiso, the putative coding regions of each transcript were identified with ORF-predictor [75]. A similarity tree of codon bias was built from correlation scores obtained between microalgae and drawn with the Darwin software (http://darwin.cirad.fr/) using the hierarchical clustering method WPGMA [76].

Annotation of Tiso trancriptome

Research of putative function was undertaken by BLAST analysis [73] on the NCBI database (nr bank; blastx with alignment length greater than 100 amino acids and identity score greater than 30%). Transcripts were sorted automatically for major cell functions. For the transcripts showing a polymorphism between Tiso-Wt and Tiso-S2M2, expert annotation was done. A second BLAST on SwissProt database (v. May2012) and functional domain was performed with InterProScan [77]. Consensus of annotation results was manually attributed for putative function for each transcript.

Differential expression analysis

After read alignment for each strain on the reference transcriptome, to search differential expression we normalized the libraries themselves as a function of mean depth per putative transcript (324× for Tiso-Wt and 389× for Tiso-S2M2). The number of read pairs aligned per transcript was compared between strains. Differential expression was considered if the difference between the 2 strains was up to 20-fold (log2) and if for both the depth was superior to 50 reads. Differential expressions were also confirmed with the recent GFOLD algorithm [78] with the p-value fixed at 0.001.

Positional polymorphism identification and estimation of genetic diversity

Positional polymorphisms, i.e. Single Nucleotide Polymorphism (SNP) and short insertion/deletions (indel), were identified between transcripts of Tiso-Wt and Tiso-S2M2. We used Freebayes software (Erik Garrison, Marth lab) and validated polymorphisms if depth was superior to 50× for each individual and if a minority of alleles had a depth superior to 10×.

Genetic diversity was measured for each strain (Tiso-Wt and Tiso-S2M2). Polyallelic *loci* were detected in transcriptomes of each strain. A *locus* was considered polyallelic if the depth of reads was greater than 50×. For each polyallelic *locus*, frequency of each allele was measured (number of alleles divided by sum of total alleles). Comparison of global genetic diversity between the both strains was estimate with D of Jost [31] using SPADE program [79].

Sequence data for this article have been deposited in the National Center for Biotechnology Information and are accessible in: SRR823264 for Tiso-S2M2 and SRR824147 for Tiso-Wt.

Supporting Information

Data S1 Methods and results of RT-qPCR approach for the candidates *TisoTranscripts-288* and *TisoTranscripts-160*.

Figure S1 Enzymes of lipid pathways identified in universal KEGG. Each enzyme with a colour corresponds to a transcript annotated in KEEG. Figure A corresponds at lipid synthesis and Figure B catabolism of lipid.

Figure S2 Comparison of G+C content in transcriptomes of Tiso and reference microalgae. Colour code: green for chlorophytes, red for rhodophyte, orange for glaucophyte, blue for dinoflagellate, yellow for diatoms and purple for haptophytes.

Figure S3 Codon bias analysis. Figure 3A shows codon bias of Tiso and 3B shows comparison of codon bias between reference microalgae. The tree was built using a correlation score matrix of codon bias between microalgae with the hierarchical clustering method WPGMA. Color code: green for chlorophytes, red for rhodophyte, orange for glaucophyte, blue for dinoflagellate, yellow for diatoms and purple for haptophytes.

Figure S4 Distribution of positional polymorphisms (SNPs and indels) between Tiso-Wt and Tiso-S2M2 per transcripts.

Figure S5 Growth of Tiso strains and lipid accumulation. Cells were counted with a Malassez counting cell and image analysis (SAMBA software). Lipid accumulation was estimated by measurement of Nile Red fluorescence by spectrofluorimetry as described by Bougaran *et al.*, [22]. Red color for Tiso-Wt and black color for Tiso-S2M2. The yellow color shows sampling events.

Table S1 List of transcripts contained polymorphism identified between Tiso-Wt and Tiso-S2M2.

Acknowledgments

Christophe Caron from the ABiMS platform and Laure Quintric from IFREMER-Caparmor platform for the bioinformatic server maintenance. We acknowledge the Genoscope platform for the sequencing.

Author Contributions

Conceived and designed the experiments: JPC MG GC GB BSJ. Performed the experiments: GC MG BSJ. Analyzed the data: GC. Contributed reagents/materials/analysis tools: EC IP CDV LLC. Wrote the paper: GC MG BSJ JPC LLC EC IP.

References

1. Spolaore P, Joannis-Cassan C, Duran E, Isambert A (2006) Commercial applications of microalgae. Journal of Bioscience and Bioengineering 101: 87–96. doi:10.1263/jbb.101.87.
2. Sili C, Torzillo G, Vonshak A (2012) Arthrospira (Spirulina). In: Whitton BA, editor. Ecology of Cyanobacteria II. Springer Netherlands. pp. 677–705.
3. Gong N, Shao K, Feng W, Lin Z, Liang C, et al. (2011) Biotoxicity of nickel oxide nanoparticles and bio-remediation by microalgae Chlorella vulgaris. Chemosphere 83: 510–516. doi:10.1016/j.chemosphere.2010.12.059.
4. Campo JAD, García-González M, Guerrero MG (2007) Outdoor cultivation of microalgae for carotenoid production: current state and perspectives. Appl Microbiol Biotechnol 74: 1163–1174. doi:10.1007/s00253-007-0844-9.
5. Cadoret J-P, Garnier M, Saint-Jean B (2012) Chapter Eight - Microalgae, Functional Genomics and Biotechnology. In: Gwenaël Piganeau, editor. Advances in Botanical Research. Academic Press, Vol. Volume 64. pp. 285–341. Available: http://www.sciencedirect.com/science/article/pii/B9780123914996000086.
6. Saei A, Ghanbari P, Barzegari A (2012) Haematococcus as a promising cell factory to produce recombinant pharmaceutical proteins. Mol Biol Rep 39: 9931–9939. doi:10.1007/s11033-012-1861-z.
7. Cadoret J-P, Bernard O (2008) Lipid biofuel production with microalgae: potential and challenges. Journal de la Société de Biologie 202: 201–211. doi:10.1051/jbio:2008022.
8. Cadoret J-P, Bardor M, Lerouge P, Cabigliera M, Henriquez V, et al. (2008) Les microalgues : Usines cellulaires productrices de molécules commerciales recombinantes. Medecine Science 24: 375–382.
9. Wijffels RH, Barbosa MJ (2010) An Outlook on Microalgal Biofuels. Science 329: 796–799.
10. Larkum AWD, Ross IL, Kruse O, Hankamer B (2012) Selection, breeding and engineering of microalgae for bioenergy and biofuel production. Trends in Biotechnology 30: 198–205. doi:10.1016/j.tibtech.2011.11.003.
11. Bendif E, Probert I, Schroeder D, De Vargas C (2013) The description of *Tisochrysis lutea* gen. nov. sp. nov. and *Isochrysis nuda* sp. nov. in the Isochrysidales, and the transfer of Dicrateria to the Prymnesiales (Haptophyta). Journal of Applied Phycology 25: 1763–1776. doi:10.1007/s10811-013-0037-0
12. Mata TM, Martins AA, Caetano NS (2010) Microalgae for biodiesel production and other applications: A review. Renewable and Sustainable Energy Reviews 14: 217–232. doi:10.1016/j.rser.2009.07.020.
13. Mardis E (2008) The impact of next-generation sequencing technology on genetics. Trends in genetics : TIG 24: 133–141. doi:10.1016/j.tig.2007.12.007.
14. Liu H, Probert I, Uitz J, Claustre H, Aris-Brosou S, et al. (2009) Extreme diversity in noncalcifying haptophytes explains a major pigment paradox in open oceans. Proceedings of the National Academy of Sciences 106: 12803–12808.
15. Houdan A, Billard C, Marie D, Not F, Sáez AG, et al. (2004) Holococcolithophore-heterococcolithophore (Haptophyta) life cycles: Flow cytometric analysis of relative ploidy levels. Systematics and Biodiversity 1: 453–465. doi:10.1017/S1477200003001270.
16. Jordan RW, Cros L, Young JR (2004) A revised classification scheme for living haptophytes. Micropaleontology 50: 55–79.
17. Vargas C de, Probert I (2004) New Keys to the Past: Current and Future DNA Studies in Coccolithophores. Micropaleontology 50: 45–54. doi:10.2307/4097102.
18. Chepurnov VA, Chaerle P, Roef L, Meirhaeghe A, Vanhoutte K (2011) Classical Breeding in Diatoms: Scientific Background and Practical Perspectives. In: Seckbach J, Kociolek P, editors. Cellular Origin, Life in Extreme Habitats and Astrobiology. Springer Netherlands, Vol. 19. pp. 167–194. Available: http://dx.doi.org/10.1007/978-94-007-1327-7_7.
19. Kim S, Lee H, Lee Y, Yim J (2008) Mutant selection of Hahella chejuensis KCTC 2396 and statistical optimization of medium components for prodigiosin yield-up. The Journal of Microbiology 46: 183–188. doi:10.1007/s12275-008-0037-y.

20. Mendoza H, De la Jara A, Freijanes K, Carmona L, Ramos AA, et al. (2009) Characterization of Dunaliella salina strains by flow cytometry: a new approach to select carotenoid hyperproducing strains.
21. Doan TTY, Obbard JP (2012) Enhanced intracellular lipid in Nannochloropsis sp. via random mutagenesis and flow cytometric cell sorting. Algal Research 1: 17–21. doi:10.1016/j.algal.2012.03.001.
22. Bougaran G, Rouxel C, Dubois N, Kaas R, Grouas S, et al. (2012) Enhancement of neutral lipid productivity in the microalga Isochrysis affinis Galbana (T-Iso) by a mutation-selection procedure. Biotechnol Bioeng. doi:10.1002/bit.24560.
23. Mendoza Guzmán H, Jara Valido A, Freijanes Presmanes K, Carmona Duarte L (2012) Quick estimation of intraspecific variation of fatty acid composition in Dunaliella salina using flow cytometry and Nile Red. J Appl Phycol 24: 1237–1243. doi:10.1007/s10811-011-9768-y.
24. Evans KM, Chepurnov VA, Sluiman HJ, Thomas SJ, Spears BM, et al. (2009) Highly Differentiated Populations of the Freshwater Diatom Sellaphora capitata Suggest Limited Dispersal and Opportunities for Allopatric Speciation. Protist 160: 386–396. doi:10.1016/j.protis.2009.02.001.
25. Evans KM, Kühn SF, Hayes PK (2005) High levels of genetic diversity and low levels of genetic differentiation in north sea pseudo-nitzschia pungens (bacillariophyceae) populations. Journal of Phycology 41: 506–514. doi:10.1111/j.1529-8817.2005.00084.x.
26. Rouxel C, Bougaran G, Doulin Grouas S, Dubois N, Cadoret J-P (2011) novel isochrysis sp. Thahitan clone and uses therefore. EP 11006712.1
27. Wang Z, Gerstein Mark, Snyder Michael (2009) RNA-Seq: a revolutionary tool for transcriptomics. Nat Rev Genet 10: 57–63.
28. Cracraft J, Donoghue MJ (2004) Assembling the Tree of Life. Oxford University Press. 594 p.
29. Fox JM, Erill I (2010) Relative Codon Adaptation: A Generic Codon Bias Index for Prediction of Gene Expression. DNA Res 17: 185–196. doi:10.1093/dnares/dsq012.
30. Wuitschick JD, Karrer KM (1999) Analysis of Genomic G+C Content, Codon Usage, Initiator Codon Context and Translation Termination Sites In Tetrahymena Thermophila. Journal of Eukaryotic Microbiology 46: 239–247. doi:10.1111/j.1550-7408.1999.tb05120.x.
31. Chao A, Jost L, Chiang SC, Jiang Y-H, Chazdon RL (2008) A Two-Stage Probabilistic Approach to Multiple-Community Similarity Indices. Biometrics 64: 1178–1186. doi:10.1111/j.1541-0420.2008.01010.x.
32. Misra N, Panda PK, Parida BK, Mishra BK, Misra N, et al. (2012) Phylogenomic Study of Lipid Genes Involved in Microalgal Biofuel Production-Candidate Gene Mining and Metabolic Pathway Analyses. Evolutionary Bioinformatics 8: 545. doi:10.4137/EBO.S10159.
33. Bendif EM, Probert I, Hervé A, Billard C, Goux D, et al. (2011) Integrative Taxonomy of the Pavlovophyceae (Haptophyta): A Reassessment. Protist 162: 738–761. doi:10.1016/j.protis.2011.05.001.
34. Sáez AG, Probert I, Geisen M, Quinn P, Young JR, et al. (2003) Pseudo-cryptic speciation in coccolithophores. Proceedings of the National Academy of Sciences 100: 7163–7168.
35. Hamblin MT, Di Rienzo A (2000) Detection of the Signature of Natural Selection in Humans: Evidence from the Duffy Blood Group Locus. The American Journal of Human Genetics 66: 1669–1679. doi:10.1086/302879.
36. Singh RS, Krimbas CB (2000) Evolutionary Genetics: From Molecules to Morphology. Cambridge University Press. 736 p.
37. Hermisson J, Redner O, Wagner H, Baake E (2002) Mutation–Selection Balance: Ancestry, Load, and Maximum Principle. Theoretical Population Biology 62: 9–46. doi:10.1006/tpbi.2002.1582.
38. Reed DH (2009) When it comes to inbreeding: slower is better. Molecular Ecology 18: 4521–4522. doi:10.1111/j.1365-294X.2009.04367.x.
39. Jump AS, Marchant R, Peñuelas J (2009) Environmental change and the option value of genetic diversity. Trends in Plant Science 14: 51–58. doi:10.1016/j.tplants.2008.10.002.
40. Kristensen TN, Sørensen AC, Sorensen D, Pedersen KS, Sørensen JG, et al. (2005) A test of quantitative genetic theory using Drosophila– effects of inbreeding and rate of inbreeding on heritabilities and variance components. Journal of Evolutionary Biology 18: 763–770. doi:10.1111/j.1420-9101.2005.00883.x.
41. Demontis D, Pertoldi C, Loeschcke V, Mikkelsen K, Axelsson T, et al. (2009) Efficiency of selection, as measured by single nucleotide polymorphism variation, is dependent on inbreeding rate in Drosophila melanogaster. Molecular Ecology 18: 4551–4563. doi:10.1111/j.1365-294X.2009.04366.x.
42. Taylor R, Fletcher RL (1998) Cryopreservation of eukaryotic algae – a review of methodologies. Journal of Applied Phycology 10: 481–501. doi:10.1023/A:1008094622412.
43. Klug A (2010) The Discovery of Zinc Fingers and Their Applications in Gene Regulation and Genome Manipulation. Annual Review of Biochemistry 79: 213–231. doi:10.1146/annurev-biochem-010909-095056.
44. Wright EM, Turk E (2004) The sodium/glucose cotransport family SLC5. Pflugers Arch - Eur J Physiol 447: 510–518. doi:10.1007/s00424-003-1063-6.
45. Wright EM, Loo DDF, Hirayama BA (2011) Biology of Human Sodium Glucose Transporters. Physiol Rev 91: 733–794. doi:10.1152/physrev.00055.2009.
46. Fang Z, Cui X (2011) Design and validation issues in RNA-seq experiments. Brief Bioinformatics 12: 280–287. doi:10.1093/bib/bbr004.
47. Al-Khodor S, Price CT, Kalia A, Abu Kwaik Y (2010) Functional diversity of ankyrin repeats in microbial proteins. Trends in Microbiology 18: 132–139. doi:10.1016/j.tim.2009.11.004.
48. Müller A, Klöppel C, Smith-Valentine M, Van Houten J, Simon M (2012) Selective and programmed cleavage of GPI-anchored proteins from the surface membrane by phospholipase C. Biochimica et Biophysica Acta (BBA) - Biomembranes 1818: 117–124. doi:10.1016/j.bbamem.2011.10.009.
49. Groot PH, Scholte HR, Hülsmann WC (1976) Fatty acid activation: specificity, localization, and function. Adv Lipid Res 14: 75–126.
50. Watkins PA (1997) Fatty acid activation. Prog Lipid Res 36: 55–83.
51. Akoh C, Lee G, Liaw Y, Huang T, Shaw J (2004) GDSL family of serine esterases/lipases. Prog Lipid Res 43: 534–552. doi:10.1016/j.plipres.2004.09.002.
52. Kroeker KJ, Kordas RL, Crim RN, Singh GG (2010) Meta-analysis reveals negative yet variable effects of ocean acidification on marine organisms. Ecology Letters 13: 1419–1434. doi:10.1111/j.1461-0248.2010.01518.x.
53. Morgan TH (1917) The Theory of the Gene. The American Naturalist 51: 513–544. doi:10.2307/2456204.
54. Darwin C (1859) On the origin of species by means of natural selection, or the preservation of favoured races in the struggle for life.
55. Chepurnov VA, Mann DG, Sabbe K, Vyverman W (2004) Experimental Studies on Sexual Reproduction in Diatoms. International Review of Cytology. Academic Press, Vol. Volume 237. pp. 91–154. Available: http://www.sciencedirect.com/science/article/pii/S0074769604370038.
56. Mouget J-L, Gastineau R, Davidovich O, Gaudin P, Davidovich NA (2009) Light is a key factor in triggering sexual reproduction in the pennate diatom Haslea ostrearia. FEMS Microbiology Ecology 69: 194–201. doi:10.1111/j.1574-6941.2009.00700.x.
57. Walne P (1966) Experiments in the large-scale culture of the larvae of Ostrea edulis L. 2. (London): Minist. Agric. Fish. Invest., Vol. 25. p. 53970. Available: http://openlibrary.org/b/OL5568472M/Experiments_in_the_large-scale_culture_of_the_larvae_of_Ostrea_edulis_L.
58. Ewing B, Green P (1998) Base-Calling of Automated Sequencer Traces UsingPhred. II. Error Probabilities. Genome Res 8: 186–194. doi:10.1101/gr.8.3.186.
59. Martin M (2011) Cutadapt removes adapter sequences from high-throughput sequencing reads. EMBnet.journal; Vol 17, No 1: Next Generation Sequencing Data Analysis. Available: http://journal.embnet.org/index.php/embnetjournal/article/view/200.
60. Carrier G, Le Cunff L, Dereeper A, Legrand D, Sabot F, et al. (2012) Transposable Elements Are a Major Cause of Somatic Polymorphism in Vitis vinifera L. PLoS ONE 7: e32973. doi:10.1371/journal.pone.0032973.
61. Dohm JC, Lottaz C, Borodina T, Himmelbauer H (2008) Substantial biases in ultra-short read data sets from high-throughput DNA sequencing. Nucleic Acids Research 36: e105–e105.
62. Grabherr MG, Haas BJ, Yassour M, Levin JZ, Thompson DA, et al. (2011) Full-length transcriptome assembly from RNA-Seq data without a reference genome. Nat Biotech: 1087–0156.
63. Zhao Q-Y, Wang Y, Kong Y-M, Luo D, Li X, et al. (2011) Optimizing de novo transcriptome assembly from short-read RNA-Seq data: a comparative study. BMC Bioinformatics 12: S2. doi:10.1186/1471-2105-12-S14-S2.
64. Fu L, Niu B, Zhu Z, Wu S, Li W (2012) CD-HIT: accelerated for clustering the next generation sequencing data. Bioinformatics. Available: http://bioinformatics.oxfordjournals.org/content/early/2012/10/11/bioinformatics.bts565.abstract.
65. Puerta MVS, Bachvaroff TR, Delwiche CF (2004) The Complete Mitochondrial Genome Sequence of the Haptophyte Emiliania huxleyi and its Relation to Heterokonts. DNA Research 11: 1–10. doi:10.1093/dnares/11.1.1.
66. Bowler C, Allen AE, Badger JH, Grimwood J, Jabbari K, et al. (2008) The Phaeodactylum genome reveals the evolutionary history of diatom genomes. Nature 456: 239–244. doi:10.1038/nature07410.
67. Armbrust EV, Berges JA, Bowler C, Green BR, Martinez D, et al. (2004) The Genome of the Diatom Thalassiosira Pseudonana: Ecology, Evolution, and Metabolism. Science 306: 79–86.
68. Blanc G, Duncan G, Agarkova I, Borodovsky M, Gurnon J, et al. (2010) The Chlorella variabilis NC64A Genome Reveals Adaptation to Photosymbiosis, Coevolution with Viruses, and Cryptic Sex. The Plant Cell Online 22: 2943–2955.
69. Merchant SS, Prochnik SE, Vallon O, Harris EH, Karpowicz SJ, et al. (2007) The Chlamydomonas Genome Reveals the Evolution of Key Animal and Plant Functions. Science 318: 245–250.
70. Jaeckisch N, Yang I, Wohlrab S, Glöckner G, Kroymann J, et al. (2011) Comparative Genomic and Transcriptomic Characterization of the Toxigenic Marine Dinoflagellate Alexandrium ostenfeldii. PLoS ONE 6: e28012. doi:10.1371/journal.pone.0028012.
71. Nozaki H, Maruyama S, Matsuzaki M, Nakada T, Kato S, et al. (2009) Phylogenetic positions of Glaucophyta, green plants (Archaeplastida) and Haptophyta (Chromalveolata) as deduced from slowly evolving nuclear genes. Molecular Phylogenetics and Evolution 53: 872–880. doi:10.1016/j.ympev.2009.08.015.
72. Matsuzaki M, Misumi O, Shin-i T, Maruyama S, Takahara M, et al. (2004) Genome sequence of the ultrasmall unicellular red alga Cyanidioschyzon merolae 10D. Nature 428: 653–657. doi:10.1038/nature02398.

73. Altschul SF, Gish W, Miller W, Myers EW, Lipman DJ (1990) Basic local alignment search tool. Journal of Molecular Biology 215: 403–410. doi:10.1016/S0022-2836(05)80360-2.
74. Stothard P (2000) The sequence manipulation suite: JavaScript programs for analyzing and formatting protein and DNA sequences. BioTechniques 28. Available: http://view.ncbi.nlm.nih.gov/pubmed/10868275.
75. Min XJ, Butler G, Storms R, Tsang A (2005) OrfPredictor: predicting protein-coding regions in EST-derived sequences. Nucleic Acids Research 33: W677–W680.
76. Perrier X, Flori A, Bonnot F (2003) Data analysis methods. Genetic diversity of cultivated tropical plants. Enfield, Science Publishers. p. pp 43–76.
77. Hunter S, Jones P, Mitchell A, Apweiler R, Attwood TK, et al. (2012) InterPro in 2011: new developments in the family and domain prediction database. Nucleic Acids Research 40: D306–D312.
78. Feng J, Meyer CA, Wang Q, Liu JS, Liu XS, et al. (2012) GFOLD: a generalized fold change for ranking differentially expressed genes from RNA-seq data. Bioinformatics 28: 2782–2788. doi:10.1093/bioinformatics/bts515.
79. Chao A, Shen T-J (2010) SPADE (Species Prediction And Diversity Estimation).

20

Exploring Germplasm Diversity to Understand the Domestication Process in *Cicer* spp. Using SNP and DArT Markers

Manish Roorkiwal[1], Eric J. von Wettberg[2,3], Hari D. Upadhyaya[1], Emily Warschefsky[2,3], Abhishek Rathore[1], Rajeev K. Varshney[1]*

1 International Crops Research Institute for the Semi-Arid Tropics (ICRISAT), Hyderabad, Andhra Pradesh, India, **2** Department of Biological Sciences, Florida International University, Miami, Florida, United States of America, **3** Center for Tropical Plant Conservation, Fairchild Tropical Botanic Garden, Miami, Florida, United States of America

Abstract

To estimate genetic diversity within and between 10 interfertile *Cicer* species (94 genotypes) from the primary, secondary and tertiary gene pool, we analysed 5,257 DArT markers and 651 KASPar SNP markers. Based on successful allele calling in the tertiary gene pool, 2,763 DArT and 624 SNP markers that are polymorphic between genotypes from the gene pools were analyzed further. STRUCTURE analyses were consistent with 3 cultivated populations, representing kabuli, desi and pea-shaped seed types, with substantial admixture among these groups, while two wild populations were observed using DArT markers. AMOVA was used to partition variance among hierarchical sets of landraces and wild species at both the geographical and species level, with 61% of the variation found between species, and 39% within species. Molecular variance among the wild species was high (39%) compared to the variation present in cultivated material (10%). Observed heterozygosity was higher in wild species than the cultivated species for each linkage group. Our results support the Fertile Crescent both as the center of domestication and diversification of chickpea. The collection used in the present study covers all the three regions of historical chickpea cultivation, with the highest diversity in the Fertile Crescent region. Shared alleles between different gene pools suggest the possibility of gene flow among these species or incomplete lineage sorting and could indicate complicated patterns of divergence and fusion of wild chickpea taxa in the past.

Editor: Tianzhen Zhang, Nanjing Agricultural University, China

Funding: This study was funded by grants from Australian Indo Strategic Research Fund (AISRF) of Ministry of Science & Technology and CGIAR Generation Challenge Programme (GCP), Mexico. This work has been undertaken as part of the CGIAR Research Program on Grain Legumes. ICRISAT is a member of CGIAR Consortium. Thanks are also due to several colleagues at ICRISAT and partners in collaborating centres. EvW received support from HHMI award #52006924 to FIU's discipline-based education research group, and USDA-NIFA- Hispanic Serving Institutions Grant 2011-38422-30804 to FIU's agroecology program. EW is supported by the Fairchild Challenge of Fairchild Tropical Botanic Garden. The funders had no role in study design, data collection and analysis, decision to publish, or preparation of the manuscript.

Competing Interests: The authors have declared that no competing interests exist.

* Email: r.k.varshney@cgiar.org

Introduction

Many crops that are grown across multiple regions have limited genetic diversity due to bottlenecks from domestication, selective breeding and in some taxa, natural processes [1–4]. Recurrent selection of improved cultivars over multiple generations results in an increasingly narrow genetic base for a crop, making it more vulnerable to disease and limiting its adaptability. Such genetically depauperate crops could have disastrous consequences in the face of emerging diseases and climate change [5,6]. Recent applications of genome mapping suggest that the genetic diversity stored in germplasm banks can be utilized with a much higher level of efficiency than previously imagined [6,7]. This is particularly true for self-pollinated crops like chickpea (*Cicer arietinum*). During the past few decades, our understanding of the importance of plant genetic resources and the need to conserve them has grown [8], and wild relatives are now commonly seen as a key source of genetic diversity that can be used to increase diversity in breeding material [7,9]. Diversity estimates of germplasm collections have not been universally performed to assess the scope of diversity available in existing collections. Such estimates are critical for providing insight into efforts to introgress wild germplasm into elite lines, and for guiding future collections of wild germplasm [10].

In order to make more efficient use of wild relatives, we need improved classifications of their relationship to crop material and to other wild species [11]. Characterizing patterns of diversity within the secondary and tertiary gene pools [12] can provide insight into which subdivisions of germplasm collections contain wild material that is most likely to increase diversity and can guide the use of wild material in breeding efforts. Although wild material is rarely used in breeding programs due to agronomically poor traits, it remains a chief reservoir for many disease and abiotic stress resistance traits. Effective characterization of wild material can facilitate its more effective use [13].

Chickpea is an important crop in semi-arid tropical regions such as South Asia and Eastern & Southern Africa, Mediterranean regions, and cool temperate areas [14]. Globally, chickpea is the second most widely consumed legume after beans (*Phaseolus*) [15]. Lack of genetic diversity has long been a critical problem for chickpea breeding [16], limiting efforts to improve resistance to diseases like *Ascochyta* blight and *Fusarium* wilt, pod borer insects, and tolerance to abiotic stresses like terminal drought, high and low temperatures [17,18]. Chickpea reference set has also been used to understand the available diversity for stress responsive genes [19]. Widening the genetic diversity of cultivated chickpea is dependent on the introduction of alleles controlling the traits of interest from wild germplasm [1]. Currently chickpea's immediate ancestor, *C. reticulatum*, and its interfertile sister species *C. echinospermum*, is the main source of new variation, although introgression is possible from the more distantly related gene pools with greater effort [20].

Cultivated chickpea first appears in the archaeological record some 6.6–7.2 thousand years ago in Syria [21,22]. The immediate wild relatives (*C. reticulatum* and *C. echinospermum*) of chickpea are restricted to southeastern Turkey [1]. Domestication is thought to have happened earlier, as much as 10.5 thousand years ago, concurrent with or soon after the domestication of other Fertile Crescent crops such as wheat, barley, pea, and lentil. Domesticated chickpea was likely brought to Syria about 7,000 years ago, while records for the dates of introduction into East Africa and the Indian subcontinent are limited [22]. Abbo and co-workers [1,23] have speculated that chickpea is particularly genetically depauperate because it may have gone through four distinct bottlenecks: modern breeding, domestication, a shift early in its cultivation from a winter annual phenology to a spring phenology, and wild relatives (particularly *C. reticulatum* and *C. echinospermum*) that have a narrow geographic distribution compared to other crops domesticated in the Fertile Crescent. The shift in phenology may have accompanied the introduction of other crops such as sesame and sorghum that are summer annuals [24]. Breeding for preferred phenotypes, such as seed colour and shape, may exacerbate chickpea's narrow genetic base and may be one of the key reasons for slow progress in yield improvement and increased tolerance to various biotic and abiotic stresses. Based on seed shape, size and colour, chickpea is classified into two seed types, kabuli and desi. The kabuli chickpea is characterized by a larger, cream-coloured seed with a thin seed coat, while the desi seed type has a smaller, darker coloured seed with a thick seed coat. In addition, a third seed type, designated as intermediate or pea-shaped, is characterized by medium to small size and round, pea-shaped seeds [25].

Single nucleotide polymorphism (SNP) markers have become the markers of choice for various genome wide analyses because they are widespread across genomes, accurate and reproducible, and well suited to automated detection [26]. A range of low- to high-throughput SNP genotyping platforms have become available to make SNP genotyping cost-effective such as BeadXpress, KBioscience Competitive Allele-Specific Polymerase chain reaction (KASPar) assays, and GoldenGate assays from Illumina Inc. [27,28]. In addition, another high-throughput marker system, Diversity arrays technology (DArT), has proven useful for screening large numbers of loci in crops with low genetic diversity, and DArT markers for chickpea have recently been developed [29].

The present study is focused on the assessment of relationships in a diversity panel of chickpea which includes breeding material from the three seed types (kabuli, desi, and pea-shaped) and wild species from the primary, secondary, and tertiary gene pools using KASPar technology and hybridization based DArT arrays for high-throughput SNP genotyping. We examined the level of genetic differentiation among these groups of genotypes and assessed how segregating variation is spread across the genome of chickpea.

Materials and Methods

Germplasm and DNA isolation

A diverse set of 94 chickpea genotypes (Table S1) including 66 cultivars and landraces (23 desi, 41 kabuli, and 2 pea-shaped seed type genotypes) and 28 genotypes from 9 wild species including genotypes from primary, secondary and tertiary gene pool was selected as a diversity panel for assessment from the ICRISAT germplasm collection [30].

Total genomic DNA was isolated from 10–12 leaves of two week old plants following a modified CTAB protocol as described in Cuc et al. [31]. Only one plant per accession was used for DNA isolation. DNA quality and quantity for each sample was assessed on 0.8% agarose gel.

Genotyping

SNPs were identified using four different approaches: Solexa/Illumina sequencing, mining of Sanger Expressed Sequence Tags (ESTs), allele-specific sequencing of candidate genes, and allele-specific sequencing of tentative orthologous genes (TOGs) as described by Hiremath et al. [28]. In total, 2,486 SNPs were used for validation and development of KASPar assays by KBioscience, of which 2,005 (80.6%) assays could be validated and designated as Chickpea KASPar Assay Markers (CKAMs) [28]. A subset of highly polymorphic 651 CKAMs was used for genotyping using KASPar assays. In addition, this diverse set was also genotyped with high-density DArT array with 15,360 DArT clones as described in Thudi et al. [29].

Data Analysis

The germplasm was divided into three different clusters based on geographical origin, namely the Fertile Crescent, Central and South Asia, and Ethiopian Highlands (Figure 1). Additionally, germplasm was classified based on gene pools (primary, secondary, and tertiary) [32], seed type (desi, kabuli, and pea-shaped) and wild vs. cultivated species. The purpose of these different divisions of the data was to determine the scale over which genetic variation is present in the germplasm collection. In order to assess hierarchical levels of variation within and between different sub-groups, DArT and SNP genotyping data were analyzed separately. AMOVA was conducted on the DArT markers based on the hierarchical model and permutational procedures of Excoffier et al. [33] to assess the level of variation among these wild and domesticated groups. We implemented AMOVA in GenAlEx 6.5 [34,35] and Arlequin [36]. AMOVA analysis with populations nested within regions was also performed to examine the distribution of variation and differential connectivity among populations (PhiPT; an analogue of Fst, i.e., genetic diversity among populations). In addition, Shannon information index (measure of species diversity in a population) was calculated for all the population using GenAlEx 6.5. This index provides important information about rarity and commonness of species in a community by taking relative abundances of different species into account [34,37].

A separate AMOVA was performed on the SNP data to assess variation within and among desi, kabuli, and pea-shaped seed types. In both AMOVAs, we assessed genetic variation within groups (Fct), within populations (Fst), between populations within a group (Fsc), population polymorphism, and Nei's genetic distance and gene flow (Nm) using GenAlEx v.6.41 [34,35] and

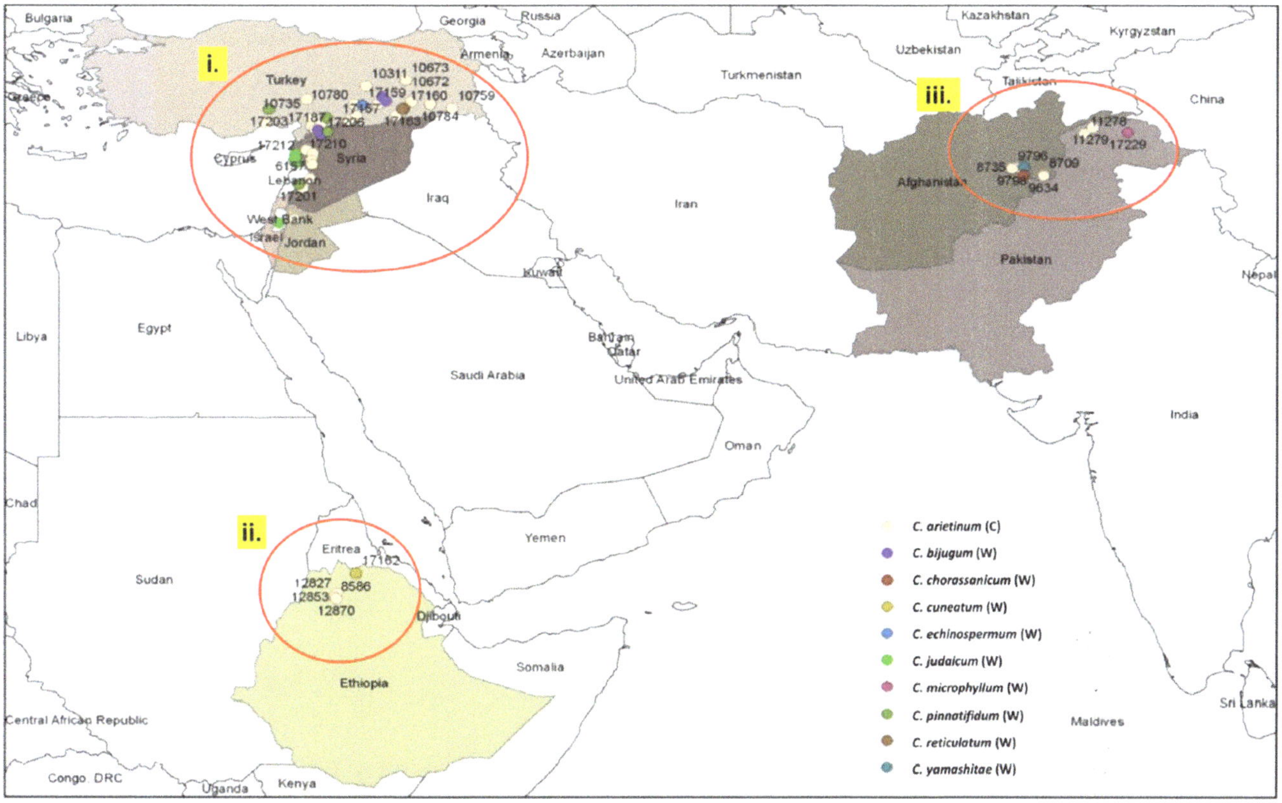

Figure 1. Geographic locations of cultivated and wild *Cicer* species collection sites (C: Cultivated; W: Wild) i. Fertile Crescent; ii. Ethiopia; iii. Central Asia.

Arlequin [36]. For each group presence of private alleles (np), percentage of polymorphic loci (%p), the average number of alleles per locus (k), the expected heterozygosity (He), and unbiased expected heterozygosity (UHe) across different subgroups (i.e., wild species *vs* cultivated with the DArT markers and seed type with the SNP markers) was calculated. The polymorphism information content (PIC) values for SNP and DArT markers across 94 diverse genotypes were calculated by using Power-Marker software [38].

STRUCTURE 2.3 [39] was used to estimate the number of natural genetic groups (K), the distribution of individuals among these groups, and to assign individual genotypes to a specified number of groups "K" based on membership coefficients calculated from the genotype data. This approach is an important complement to the hierarchical division of the germplasm (see above), as it can determine the number of groups best supported by the DArT and SNP data. DArT data was converted in to psuedo-diploid format by assigning a row of missing data to each individual so that it could be analysed with STRUCTURE. We assessed a range of population numbers from K = 1 to K = 15 using a burn-in period of 50,000 steps followed by 500,000 MCMC (Monte Carlo Markov Chain) replicates with 3X iterations, assuming admixture and correlated allele frequencies. Due to missing SNP calls in the wild material, data from wild material was separated from that of cultivated material and a separate STRUCTURE analysis of cultivated material alone was performed using SNP markers. In order to compliment the STRUCTURE analyses, pair-wise genetic differentiation between individuals was calculated from the DarT markers, which was used in principal coordinate analysis (PCoA), implemented in GenAlEx 6.5. These analyses labelled the material based on its source region: the Fertile Crescent, Central Asia, and the Ethiopian highlands.

A complementary approach to assessing relationships among taxa is a phylogenetic analysis. Distance-based phylogenetic analysis of SNP data was performed using the software package Geneious v. 7.0.6 (Biomatters) (http://www.geneious.com). A cladogram was produced using unweighted pair-group method with arithmetic mean (UPGMA) cluster analysis under the Jukes-Cantor genetic distance model with 100 bootstrap replications. The consensus tree was then rooted with the clade of individuals from the tertiary gene pool.

Results

Marker attributes

In total, 651 SNP markers using KASPar assays and DArT arrays were used for genotyping the set of 94 diverse chickpea genotypes. This set includes 66 cultivated chickpea genotypes and 27 wild relatives representing eight wild *Cicer* species from primary, secondary, and tertiary gene pools along with one perennial wild chickpea genotype. The genotypes were carefully selected to represent geographical areas with the most phenotypic diversity: the Fertile Crescent, Central Asia, and the Ethiopian highlands (Figure 1). SNP markers were highly polymorphic across this diverse set and a total of 611 SNPs were found polymorphic. The polymorphic information content (PIC) value ranged from 0.02 to 0.50 across these 94 genotypes with mean PIC value of 0.23 (Figure 2a). Although these SNPs were highly polymorphic, in many cases SNPs could not be called for wild chickpea genotypes (Table S2). SNPs were developed using cultivated chickpea and

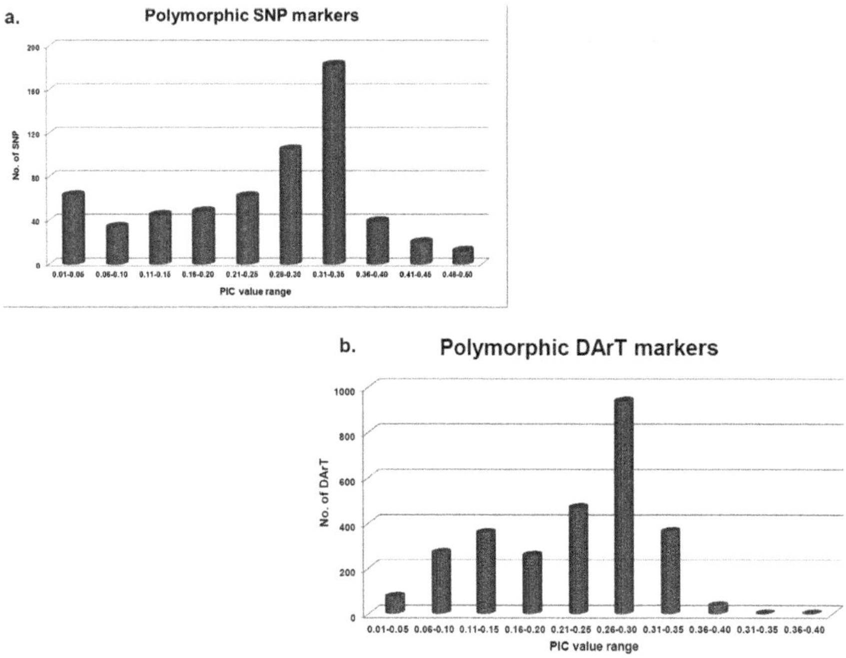

Figure 2. Polymorphism information content (PIC) value of markers used in study. a. PIC value of SNP markers used for diversity analysis. b. PIC value of DArT markers used for diversity analysis.

later used for genotyping the wild species, which may account for the greater number of missing loci in the tertiary gene pool and the bimodal distribution of PIC values.

To overcome the issue of missing data in the wild material and to compliment the SNP data, the set was genotyped using high density DArT arrays with 15,360 clones [29]. A total, 5,257 DArT markers were polymorphic across 94 lines. Of these, a subset of 2,763 markers was selected for use in the present study based on the presence of the allele in wild chickpea (tertiary gene pool). PIC for these 2,763 DArT markers ranged from 0.02 to 0.37, with an average of 0.22 across the 94 genotypes (Figure 2b) (Table S3).

Differences among the wild species and cultivated germplasm

The chickpea diversity panel used in the present study is comprised of 94 genotypes from 9 wild species (8 annual and 1 perennial) and cultivated species (*C. arietinum*). DArT data was used to understand the diversity and genetic architecture of the germplasm. As expected, wild species genotypes had higher levels of polymorphic markers (99.60%) compared to cultivated genotypes (35.79%) (Table 1). A UPGMA tree was constructed based on pairwise genetic distances using the SNP markers to understand the relationships between the genotypes from wild and cultivated species (Figure 3). Two major groups were identified by this analysis, separating wild from cultivated genotypes. Cultivated and wild species genotypes from the primary gene pool were grouped in one cluster (Figure 3). However, genotypes from the chickpea ancestor, *C. reticulatum,* were interspersed with those from cultivated individuals, consistent with a close relationship between ancestral and cultivated chickpea. Genotypes from the secondary gene pool species were found to cluster together, as were genotypes from the tertiary gene pool.

In parallel, STRUCTURE was also used to understand the clustering between cultivated and wild species genotypes. With the DArT data, STRUCTURE resolved four clusters using the Evanno method (Figure 4a). This grouping indicates a substantial difference between wild and cultivated material, as well as major differences within the wild material. These results suggest that there are three major groups among the wild material (Figure 4a), corresponding to different gene pools. Individuals in the tertiary gene pool are represented largely as one cluster with admixture; although these individuals represent several species (with the capacity to hybridize) and are certainly not a homogenous group, they do cluster together. The perennial species in the tertiary gene pool, *C. microphyllum,* appears admixed with the primary gene pool. However, this could be due to its closer phylogenetic relationship to *C. reticulatum* or accidental gene flow in the germplasm collection. The secondary gene pool, with the closely related and interfertile species of *C. pinnatifidum*, *C. bijugum* and *C. judaicum* formed one tight cluster. The immediate ancestors of the crop, *C. reticulatum* and *C. echinospermum,* show up as a group with substantial admixture with the cultivated individuals. This could represent the derivation of the crop, and could also represent introgression from the crop to the wild species (or artefacts of maintenance in germplasm facilities). The cultivated accessions of *C. arietinum* showed little admixture with the wild material in this analysis.

In addition, principal coordinate analysis, which was performed as a complementary approach to display clustering of genotypes, separated cultivated genotypes from wild species genotypes. Few genotypes of the wild chickpea clustered with cultivated material. Those wild genotypes that did cluster were *C. reticulatum,* the likely progenitor of cultivated chickpea (Figure 4b). The PCoA showed substantial differences among the wild material; *C. reticulatum* and *C. echinospermum* genotypes clustered with closely related cultivated material (Figure 4b). However, the closely related species from the secondary gene pool clustered individually rather than all clustering together. Furthermore, genotypes from a species in the tertiary gene pool, *C. yamashatae,* clustered more closely with the primary gene pool than did the species of the secondary gene pool. AMOVA partitioned 39% of variation between wild and

Table 1. Assessment of genetic diversity across groups of wild and cultivated chickpea using DArT markers.

	Polymorphic marker (%)	N	Na	Ne	I	He	UHe
Cultivated	35.79	63.401±0.04	1.219±0.013	1.096±0.004	0.113±0.004	0.068±0.002	0.068±0.002
Wild	99.60	27.143±0.023	1.996±0.001	1.766±0.005	0.607±0.002	0.421±0.002	0.429±0.002
Mean	67.70	45.272±0.245	1.607±0.008	1.431±0.005	0.360±0.004	0.244±0.003	0.249±0.003

No. of polymorphic alleles (N), No. of Different Alleles (Na), No. of Effective Alleles (Ne, = 1/(Sum pi^2)), Shannon's Information Index (I = −1 * Sum (pi * Ln (pi))), Expected Heterozygosity (He = 1 − Sum pi^2) and Unbiased Expected Heterozygosity (UHe = (2N/(2N − 1)) * He).

cultivated groups and 61% of variation segregating within groups (Figure 4c).

Genetic diversity among the genotypes from wild chickpea

The present study included analysis of 28 chickpea genotypes from nine wild species including genotypes from primary, secondary, tertiary gene pools and one individual of a perennial species, *C. microphyllum*. Genotyping using SNP markers resulted in high rates of failed SNP allele calls and null alleles. We therefore used DArT data to estimate the genetic diversity and relationships among the cultivated and wild species genotypes for primary, secondary and tertiary gene pools. AMOVA of wild species genotypes indicated that 31% of variation was found among the species while 69% of variation was observed within the species. Genetic distance between populations (primary, secondary and tertiary) was calculated based on Nei's genetic distance. As expected, higher similarity was observed between the primary and secondary gene pools (Nei's genetic distance 0.15), while greater distance was observed between primary and tertiary gene pools (Nei's genetic distance 0.69). Furthermore, a greater distance was observed between the secondary and tertiary gene pools than between the primary and secondary gene pools, which suggests that genotypes from the primary and secondary gene pools are more closely related to each other than to the tertiary gene pool. Across all wild material, numbers of effective alleles and values of heterozygosity were much higher than in the crop material. Within the wild material, the secondary gene pool had the greatest diversity, with highest effective allele estimates and highest heterozygosity (Table 2).

In the PCoA of the wild material alone (Figure S1a), a few genotypes from the primary gene pool clustered with the tertiary gene pool genotypes. Other genotypes from the primary gene pool clustered with the secondary gene pool. In parallel, we performed a STRUCTURE analysis on the 28 wild species genotypes using DArT markers. The STRUCTURE results complemented the observation from PCoA and diversity analysis (Figure S1b). We selected K = 2 based on Evanno method. The first cluster corresponds to the primary gene pool, while the second cluster corresponds to the secondary gene pool. The tertiary gene pool was admixed, likely representing the great diversity in those disparate species.

Genetic diversity among phenotypic classes of cultivated chickpea

Diversity among the 66 cultivated genotypes was assessed using both the DArT and SNP markers. These 66 genotypes were classified in three sub-groups based on seed type, *i.e.* desi, kabuli and pea-shaped. SNP markers were used in the program STRUCTURE to resolve differences among phenotypic classes of cultivated chickpea. Three groups of the cultivated material (K = 3) were observed, with most individuals demonstrating substantial admixture (Figure S2a). Genetic diversity among the phenotypic classes was also assessed using DArT and SNP markers (Table 3). The number of effective alleles (Ne) and heterozygosity (He) were very similar among the phenotypic classes (with overlapping standard deviations around their means), and all values were low (i.e., <1.1 for Ne, and <0.1 for He). Hierarchical AMOVA using both SNP and DArT data provided similar results. More than 90% of variation was observed within these phenotypic classes, while only about 10% variation was reported among these different populations (Figure S2b).

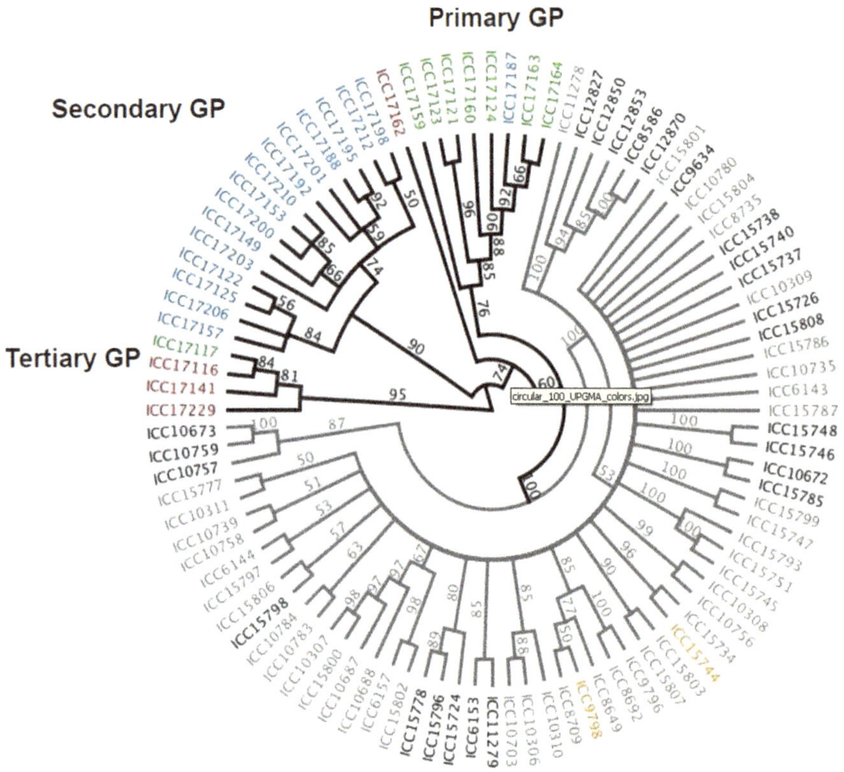

Figure 3. UPGMA tree of pairwise relatedness of cultivated (grey branches) and wild (black branches) chickpea. Genepools and seed types are represented by the following colors: primary, green; secondary, blue; tertiary, red; pea-shaped, orange; kabuli, grey; and desi, black.

Genetic diversity among the cultivars from different geographic regions

To understand the diversity in chickpea cultivars from different regions, an analysis was also performed based on the geographical distribution of cultivated and wild species genotypes. Based on geographical origin, germplasm was divided in three clusters: the Fertile Crescent, Central Asia, and the Ethiopian highlands. Substantial geographic variation was observed, with the greatest

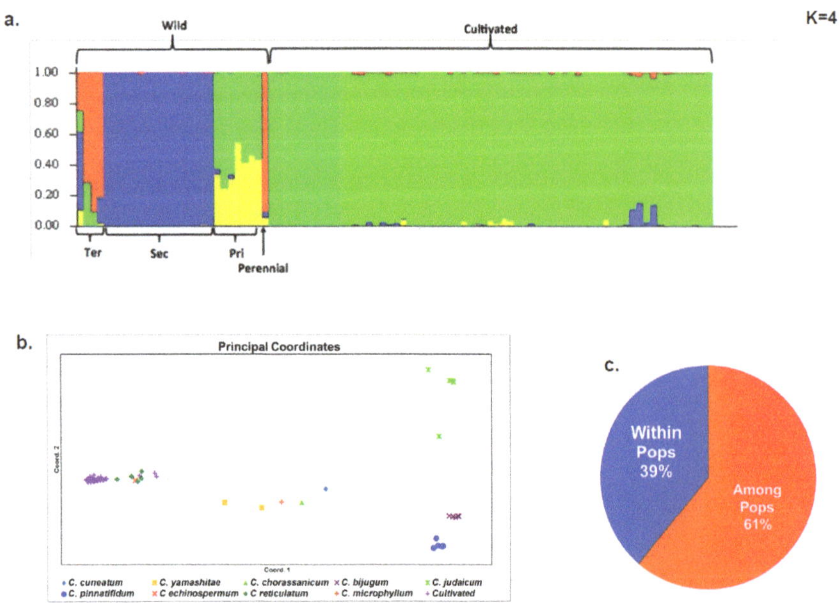

Figure 4. Population structure analysis using STRUCTURE of *Cicer* accessions. a. Structure showing distinct group of wild and cultivated species; wild further classified in primary (Pri), secondary (Sec) and tertiary (Ter) gene pool species. b. Principal coordinates analysis among wild and cultivated species. c. Analysis of molecular variance between and among wild and cultivated species genotypes.

Table 2. Assessment of genetic diversity across wild germplasm using DArT markers.

	Polymorphic marker (%)	N	Na	Ne	I	He	UHe
Primary GP	74.34	6.915±0.006	1.743±0.008	1.521±0.007	0.429±0.005	0.294±0.004	0.317±0.004
Secondary GP	95.04	15.519±0.014	1.935±0.006	1.63±0.007	0.521±0.004	0.353±0.003	0.365±0.003
Tertiary GP	10.82	3.754±0.011	0.86±0.011	1.072±0.004	0.061±0.003	0.041±0.002	0.048±0.003
Mean	60.07	8.729±0.055	1.513±0.007	1.407±0.004	0.337±0.003	0.229±0.002	0.243±0.002

No. of polymorphic alleles (N), No. of Different Alleles (Na), No. of Effective Alleles (Ne, = 1/(Sum pi^2)), Shannon's Information Index (I = −1* Sum (pi * Ln (pi))), Expected Heterozygosity (He = 1 − Sum pi^2) and Unbiased Expected Heterozygosity (UHe = (2N/(2N − 1)) * He).

Table 3. Assessment of genetic diversity across chickpea germplasm based on seed type.

Marker Type	Seed type	Polymorphic marker (%)	N	Na	Ne	I	He	UHe
SNP	**kabuli**	15.22	41.747±0.026	1.152±0.014	1.046±0.006	0.052±0.006	0.031±0.004	0.032±0.004
	desi	4.81	23.897±0.014	1.048±0.009	1.031±0.006	0.025±0.005	0.017±0.003	0.018±0.003
	pea	2.56	1.901±0.012	1.026±0.006	1.026±0.006	0.018±0.004	0.013±0.003	0.017±0.004
	Total	7.53	22.515±0.377	1.075±0.006	1.034±0.004	0.032±0.003	0.02±0.002	0.022±0.002
DArT	**kabuli**	18.64	39.101±0.028	0.987±0.012	1.06±0.003	0.065±0.003	0.04±0.002	0.041±0.002
	desi	26.89	22.348±0.019	1.094±0.013	0.11±0.004	0.109±0.004	0.069±0.003	0.071±0.003
	pea	0.54	1.952±0.004	0.725±0.009	1.004±0.001	0.003±0.001	0.002±0.001	0.003±0.001
	Total	15.36	21.134±0.167	0.936±0.007	1.058±0.002	0.059±0.002	0.037±0.001	0.038±0.001

No. of polymorphic alleles (N), No. of Different Alleles (Na), No. of Effective Alleles (Ne, = 1/(Sum pi^2)), Shannon's Information Index (I = −1* Sum (pi * Ln (pi))), Expected Heterozygosity (He = 1 − Sum pi^2) and Unbiased Expected Heterozygosity (UHe = (2N/(2N − 1)) * He).

diversity found in the Fertile Crescent and much lower diversity in the Ethiopian highlands and central Asia (Table 4). In parallel, PCoA was also performed (Figure 5). Outside of the Fertile Crescent, wild and cultivated material did not cluster together, which is consistent with a single domestication in the Fertile Crescent followed by dispersal to Central and South Asia and the East African highlands and subsequent divergence (Figure 5).

Discussion

Chickpea is believed to have been domesticated 10,000 years ago in southeastern Turkey and adjoining Syria [40–42]. The crop suffers from a narrow genetic base among the cultivated germplasm, which may be due to four population bottlenecks the crop has experienced [1]. This low genetic diversity makes the crop more susceptible to a range of diseases and pests [1,17]. Recently, Varshney et al. [43] also confirmed the problem of narrow diversity in elite chickpea using whole genome re-sequencing of 90 chickpea lines. Wild relatives of chickpea could serve an important role in enhancing the genetic base of cultivated material. In an effort to understand the genetic diversity available in cultivated and wild gene pools, the present study was undertaken using SNP and DArT markers. Genetic diversity was analyzed for these loci across a panel of domesticated and wild germplasm in the ICRISAT collection [30].

Understanding the available genetic diversity in the germplasm collection is a pre-requisite to adopt effective conservation and management strategies to use these genetic resources in crop improvement. Understanding patterns of genetic diversity can complement efforts to match collections from differing climatic regions to planting zones differing in climate [24]. The present study focuses on exploration of the genetic diversity and population structure of this diverse set of chickpea that includes cultivated and wild species genotypes ranging from primary to tertiary gene pools [12]. Global research efforts have resulted in the development of a large number of markers (SSR, SNPs, DArT) and genotyping platforms that can be used to study genetic diversity and explore the diverse germplasm for the traits to use in chickpea improvement programs [44]. KASPar assay from KBiosciences (Hertfordshire, UK) (http://www.kbioscience.co.uk) provides flexibility in use and have been proven successful for molecular breeding applications involving only few markers for genotyping a large number of segregating lines [45–47]. In the case of chickpea, more than 2,000 KASPar assay [28] and high density DArT array with 15,360 DArT clones have been developed [29]. The present study used a subset of 651 SNPs along with DArT arrays for genotyping. SNP genotyping data was used for cultivated germplasm as alleles could not be called for most of the wild species genotypes. SNPs used in the present study were designed from cultivated chickpea, which may be the reason they could not be amplified in wild species and could contribute to the biomodel PIC values. SNPs, although powerful as a marker due to their declining costs and high number [28], can be biased by being developed from a small number of individuals. This bias can skew the pool towards older and more intermediate frequency SNPs [48,49]. The benefit remains the large number of low cost markers. We minimized any effect of SNP bias by restricting its usage in the wild *Cicer* material where it lacks the information needed to separate patterns of relationships and complemented our analysis with the inclusion of independent DArT data that lacks such bias. In particular, focusing our analysis of the wild material on the DArT data should avoid the skew that SNP data can introduce.

Table 4. Genetic variation across the three primary regions of diversity: Fertile Crescent, Central Asia, and the Ethiopian highlands using DArT markers.

	Polymorphic marker (%)	N	Na	Ne	I	He	UHe
Ethiopia	49.84%	5.895±0.007	1.467±0.011	1.206±0.004	0.232±0.004	0.145±0.003	0.175±0.003
Central Asia	54.11%	12.840±0.008	1.527±0.010	1.277±0.006	0.267±0.005	0.174±0.003	0.189±0.004
Fertile Crescent	98.62%	71.810±0.047	1.986±0.002	1.415±0.004	0.439±0.003	0.277±0.002	0.281±0.002

No. of polymorphic alleles (N), No. of Different Alleles (Na), No. of Effective Alleles (Ne, = 1/(Sum pi^2)), Shannon's Information Index (I = −1* Sum (pi * Ln (pi))), Expected Heterozygosity (He = 1 − Sum pi^2) and Unbiased Expected Heterozygosity (UHe = (2N/(2N − 1)) * He).

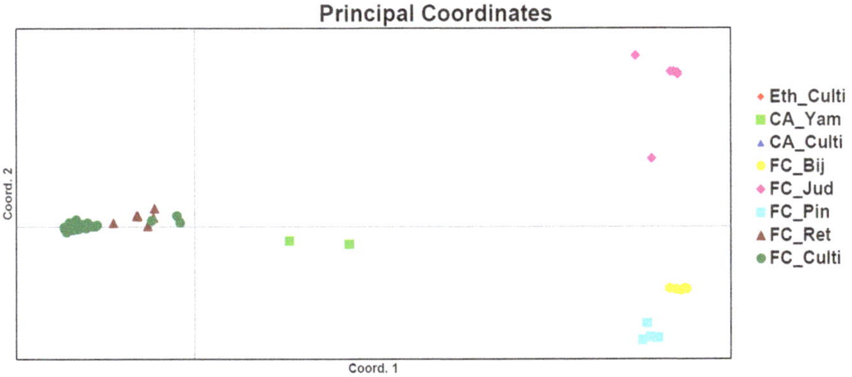

Figure 5. Principal coordinates analysis of wild and cultivated species of chickpea based on their geographical distribution (Eth_Culti: Cultivated chickpea from Ethiopia; CA_Yam: *Cicer yamashatae* from Central Asia; CA_culti: Cultivated chickpea from Central Asia; FC_Bij:, *C. bijugum*; from Fertile Crescent; FC_Jud: *C. judaicum* from Fertile Crescent; FC_Pin: *C. pinnatifidum* from Fertile Crescent; FC_Ret: *C. reticulatum* from Fertile Crescent and FC_Culti: Cultivated chickpea from Fertile Crescent).

In many crops that are deficient in genetic variation, wild relatives remain a critical resource. As is the case in other crops [4,47,50], higher levels of genetic variation were observed across all of the wild species. Significant genetic variation was observed in *C. reticulatum*, the immediate progenitor of cultivated chickpea, but genotypes of this species were less diverse than other *Cicer* species. Our results will allow the most genetically distinct of the existing accessions of these species to be used in breeding to maximize the diversity introgression into cultivated forms. However, as international germplasm collections contain only 18 unique *C. reticulatum* accessions [51], our results suggest that further collecting of *C. reticulatum*, particularly beyond the Mardin region of southeastern Anatolia where most existing collections were made, would be greatly beneficial. Relatively higher levels of genetic variation were present in the wild species of the secondary and tertiary gene pools, which span a far greater ecological range than *C. reticulatum*, which is restricted to oak savannas and disturbed pastures in southeastern Anatolia. However, the levels of genetic variation were still not all that high, consistent with the high probabilities on the assignment tests and the primarily selfing reproductive system of most *Cicer* species. Traits of wild species that are beneficial in a Mediterranean climate, such as vernalization, can hinder efforts to breed chickpea for cultivation in subtropical climates. Therefore, wild species from different regions, such as the African highlands or Central Asia could provide climatically adaptive traits for chickpea production in non-Mediterranean climates. For instance, species from outside the Fertile Crescent, such as *C. cuneatum* from Ethiopia and *C. microphyllum* from Central Asia (Pakistan and Afghanistan) could be exploited as sources of adaptive variation for those regions. Furthermore, wild species from more arid environments, such as *C. judaicum* and *C. pinnatifidum*, could be useful in expanding the resistance of cultivated chickpea to important biotic stresses like *Ascochyta*, *Helicoverpa*, *Fusarium* and *Botrytis* Gray Mold [20].

Based on seed type, chickpea has been subdivided in to three groups: desi, kabuli and pea-shaped. Significant differentiation among desi and kabuli seed type cultivars was observed, although far less than exists between wild species. The distinction could be due to a relatively recent evolution of kabuli seed type from a desi seed type ancestor that closely resembled the wild species, as previously speculated [16], but could just as easily represent artificial population structure generated by breeders [52]. Regardless, the division between the phenotypic classes of seed type appears to be weak and likely of recent origin. The dearth of desi seed type genotypes from the Fertile Crescent could suggest that kabuli seed types were favoured in this region, potentially as a means to prevent introgression from *C. reticulatum* and *C. echinospermum*, which have seed and flower colours similar to desi seed types.

Germplasm collections contain relatively low numbers of wild relatives of crops [6]. Although often several individual lines of a wild species are available, rarely has collecting been aimed at understanding patterns of variation in populations of wild relatives [53,54]. Our results indicate that collecting diverse population samples of several *Cicer* species spanning ecologically meaningful gradients in abiotic or biotic factors such as moisture, soil fertility or pathogen distribution would be extremely useful. Analysis of variation across these gradients in wild relatives could show how natural selection has adapted populations of wild relatives to these localized conditions, giving us natural targets for breeding.

Supporting Information

Figure S1 a. Principal coordinates analysis of wild species of chickpea based on primary, secondary and tertiary gene pool. b. Population structure analysis across wild chickpea accessions to understand the distribution of primary, secondary and tertiary gene pool species.

Figure S2 a. Population structure analysis across cultivated chickpea accessions based on seed type. b. Analysis of molecular variance within and among cultivated population based on seed type.

Table S1 Details about the Cicer accessions used in the study.

Table S2 Summary of the genotyping data generated using 651 CKAM markers on 94 Cicer accessions.

Table S3 Summary of the genotyping data generated using DArT markers.

Author Contributions

Conceived and designed the experiments: RKV. Performed the experiments: RKV MR. Analyzed the data: MR EvW EW AR. Contributed reagents/materials/analysis tools: RKV HDU. Wrote the paper: RKV MR EvW EW.

References

1. Abbo S, Berger J, Turner NC (2003) Evolution of cultivated chickpea: four bottlenecks limit diversity and constraint adaptation. Funct Plant Biol 30: 1081–1087.
2. Gepts P (2004) Who owns biodiversity, and how should the owners be compensated? Plant physiology 134: 1295–1307.
3. Fu YB (2006) Impact of plant breeding on genetic diversity of agricultural crops: searching for molecular evidence. Plant Genetic Resources: Characterization and Utilization 4: 71–78.
4. Gross BL, Olsen KM (2010) Genetic perspectives on crop domestication, Trends in plant science 15: 529–537.
5. Tanksley SD, McCouch SR (1997), Seed banks and molecular maps: unlocking genetic potential from the wild. Science 277: 1063–1066.
6. McCouch S, Baute GJ, Bradeen J, Bramel P, Bretting PK, et al. (2013) Agriculture: Feeding the future. Nature 499: 23–24.
7. Hajjar R, Hodgkin T (2007) The use of wild relatives in crop improvement: a survey of developments in the last 20 years. Euphytica 156: 1–13.
8. Roa V, Hodgkin T (2002) Genetic diversity and conservation and utilization of plant genetic resources. Plant Cell Tiss Org 68: 1–19.
9. Dempewolf H, Eastwood RJ, Guarino L, Khoury CK, Müller JV, et al. (2014) Adapting agriculture to climate change: A global initiative to collect, conserve, and use crop wild relatives. Agroecology and Sustainable Food Systems 38: 369–377.
10. Tester M, Langridge P (2010) Breeding technologies to increase crop production in a changing world. Science 327: 818–822.
11. Miflin B (2000) Crop improvement in the 21st century. J Exp Bot 51: 1–8.
12. Harlan JR, Wet JMJd (1971) Toward a rational classification of cultivated plants. Taxon 20: 509–517.
13. Haussmann BIG, Parzies HK, Presterl T, Sušic Z, Miedaner T (2004) Plant genetic resources in crop improvement. Plant Genetic Resources 2: 3–21.
14. Berger JD, Milroy SP, Turner NC, Siddique KHM, Imtiaz M, et al. (2011) Chickpea evolution has selected for contrasting phenological mechanisms among different habitats. Euphytica 180: 1–15.
15. Akibode S, Maredia M (2011) Global and regional trends in production, trade and consumption of food legume crops. CGIAR Draft Report. Available: http://impact.cgiar.org/sites/default/files/images/Legumetrendsv2.pdf.
16. Ladizinsky G (1985) Founder effect in crop plant evolution. Economic Botany 39: 191–199.
17. Millan T, Clarke HJ, Siddique KHM, Buhariwalla HK, Gaur PM, et al. (2006) Chickpea molecular breeding: new tools and concepts. Euphytica 147: 81–103.
18. Berger JD, Turner NC (2007) The Ecology of Chickpea. *In:* Chickpea breeding and management, Yadav SS, Redden RJ, and Sharma (eds). CAB International, Oxfordshire, UK.
19. Roorkiwal M, Nayak SN, Thudi M, Upadhyaya HD, Brunel D, et al. (2014) Allelic diversity and association analysis for candidate abiotic stress responsive genes with drought tolerance in chickpea. Front Plant Sci 5: 248 doi:10.3389/fpls.2014.00248.
20. Mallikarjuna N, Coyne C, Cho S, Rynearson S, Rajesh PN, et al. (2011) *Cicer*. Kole C (ed.), Wild Crop Relatives: Genomic and Breeding Resources, Legume Crops and Forages, DOI 10.1007/978-3-642-14387-8_4, Springer-Verlag Berlin Heidelberg.
21. Hillman GC (1975) The plant remains from Tell Abu Hureya in Syria: a preliminary report. In: Moore AMT (ed.) The Excavation of Tell Abu Hureya in Syria: A Preliminary Report. Proceedings of the Prehistory Society 41: 70–73.
22. Redden RJ, Berger JD (2007) History and Origin of Chickpea. In *Chickpea breeding and management*, Yadav SS, Redden RJ, and Sharma (eds). CAB International, Oxfordshire, U.K.
23. Abbo S, Saranga Y, Peleg Z, Kerem Z, Lev-Yadun S, et al. (2009) Reconsidering domestication of legumes versus cereals in the ancient near east. Q Rev Biol 84: 29–50.
24. Berger JD, Hughes S, Snowball R, Redden B, Bennett SJ, et al. (2013) Strengthening the impact of plant genetic resources through collaborative collection, conservation, characterisation, and evaluation: a tribute to the legacy of Dr Clive Francis. Crop and Pasture Science 64: 300–311.
25. Sharma S, Upadhyaya HD, Roorkiwal M, Varshney RK, Gowda CLL (2013) Chickpea. *In:* Genetic and Genomic Resources of Grain Legume Improvement to production. doi:http://dx.doi.org/10.1016/B978-0-12-397935-3.00001-3.
26. Varshney RK, Hiremath PJ, Lekha P, Kashiwagi J, Balaji J, et al. (2009) A comprehensive resource of drought- and salinity- responsive ESTs for gene discovery and marker development in chickpea (*Cicer arietinum* L.). BMC Genomics 10: 523.
27. Roorkiwal M, Sawargaonkar SL, Chitikineni A, Thudi M, Saxena RK, et al. (2013) Single nucleotide polymorphism genotyping for breeding and genetics applications in chickpea and pigeonpea using the BeadXpress platform. The Plant Genome 6 doi:10.3835/plantgenome2013.05.0017.
28. Hiremath PJ, Kumar A, Penmetsa RV, Farmer A, Schlueter JA, et al. (2012) Large-scale development of cost-effective SNP marker assays for diversity assessment and genetic mapping in chickpea and comparative mapping in legumes. Plant Biotechnol J 10: 716–732.
29. Thudi M, Bohra A, Nayak SN, Varghese N, Shah TM, et al. (2011) Novel SSR markers from BAC-end sequences, DArT arrays and a comprehensive genetic map with 1,291 marker loci for chickpea (*Cicer arietinum* L.). PLoS One 6: e27275.
30. Gowda CLL, Upadhyaya HD, Sharma S, Varshney RK, Dwivedi SL (2013) Exploiting genomic resources for efficient conservation and use of chickpea, groundnut, and pigeonpea collections for crop improvement. The Plant Genome 6: doi:10.3835/plantgenome2013.05.0016.
31. Cuc LM, Mace ES, Crouch JH, Quang VD, Long TD, et al. (2008) Isolation and characterization of novel microsatellite markers and their application for diversity assessment in cultivated groundnut (*Arachis hypogaea*). BMC Plant Biology 8: 55.
32. Ladizinsky G, Adler A (1976) Genetic relationships among the annual species of *Cicer* L. Theor Appl Genet 48: 197–203.
33. Excoffier L, Smouse PE, Quattro JM (1992) Analysis of molecular variance inferred from metric distances among DNA haplotypes: application to human mitochondrial DNA restriction data. Genetics 131: 479–491.
34. Peakall R, Smouse PE (2006) GENALEX 6: genetic analysis in Excel. Population genetic software for teaching and research. Molecular Ecology Notes 6: 288–295.
35. Peakall R, Smouse PE (2012) GenAlEx 6.5: genetic analysis in Excel. Population genetic software for teaching and research-an update. Bioinformatics 28: 2537–2539.
36. Excoffier L, Estoup A, Cornuet JM (2005) Bayesian analysis of an admixture model with mutations and arbitrarily linked markers. Genetics 169: 1727–1738.
37. Shannon CE (1948) A mathematical theory of communication. The Bell System Technical Journal 27: 379–423 and 623–656.
38. Liu K, Muse SV (2005) PowerMarker: an integrated analysis environment for genetic marker analysis. Bioinformatics 21: 2128–2129.
39. Pritchard JK, Stephens M, Donnelly P (2000) Inference of population structure using multilocus genotype data. Genetics 155: 945–959.
40. van der Maesen LJG (1987) Origin, history and taxonomy of chickpea. In: Saxena MC, Singh RB (eds) The chickpea. CABI, Wallingford, UK, 139–156.
41. Lev-Yadun S, Gopher A, Abbo S (2000) Archaeology. The cradle of agriculture. Science 288: 1602–1603.
42. Zohary D, Hopf M (2000) Domestication of plants in the old world, 3rd edn. Oxford University Press, New York, USA.
43. Varshney RK, Song C, Saxena RK, Azam S, Yu S, et al. (2013) Draft genome sequence of chickpea (*Cicer arietinum*) provides a resource for trait improvement. Nat Biotechnol 31: 240–246.
44. Varshney RK, Mohan SM, Gaur PM, Gangarao NV, Pandey MK, et al. (2013) Achievements and prospects of genomics-assisted breeding in three legume crops of the semi-arid tropics. Biotechnol Adv 31: 1120–1134.
45. Allen AM, Barker GL, Berry ST, Coghill JA, Gwilliam R, et al. (2011) Transcript-specific, single-nucleotide polymorphism discovery and linkage analysis in hexaploid bread wheat (*Triticum aestivum* L.). Plant Biotechnol J 9: 1086–1099.
46. Cortés AJ, Chavarro MC, Blair MW (2011) SNP marker diversity in common bean (*Phaseolus vulgaris* L.). Theor Appl Genet 123: 827–845.
47. Saxena RK, von Wettberg E, Upadhyaya HD, Sanchez V, Songok S, et al. (2014) Genetic Diversity and Demographic History of *Cajanus* spp. Illustrated from Genome-Wide SNPs. PLoS One 9(2): e88568.
48. Albrechtsen A, Nielsen FC, Nielsen R (2010) Ascertainment biases in SNP chips affect measures of population divergence. Mol Biol Evol 27: 2534–2547.
49. Lachance J, Tishkoff SA (2013) SNP ascertainment bias in population genetic analyses: why it is important, and how to correct it. BioEssays 35: 780–786.
50. Kassa MT, VarmaPenmetsa R, Carrasquilla-Garcia N, Sarma BK, Datta S, et al. (2012) Genetic patterns of domestication in pigeonpea (*Cajanus cajan* (L) Millsp) and wild *Cajanus* relatives. PloS one 7: 6.
51. Berger J, Abbo S, Turner NC (2003) Ecogeography of annual wild *Cicer* species: The poor state of the world collection. Crop Sci 41: 1976–1090.
52. Moreno MT, Cubero JI (1978) Variation in *Cicer arietinum*. Euphytica 27: 465–485.
53. Gepts P, Famula TR, Bettinger RL, Brush SB, Damania AB, et al. (eds) (2012) Biodiversity in agriculture: domestication, evolution, and sustainability. Cambridge University Press, Cambridge, U.K.
54. Hufford MB, Xu X, van Heerwaarden J, Pyhäjärvi T, Chia JM, et al. (2012) Comparative population genomics of maize domestication and improvement. Nat Genet 44: 808–811.

21

Genetic Structure and Diversity of the Endangered Fir Tree of Lebanon (*Abies cilicica* Carr.): Implications for Conservation

Lara Awad[1]*, Bruno Fady[2], Carla Khater[3], Anne Roig[2], Rachid Cheddadi[1]

1 University of Montpellier II, Institute of Evolutionary Sciences, CNRS UMR 5554, Montpellier, France, **2** INRA, UR629, Ecologie des Forêts Méditerranéennes (URFM), Domaine St Paul, Avignon, France, **3** Center for Remote Sensing, National Council for Scientific Research-Lebanon, Bir Hassan, Beirut, Lebanon

Abstract

The threatened conifer *Abies cilicica* currently persists in Lebanon in geographically isolated forest patches. The impact of demographic and evolutionary processes on population genetic diversity and structure were assessed using 10 nuclear microsatellite loci. All remnant 15 local populations revealed a low genetic variation but a high recent effective population size. F_{ST}-based measures of population genetic differentiation revealed a low spatial genetic structure, but Bayesian analysis of population structure identified a significant Northeast-Southwest population structure. Populations showed significant but weak isolation-by-distance, indicating non-equilibrium conditions between dispersal and genetic drift. Bayesian assignment tests detected an asymmetric Northeast-Southwest migration involving some long-distance dispersal events. We suggest that the persistence and Northeast-Southwest geographic structure of *Abies cilicica* in Lebanon is the result of at least two demographic processes during its recent evolutionary history: (1) recent migration to currently marginal populations and (2) local persistence through altitudinal shifts along a mountainous topography. These results might help us better understand the mechanisms involved in the species response to expected climate change.

Editor: Giovanni G. Vendramin, CNR, Italy

Funding: This paper represents a part of the doctoral research of Lara Awad who acknowledges a scholarship from the Institut Ecologie et Environnement (INEE) of the Centre National de la Recherche Scientifique (CNRS) (Contract number 245303) (http://www.cnrs.fr/inee/) and (http://www.cnrs.fr/index.php). Fieldwork in Lebanon was facilitated by a financial support from the Grant Research Program (GRP) of the National Council for Scientific Research-Lebanon (NCSR) (http://www.cnrs.edu.lb/). The stay of Lara Awad at INRA-URFM for the laboratory work was facilitated by a financial support from the EcoChange project (http://www.ecochange-project.eu/ecochange-project). The funders had no role in study design, data collection and analysis, decision to publish, or preparation of the manuscript.

Competing Interests: The authors have declared that no competing interests exist.

* E-mail: lara.awad@univ-montp2.fr

Introduction

Geographic patterns of genetic variation in conifers have long been studied at the species and populations levels [1–9]. Geographic range size, geographic isolation and biogeographic position may play a crucial role in shaping patterns of genetic variation [10]. Widely distributed woody plant species with high longevity and high fecundity generally retain high genetic diversity within populations but low genetic differentiation among populations [11], [12]. The opposite is generally found for endemic and narrowly distributed woody plants with the same combination of traits [12]. However, the low genetic variation found in rare species may be an overgeneralization [13]. For instance, high gene diversity and population structure can be found in endemic species with restricted geographic range [14]. The imprint of geographic range size on patterns of genetic diversity may also be indirectly driven by species' dispersal ability. At the end of the Pleistocene, the rapid geographical range expansion of many tree species was characterized by the occurrence of long-distance dispersal events [15]. Consequences of such range expansions on spatial genetic patterns and on the maintenance or erosion of genetic diversity may respectively depend on the forms of dispersal [16] and on the number of long-distance dispersal events [17]. Gene flow arising from dispersal can either impede or promote evolution depending on the geographic structure of natural populations [18]. For instance, local adaptation may be achieved despite high gene flow and long-distance dispersal in expanding populations [19], [20].

The genetic imprint of geographic isolation, or fragmentation, on natural populations can vary with respect to their demographic structure. Relict populations in fragmented landscapes predictably show low genetic variation but high genetic differentiation due to spatial isolation [21]. Demographic bottlenecks resulting from fragmentation may not be accompanied by genetic bottlenecks due to temporal fluctuation in population sizes, mating system dynamics, or metapopulation structure [22], [23], [24]. Short-range dispersal occurring between neighboring populations should contribute to the genetic differentiation of remote populations via isolation-by-distance [25]. Over time, the isolation-by-distance genetic pattern should be pronounced at equilibrium between dispersal and genetic drift in long-established populations [26]. In demographically unstable populations, however, fragmentation can either increase or decrease the genetic variance among populations. The outcome may depend on the mode of colonization [27], on temporal and spatial variation and co-variation in demographic parameters [28], or on the rates of extinction and recolonization relative to the effective population size [18] or relative to the effective number of migrants [27], [29].

The biogeographic position within a geographic range or a biome may impact genetic diversity within and among populations. In the Mediterranean basin, genetic diversity is higher than other regions of the world for conifers [30] and the rear-edge populations of temperate tree species often display large regional genetic diversity and unique genetic variants [31]. Most rear-edge populations have not been involved in major postglacial recolonization events [32]. Instead, they have persisted locally with small altitudinal shifts in range [33], [34]. Longitude and underlying demographic processes such as climate-related effective population size changes and migration can also explain genetic diversity gradients. For instance, a pattern of decreasing within-population genetic diversity is observable from east to west across the Mediterranean basin in a wide range of taxa [10], [35].

In this study, we have explored the imprint of geographic range size, geographic isolation and biogeographic position and that of demographic processes such as dispersal ability and viability on the modern genetic diversity of *Abies cilicica* Carr. in Lebanon (Cilician fir). *Abies cilicica* is a long-lived and outcrossed conifer species. It is endemic to the Eastern Mediterranean in the Taurus, Anti-Taurus and Amanus mountains of Turkey, the Alawi mountains in Syria and the Northern Mount Lebanon range in Lebanon (Figure 1). However, its range is not continuous, and in Lebanon, noticeably, it is geographically restricted and geographically isolated from the rest of its range with more than a 150 km distance from *Abies cilicica* in Syria (Figure 1). In Lebanon, *Abies cilicica* is geographically marginal as it currently occupies the southernmost geographical position of its range (Figure 1). The varied topography of Mount Lebanon shelters geographically isolated forest patches. Historically, Ancient Egyptian texts attest the use of *Abies cilicica* for its timber, during the Ancient Egyptian New Kingdom, as a sign of power of the Pharaoh, particularly in the construction of flagpoles for temple pylons, in shipbuilding, and in the construction of the Barque of Amun [36]. During World War II, the wood of *Abies cilicica* was cut mainly from Qammouaa to build the railroad between Tripoli and Haifa [37]. Major current anthropogenic disturbances include illegal logging and overgrazing activities. The IUCN status of *Abies cilicica* is Near Threatened overall [38], but IUCN considers it critically endangered at national level in Lebanon.

Because genetic diversity is potentially high in the Eastern Mediterranean and for mountain species [10], and because its distribution range is patchy and some populations display low census size, *Abies cilicica* in Lebanon offers the opportunity to answer the following questions for which accumulating experimental evidence remains central: (1) Is biogeographic position affecting genetic diversity within and among populations?, (2) Is habitat fragmentation increasing spatial genetic structure?, (3) Is geographic isolation limiting dispersal?, and (4) Is geographic marginality affecting persistence?

Material and Methods

Ethics Statement

All the field sampling of *Abies cilicica* in Lebanon was carried out after obtaining mission permits from CNRS in France and NCSR in Lebanon (number L03504, L03701, 2938/w and 5170/w) and following the approval in Lebanon of corresponding municipalities and nature reserves organizational teams. Sampling of *Abies marocana* Trab. (also known as *Abies pinsapo* subsp. *marocana* Trab.) was permitted by the Talassemtane National Park, Morocco. Needle samples of *Abies bornmuelleriana* Mattf., *Abies cephalonica* Loud. and *Abies alba* Mill. were provided by INRA-URFM.

Sampling Area and Plant Material

Abies cilicica is currently distributed in Lebanon across 15 remnant populations located between ca. 1000–1800 m.a.s.l. on the western slopes of northern Mount Lebanon. Moist airstreams off the Mediterranean approaching over Mount Lebanon result in a classic orographic precipitation pattern. Rain falling on the western side of Mount Lebanon becomes snow at higher elevations during winter. The species currently occupies humid bioclimatic zones with a mean annual precipitation over 800 mm. The varied topography of Mount Lebanon allows it to inhabit both the Supra-Mediterranean (1000–1500 m.a.s.l.) and Mountain-Mediterranean (1500–2000 m.a.s.l.) vegetation zones. Currently, it occupies a patchy distribution covering 1613 ha equivalent to 1.2% of the total forest cover of Lebanon, compared to 1.58% for *Cedrus libani* A. Rich. (Cedar of Lebanon) [39].

We have exhaustively sampled all the 15 remnant natural populations of *A. cilicica* in Lebanon (Table 1, Table S1). Adult trees were sampled randomly at distance intervals of 20 m approximately keeping sampled areas to a minimum. A total of 351 leaf samples of *A. cilicica* from Lebanon were collected and then conserved at −20°C prior to DNA extraction.

DNA Extraction

Total genomic DNA was successfully extracted and purified from 45 mg of needles per sample using the QIAGEN DNeasy 96 plant kit following manufacturer's protocol. Furthermore, 48 needle samples from four Mediterranean *Abies* species were also used for DNA extraction to serve as reference for the genetic analyses. These included 6 samples of *A. marocana* from the Talassemtane National Park in Morocco, 8 samples representing a mixture of 6 populations of *A. bornmuelleriana* from Turkey, 8 samples representing a mixture of 8 populations of *A. cephalonica* from Greece, 10 samples representing a mixture of 10 populations of *A. alba* from the French Pyrenees and 16 samples representing a mixture of 4 populations of *A. alba* from the French Southern Alps (Table S1).

Microsatellite Analysis

Microsatellites are efficient molecular markers for describing the demographic history of populations [40], and therefore addressing specific questions related to their evolutionary ecology and conservation [41]. Genomic DNA amplification was performed using 10 nuclear microsatellites: NFF3, NFH15, NFH3 and NFF7 developed for *A. nordmanniana* Stev. [42], and SF1, SFb4, SFb5, SF50, SF78 and SF333 developed for *A. alba* Mill. [43]. The forward primer was 5′ end labeled by one of fluorescent dyes 6-FAM (blue), HEX (green), TAMRA (yellow) or ATTO565 (red) (Eurofins MWG Operon). Two multiplex microsatellite sets (NFF3, NFH15, NFH3, NFF7 and SF1, SFb4, SFb5, SF50, SF78, SF333) for which allele drop-out was controlled, were amplified for all individuals by polymerase chain reaction (PCR) using Qiagen Multiplex PCR Kit. Multiplex PCR was run in an 11 µL reaction on a Mastercycler EP Gradient S96 thermal cycler (Eppendorf). The PCR mix contained 2× Qiagen Multiplex PCR Master Mix (1× final concentration), 0.2 µM of each forward and reverse primers, 5× Q-Solution (0.5× final concentration), RNase-free water (Qiagen) qs to 9 µL and 2 µL (about 20 ng) of genomic DNA. The thermal cycling consisted of initial denaturation step at 95°C for 15 minutes; 3-step cycling repeated 30 times and consisting of denaturation at 94°C for 30 seconds, annealing at 57°C for 90 seconds and extension at 72°C for 60 seconds; and a final extension step at 72°C for 10 minutes.

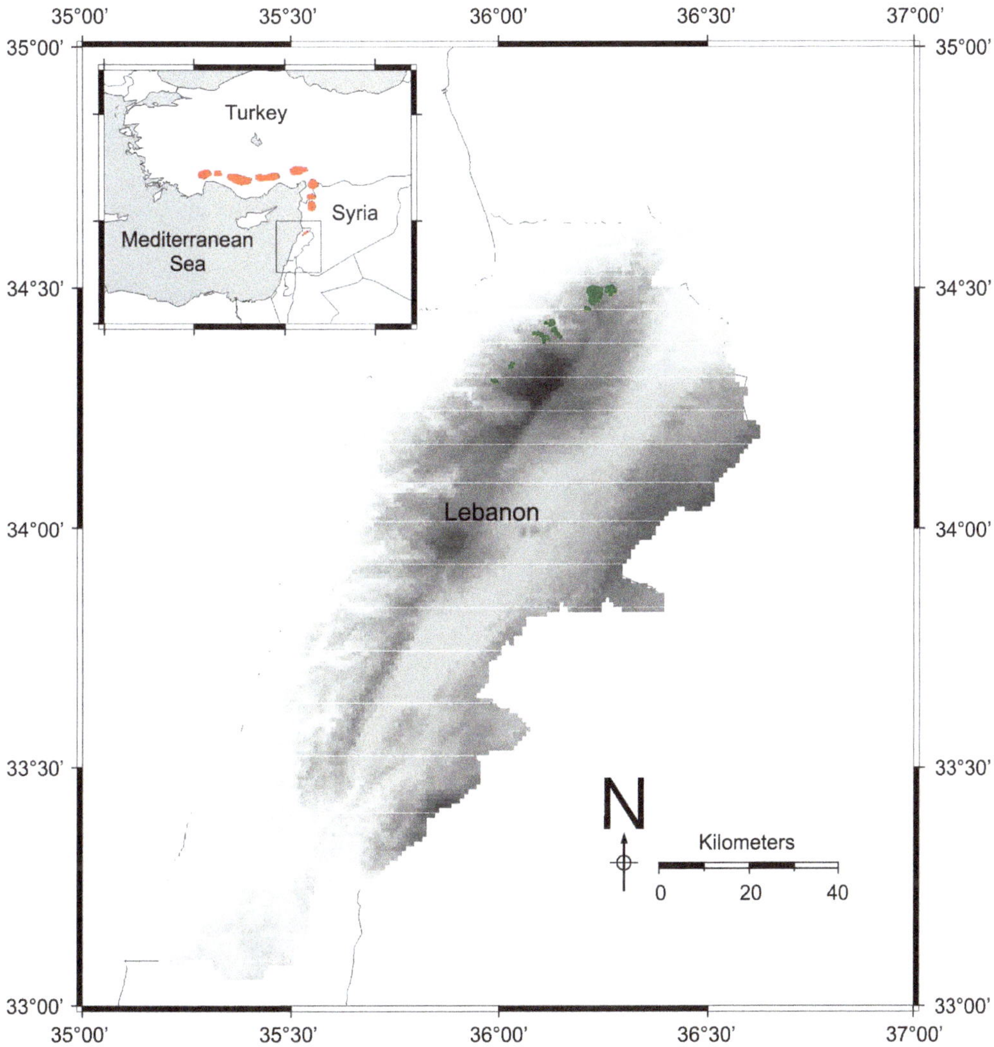

Figure 1. Range of distribution of *Abies cilicica* in the Eastern Mediterranean. Remnant populations of *Abies cilicica* in Turkey, Syria and Lebanon are plotted in red onto the map of the Eastern Mediterranean. Remnant populations of *Abies cilicica* in Lebanon are plotted in green onto the map of Lebanon where they are currently found on the western slopes of the northern part of Mount Lebanon. Geographical position of *Abies cilicica* in Turkey and Syria were plotted from a EUFORGEN map [105] while the geographical position of *Abies cilicica* in Lebanon corresponds to our sampling data. Map was prepared using QGIS version 1.8.0 [106] and the GMT package [107] within R [108].

Capillary electrophoresis of multiplexed PCR products was performed on an ABI 3730XL automated sequencer (Applied Biosystems) at INRA Avignon, with the GeneScan™ - 500 LIZ™ size standard (Applied Biosystems) run with each sample. In order to decrease the risks of genotyping errors related to automated or arbitrary decisions in allelic binning [44], [45], allelic binning and scoring of genotypes were performed manually by two different persons using the software GeneMapper 4.1 (Applied Biosystems) and compared to generate the final data set. The genotype dataset for the 10 microsatellites of 399 individuals of *A. cilicica*, *A. cephalonica*, *A. bornmuelleriana*, *A. marocana* and *A. alba* is publicly deposited in the DRYAD repository.

Data Analysis

Hardy-Weinberg Equilibrium, Linkage Disequilibrium and Null Alleles

Exact tests for deviations from Hardy–Weinberg equilibrium (HWE) were performed using the heterozygote deficiency and heterozygote excess tests implemented in GENEPOP 4.1.3 [46], [47]. Testing for linkage disequilibrium (LD) was also performed in GENEPOP 4.1.3 by using the log likelihood ratio statistic test (*G*-test). Exact P-values for the individual population or locus tests were estimated using the Markov Chain algorithm implemented in GENEPOP 4.1.3 with 10000 dememorization steps, 100 batches and 5000 iterations and were then corrected both using the sequential Bonferroni procedure [48] and the Benjamini-Hochberg method [49] to ensure controlling both the Family Wise Error Rate (FWER) and the False Discovery Rate (FDR). The presence and frequency of null alleles were examined using FreeNA [50] following the Expectation Maximization (EM) algorithm [51] which was found to provide unbiased and low variance estimates of null allele frequencies [50]. Given that the presence of null alleles result in an overestimation of population genetic differentiation [50], FreeNA was used to compute the F_{ST} statistic [52] both using and without using the *ENA* (Excluding Null Alleles) correction method, i.e. by excluding or not excluding null allelic

Table 1. Origin, location and sample size of *Abies cilicica* populations in Lebanon.

Code	Location	Longitude	Latitude	Altitude*	Sample size
KCHB[1]	Kobayat	36° 15.720′ E	34° 29.881′ N	1582–1758	30
BTJF[1]	Hermel	36° 16.643′ E	34° 29.731′ N	1538–1593	30
CHIR[1]	Qammouaa	36° 14.533′ E	34° 29.790′ N	1553–1635	28
JBMD[1]	Qammouaa	36° 13.594′ E	34° 29.099′ N	1394–1581	30
MKTF[1]	Qammouaa	36° 13.580′ E	34° 28.101′ N	1548–1597	30
NBSH[1]	Qammouaa	36° 12.626′ E	34° 27.404′ N	1460–1545	30
QARS[2]	Qarsita	36° 06.849′ E	34° 25.670′ N	1037–1140	11
QJRN[1]	Jaïroun	36° 07.610′ E	34° 25.549′ N	1304–1439	28
QEMN[1]	Wadi Jhanam	36° 08.056′ E	34° 24.846′ N	1102–1191	28
HNKT[1]	Kfarbnine	36° 05.010′ E	34° 24.353′ N	1389–1453	12
HNKF[1]	Kfarbnine	36° 06.300′ E	34° 24.081′ N	1422–1505	21
GHAB[1]	Kfarbnine	36° 06.484′ E	34° 23.953′ N	1480–1597	24
HKYT[2]	Bqaa Safrine	36° 01.886′ E	34° 20.649′ N	1551–1695	12
KMHR[2]	Karm El Mohr	36° 01.585′ E	34° 20.186′ N	1639–1704	18
EHDN[2]	Ehden	35° 59.533′ E	34° 18.460′ N	1447–1613	19

*Elevation in meters above sea level.
[1]membership of the local population to deme 1 (Northeastern ridge).
[2]membership of the local population to deme 2 (Southwestern ridge).

states from the F_{ST} computation [50]. The bootstrap 95% confidence intervals (CI) for the global F_{ST} values were calculated using 50000 replicates over loci.

Within-Population Genetic Diversity

GenAlEx v6 [53] and FSTAT version 2.9.3.2 [54] were used to estimate observed heterozygosity H_O, expected heterozygosity H_E [55], [56], rarefied allelic richness A_R [57], [58] for a sample size of n = 20 gene copies (10 diploid trees), and inbreeding coefficient F_{IS} [59]. The bootstrap 99% CI for the global F_{IS} value over loci was calculated in FSTAT version 2.9.3.2 using 15000 replicates. Rarefied private allelic richness Π_s was computed using the rarefaction method [57], [60] implemented in HP-Rare 1.1 [61] with a minimum sample size of n = 20 gene copies. Statistical significance at the 5% nominal level of the difference between the mean value of H_O and H_E across all samples for all loci was verified using a Student's *t*-test. Differences in the within-population genetic diversity and allelic richness among *A. cilicica*, *A. marocana*, *A. bornmuelleriana*, *A. cephalonica* and *A. alba* were assessed using a permutation test implemented in FSTAT version 2.9.3.2 [54]. Statistical significance at the 5% nominal level was calculated after 10000 permutations of populations among groups.

Among-Population Genetic Differentiation

The differentiation between populations was computed in SPAGeDi [62] based on allele identity and allele size by respectively using F_{ST} [59] and R_{ST} [63], [64], and their significance was determined by 10000 permutations. The presence of a phylogeographic pattern was also assessed, using the same software, after 10000 randomizations of allele sizes among alleles within loci [65], by testing whether R_{ST} computed before randomization was larger than R_{ST} obtained after allele size permutation (pR_{ST}). The randomization test also infers whether or not strict stepwise-like mutations contributed to population differentiation [65].

Population genetic structure was also assessed using two Bayesian clustering approaches with different underlying assumptions concerning the units of clustering (individual, population), the types of clustering (non-spatial, spatial), and the computation algorithm (Markov chain Monte Carlo (MCMC), stochastic optimization). The first Bayesian clustering approach, implemented in the software STRUCTURE 2.3.3 [66], identifies the number of distinct genetic clusters (K) and probabilistically assigns individual multilocus genotypes to these clusters. To infer posterior probabilities of K, we ran a series of independent runs with different user-defined values of K clusters ranging from 1 to 15. Assuming that individuals may have mixed ancestry and that the different populations have correlated allele frequency [67], analysis was run ten times for 10^6 iterations of MCMC after a burn-in period of 30000. No prior information on the sampling locations of each individual was used as recommended in the software documentation [68]. STRUCTURE results were then processed in a web-based program, STRUCTURE HARVESTER [69], which implements the Evanno method [70] to estimate a mode at the true K. The second Bayesian approach, implemented in the software BAPS 6 [71], estimates the hidden population substructure by clustering sampled populations into panmictic groups [72]. Analysis was performed at the population level by using a spatial mixture model with prior information on the geographical position of the sampled populations, to prevent both over-smoothing in the case of a weak spatial structure and the emergence of spurious genetic structure as a result of weak stochastic fluctuations in the allele frequencies [73], [74]. Spatial genetic mixture analysis was run ten times after choosing a set of possible values of K clusters ranging from 1 to 15. To find the optimal partition, the program uses a stochastic optimization algorithm, which is less computationally intensive than the MCMC algorithm used in STRUCTURE [75]. A joint posterior distribution of partitions of the sampled populations into genetically divergent groups was produced and then plotted onto the map of Lebanon.

FSTAT version 2.9.3.2 [54] was used to compute, for the identified clusters, the observed (H_O) and expected (H_E) proportion of heterozygotes and the overall gene diversity (H_T) according to [76], F_{IS} and F_{ST} according to [59], and rarefied allelic richness (A_R) for a sample size of n = 102 gene copies or 51 diploid individuals [57], [58]. Statistical significance at the 5% significance level was obtained after 10000 permutations of populations among demes. R_{ST} [63], [64] was also computed among populations of each of the identified clusters using the software SPAGeDi [62], as well as the occurrence of a phylogeographic pattern as described above.

Direct estimates of recent migration rates within and between the current clusters were generated in the software GENECLASS2 [77] through assignment tests allowing the detection of first-generation (F_0) migrants. The likelihood (L) that an individual originates from a given population sample was computed using the statistical criterion $\frac{L\_home}{L\_max}$, i.e. the ratio of the likelihood of finding a given individual in the population in which it was sampled to the greatest likelihood among all sampled populations [78]. Given that all the populations of *A. cilicica* in Lebanon have been extensively sampled, this criterion was considered to be informative on recent dispersal [77], [78]. The likelihood estimation was computed using the likelihood-based Bayesian method of [79], which has been shown to perform better than frequency-based and distance-based methods [80]. The probability computation, at the 1% significance level, was assessed by using the Monte Carlo resampling method of [78] with 10000 simulated individuals instead of the Monte Carlo resampling methods of [79], [80] to better control type I error rates.

Isolation-By-Distance

Isolation-by-distance (IBD) [25] was analyzed using IBDWS version 3.23 [81] by assessing the statistical significance, in a Mantel test [82] with 30000 permutations, of the relationship between the measure of similarity $\hat{M} = \frac{\frac{1}{F_{ST}-1}}{4}$ [26] among population pairs and the comparable matrix of geographic distances in a two-dimensional stepping stone model. This approach was chosen as it has been shown to distinguish between equilibrium and non-equilibrium populations under dispersal and genetic drift [26]. Both $\hat{M}$ and the geographic distance were log-transformed as suggested in [26]. The strength of IBD was quantified using a reduced major axis (RMA) regression, which is more appropriate for this purpose than a standard ordinary least squares (OLS) regression [83].

Genetic Bottlenecks

Three complementary methods were used to detect whether or not *A. cilicica* in Lebanon had experienced recent and severe genetic bottlenecks. The first method, implemented in the software M-P-Val and Critical_M [84], is based on the computation of the parameter $M = \frac{k}{r+1}$, i.e. the mean ratio of the number of alleles (k) to the range in allele size (r), from a population sample of microsatellite loci [84], [85]. Along with the fact that when effective population size decreases, genetic drift intensifies and alleles are subsequently lost, this ratio is expected to be smaller in recently and severely bottlenecked populations than in populations under mutation-drift equilibrium [84]. This method can detect recent genetic bottlenecks over a period of time of 100 generations [84]. The analysis was run assuming a Two-Phase Mutation Model (TPM) [86] of microsatellite loci with parameters p_s (proportion of one-step mutations) = 0.88 and Δ_g (average size of non one-step mutations) = 2.8 as reviewed in [84]. In M-P-Val, the average observed value of M calculated from the microsatellite dataset was compared with a simulated equilibrium distribution of M for different values of the population parameter $\theta = 4N_e\mu$ (where N_e is the pre-bottleneck effective population size, and μ is the mutation rate of microsatellites per locus per generation). In Critical_M, values of the critical M threshold (M_c), below which a bottleneck is evident, were also calculated for the different θ values. Considering an average mutation rate per locus per generation equal to 0.0005 [84], [87], statistical significance was assessed with 10000 simulations carried out with values of $\theta \in \{10; 1; 0.5; 0.1\}$ corresponding respectively to a pre-bottleneck effective population size of 5000, 500, 250 and 50.

The second method makes use of a Wilcoxon's signed rank test to compare the distribution of the heterozygosity expected under mutation-drift equilibrium and under HWE and can detect recent genetic bottlenecks over the past 4 N_e or 2 N_e generations [88]. The third method is a mode-shift test, which searches for bottleneck-induced and transient distortions of allele frequency distributions over the past few dozen generations [89]. The last two methods were performed using the software BOTTLENECK [90] and a Two-Phase Mutation Model (TPM) [86] consisting of 95% of Stepwise Mutation Model (SMM) [91] and 5% of Infinite Allele Model (IAM) [92] with a variance of 12% among multi-steps, as recommended in [90], and 10000 iterations.

Results

Hardy-Weinberg Equilibrium, Linkage Disequilibrium and Null Alleles

Out of a total of 149 tests for heterozygote deficiency, only the population KCHB for the locus SFb5 showed heterozygote deficiency (P<0.05), after controlling both the FWER and the FDR. Heterozygosity excess was never detected in any population and any locus. Further, the LD test showed no statistical significance for all the 666 tests between pairs of loci indicating an independent segregation of loci. The locus SFb5 for the population KCHB showed a relatively high percentage of null alleles (21%). However, the percentage of null alleles for SFb5 on average across the 15 local populations was only 4.5%. In total, null allele frequency estimates ranged from 0.6% to 4.5% with 2.5% on average across loci. The estimation of F_{ST} both using and without using the ENA correction method gave equal results; F_{ST} $^{\text{not using } ENA} = F_{ST}$ $^{\text{using } ENA} = 0.0524$ with the respective 95% CI [0.039–0.072] and [0.038–0.073]. Classical measures of population differentiation are only slightly biased with a null allele frequency ranging between 5% and 8% on average across loci [50], [93]. Given that the average percentage of null allele frequency across loci (2.5%) is lower than 5%, and that the estimation of F_{ST} did not vary after excluding null alleles, all loci were kept for further statistical analyses.

Within-Population Genetic Diversity

In total, 109 different alleles were scored in the 351 genotyped individuals. Mean rarefied allelic richness per locus and per population averaged 4.14 (±0.02 SE) for a standardized sample size of n = 20 gene copies and 2.77 (±0.01 SE) alleles per locus and per population for a standardized sample size of n = 6 gene copies (Table 2). Mean rarefied private allelic richness per locus and per population averaged 0.15 (±0.005 SE) (Table 2). The mean observed heterozygosity (H_O = 0.578±0.017 SE) was not significantly different from the mean expected heterozygosity under HWE (H_E = 0.595±0.016 SE) (P>0.05) (Table 2). The overall mean inbreeding coefficient (F_{IS} = 0.029) was not significantly different from zero with the 99% CI [−0.024–0.090] (Table 2). Permutation tests showed that *A. cilicica* was significantly less genetically diverse (P<0.001) and had significantly less allelic richness (P<0.001) than the studied northern Mediterranean fir species (*A. alba*, *A. bornmuelleriana* and *A. cephalonica*) (Figure 2, Table 2). Despite being significantly less genetically diverse (P<0.01) and significantly less rich in alleles (P<0.01) than *A. alba* populations from the French Pyrenees and the French Southern Alps, no significant difference was observed for gene diversity and rarefied allelic richness between populations of *A. alba* from the Pyrenees and *A. cilicica* from Lebanon (P>0.05) (Figure 2, Table 2). Similarly, *A. cilicica* and *A. marocana* were not significantly different in both gene diversity and rarefied allelic richness (P>0.05) (Figure 2, Table 2).

Among-Population Genetic Differentiation

The among-population genetic differentiation based on allele identities (F_{ST}) and on allele sizes (R_{ST}) averaged 0.0536 (P = 0) and 0.0374 (P = 0), respectively, across all loci. Allele size permutation tests showed that the observed value of R_{ST} computed before randomization was not significantly different from the R_{ST} obtained after allele size permutation (pR_{ST}) (P = 0.43). Therefore, strict stepwise mutations did not contribute to among-population differentiation and no phylogeographic pattern was detected.

The Bayesian analysis of population structure in STRUCTURE and calculation of ΔK from the STRUCTURE output produced a modal value near 100 of the statistic at K = 2, thus indicating that the uppermost hierarchical level detected by STRUCTURE was two distinct genetic clusters. The Bayesian analysis of population structure in BAPS split the 15 local populations into two genetically distinct groups with a posterior probability of 0.95: 11 populations (KCHB, BTJF, CHIR, JBMD, MKTF, NBSH, QJRN, QEMN,

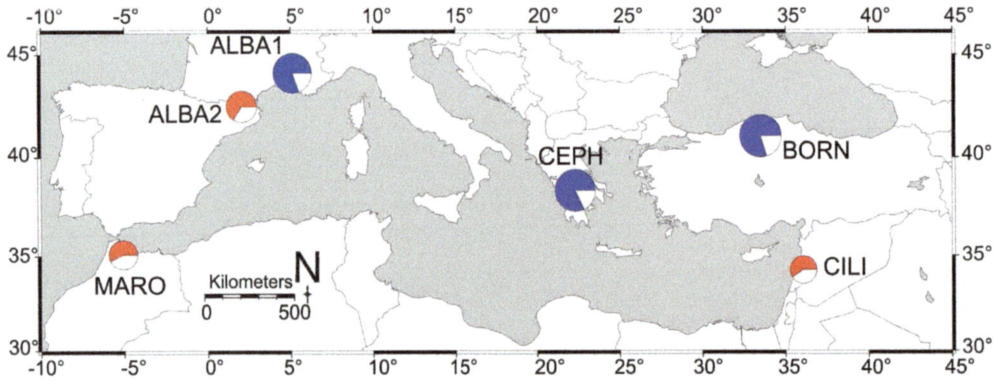

Figure 2. Genetic variation of *Abies cilicica* and other Mediterranean *Abies* species. Pie charts are used to depict the gene diversity corrected for sample size [55], [56] and the rarefied allelic richness [57], [58], averaged over 10 nuclear microsatellite loci, for *A. cilicica* in Lebanon (CILI), *A. bornmuelleriana* in Turkey (BORN), *A. cephalonica* in Greece (CEPH), *A. alba* in the French Southern Alps (ALBA1), *A. alba* in the French Pyrenees (ALBA2), and *A. marocana* in Morocco (MARO). Diameter of pie charts is relative to the correspondent rarefied allelic richness of the species. Proportion of red or blue color within pie charts is relative to the correspondent gene diversity of the species. Blue and red colors are employed to separate the studied fir populations into a group with significant lower genetic variation (red pie charts), and a group with significant higher genetic variation (blue pie charts) based on 10000 permutations of populations among groups performed in FSTAT version 2.9.3.2 [54].

Table 2. Genetic variation and genetic distinctiveness of *Abies cilicica* and other Mediterranean *Abies* species.

Code	H_O	H_E	$A_{R\ (20)}$	$A_{R\ (6)}$	Π_s	F_{IS}
KCHB	0.568	0.616	4.95	3.04	0.18	0.079
BTJF	0.582	0.607	4.40	2.9	0.24	0.042
CHIR	0.629	0.631	4.57	2.95	0.05	0.005
JBMD	0.615	0.629	4.57	2.93	0.14	0.023
MKTF	0.561	0.612	4.32	2.85	0.22	0.086
NBSH	0.567	0.572	4.02	2.67	0.10	0.010
QARS	0.448	0.497	3.23	2.34	0.24	0.103
QJRN	0.609	0.598	4.39	2.83	0.35	−0.019
QEMN	0.603	0.585	4.33	2.78	0.18	−0.031
HNKT	0.665	0.611	4.09	2.77	0.08	−0.094
HNKF	0.621	0.619	4.47	2.91	0.20	−0.004
GHAB	0.579	0.625	4.45	2.90	0.22	0.075
HKYT	0.558	0.600	3.21	2.56	0	0.072
KMHR	0.621	0.599	3.45	2.67	0	−0.037
EHDN	0.446	0.522	3.67	2.53	0.01	0.149
Mean	0.578	0.595	4.14	2.77	0.15	0.029
MARO	0.405	0.572	-	2.86	-	-
BORN	0.668	0.799	-	4.22	-	-
CEPH	0.713	0.821	-	4.17	-	-
ALBA1	0.663	0.808	-	3.93	-	-
ALBA2	0.640	0.662	-	3.08	-	-

H_O observed heterozygosity per population averaged over loci; H_E expected heterozygosity corrected for sample size averaged over loci [55], [56]; $A_{R\ (20)}$ rarefied allelic richness for a sample size of 20 gene copies averaged over loci [57], [58]; $A_{R\ (6)}$ rarefied allelic richness for a sample size of 6 gene copies averaged over loci [57], [58]; Π_s rarefied private allelic richness for a sample size of 20 gene copies averaged over loci [57], [60]; F_{IS} inbreeding coefficient [59]. MARO refers to *Abies marocana* in Morocco, BORN to *Abies bornmuelleriana* in Turkey, CEPH to *Abies cephalonica* in Greece, ALBA1 to *Abies alba* in the French Southern Alps and ALBA2 to *Abies alba* in the French Pyrenees. Other codes refer to *Abies cilicica* in Lebanon (See Tables 1 and S1).

HNKT, HNKF and GHAB) were grouped into a single deme (deme 1) while the other 4 populations (QARS, HKYT, KMHR and EHDN) formed another deme (deme 2). Because of their geographic distribution, throughout this paper, the demes 1 and 2 are referred to as "Northeastern ridge" and "Southwestern ridge" respectively.

The permutation test of populations performed in FSTAT revealed significant differences among demes. Notably, the Southwestern ridge had significantly lower allelic richness ($P<0.001$), had significantly lower gene diversity ($P<0.01$), but significantly higher differentiation ($P<0.05$) than the Northeastern ridge (Table 3). The total genetic diversity of each deme, however, was not significantly different and neither was F_{IS} ($P>0.05$) (Table 3). Additionally, R_{ST} was not significantly different from pR_{ST} for each deme ($P>0.05$), thus, no phylogeographic pattern was detected within the Northeastern ridge and within the Southwestern ridge (Table 3).

The likelihood-based Bayesian assignment test performed within GENECLASS2 allowed the detection of 11 first-generation migrants among the 15 remnant populations of *A. cilicica* in Lebanon ($P<0.01$) (Table 4). Gene flow within the Northeastern ridge was pronounced with a total of 6 F_0 migrants, although only one F_0 migrant was detected within the Southwestern ridge. Among-deme gene flow was asymmetrical: more gene flow from the Northeastern ridge towards the Southwestern ridge (3 F_0 migrants) and less gene flow from the Southwestern ridge towards the Northeastern ridge (1 F_0 migrant) (Figure 3). Dispersal occurred not only between neighboring populations, but also between distant populations at distance intervals comprised between 10 Km and 20 Km. Dispersal was not observed between the two most distant local populations, i.e. KCHB and EHDN.

Isolation-By-Distance

Isolation-by-distance (IBD) assessed over all populations was significant (Mantel test, $P<0.01$). However, the strength of the IBD was weak, with the RMA explaining only 5% of the variance in the whole distribution area ($r=-0.2248$; $R^2=0.05$) (Figure 4), thus implying non-equilibrium conditions between dispersal and genetic drift. The weak strength of IBD and the lack of phylogeographic patterns further implied a recent isolation of local populations. No pattern of IBD was detected within the

Table 3. Genetic variation, population structure and phylogeography in the Northeastern and Southwestern ridges.

	H_O	H_E	A_R	H_T	F_{IS}	F_{ST}	R_{ST}	pR_{ST}
Northeastern ridge	0.595 **	0.609**	4.413***	0.631^{NS}	0.023^{NS}	0.038*	0.018^{NS}	0.024^{NS}
Southwestern ridge	0.521 **	0.556**	3.389***	0.594^{NS}	0.064^{NS}	0.081*	0.0731^{NS}	0.0858^{NS}

H_O observed heterozygosity per deme [76]; H_E expected heterozygosity per deme and corrected for sample size [76]; A_R rarefied allelic richness per deme for a sample size of 102 gene copies [57], [58]; H_T overall gene diversity per deme [76]; F_{IS} inbreeding coefficient per deme [59]; F_{ST} genetic differentiation per deme based on allele identity [59]; R_{ST} genetic differentiation per deme based on allele size [63], [64]; pR_{ST} permuted R_{ST} after allele size permutation test [65]. Columns 1 to 6 indicate the statistical differences in genetic variation and population structure between populations of the Northeastern ridge and populations of the Southwestern ridge. Columns 7 and 8 indicate the results of the test of phylogeographic patterns corresponding to the statistical differences between R_{ST} and pR_{ST} within the Northeastern ridge and within the Southwestern ridge. * P<0.05; ** P<0.01; *** P<0.001. ^{NS}Non-Significant test.

Northeastern ridge (P = 0.1568) and within the Southwestern ridge (P = 0.0834).

Genetic Bottlenecks

The observed values of the Garza-Williamson indices ranged between 0.54 and 0.67. However, there was no statistical evidence of a recent and severe genetic bottleneck. First, the observed *M*-ratios were not significantly lower than the simulated equilibrium distribution of *M* for all the different pre-bottleneck θ values (P>0.05). Second, the bottleneck threshold values, calculated for all the pre-bottleneck θ values, were significantly lower than the observed *M*-ratios (P<0.05) (Figure 5). The Wilcoxon's signed rank test revealed recent heterozygosity excess in only 4 populations (MKTF, P<0.05; QARS, P<0.01; HNKT, P<0.001; and KMHR, P<0.05). However, for each local population, the mode-shift test detected the typical L-shaped allele frequency distribution of non-bottlenecked populations at mutation-drift equilibrium (Figure 6). We concluded, for the 3 different tests, that no recent and severe genetic bottleneck was evident for any local population.

Discussion

A. cilicica in Lebanon revealed low within-population genetic diversity (0.595±0.016 SE), low allelic richness (2.77±0.01 SE), compared to other *Abies* species, and low among-population

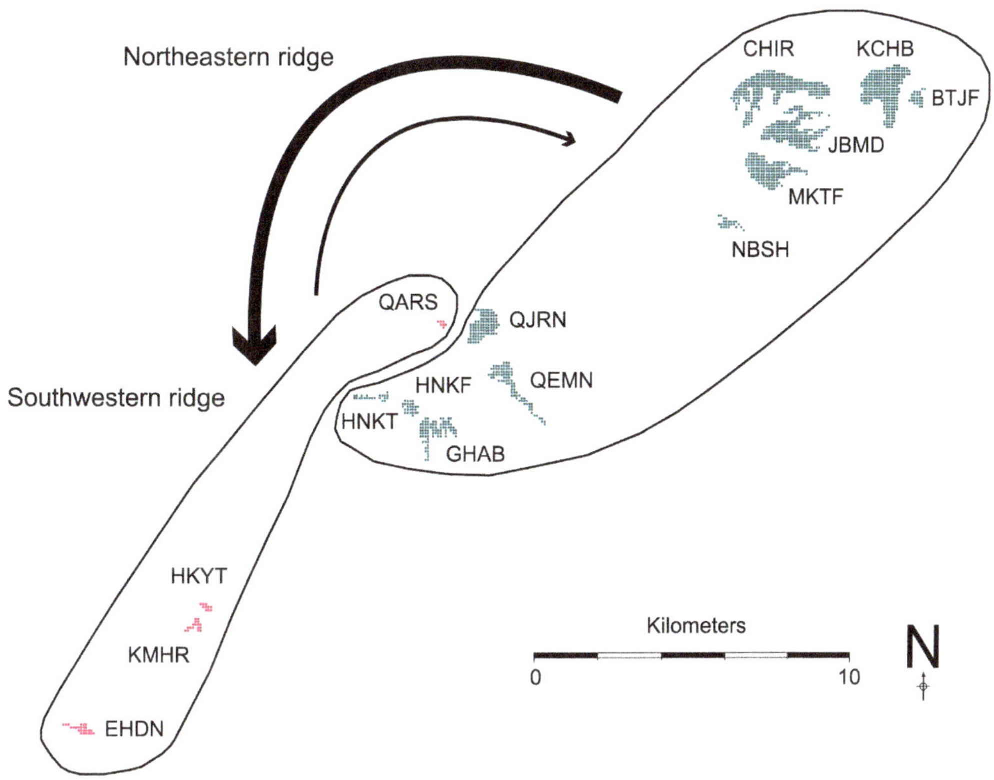

Figure 3. Asymmetric Northeast-Southwest migration of *Abies cilicica* in Lebanon. The Northeastern and Southwestern ridges correspond to the two genetically-distinct demes assigned by BAPS [71]. Populations of the Northeastern ridge are plotted in green color while those of the Southwestern ridge are plotted in magenta color. A likelihood-based Bayesian assignment test [79] performed in GENECLASS2 [77] allowed the detection, in the Southwestern ridge, of 3 F_0 migrants originating from the Northeastern ridge. It has also allowed the detection, in the Northeastern ridge, of 1 F_0 migrant originating from the Southwestern ridge, suggesting an asymmetric Northeast-Southwest migration.

Table 4. First-generation migrants among populations and among demes.

F_0 migrant	Sampled population	Population of origin	P-value
1	CHIR[1]	HNKF[1]	0.0084
2	MKTF[1]	HNKF[1]	0.0003
3	NBSH[1]	HNKT[1]	0.0098
4	KCHB[1]	BTJF[1]	0.0045
5	BTJF[1]	KCHB[1]	0.0049
6	QEMN[1]	GHAB[1]	0.0021
7	KMHR[2]	HKYT[2]	0.0003
8	QARS[2]	BTJF[1]	0.0000
9	EHDN[2]	QEMN[1]	0.0010
10	EHDN[2]	GHAB[1]	0.0091
11	QEMN[1]	EHDN[2]	0.0025

This table corresponds to the results of the likelihood-based Bayesian assignment test [79] performed within GENECLASS2 [77] allowing the detection of first-generation (F_0) migrants within and between the Northeastern and Southwestern ridges. The probability computation at the 1% significance level was performed using the Monte Carlo resampling method of [78] with 10000 simulated individuals. Rows 1 to 6 correspond to recent dispersal between populations of the Northeastern ridge (6 F_0 migrants) while row 7 corresponds to recent dispersal between populations of the Southwestern ridge (1 F_0 migrant). Recent dispersal between both the Northeastern and Southwestern ridges is represented in rows 8 to 11 (4 F_0 migrants). See Tables 1 and S1 for codes.
[1]membership of the local population to deme 1 (Northeastern ridge).
[2]membership of the local population to deme 2 (Southwestern ridge).

genetic differentiation (F_{ST} = 0.0536; R_{ST} = 0.0374). Populations were not inbred and showed no signal of a recent and severe genetic bottleneck. Despite the low spatial genetic structure inferred from classical genetic differentiation parameters, populations showed a significant genetic divergence, forming two demes, the Northeastern and Southwestern ridges. Additionally, despite their isolation-by-distance, populations were capable of long-distance dispersal. We discuss in the following sections how these contradictory features may have emerged and their effect on the persistence of *A. cilicica* in Lebanon. We finally discuss the implications of these results for conservation.

A Lower Within-Population Genetic Diversity than Expected

Previous comparative studies between *Abies* species in the Mediterranean, using allozymes and chloroplast microsatellites [2], [3], have respectively shown that *A. cilicica* in Turkey and Syria harbor high values of heterozygosity. High levels of genetic variation have also been found for *A. cephalonica* and *A. bornmuelleriana* in numerous studies [2–5], and were further supported in our study. Contrary to its geographical position in the Eastern Mediterranean and its life-history traits as a conifer species, our results have shown that *A. cilicica* in Lebanon revealed unexpected low within-population genetic diversity and low allelic richness. Particularly, this low genetic variation was not significantly different from the genetic variation of *A. marocana* in the Moroccan Rif and *A. alba* in the French Pyrenees. Previous studies have also reported low genetic variation in *A. marocana* [2], [3]. The significant lower genetic variation found in *A. alba* in the French Pyrenees with respect to *A. alba* in the French Southern Alps is also consistent with a previous study [6]. Particularly, *A. alba* in the Pyrenees is isolated within a glacial refugium and highly differentiated from the rest of its range [7]. The low genetic variation detected in *A. alba* in the Pyrenees might be the result of their exposure to a historic severe genetic bottleneck [8].

The low genetic variation of *A. cilicica* in Lebanon might be explained by the historical timber use during the Ancient Egyptian New Kingdom's rule over Phoenicia [36] and more recently during World War II [37], and by current severe illegal logging and overgrazing activities, resulting in strong habitat fragmentation. In this context, the absence of signatures of recent and severe genetic bottlenecks is unexpected and so is the high recent effective population size of *A. cilicica* in Lebanon. The low genetic variation might have been shaped over time by a rather long-term than a short-term low effective population size. One probable cause of this might be the temporal fluctuation in population sizes [22], [23], which can be typical of populations situated at the margins of geographic distributions [94]. It might also be the result of temporal variation in mating system dynamics, which can be typical of conifer species with dynamic mating systems [95], [96]. It might as well be the result of the founder effect mediated by local extinction and recolonizations in a metapopulation context

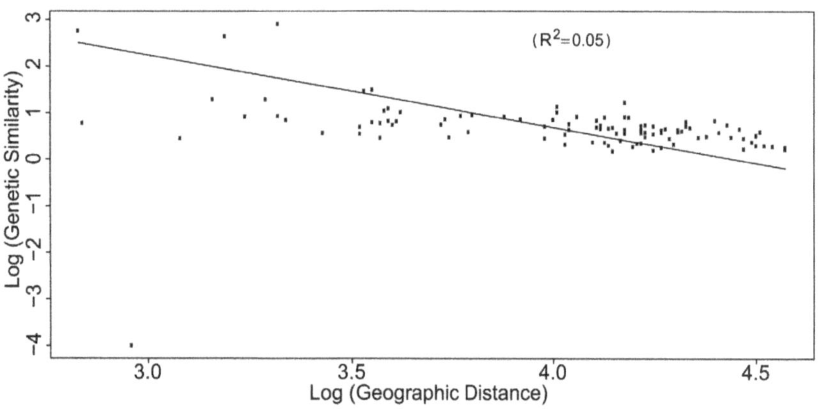

Figure 4. Isolation-by-distance and non-equilibrium conditions between dispersal and genetic drift. The logarithm of genetic similarity is plotted against the logarithm of geographic distance among population pairs [26]. Estimations of slope, intercept and R^2 of the relationships were calculated using Reduced Major Axis (RMA) regression [83]. The RMA regression explained only 5% of the variance in the whole distribution area of *Abies cilicica* in Lebanon. Analysis was performed in IBDWS [81].

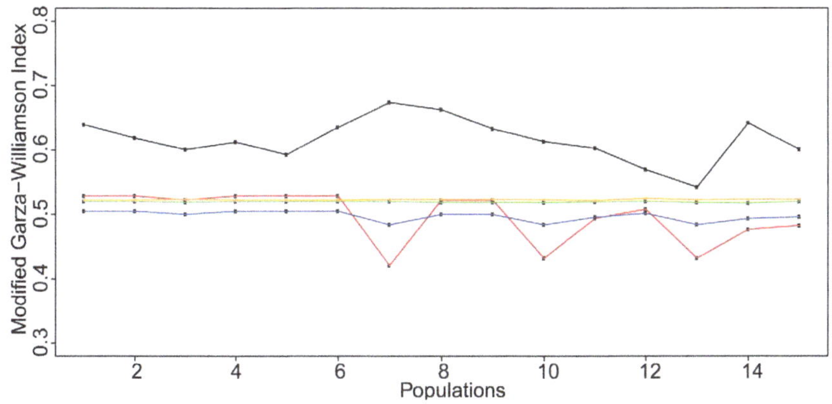

Figure 5. Observed and simulated values of the modified Garza-Williamson index for *Abies cilicica* populations. This figure shows the different observed values of the *M*-ratios calculated for the microsatellite loci and the values of the simulated critical *M* threshold (M_c) below which a bottleneck is evident. The observed *M*-ratio values are displayed in black solid line. The M_c values calculated for $\theta = 10$, $\theta = 1$, $\theta = 0.5$, and $\theta = 0.1$ are respectively displayed in red, blue, green and orange solid lines. All the observed *M*-ratio values were significantly higher than the simulated M_c values for different pre-bottleneck θ values. Values of the *M*-ratio and the M_c parameters were obtained using M-P-val and critical_M [84].

[24]. Nevertheless, the genetic imprint of recent founder events within these populations might have been rapidly masked by gene flow [28], and particularly by long-distance dispersal.

A Significant Genetic Structure despite Low F_{ST}

Despite the low observed value of F_{ST}, the Bayesian clustering methods detected two genetically distinct, but highly admixed demes, the Northeastern and Southwestern ridges. Populations of the Southwestern ridge inhabited the smallest habitat patches and occupied the southernmost part of the distribution area. Populations of the Southwestern ridge were significantly less genetically diverse, had significantly less allelic richness, but were significantly more genetically differentiated than populations of the Northeastern ridge. Despite these differences, the Northeastern and Southwestern ridges shared similar ecological attributes. Both demes occupied a similar elevation range, between ca. 1000 and 1800 m.a.s.l. Neither deme showed inbreeding or sub-population structure. Both demes preserved the same amount of total genetic diversity (H_T), and were capable of short-distance and long-distance dispersal. The recent large effective population size and the weak IBD pattern suggest an outperformance of migration over genetic drift. Consequently, genetic drift may not be the leading evolutionary force for population differentiation in *A. cilicica* from Lebanon. In the latter case, F_{ST} may not be accurate [97] and molecular measures of genetic diversity may weakly predict quantitative genetic variation [98].

High Dispersal Ability and Persistence despite Geographic Marginality

Similarly to many tree species, the genetic structure of *A. cilicica* in Lebanon may result from rapid colonization where both long-distance and short-distance dispersal are key demographic processes [15], [99], [100]. We define the observed migration type found in *A. cilicica* from Lebanon as a recent rear-edge migration with both recent colonization balancing local extinctions and recent range expansion along altitudinal gradients. Considering that brief episodes of range expansions are thought to leave strong and persistent signature on the genetic structure of natural populations [101], the Northeast-Southwest structure of *A. cilicica* in Lebanon suggests two plausible demographic processes promoting persistence.

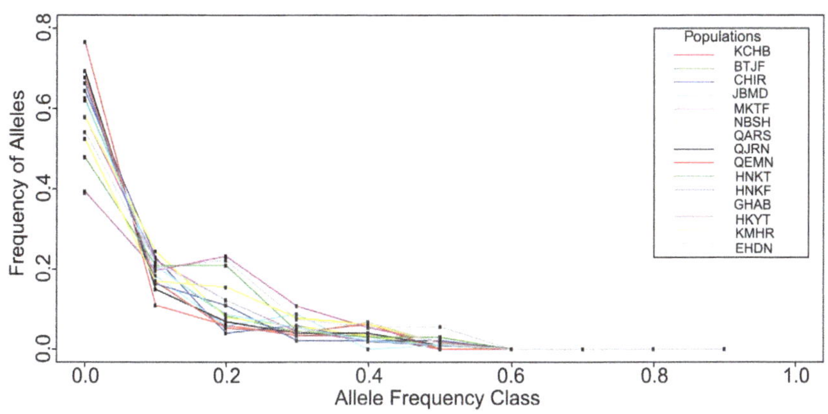

Figure 6. Allele frequency distribution over different allele frequency classes for each population of *Abies cilicica*. This figure shows the distribution of allele frequencies across the 15 local populations. The values were obtained using the mode-shift test [89]. The distribution of allele frequencies across the 15 local populations is plotted using different colors. All populations showed a higher abundance of rare alleles (frequency<0.1) than alleles at intermediate frequencies (0.1–0.2).

First, the Northeast-Southwest structure might result from recent colonization balancing local extinctions. The likelihood-based Bayesian assignment test detected an emigration rate from the Northeastern ridge exceeding three-fold that from the Southwestern ridge (Figure 3). Although, this asymmetric migration might be a simple consequence of the smaller size of the Southwestern ridge, gene flow from more genetically diverse populations may counteract drift and contribute to restoring genetic diversity in populations of the Southwestern ridge (e.g. QARS and EHDN), thus making persistence possible.

Second, the Northeast-Southwest structure might also result from recent range expansion along altitudinal gradients. Private alleles in both the Northeastern and Southwestern ridges could indicate the presence of different glacial microrefugia in both demes (Table 2). In the Northeastern ridge, more or less all local populations harbored private alleles, both at low and high elevations, suggesting multiple low-elevation and high-elevation microrefugia. The higher genetic diversity of Northeastern ridge populations might result from the mixing of these different microrefugial lineages along mountain slopes during recolonization. In the Southwestern ridge, high private allelic richness was only found in low-elevation population QARS. Other Southwestern ridge populations, located at higher elevations, harbored either no private alleles (HKYT and KMHR) or very few private alleles (EHDN), suggesting a postglacial colonization from QARS upwards. Range expansion along altitudinal gradients may have occurred concomitantly or at different time scales from the Northeastern-Southwestern asymmetric migration described earlier. Microrefugial populations may be crucial for predicting climate-change-related range shifts of tree species [102], and particularly, altitudinal range shifts may be an efficient mechanism for populations to cope with changing environmental and/or climatic conditions. A recent meta-analysis study showed that the distribution of species had recently shifted to higher elevations at an average rate of 11 meters per decade [103].

Implications for Conservation

Because they contain most of the genetic diversity of *A. cilicica* in Lebanon, and have shaped current diversity via long-distance migration and altitudinal range shifts, all local populations of the Northeastern ridge as well as QARS of the Southwestern ridge have disproportionate conservation importance. The very low census size of QARS and its location on the lowest elevation warrant special attention. Conserving other Southwestern ridge populations (such as EHDN) would also be useful as they are stepping stones for gene flow between the two demes. EHDN is an interesting example for highlighting conservation challenges. A mixed fir/cedar population, EHDN has been managed as a nature reserve since 1992. The low number of surviving fir trees in this nature reserve as well as their low genetic variation suggest that a change of strategy is needed if conserving the genetic diversity of *A. cilicica* is a target for management. Dynamic conservation of genetic diversity [104] *in situ* requires that a particular attention should be given to regeneration and to avoid introducing seeds or seedlings from exotic locations outside demes or farther away. Attention should be given to keeping gene flow between the Northeastern and Southwestern ridges at reasonable rates. This can be potential through maintaining historic-likewise levels of gene flow by (1) preserving the natural ecological corridor between the Northeastern and Southwestern ridges, and (2) preserving a buffer zone across each local population to facilitate the natural recolonization of local populations over unoccupied habitat patches. For local populations with very low census size and genetic diversity, *ex situ* conservation (e.g. establishing seed banks and clonal archives) or assisted migration to suitable locations where future climate should be favorable to *A. cilicica* is strongly advisable.

Conclusions

The population genetic diversity and structure of *A. cilicica* in Lebanon is at odds with biogeographical and life history trait expectations. Rather, demographic processes linked to current habitat fragmentation and geographic range structure, and to original historical range contraction and recolonization better explain genetic diversity. *A. cilicica* in Lebanon is made of two demes moderately connected by geneflow shaping a weak isolation-by-distance pattern. Persistence of *A. cilicica* in Lebanon was linked to two plausible demographic processes, occurring concurrently or sequentially, which we called rear-edge type migration: (1) alternating colonization and extinction events and (2) range shifts along altitudinal gradients. These two demographic events might help us better understand the species response to expected climate change. Populations of the Northeastern and Southwestern ridges might harbor a large evolutionary potential, and should be consequently prioritized for conservation.

Supporting Information

Table S1 Origin, location sample size of *Abies cilicica* and other Mediterranean *Abies* species. Legend: Origin, location, and sample size of the 15 populations of *Abies cilicica* in Lebanon and other Mediterranean *Abies* species. Altitude is in meters above sea level (m.a.s.l.) and the surface of the sampled populations of *Abies cilicica* in Lebanon is in hectares. N.B. Surface is not indicative of population density.

Acknowledgments

The authors would like to thank all of those who assisted in sampling the fir populations in Lebanon and Morocco. We are grateful to the National Council for Scientific Research-Lebanon as well as the municipalities of Fnaydeq and Bqaa Safrine in Lebanon for the logistics support. We also would like to thank Matthieu Lingrand (INRA-URFM) for the technical support in the laboratory. This is an ISEM contribution number ISEM 2013-182.

Author Contributions

Conceived and designed the experiments: RC CK LA. Performed the experiments: LA AR. Analyzed the data: LA BF RC. Contributed reagents/materials/analysis tools: RC LA BF. Wrote the paper: LA BF RC. Performed the field sampling: LA CK RC. Read and approved the manuscript: LA BF RC CK AR.

References

1. Billington HL (1991) Effect of population size on genetic variation in a dioecious conifer. Conserv Biol 5: 115–119.
2. Scaltsoyiannes A, Tsaktsira M, Drouzas AD (1999) Allozyme differentiation in the Mediterranean firs (*Abies, Pinaceae*). A first comparative study with phylogenetic implications. Plant Syst Evol 216: 289–307.
3. Liepelt S, Mayland-Quellhorst E, Lahme M, Ziegenhagen B (2010) Contrasting geographical patterns of ancient and modern genetic lineages in Mediterranean *Abies* species. Plant Syst Evol 284: 141–151.
4. Fady B, Conkle MT (1993) Allozyme variation and possible phylogenetic implications in Abies cephalonica Loudon and some related Eastern Mediterranean firs. Silvae Genet 42: 351–359.

5. Parducci L, Szmidt AE, Madaghiele A, Anzidei M, Vendramin GG (2001) Genetic variation at chloroplast microsatellites (cpSSRs) in *Abies nebrodensis* (Lojac.) Mattei and three neighboring *Abies* species. Theor Appl Genet 102: 733–740.
6. Fady B, Forest I, Hochu I, Ribiollet A, de Beaulieu J-L, et al. (1999) Genetic differentiation in *Abies alba* Mill. populations from Southeastern France. Forest Genetics 6: 129–138.
7. Konnert M, Bergmann F (1995) The geographical distribution of genetic variation of silver fir (*Abies alba, Pinaceae*) in relation to its migration history. Plant Syst Evol 196: 19–30.
8. Vendramin GG, Degen B, Petit RJ, Anzidei M, Madaghiele A et al. (1999) High level of variation at *Abies alba* chloroplast microsatellite loci in Europe. Mol Ecol 8: 1117–1126.
9. Pluess AR (2011) Pursuing glacier retreat: genetic structure of a rapidly expanding *Larix decidua* population. Mol Ecol 20: 473–485.
10. Conord C, Gurevitch J, Fady B (2012) Large-scale longitudinal gradients of genetic diversity: a meta-analysis across six phyla in the Mediterranean basin. Ecol Evol 2: 2600–2614.
11. Hamrick JL, Godt MJW (1989) Allozyme diversity in plant species. In: Brown AHD, Clegg MT, Kahler AL, Weir BS, editors. Plant population genetics, breeding and genetic resources. Sunderland, Massachusetts: Sinauer Press. pp. 43–63.
12. Hamrick JL, Godt MJW, Sherman-Broyles SL (1992) Factors influencing levels of genetic diversity in woody plant species. New Forests 6: 95–124.
13. Gitzendanner MA, Soltis PS (2000) Patterns of genetic variation in rare and widespread plant congeners. Am J Bot 87: 783–792.
14. Eliades N-GH, Gailing O, Leinemann L, Fady B, Finkeldey R (2011) High genetic diversity and significant population structure in *Cedrus brevifolia* Henry, a narrow endemic Mediterranean tree from Cyprus. Plant Syst Evol 294: 185–198.
15. Clark JS (1998) Why trees migrate so fast: confronting theory with dispersal biology and the paleorecord. Am Nat 152: 204–224.
16. Ibrahim KM, Nichols RA, Hewitt GM (1996) Spatial patterns of genetic variation generated by different forms of dispersal during range expansion. Heredity 77: 282–291.
17. Bialozyt R, Ziegenhagen B, Petit RJ (2006) Contrasting effects of long distance seed dispersal on genetic diversity during range expansion. J Evol Biol 19: 12–20.
18. Slatkin M (1987) Gene flow and the geographic structure of natural populations. Science 236: 787–792.
19. Mimura M, Aitken SN (2007) Adaptive gradients and isolation-by-distance with postglacial migration in *Picea sitchensis*. Heredity 99: 224–232.
20. Mimura M, Aitken SN (2010) Local adaptation at the range peripheries of Sitka spruce. J Evol Biol 23: 249–258.
21. Young A, Boyle T, Brown T (1996) The population genetic consequences of habitat fragmentation for plants. Tree 11: 413–418.
22. Pimm SL, Gittleman JL, McCracken GF, Gilpin ME (1989) Plausible alternatives to bottlenecks to explain reduced genetic diversity. Trends Ecol Evol 4: 176–178.
23. Cornuet J-M, Luikart G (1996) Description and power analysis of two tests for detecting recent population bottlenecks from allele frequency data. Genetics 144: 2001–2014.
24. Gilpin M (1991) The genetic effective size of a metapopulation. Biol J Linn Soc Lond 42: 165–175.
25. Wright S (1943) Isolation by distance. Genetics 38: 114–138.
26. Slatkin M (1993) Isolation by distance in equilibrium and non-equilibrium populations. Evolution 47: 264–279.
27. Wade MJ, McCauley DE (1988) Extinction and recolonization: their effects on the genetic differentiation of local populations. Evolution 42: 995–1005.
28. Whitlock MC (1992) Temporal fluctuations in demographic parameters and the genetic variance among populations. Evolution 46: 608–615.
29. Lande R (1992) Neutral theory of quantitative genetic variance in an island model with local extinction and colonization. Evolution 46: 381–389.
30. Fady-Welterlen B (2005) Is there really more biodiversity in Mediterranean forest ecosystems?.Taxon 54: 905–910.
31. Petit RJ, Hampe A, Cheddadi R (2005) Climate changes and tree phylogeography in the Mediterranean. Taxon 54: 877–885.
32. Petit RJ, Aguinagalde I, de Beaulieu J-L, Bittkau C, Brewer S, et al. (2003) Glacial refugia: hotspots but not melting pots of genetic diversity. Science 300: 1563–1565.
33. Hewitt G (2000) The genetic legacy of the Quaternary ice ages. Nature 405: 907–913.
34. Hampe A, Petit RJ (2005) Conserving biodiversity under climate change: the rear edge matters. Ecol Lett 8: 461–467.
35. Fady B, Conord C (2010) Macroecological patterns of species and genetic diversity in vascular plants of the Mediterranean basin. Diversity and Distributions 16: 53–64.
36. Bardinet T (2008) Relations économiques et pressions militaires en Méditerranée orientale et en Libye au temps des Pharaons. Histoire des importations égyptiennes des résines et des conifères du Liban et de la Libye depuis la période archaïque jusqu'à l'époque ptolémaïque. In: Cybele, editor. Paris: Etudes et mémoires d'égyptologie. 337p.
37. Mikesell MW (1969) The deforestation of Mount Lebanon. Geogr Rev 59: 1–28.
38. Gardner M, Knees S (2013) *Abies cilicica*. In: IUCN 2013. IUCN Red List of Threatened Species Version 2013.1. Available: http://bit.ly/1bKIa6C. Accessed 18 September 2013.
39. MOA DRDNR (2005) National forest and tree assessment and inventory TCP/LEB/2903. FAO. Available: http://bit.ly/1a9yJkQ. Accessed 19 March 2013.
40. Selkoe KA, Toonen RJ (2006) Microsatellites for ecologists: a practical guide to using and evaluating microsatellite markers. Ecol Lett 9: 615–629.
41. Balloux F, Lugon-Moulin N (2002) The estimation of population differentiation with microsatellite markers. Mol Ecol 11: 155–165.
42. Hansen OK, Vendramin GG, Sebastiani F, Edwards KJ (2005) Development of microsatellite markers in *Abies nordmanniana* (Stev.) Spach and cross-species amplification in the *Abies* genus. Mol Ecol Notes 5: 784–787.
43. Cremer E, Liepelt S, Sebastiani F, Buonamici A, Michalczyk IM, et al. (2006) Identification and characterization of nuclear microsatellite loci in *Abies alba* Mill. Mol Ecol Notes 6: 374–376.
44. Amos W, Hoffman JI, Frodsham A, Zhang L, Best S, et al. (2007) Automated binning of microsatellite alleles: problems and solutions. Mol Ecol Notes 7: 10–14.
45. Guichoux E, Lagache L, Wagner S, Chaumeil P, Léger P, et al. (2011) Current trends in microsatellite genotyping. Mol Ecol Resour 11: 591–611.
46. Raymond M, Rousset F (1995) GENEPOP (version 1.2): Population genetics software for exact tests and ecumenicism. J Hered 86: 248–249.
47. Rousset F (2008) GENEPOP'007: a complete re-implementation of the genepop software for Windows and Linux. Mol Ecol Resour 8: 103–106.
48. Rice WR (1989) Analyzing tables of statistical tests. Evolution 43: 223–225.
49. Benjamini Y, Hochberg Y (1995) Controlling the False Discovery Rate: a practical and powerful approach to multiple testing. J R Stat Soc Series B Stat Methodol 57: 289–300.
50. Chapuis M-P, Estoup A (2007) Microsatellite null alleles and estimation of population differentiation. Mol Biol Evol 24: 621–631.
51. Dempster AP, Laird NM, Rubin DB (1977) Maximum Likelihood from incomplete data via the EM algorithm. J R Stat Soc Series B Stat Methodol 39: 1–38.
52. Weir BS (1996) Genetic data analysis II. Sunderland, Massachusetts: Sinauer Associates, Inc. 445 p.
53. Peakall R, Smouse PE (2006) GENALEX 6: genetic analysis in Excel. Population genetic software for teaching and research. Mol Ecol Notes 6: 288–295.
54. Goudet J (1995) FSTAT (Version 1.2): A computer program to calculate F-statistics. J Hered 86: 485–486.
55. Nei M (1978) Estimation of average heterozygosity and genetic distance from a small number of individuals. Genetics 89: 583–590.
56. Hartl DL, Clark AG (1997) Principles of population genetics, 3rd edition. Sunderland, Massachusetts: Sinauer Associates, Inc. 542 p.
57. Hurlbert SH (1971) The nonconcept of species diversity: a critique and alternative parameters. Ecology 52: 577–586.
58. El Mousadik A, Petit RJ (1996) High level of genetic differentiation for allelic richness among populations of the argan tree [*Argania spinosa* (L.) Skeels] endemic to Morocco. Theor Appl Genet 92: 832–839.
59. Weir BS, Cockerham CC (1984) Estimating F-statistics for the analysis of population structure. Evolution 38: 1358–1370.
60. Kalinowski ST (2004) Counting alleles with rarefaction: private alleles and hierarchical sampling designs. Conserv Genet 5: 539–543.
61. Kalinowski ST (2005) HP-RARE 1.0: a computer program for performing rarefaction on measures of allelic richness. Mol Ecol Notes 5: 187–189.
62. Hardy OJ, Vekemans X (2002) SPAGEDI: a versatile computer program to analyse spatial genetic structure at the individual or population levels. Mol Ecol Notes 2: 618–620.
63. Slatkin M (1995) A measure of population subdivision based on microsatellite allele frequencies. Genetics 139: 457–462.
64. Michalakis Y, Excoffier L (1996) A generic estimation of population subdivision using distances between alleles with special reference for microsatellite loci. Genetics 142: 1061–1064.
65. Hardy OJ, Charbonnel N, Fréville H, Heuertz M (2003) Microsatellite allele sizes: a simple test to assess their significance on genetic differentiation. Genetics 163: 1467–1482.
66. Pritchard JK, Stephens M, Donnelly P (2000) Inference of population structure using multilocus genotype data. Genetics 155: 945–959.
67. Falush D, Stephens M, Pritchard JK (2003) Inference of population structure using multilocus genotype data: linked loci and correlated allele frequencies. Genetics 164: 1567–1587.
68. Pritchard JK, Wen X, Falush D (2010) Documentation for STRUCTURE software: version 2.3. Available: http://stanford.io/1dGRJYQ. Accessed 26 April 2012.
69. Earl DA, vonHoldt BM (2011) STRUCTURE HARVESTER: a website and program for visualizing STRUCTURE output and implementing the Evanno method. Conserv Genet Resour 4: 359–361.
70. Evanno G, Regnaut S, Goudet J (2005) Detecting the number of clusters of individuals using the software STRUCTURE: a simulation study. Mol Ecol 14: 2611–2620.
71. Corander J, Cheng L, Marttinen P, Sirén J, Tang J (2013) BAPS: Bayesian Analysis of Population Structure. Manual v.6.0. Available: http://bit.ly/1eAKi4e. Accessed 20 February 2013.

72. Corander J, Waldmann P, Sillanpää MJ (2003) Bayesian analysis of genetic differentiation between populations. Genetics 163: 367–374.
73. Corander J, Sirén J, Arjas E (2008) Bayesian spatial modeling of genetic population structure. Computational Statistics 23:111–129.
74. Rodríguez-Ramilo ST, Toro MA, Fernández J (2009) Assessing population genetic structure via the maximisation of genetic distance. Genet Sel Evol 41:49.
75. Corander J, Marttinen P, Mäntyniemi S (2006) Bayesian identification of stock mixtures from molecular marker data. Fishery Bulletin 104: 550–558.
76. Nei M (1987) Molecular evolutionary genetics. New York: Columbia University Press. 512 p.
77. Piry S, Alapetite A, Cornuet J-M, Paetkau D, Baudouin L, et al. (2004) GENECLASS2: a software for genetic assignment and first-generation migrant detection. J Hered 95: 536–539.
78. Paetkau D, Slade R, Burden M, Estoup A (2004) Genetic assignment methods for the direct, real-time estimation of migration rate: a simulation-based exploration of accuracy and power. Mol Ecol 13: 55–65.
79. Rannala B, Mountain JL (1997) Detecting immigration by using multilocus genotypes. Proc Natl Acad Sci U S A 94: 9197–9201.
80. Cornuet J-M, Piry S, Luikart G, Estoup A, Solignac M (1999) New methods employing multilocus genotypes to select or exclude populations as origins of individuals. Genetics 153: 1989–2000.
81. Jensen JL, Bohonak AJ, Kelley ST (2005) Isolation by distance, web service. BMC Genetics 6: 13.
82. Manly BFJ (1994) Multivariate statistical methods: a primer, 2nd edition. New York: Chapman and Hall.
83. Hellberg ME (1994) Relationships between inferred levels of gene flow and geographic distance in a philopatric Coral, *Balanophyllia elegans*. Evolution 48: 1829–1854.
84. Garza JC, Williamson EG (2001) Detection of reduction in population size using data from microsatellite loci. Mol Ecol 10: 305–318.
85. Excoffier L, Estoup A, Cornuet J-M (2005) Bayesian analysis of an admixture model with mutations and arbitrarily linked markers. Genetics 169: 1727–1738.
86. Di Rienzo A, Peterson AC, Garza JC, Valdes AM, Slatkin M, et al. (1994) Mutational processes of simple-sequence repeat loci in human populations. Proc Natl Acad Sci U S A 91: 3166–3170.
87. Estoup A, Angers B (1998) Microsatellites and minisatellites for molecular ecology: theoretical and empirical considerations. In: Carvalho G, editor. Advances in molecular ecology. Amsterdam: IOS Press. pp. 55–86.
88. Luikart G (1997) Usefulness of molecular markers for detecting population bottlenecks and monitoring genetic change. Ph. D. Thesis. Missoula: University of Montana.
89. Luikart G, Allendorf FW, Cornuet J-M, Sherwin WB (1998) Distortion of allele frequency distributions provides a test for recent population bottlenecks. J Hered 89: 238–247.
90. Piry S, Luikart G, Cornuet J-M (1999) BOTTLENECK: a computer program for detecting recent reductions in the effective population size using allele frequency data. J Hered 90: 502–503.
91. Ohta T, Kimura M (1973) A model of mutation appropriate to estimate the number of electrophoretically detectable alleles in a finite population. Genet Res 22: 201–204.
92. Kimura M, Crow JF (1964) The number of alleles that can be maintained in a finite population. Genetics 49: 725–738.
93. Dakin EE, Avise JC (2004) Microsatellite null alleles in parentage analysis. Heredity 93: 504–509.
94. Cuervo JJ, Møller AP (2013) Temporal Variation in Population Size of European Bird Species: Effects of Latitude and Marginality of Distribution. PLoS ONE 8(10): e77654. doi:10.1371/journal.pone.0077654.
95. Cheliak WM, Dancik BP, Morgan K, Yeh FCH, Strobeck C (1985) Temporal variation of the mating system in a natural population of Jack Pine. Genetics 109: 569–584.
96. Mitton JB (1992) The dynamic mating systems of conifers. New Forests 6: 197–216.
97. Neigel JE (2002) Is F_{ST} obsolete?. Conserv Genet 3: 167–173.
98. Reed DH, Frankham R (2001) How closely correlated are molecular and quantitative measures of genetic variation? A meta-analysis. Evolution 55: 1095–1103.
99. Le Corre V, Machon N, Petit RJ, Kremer A (1997) Colonization with long-distance seed dispersal and genetic structure of maternally inherited genes in forest trees: a simulation study. Genet Res 69: 117–125.
100. Roques L, Hamel F, Fayard J, Fady B, Klein EK (2010) Recolonization by diffusion can generate increasing rates of spread. Theor Popul Biol 77: 205–212.
101. Petit RJ (2011) Early insights into the genetic consequences of range expansions. Heredity 106: 203–204.
102. Feurdean A, Bhagwat SA, Willis KJ, Birks HJB, Lischke H, et al. (2013) Tree Migration-Rates: Narrowing the Gap between Inferred Post-Glacial Rates and Projected Rates. PLoS ONE 8(8): e71797. doi:10.1371/journal.pone.0071797
103. Chen I-C, Hill JK, Ohlemüller R, Roy DB, Thomas CD (2011) Rapid range shifts of species associated with high levels of climate warming. Science 333: 1024–1026.
104. Koskela J, Lefèvre F, Schueler S, Kraigher H, Olrik DC, et al. (2013) Translating conservation genetics into management: Pan-European minimum requirements for dynamic conservation units of forest tree genetic diversity. Biol Conserv 157: 39–49.
105. Alizoti PG, Fady B, Prada MA, Vendramin GG (2011) EUFORGEN Technical Guidelines for genetic conservation and use of Mediterranean firs (*Abies* spp.) Bioversity International, Rome, Italy. 6 p. Available: http://bit.ly/1f6VcDo. Accessed 5 September 2013.
106. QGIS Development Team (2012) QGIS Geographic Information System. Open Source Geospatial Foundation Project. Available: http://bit.ly/1d3UwLH. Accessed 19 July 2012.
107. Wessel P, Smith WHF (2006) New, improved version of generic mapping tools released. Eos, Transactions American Geophysical Union 79: 579–579.
108. R Development Core Team (2008) R: A language and environment for statistical computing. R Foundation for Statistical Computing, Vienna, Austria. Available: http://bit.ly/19WExR5. Accessed 10 January 2013.

Genetic Pool Information Reflects Highly Suitable Areas: The Case of Two Parapatric Endangered Species of Tuco-tucos (Rodentia: Ctenomiydae)

Daniel Galiano[1]*, Jorge Bernardo-Silva[2], Thales R. O. de Freitas[1,2]

1 Programa de Pós-Graduação em Biologia Animal, Universidade Federal do Rio Grande do Sul, Porto Alegre, RS, Brazil, **2** Departamento de Genética, Universidade Federal do Rio Grande do Sul, Porto Alegre, RS, Brazil

Abstract

Conservation of small mammals requires knowledge of the genetically and ecologically meaningful spatial scales at which species respond to habitat modifications. Conservation strategies can be improved through the use of ecological niche models and genetic data to classify areas of high environmental suitability. In this study, we applied a Maxent model integrated with genetic information (nucleotide diversity, haplotype diversity and Fu's *Fs* neutrality tests) to evaluate potential genetic pool populations with highly suitable areas for two parapatric endangered species of tuco-tucos (*Ctenomys minutus* and *C. lami*). Our results demonstrated that both species were largely influenced by vegetation and soil variables at a landscape scale and inhabit a highly specific niche. *Ctenomys minutus* was also influenced by the variable altitude; the species was associated with low altitudes (sea level). Our model of genetic data associated with environmental suitability indicate that the genetic pool data were associated with highly suitable areas for *C. minutus*. This pattern was not evident for *C. lami*, but this outcome could be a consequence of the restricted range of the species. The preservation of species requires not only detailed knowledge of their natural history and genetic structure but also information on the availability of suitable areas where species can survive, and such knowledge can aid significantly in conservation planning. This finding reinforces the use of these two techniques for planning conservation actions.

Editor: Jesus E. Maldonado, Smithsonian Conservation Biology Institute, United States of America

Funding: Funding was provided by Coordenação de Aperfeiçoamento de Pessoal de Nível Superior (Capes) and Conselho Nacional de Desenvolvimento Cientifico e Tecnológico (CNPq), Brazilian Ministry of Science and Technology. The funders had no role in study design, data collection and analysis, decision to publish, or preparation of the manuscript.

Competing Interests: The authors have declared that no competing interests exist.

* E-mail: galiano3@hotmail.com

Introduction

Subterranean rodents of the genus *Ctenomys* inhabit the southern part of the Neotropical region from the extreme south up to southern Peru, including the entire Patagonian region and showing a wide latitudinal variation. The range of the genus extends from sea level to 4,000 m in the Andean region [1]. These small subterranean herbivores are among the most geographically variable mammals and are also the most speciose of all subterranean rodent groups [2]. Presently, the genus includes approximately 60 species [2]. South American tuco-tucos (*Ctenomys*) have attracted special interest in terms of speciation and evolution by virtue of their patchy distributions, low vagility, territoriality, and extensive karyotypic variation [1].

In the coastal plain of southern Brazil, the species with the widest geographic distribution is *Ctenomys minutus*, which inhabits sandy fields and dunes and has a range extending from Jaguaruna Beach in the state of Santa Catarina to the town of São José do Norte in the state of Rio Grande do Sul. *Ctenomys lami* is an endemic species inhabiting a sandy region named 'Coxilha das Lombas' along a narrow line of old dunes that extends from north of Guaiba Lake to the northwestern sandbanks of Barros Lake [3,4]. A hybrid zone between *C. minutus* and *C. lami* is also present. It is probable that this zone was formed due to habitat alterations. A wide humid zone once separated both species and represented a barrier [5]. *Ctenomys lami* is cited as vulnerable in the Red List of the International Union for Conservation of Nature [6], and *C. minutus* is listed as data deficient. However, given the current state of information about tuco-tucos in Brazil, Fernandes et al. [7] have affirmed that the vulnerability of these subterranean rodents is greater than presently supposed and that any conservation effort should be based on consistent and detailed studies of habitat occupation.

Ecological niche models (ENMs), based on the use of multiple strategies in conservation plans, are an important tool for determining the distribution of threatened species for conservation purposes [8]. In particular cases of restricted distribution or sparse populations of rare or endangered species, the use of ENMs may be necessary to identify a set of areas for protection [9]. Functionally, ENMs could be considered to furnish a specification of the relationship between the points of occurrence of a species and a set of multivariate environmental data [10]. The need to base strategies of ENM development on niche theory facilitates the interpretation and discussion of the resulting models relative to conservation issues. Conservation of small mammals requires knowledge of the genetically and ecologically meaningful spatial scales at which species respond to habitat modification. Understanding this information is important because small mammals represent a major assemblage of species occupying the most varied environments [11]. Large-scale, presence-only models are vital to complement the information obtained by local studies. They reveal otherwise-overlooked ecological requirements by identifying the environmental parameters that influence species distributions on a broad geographical scale [12].

Ctenomys minutus and *C. lami* inhabit the coastal plain of southern Brazil. Several studies have addressed the degradation of this coastal environment, identifying both natural and human-induced factors contributing to long- and short-term change [13,14]. Anthropogenic changes, such as urbanization in active dune areas [14,15], have modified the natural landscape over the years. As a result of human activity, native habitats have become increasingly fragmented or destroyed, changing the patterns of gene flow between populations and modifying the levels of genetic diversity. In our particular case, conservation strategies can be improved with the use of ENMs and genetic data to classify areas of high environmental suitability and to investigate how genetic characteristics diverge in different areas.

In this study, we applied ENMs integrated with genetic information to evaluate potential genetic pool populations with highly suitable potential areas. We developed maximum-entropy (Maxent, Phillips et al. [10]) presence-only distribution models for *C. lami* and *C. minutus* based on the following goals:

1. To conduct the first geographical distribution analysis for both species over their entire distributional range;
2. To determine which ecological factors may be limiting the distributions of the species;
3. To develop an integrated framework to evaluate geographic patterns of genetic diversity within potential genetic pools;
4. To discuss conservation issues in the light of this concatenated approach.

Methods

Study area

The study was conducted on the southern coast of Brazil. The environment consists of sand dunes and relict 'restinga' forests [16]. The landscape of the coastal plain of southern Brazil is characterized by lakes, lagoons, rivers, and dunes that represent natural geographical barriers for different populations of *Ctenomys* [4,17] (Figure 1).

Species records

Current knowledge about the distributions of *C. minutus* and *C. lami* is based on records from the collection of the Laboratório de Citogenética e Evolução of the Universidade Federal do Rio Grande do Sul and collected by researchers [3–5,18–21]. All of the coordinates were recorded using a GPS at the exact point of collection or observation. We considered a total of 74 records for *C. minutus* (n = 45) and *C. lami* (n = 29), each representing an established population of these tuco-tucos, to generate the ENMs (Figure S1).

Environmental data

We used 19 Worldclim bioclimatic variables obtained through interpolated data from derived rainfall and temperature, with a resolution of 2.5 arc-minutes (five kilometers); one altitude variable [22], available at http://www.worldclim.org/download; and two categorical variables for soil and vegetation composition [23] available at http://mapas.mma.gov.br/i3geo/datadownload.htm. We included those two categorical variables due to their recognized ecological importance to the study species and genus [24–27]. The vegetation and soil maps were converted to rasters at the same level of resolution as the bioclimatic variables using ArcMap v. 10.0. We generated a matrix with the values of each climatic variable for the entire study area. We performed a Principal Components Analysis (PCA) on this matrix to identify correlations between variables, selecting the axes that explained ≥ 95% of the correlation structure. From this result, we selected variables with the highest absolute coefficient in each axis. This procedure yielded seven variables for *C. minutus* (mean temperature of coldest quarter, precipitation seasonality, mean temperature of warmest quarter, mean diurnal range (mean of monthly (max temp - min temp)), soil, vegetation, and altitude) and five variables for *C. lami* (maximum temperature of warmest month, minimum temperature of coldest month, mean temperature of warmest quarter, soil, and vegetation). To avoid information loss, we included all variables represented in the PCAs from both species to generate models for the two species. The models were constructed with a total of nine variables because equal variables were selected for the two species.

Modeling

The models were constructed with Maxent version 3.3.3k [10], available at (http://www.cs.princeton.edu/~schapire/maxent/). Maxent estimates the ecological niche of a species by determining the distribution of maximum entropy (ME), subject to the constraint that the expected value of each environmental variable under this estimated distribution matches its empirical average. The technique was designed as a machine learning algorithm [10,28,29]. The default parameters in Maxent were used to construct the models: automatic features election, regularization multiplier at unity, maximum iterations 500 and convergence threshold 10^{-5}. We produced maps of the potential distribution of the species using the logistic output format [29]. This format is an attempt to ensure the closest possible approach to an estimate of the probability that the species is present given the environment [28]. We chose Maxent because it requires only presence data, because it can process categorical data and model interactions, and because it has performed favorably when compared with alternative approaches [30,31]. The models were validated by calculating the area under the curve (AUC) from a receiver operating characteristic curve (ROC). The relative importance of the variables was assessed with Maxent's built-in Jackknife functionality.

Genetic data

Our basic data consisted of mitochondrial DNA (mtDNA) sequences corresponding to the cytochrome oxidase I gene and the hypervariable control region. The data were obtained from Lopes et al. [20] for *C. minutus* and Lopes and Freitas [19] for *C. lami*, where a detailed description of the genetic data can be obtained. For *C. minutus*, a total of 30 localities were sampled across the entire distribution of the species, and seven principal clades were highlighted in the Bayesian phylogenetic tree. We focused on these seven main genetic haplogroups described by Lopes et al. [20] (North 1, North 2, Coast, Barros Lake, Mostardas, Tavares, and South; see Lopes et al. [20] for details). For *C. lami*, 28 locations were sampled, and the data are presented by the authors based on the four karyotypic blocks proposed for the species (Block A, B, C and D; see Freitas [4], Lopes and Freitas [19]). We chose to present our results for this concatenated approach (genetic data and modeling) based on the haplogroups for *C. minutus* and karyotipic blocks for *C. lami* since we did not directly compare these two different units, but discuss the results independently for each species. We believe that the use of this approach may not overestimate the interpretation of the results.

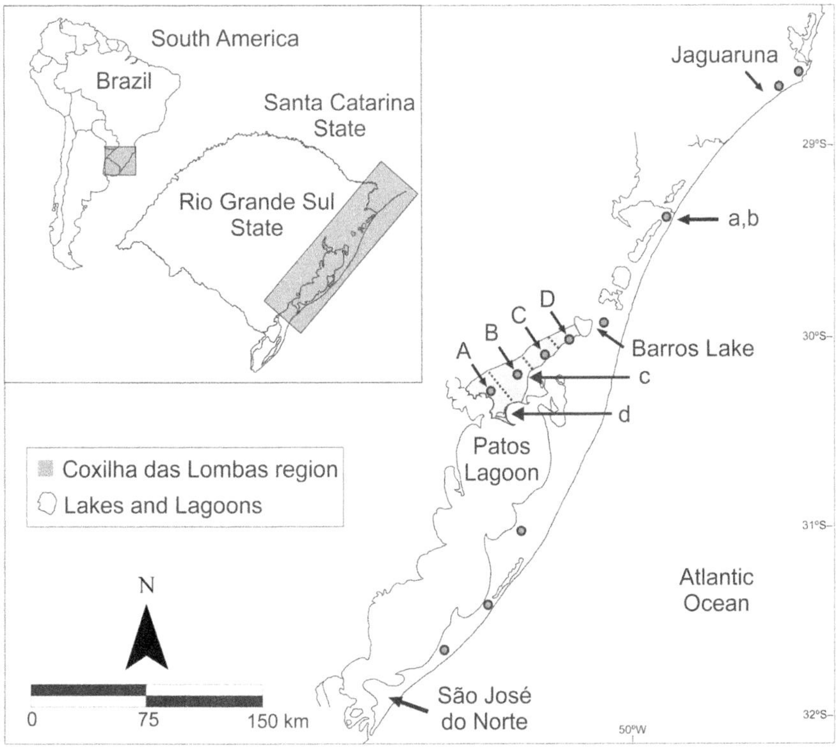

Figure 1. Coastal plain of southern Brazil with geographic details. The most important lakes and rivers are shown. Points (●) represent geographical regions for which genetic data are available from Lopes and Freitas (2012) and Lopes et al. (2013). In the region of Coxilha das Lombas are displayed the four karyotypic blocks described by Freitas (2007) (A, B, C and D). Conservation areas within the distribution of each species are also displayed (a, b: Parque de Itapeva and Parque da Guarita; c: Parque de Itapuã; d: Refugio de Vida Silvestre Banhado dos Pachecos).

Briefly, the geographical genetic structure of *C. minutus* was characterized by examining 340 individuals over the entire distributional range and using information from cytochrome oxidase I gene and control region sequences. For *C. lami*, a total of 178 specimens were sampled, using information from the same mitochondrial DNA markers. Our analyses were based primarily on three different genetic parameters obtained from genetic data estimated by these authors [19,20]. The first parameter, nucleotide diversity (π), is defined by the equation

$$\pi = \sum_{ij} x_i x_j \pi_{ij},$$

where π_{ij} is the proportion of nucleotide differences between the *ith* and *jth* types of DNA sequences and x_i and x_j are the respective frequencies of these sequences. Nucleotide diversity is a measure of genetic variation. The second parameter, haplotype diversity (H_d), is a measure of the uniqueness of a particular haplotype in a given population. H_d is defined by the equation

$$Hd = \frac{N}{N-1}(1 - \sum i x^2 i),$$

where x_i is the (relative) haplotype frequency of each haplotype in the sample and N is the sample size. Haplotype diversity is given for each sample. The third parameter, Fu's *Fs*, is the probability of observing a random sample with a number of alleles equal to or smaller than the observed value given the observed level of diversity and the assumption that all of the alleles are selectively neutral. If we call this probability $\hat{S}$, then

$$Fs = \ln\left(\frac{\hat{S}}{1-\hat{S}}\right),$$

so that a negative value of *Fs* would be expected from a recent population expansion or from genetic hitchhiking. A positive value of *Fs* is evidence for a deficiency of alleles, as would be expected from a recent population bottleneck or from overdominant selection [32]. To display the genetic information on a map, we established a geographic coordinate associated with each haplogroup for *C. minutus* and with each karyotypic block for *C. lami*.

Results

The models

The models constructed with Maxent produced distributional predictions for each species. The map with the widest distributional area was generated for *C. minutus*, whereas *C. lami* presented a very narrow distribution probability (Figure 2).

According to the ME model, the occurrence of *C. lami* was most strongly associated with soil and vegetation. The variable with the highest gain when used in isolation was vegetation. For this reason, vegetation appears to be the single variable that furnishes the most useful information. The variable that decreased the gain most markedly when omitted was soil. For this reason, soil appears to furnish the greatest amount of information not present in the other variables. For *C. minutus*, occurrence was strongly associated with altitude, vegetation, and soil. The variable with highest gain when used in isolation was vegetation, and the variable that decreased the gain most markedly was also vegetation (Figure 3).

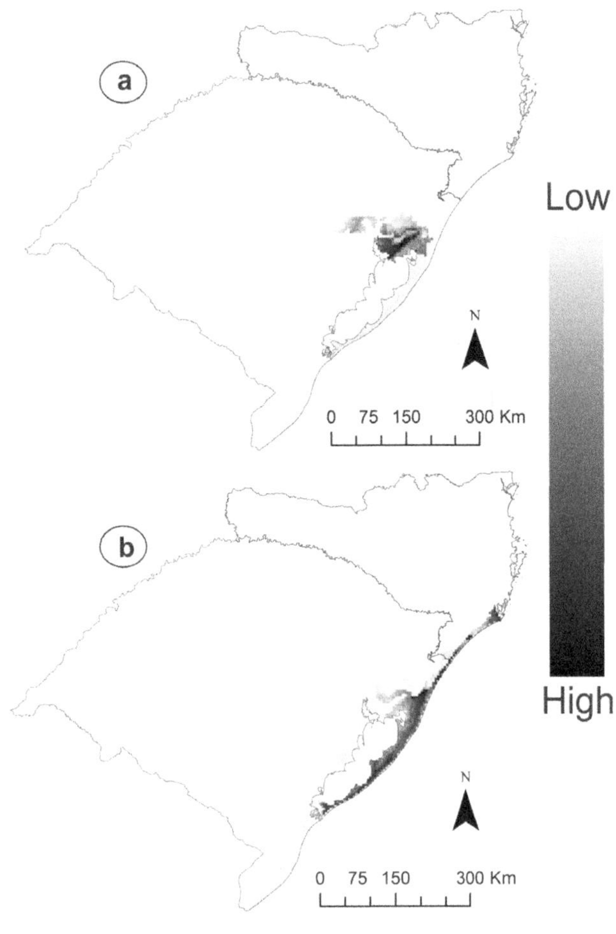

Figure 2. Suitable areas for (a) *Ctenomys lami* and (b) *Ctenomys minutus*, according to Maxent (maximum entropy) model in southern Brazilian coastal plain. Darker regions indicate greater ecological suitability.

Both species were strongly influenced by vegetation and soil variables. Model performance, defined as the area under the curve, was highly discriminative for both species (AUC values for *C. lami* = 0.997 and *C. minutus* = 0.993), indicating that they inhabit a highly specific landscape niche.

Genetic pool data

The three local genetic parameters estimated showed no apparent patterns for *C. lami*. The suitability values calculated between the blocks were the same (high), and block C showed the highest value of π; however, the same pattern was not followed by the H_d values, which were higher in blocks B and D. The *Fs* values were not significant for any of the analyzed blocks (Figure 4).

For *Ctenomys minutus*, areas of high suitability were in accordance with the seven principal genetic haplogroups found for the species (see Lopes et al. [20] for details). For π values, the northern portion of the distribution presented higher values (except for North 1). H_d presented the same trend, with the center of the distribution showing greater diversity. For the *Fs* index, only the Barros Lake haplogroup showed a significant result (−5.30), as would be expected from a recent population expansion or from genetic hitchhiking (Figure 4).

Discussion

Maximum entropy models and environmental variables

This is the first study to use ENM methods to test parameters that influence the occurrence of the genus *Ctenomys* and one of the first to use these techniques on subterranean mammals. The models presented here were successful in identifying regions that are suitable for the species. Both models showed few commission errors, such as the areas of medium environmental suitability on the east side of the Coxilha das Lombas region for *C. lami* and the areas of low suitability in the countryside along Patos Lagoon for *C. minutus*. In addition, model performance was highly discriminative for both species. Although we identified these commission errors, we must consider that is very difficult to incorporate the dispersal abilities of the species and geographical accessibility to the model (landscape configuration) at present or through history; however, this factor is necessary for the actual presence of species [33]. In our particular case, geographical accessibility is a very important issue because the tuco-tucos are characterized by limited individual mobility and a patchy distribution of local populations [25,34]. *Ctenomys lami* appears to be restricted to the region of Coxilha das Lombas and showed a potential distribution at low to medium levels of occurrence beyond the limits of this region. The species was shown to be associated with local vegetation and soil type and to be geographically limited by the presence of wetlands and geographical inaccessibility, factors that characterize this region.

The distribution of *Ctenomys minutus* was restricted to the coastal plain of southern Brazil, including the region's first line of dunes and extending to sandy fields. This species shows a sympatric zone with another tuco-tuco species (*C. flamarioni*) in a range of approximately 15 km [35]. In an area extending to the northern portion of this region, *C. minutus* inhabits the first line of dunes. Below the sympatric zone, *C. minutus* occupies only the regions of sandy fields, although the ME model indicates that it can potentially occur up to the line of dunes. The reason for this apparent discrepancy is that near the southern limits of this sympatric zone, the first line of dunes is already occupied by *C. flamarioni* up to the southern boundary of Rio Grande do Sul state. This finding is, most likely, related to the historical patterns of occupation by these species. *Ctenomys minutus* shows a pattern of north-south occupation extending from Santa Catarina to Rio Grande do Sul state [20], and *C. flamarioni* shows a pattern of occupancy from south to north [36,37]. This transition between the sand fields and dune line represents an environmental discontinuity along the range of *C. minutus*. According to Fornel et al. [38], who have found significant differences in skull morphology between specimens of *C. minutus* that inhabit sand fields and those that inhabit dunes, environmental characteristics such as soil hardness produce morphological adaptations in tuco-tucos. These authors found some variation among different populations belonging to different habitats in cranial shape, and these intraspecific phenotypic differences appear to arise as a combination of selection and drift acting as diversifying forces. However, according to our model of environmental suitability, this discontinuity may not inhibit free dispersal by specimens from different habitats. Therefore, these different environments may also be associated with distinct levels of environmental suitability, although we could not measure this difference in our models.

Although they are morphologically similar, both species studied here occupy distinct areas of the coastal plain. According to the ME models, *C. minutus* appears to show higher environmental plasticity than *C. lami*. The models proposed for these two species showed a low overlap of areas of high suitability. This

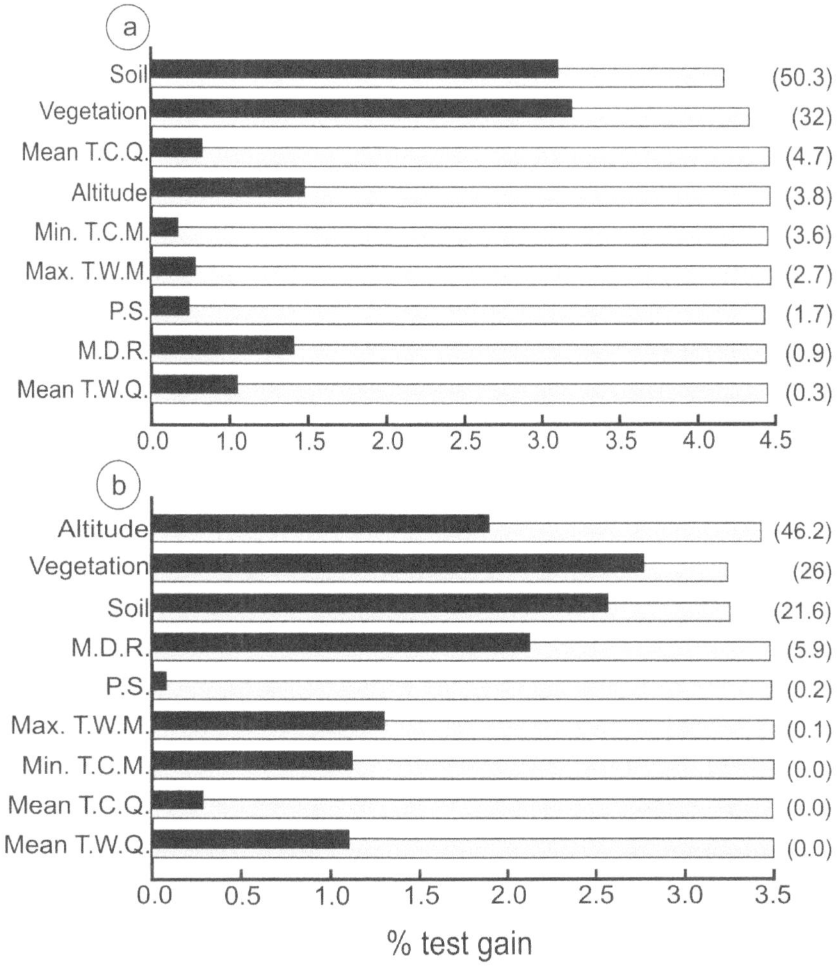

Figure 3. Jackknife analyses of the importance of environmental variables in maximum entropy modeling of *C. lami* (a) and *C. minutus* (b) occurrence. A heuristic estimate of the relative contribution of each variable to the global model is given in parentheses, with variables listed in descending order of importance. Grey bars show the performance of the global model (known as test gain) without each variable, and black bars show the influence of each variable in isolation (derived from a univariate model only). The variables are the following: Mean T.C.Q.: mean temperature of coldest quarter; Min. T.C.M.: minimum temperature of coldest month; Max. T.W.M.: maximum temperature of warmest month; P.S.: precipitation seasonality; M.D.R.: mean diurnal range (mean of monthly (max temp - min temp)); Mean T.W.Q.: mean temperature of warmest quarter.

characteristic appears to be defined primarily by the presence of the Coxilha das Lombas region and the geographical inaccessibility of this region. A hybrid zone between *C. lami* and *C. minutus* is present in the northern portion of this coxilha. This zone has been the subject of previous study. According to Gava and Freitas [5], it was formed by modifications of the environment because a humid zone that separated these species was drained for rice cultivation. Environmentally, our results suggest a medium probability of occurrence for both species in the region where they hybridize.

Moreover, edaphic conditions can be very important for the definition of a species' fundamental niche [39,40]. The occurrence of *C. lami* and *C. minutus* was strongly influenced by soil and vegetation. *C. minutus* was also influenced by altitude, as it was associated with low altitudes (sea level). According to Kubiak [35], *C. minutus* is associated with areas with higher vegetation and biomass, which strengthens the evidence for the importance of this variable to the species. Given that tuco-tucos are exclusively subterranean, the use of variables that are directly linked to the species niche demonstrated to be extremely important in building the models, regardless of soil and vegetation variables, might have suffered from their degree of generalization. It is recommended that the choice of predictors in ENMs should consider the ecological relevancy of the predictors to the target species [41]. Such considerations are extremely important to improve conservation actions.

Genetic information and conservation actions

Our analyses of the genetic data associated with the ME model for *C. lami* showed that there are differences in the levels of genetic diversity (π and H_d) between populations, although the suitability is currently the same for all of the blocks. Lopes and Freitas [19] found that the genetic structure associated separately with each of the four different karyotypic blocks was inconsistent for mtDNA and microsatellite data. Environmentally, we also found no differences among these four karyotypic blocks. Thus, the presence of chromosomal rearrangements or distinct karyotypes between individuals noted by Freitas [4] does not appear to be limited by the environment.

Lopes et al. [20] recovered a pattern of genetic structure of the sampling sites subdivided into seven main haplogroups for *C. minutus*. In the current study, these haplogroups, represented by three genetic parameters, were associated with areas of high

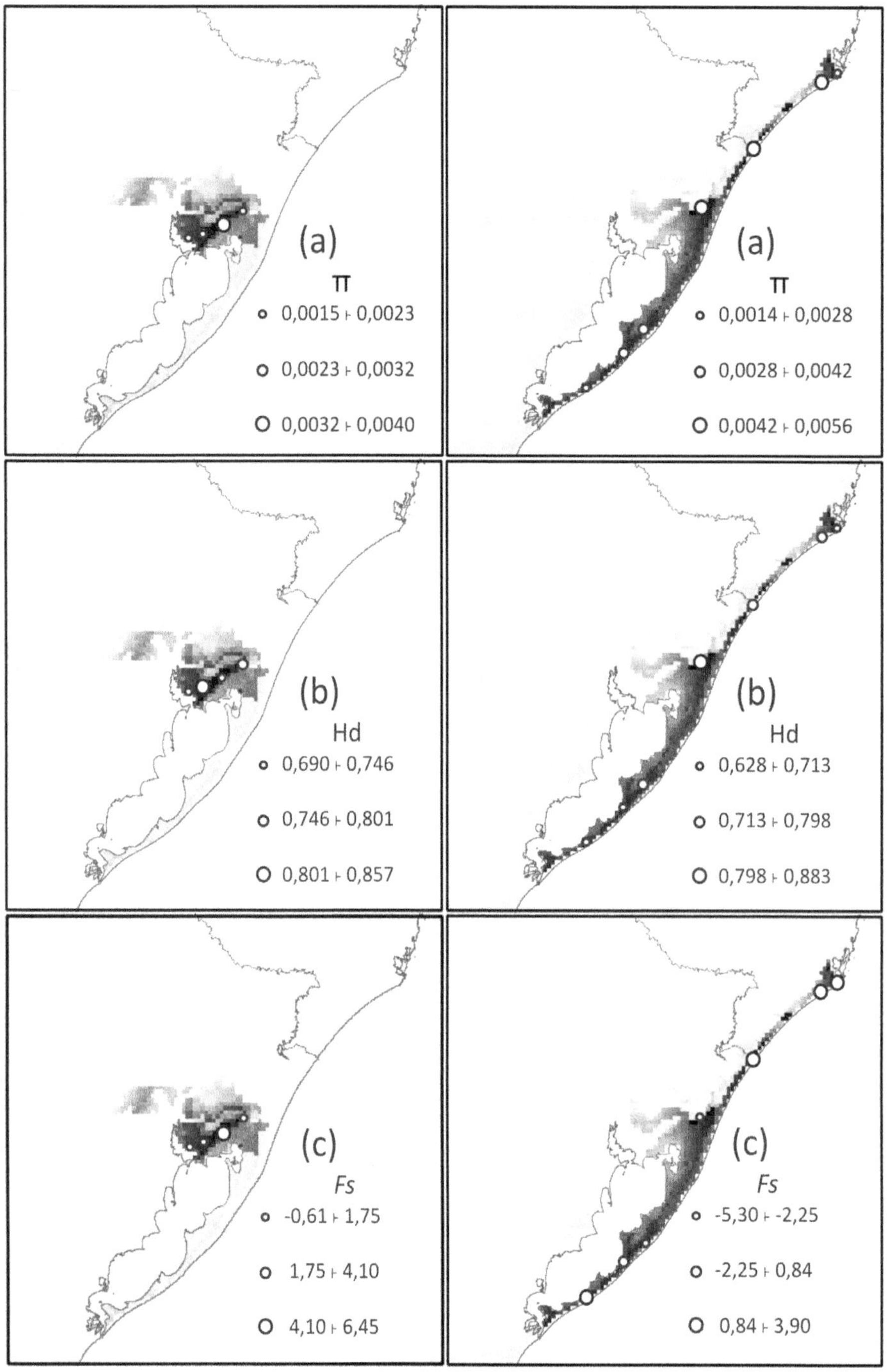

Figure 4. Geographical patterns of genetic parameters estimated within phylogeographic groups (*C. minutus*) and karyotype blocks (*C. lami*). Nucleotide diversity, Π (a); Haplotype diversity, *Hd* (b); Fu's neutrality parameter, *Fs* (c), overlapping the potential distribution of *C. lami* (left column) and *C. minutus* (right column). Darker regions indicate greater ecological suitability.

environmental suitability. This pattern may result from the occurrence of higher genetic diversity in the areas of higher suitability for the species. According to Lopes et al. [20], a relatively ancient genetic structure was present in the northern area of the geographical distribution, whereas the southern sampling sites may exhibit a founder effect of more recent occurrence. However, areas with high suitability showed different values of π and H_d, thus other extrinsic (historical or contemporary) or intrinsic (behavioral) factors could be affecting the pattern of genetic diversity and structure.

The only value of the *Fs* index that was statistically significant for *C. minutus* was from the Barros Lake haplogroup, which was associated with a highly suitable area. As this pattern was not found for *C. lami* or for the other haplogroups of *C. minutus*, it is not possible to support our hypothesis that highly suitable areas favor population expansion or genetic hitchhiking, as indicated by a significant negative value of the *Fs* index.

The preservation of species requires not only detailed knowledge of their natural history and genetic structure but also information on the availability of suitable areas where species can survive; such knowledge can aid significantly in conservation planning. Both species analyzed here have suffered intense pressure from the fragmentation and reduction of their habitats. This process results in the death or isolation of individuals because these species are extremely niche-specific. The principal threats to *C. lami* are the progressive urbanization and human settlement in its territory, associated, for example, with agricultural activities [7,19]. In the northern portion of the coast, *C. minutus* is threatened by land speculation associated with urban constructions along the coast and on the shores of the lagoon and by the fragmentation of the dune habitat. In the south, silviculture involving *Pinus* sp. directly affects the populations of the species ([7], authors' personal observations). A study by Bernardo-Silva et al. [42] focused on the conservation of two endangered species of red-bellied toads, and the authors suggested five important areas for species conservation based on a hotspot analysis. These areas included a significant number of threatened species, including *C. minutus*. Our results underscore that the areas proposed by those authors should be considered for reserve planning because they cover three highly suitable areas within the range of *C. minutus* (see Bernardo-Silva et al. [42]). Although their results did not focus on underground rodents, these hotspot areas are of great interest for *C. minutus* because considering conservation of habitats as a priority, the protection of sandy field areas with high suitability for underground tunneling is an interesting solution for protecting the species. Additionally, *C. minutus* has only two protected areas within its entire distribution (Parque de Itapeva and Parque da Guarita). These areas are in neighboring locations in the northeastern portion of the distribution. However, none of the southern populations are located in protected areas. Two state conservation units (Parque de Itapuã and Refugio de Vida Silvestre Banhado dos Pachecos) are located within the range of *C. lami*, but only parts of these areas overlap the distribution of the species. Over the past year, extensive field work has been conducted to search for *C. lami*, but few populations have been found outside these two units (authors' personal observations).

Our models of genetic data associated with environmental suitability indicated that genetic pool data were associated with highly suitable areas for *C. minutus*. This result supports the use of these two techniques to plan conservation actions. This pattern was not evident for *C. lami*, perhaps due to the restricted range of the species. Despite the current wide use of molecular markers to estimate genetic variation at different geographical scales and originating from different evolutionary processes [43,44], it is still difficult to match these data with ecological parameters at broad scales. Moreover, despite many studies testing the relationship between genetic diversity and geographical populations, it is necessary to test alternative mechanisms, such as fragmentation and human effects on species distributions, by integrating genetic and demographic data in an ecological context.

Acknowledgments

We are grateful to Tatiane Noviski for help in preparing the figures, and Paula A. Roratto, Gislene L. Gonçalves, Rodrigo Fornel, Gabriela P. Fernández and one anonymous reviewer for suggestions and for reviewing this manuscript. We also thank colleagues from the Laboratório de Citogenética e Evolução for support at various stages of this research.

Author Contributions

Conceived and designed the experiments: DG JBS TROF. Performed the experiments: DG JBS. Analyzed the data: DG JBS. Contributed reagents/materials/analysis tools: DG JBS TROF. Wrote the paper: DG JBS TROF. Acquired the data: DG TROF.

References

1. Reig OA, Busch C, Contreras MO, Ortells JR (1990) An overview of evolution, systematics, population and speciation in *Ctenomys*. In: Nevo E, Reig OA, editors. Evolution of subterranean mammals at the organismal and molecular levels. New York: Wiley-Liss.pp. 71–96.
2. Parada A, D'Elia G, Bidau CJ, Lessa EP (2011) Species groups and the evolutionary diversification of tuco-tucos, genus *Ctenomys* (Rodentia: Ctenomyidae). J Mammal 92:671–682.
3. Freitas TRO (2001) Tuco-tucos (Rodentia, Octodontidae) in Southern Brazil: *Ctenomys lami* spec. nov. separated from *C. minutus* Nehring, 1887. Stud Neotrop Fauna Environ 36:1–8.
4. Freitas TRO (2007) *Ctenomys lami*: the highest chromosomal variability in *Ctenomys* (Rodentia, Ctenomyidae) due to a centric fusion/fission and pericentric inversion system. Acta Theriol 52:171–180.
5. Gava A, Freitas TRO (2003) Inter and intra-specific hybridization in tuco-tucos (*Ctenomys*) from Brazilian coastal plains (Rodentia: Ctenomyidae). Genetica 119:11–17.
6. IUCN (2013) IUCN Red List of Threatened Species, version 2013.1. Available: http://www.iucnredlist.org. Accessed: 2013 Aug 12.
7. Fernandes FA, Fernández-Stolz GP, Lopes CM, Freitas TRO (2007) The conservation status of the tuco-tucos, genus *Ctenomys* (Rodentia: Ctenomyidae), in southern Brazil. Braz J Biol 67:839–847.
8. De Marco Júnior P, Siqueira MF (2009) Como determinar a distribuição potencial de espécies sob uma abordagem conservacionista? Megadiversidade 5:65–76.
9. Siqueira MF, Durigan G, De Marco P, Peterson AT (2009) Something from nothing: using landscape similarity and ecological niche modeling to find rare plant species. J Nat Conserv 17:25–32.
10. Phillips SJ, Anderson RP, Schapire RE (2006) Maximum entropy modeling of species geographic distributions. Ecol Model 190:231–259.
11. Manning JA, Edge WD (2004) Small mammal survival and downed wood at multiple scales in managed forests. J Mammal 85:87–96.
12. Razgour O, Hanmer J, Jones G (2011) Using multi-scale modelling to predict habitat suitability for species of conservation concern: The grey long-eared bat as a case study. Biol Conserv 144:2922–2930.
13. Esteves LS, Toldo EE, Dillenburg SR, Tomazelli LJ (2002) Long- and short-term coastal erosion in southern Brazil. J Coast Res 36:273–282.
14. Tomazelli LJ, Villwock JA (2000) O Cenozoico no Rio Grande do Sul: geologia da Planície Costeira. In: Holz M, De Ros LF, editors. Geologia do Rio Grande do Sul. Porto Alegre, Brazil: Cigo Ufrgs.pp. 375–406.
15. Fernandez-Stolz GP, Stolz JFB, Freitas TRO (2007) Bottlenecks and dispersal in the tuco-tuco-das-dunas, *Ctenomys flamarioni* (Rodentia: Ctenomyidae) in southern Brazil. J Mammal 88:935–945.
16. Hesp PA, Giannini PCF, Martinho TC, Da Silva GM, Neto NEA (2009) The Holocene Barrier Systems of the Santa Catarina Coast, Southern Brazil. In: Dillenburg SR, Hesp PA, editors. Geology and Geomorphology of Holocene Coastal Barriers of Brazil. Berlin: Springer.pp. 92–133.
17. Moreira DM, Franco MHLP, Freitas TRO, Weimer TA (1991) Biochemical polymorphisms and phenetic relationships in rodents of the genus *Ctenomys* from southern Brazil. Biochem Genet 29:601–615.
18. Freitas TRO (1995) Geographic distribution and conservation of four species of the genus *Ctenomys* in Southern Brazil. Stud Neotrop Fauna Environ 30:53–59.
19. Lopes CM, Freitas TRO (2012) Human impact in naturally patched small populations: genetic structure and conservation of the burrowing rodent, tuco-tuco (*Ctenomys lami*). J Heredity 103:672–681.
20. Lopes CM, Ximenes SSF, Gava A, Freitas TRO (2013) The role of chromosomal rearrangements and geographical barriers in the divergence of lineages in a South American subterranean rodent (Rodentia: Ctenomyidae: *Ctenomys minutus*). Heredity 1–13.
21. Marinho JR, Freitas TRO (2006) Structure of a *Ctenomys minutus* (Rodentia, Hystricognathi) population in the Coastal Plain of Rio Grande do Sul, Brazil. Acta Theriol 51:53–60.
22. Hijmans RJ, Cameron SE, Parra JL, Jones PG, Jarvis A (2005) Very high resolution interpolated climate surfaces for global land areas. Int J Climatol 25:1965–1978.

23. MMA (2012) Ministério do Meio Ambiente. Data download. Available: http://mapas.mma.gov.br/i3geo/datadownload. Accessed 2013 Dec 10.
24. Comparatore VM, Agnusdei M, Busch C (1992) Habitat relations in sympatric populations of *Ctenomys australis* and *Ctenomys talarum* (Rodentia, Octodontidae) in a natural grassland. Z für Säugetierkunde 57:47–55.
25. Lacey EA, Patton JL, Cameron GN (2000) Life underground: the biology of subterranean rodents. Chicago, Illinois: University of Chicago Press. 449 p.
26. Malizia AI, Vassallo AI, Busch C (1991) Population and habitat characteristics of two sympatric species of *Ctenomys* (Rodentia: Octodontidae). Acta Theriol 36:87–94.
27. Vassallo AI (1998) Functional morphology, comparative behaviour, and adaptation in two sympatric subterranean rodents genus *Ctenomys* (Caviomorpha: Octodontidae). J Zool Lond 244:415–427.
28. Elith J, Phillips SJ, Hastie T, Dudik M, Cheen YE, et al. (2011) A statistical explanation of Maxent for ecologists. Divers Distrib 17:43–57.
29. Phillips SJ, Dudik M (2008) Modeling of species distributions with Maxent: new extensions and a comprehensive evaluation. Ecography 31:161–175.
30. Elith J, Graham CH, Anderson RP, Dudik M, Ferrier S, et al. (2006) Novel methods improve prediction of species' distributions from occurrence data. Ecography 29:129–151.
31. Wisz MS, Hijmans RJ, Li J, Peterson AT, Graham CH, et al. (2008) Effects of sample size on the performance of species distribution models. Divers Distrib 14:763–773.
32. Fu YX (1997) Statistical tests of neutrality of mutations against population growth, hitchhiking, and background selection. Genetics 147:915–925.
33. Soberon J, Peterson AT (2005) Interpretation of models of fundamental ecological niches and species' distributional areas. Biodivers Inform 2:1–10.
34. Cutrera AP, Mora MS, Antinucci CD, Vassallo AI (2010) Intra-and interspecific variation in home-range size in sympatric tuco-tucos, *Ctenomys australis* and *Ctenomys talarum*. J Mammal 91:1425–1434.
35. Kubiak BB (2013) Relação entre a distribuição espacial e características ambientais de duas espécies de tuco-tuco (Rodentia: *Ctenomys*) em uma nova área de simpatria para o gênero no sul do Brasil. M.Sc. dissertation, Federal University of Rio Grande do Sul.
36. Freitas TRO (1994) Geographical variation of heterochromatin in *Ctenomys flamarioni* (Rodentia: Octodontidae) and its cytogenetic relationship with other species of the genus. Cytogenet Cell Genet 67:193–198.
37. Massarini AI, Freitas TRO (2005) Morphological and cytogenetics comparison in species of the mendocinus -group (genus: *Ctenomys*) with emphasis in *C.australis* and *C. flamarini* (Rodentia-Ctenomyidae). Caryologia 58:21–27.
38. Fornel R, Cordeiro-Estrela P, Freitas TRO (2010) Skull shape and size variation in *Ctenomys minutus* (Rodentia: Ctenomyidae) in geographical, chromosomal polymorphism, and environmental contexts. Biol J Linn Soc Lond 101:705–720.
39. Bertrand R, Perez V, Gégout JC (2012) Disregarding the edaphic dimension in species distribution models leads to the omission of crucial spatial information under climate change: the case of *Quercus pubescens* in France. Glob Chang Biol 18:2648–2660.
40. Tuomisto H (2006) Edaphic niche differentiation among *Polybotrya ferns* in western Amazonia: implications for coexistence and speciation. Ecography 29:273–284.
41. Elith J, Leathwick JR (2009) Species distribution models: ecological explanation and prediction across space and time. Annu Rev Ecol Evol Syst 40:677–697.
42. Bernardo-Silva J, Martins-Ferreira C, Maneyro R, Freitas TRO (2012) Identification of priority areas for conservation of two endangered parapatric species of red-bellied toads using ecological niche models and hotspot analysis. Nat Conserv 10:207–213.
43. Diniz-Filho JAF, Telles MPC, Bonatto SL, Eizirik E, Freitas TRO, et al. (2008) Mapping the evolutionary twilight zone: molecular markers, populations and geography. J Biogeogr 35:753–763.
44. Diniz-Filho JAF, Nabout JC, Bini LM, Soares TN, Telles MPC, et al. (2009) Niche modelling and landscape genetics of *Caryocar brasiliense* ("Pequi" tree: Caryocaraceae) in Brazilian Cerrado: an integrative approach for evaluating central–peripheral population patterns. Tree Genet Genomes 5:617–627.

23

Conservation Priorities when Species Interact: The Noah's Ark Metaphor Revisited

Pierre Courtois, Charles Figuieres*, Chloé Mulier

INRA-LAMETA, Montpellier, France

Abstract

This note incorporates ecological interactions into the Noah's Ark problem. In doing so, we arrive at a general model for ranking *in situ* conservation projects accounting for species interrelations and provide an operational cost-effectiveness method for the selection of best preserving diversity projects under a limited budget constraint.

Editor: David L. Roberts, University of Kent, United Kingdom

Funding: This study was supported by the ONEMA program 2013-2015. The funders had no role in study design, data collection and analysis, decision to publish, or preparation of the manuscript.

Competing Interests: The authors have declared that no competing interests exist.

* Email: figuiere@supagro.inra.fr

Introduction

Weitzman [13] is a milestone in the economic theory of biodiversity. His "Noah's Ark Problem" is not only a modeled metaphor that is helpful to organize thinking on how to face conservation trade-offs with finite resources. It also results in a practical cost-effectiveness methodology that can serve as inspiration to guide conservation policies. The idea is, for each species i, to collect information about: *i*) C_i, the cost of its protection, *ii*) ΔP_i, the increase of survival probability resulting from it, *iii*) U_i, the direct utility of how much we value the species, *iv*) D_i, its distinctiveness. From this information, each species is assigned a number R^i *via* the formula:

$$R^i = \frac{\Delta P_i}{C_i}(D_i + U_i) \,, \tag{1}$$

which indicates its rank in conservation priorities. This ranking criterion has a theoretical foundation: it is rooted in a rigorous optimization model ([13], Theorem 4, p. 1295).

This criterion sheds light on real biodiversity issues and has actually been used in several applications. Some of these have led to changes in allocation of conservation funding (e.g., in New Zealand; [9]), and variants have been used to allocate surveillance effort over space (e.g., [8]). Other applications are quoted in [5]. But it is fair to say that this approach is more appropriate for *ex situ* conservation projects - say to build a gene bank or a zoo - rather than to manage a set of interacting species in their natural habitats. This is so because formula (1) uses no information of any kind about the web of life. Yet, in ecosystems, species interact. Some of them compete to share common resources, others develop synergies and mutually enhance each other or they simply pertain to the same trophic chain. Suppose, then, that the conservation authority has information about those ecological interactions, even if it is only under the rudimentary form of survival probability interdependencies. That is, it knows that a marginal increase of survival probability of species j will have an impact r_{ij} on the survival probability of species i. Could this information be used to qualify formula (1) and increase its relevance when it comes to *in situ* conservation trade-offs?

To our knowledge, three recent articles stress the need to account for ecological interactions in Weitzman's diversity concept. They have in common: *i*) to take into account the ecological interactions *via* interdependent survival probabilities in a simplified version of the Noah's Ark metaphor with two species [1], [11] or three species [12], *ii*) to show that this consideration can reverse the conservation priorities. The key of this note is to provide a general analysis of *in situ* conservation problems considering interdependent survival probabilities. Revisiting Weitzman's optimization problem, we extend his model in order to incorporate species interactions. Our principal output is to forward a general ranking formula that could be used as a rule of thumb for deciding *in situ* conservation priorities under a limited budget constraint.

The sketch of the paper is the following. Section 2 incorporates ecological interactions in Weitzman's parable of Noah's Ark, with any arbitrary number of species. The crux of the section is to provide with a new rule for establishing *in situ* conservation priorities through the expression (12) below that encompasses formula (1) as a special case. The link between this formula and Noah's optimal policy is explained. Section 3 illustrates the relevance of this new formula within a two-species example. We check the robustness of our formula and end the paper with a discussion on the possibility of ranking reversal in relation to three stylized kinds of ecological interactions: *predation*, *mutualism* and *competition*.

Analysis

The "Noah's Ark Problem" is a parable intended to be a kind of canonical form representing how best to preserve biodiversity under a limited budget constraint. In the initial version of Weitzman's modeled allegory, Noah's decision problem is, for each species i, to choose a survival probability between a lower and an upper bound, $P_i \in [\underline{P}_i, \overline{P}_i]$, in order to maximize the sum of the *expected diversity function*:

$$W\left(\{P_i\}_{i=1}^k\right),$$

and the *expected utility of the set of species:*

$$U\left(\{P_i\}_{i=1}^k\right) = \sum_{i=1}^{k} U_i * P_i .$$

Weitzman devotes much of his paper to defining the expected diversity function $W\left(\{P_i\}_{i=1}^k\right)$ and to explaining its link with the concept of information content (see his Theorem 1, p. 1284). This function could take various specific forms, depending on the way dissimilarity is conceptualized. A precise example, from [13], is discussed in Section 4. In order for our results to remain as general as possible, we simply consider in this paper the class of C^2 functions, *i.e* whose first and second order derivative both exist and are continuous.

And we assume they admit Hessian matrices that are *nowhere* negative semi-definite, *i.e* there is no admissible $\{P_i\}_{i=1}^k$ such that the Hessian of $W\left(\{P_i\}_{i=1}^k\right)$ is negative semi-definite at $\{P_i\}_{i=1}^k$. Weitzman's expected diversity function belongs to this class. It encompasses - but is not limited to - functions W with a positive definite Hessian matrix, *i.e.* that are strictly convex functions.

Now let us take a step away from this initial metaphor, towards reality. Two modifications are brought into the formalism. First, rather than controlling directly the probability of survival P_i of each species $i=1,...,k$, Noah can exert a protection effort within an admissible range, $x_i \in [0,\bar{x}_i]$, which is interpreted as the controlled increase of survival probability P_i - say that x_i is the increase of survival probability for species i resulting from a protection effort, *e.g.* an investment in a vaccination campaign, the provision of supplementary food, the protection and enhancement of habitat [6]. It is important to distinguish the effort from the change in the survival probability because P_i is also determined by other factors, for there are ecological interactions among species. And this is where our second, most important, qualification appears: probabilities of survival are interdependent and the nature of those interactions are known. Nowadays, Noah can rely on the knowledge gained from the new and booming conservation biology literature on *species distribution models* and *population viability analysis*. See for instance[3], [14], [7], or [4] for a recent overview. Note that this literature does not take into account directly of species interactions; it just provides estimates of probabilities in space and time. From there, although applied econometric problems will have to be overcome, correlations between probabilities could be estimated.

A group of experts can measure the marginal impact, say r_{ih}, that an increase in the probability of survival of a species h can have on the probability of survival of another species i. The experts can also appraise the impact of protection efforts on these probabilities. Assume, then, that the relationships between extinction risks are linear. Put differently, a tractable approximation of all those pieces of information can be summarized by the system (2) of linear equations:

$$P_i = q_i + x_i + \sum_{h \neq i} r_{ih} P_h , \quad q_i \in [0,1[, \quad x_i \in [0,\bar{x}_i]. \tag{2}$$

There are biological and economic factors that determines eligible efforts. Formally, *admissible* ranges of efforts are $\times_{i=1}^k [0,\bar{x}_i]$. Implicitly, additional efforts beyond the threshold $\bar{x}_i$ have no effect on the survival probabilities. And we assume:

$$P_i \in \Pi_i = [\underline{\Pi}_i, \overline{\Pi}_i] \sqsubseteq [0,1], \quad \forall i , \forall x_i \in [0,\bar{x}_i] .$$

We denote $\underline{P}_i$ as the survival probability of species i without any conservation efforts, $x_i = 0, \underline{P}_i \geq \underline{\Pi}_i$. In the absence of natural interactions, which corresponds to the case studied by Weitzman, we have $r_{ih} = 0, \forall i, \forall h$. A consequence is that in the very particular case with no ecological interactions and no conservation efforts, species i has a probability of survival q_i. The survival probabilities interval, without ecological interactions, would thus take values ranging from $\underline{P}_i = q_i$ to $\overline{P}_i = \underline{P}_i + \bar{x}_i$.

Noah also has to cope with a budget constraint:

$$\sum_{i=1}^{k} \beta_i * x_i \leq B . \tag{3}$$

where B is the total budget to be allocated to conservation - metaphorically, the size of the Ark - and β_i is the cost per unit of effort to preserve species i.

It is worthwhile making three remarks about this budget constraint. Firstly, it is assumed that changes in extinction probability are a linear function of expenditure. This may be inconsistent in real world applications where the marginal expense needed to reduce extinction risks is increasing. For example, [10] documents that the marginal preservation cost of threatened Australian birds increases when probability of extinction approaches zero. Weitzman rightly defends this linearity assumption as an acceptable approximation when the variation of probability falls in a sufficiently narrow range. But clearly, if costs are non linear and convex functions of efforts, an important qualitative result of our paper could change (Theorem 1 below may not hold any longer). Secondly, as a formal matter one could retrieve Weitzman's model with a simple change of variable, $\beta_i \equiv C_i / \Delta P_i$ where C_i is the cost per unit of increase of survival probability in the range $\Delta P_i = \overline{\Pi}_i - \underline{\Pi}_i$. Thirdly, except when ecological interactions are negligible, Noah can increase the probability of survival of any species i *via* two different channels: a direct one by increasing the protection effort x_i, at a cost $\beta_i * x_i$, and an indirect one through ecological interactions, due to the protection of another species j, with a cost $\beta_j * x_j$.

Noah's Ark problem, when ecological interactions are taken into account, is then:

$$\max_{\{x_i\}_{i=1}^k \in \times_{i=1}^k [0,\bar{x}_i]} W\left(\{P_i\}_{i=1}^k\right) + U\left(\{P_i\}_{i=1}^k\right), \quad (4)$$

subject to (2) and (3).

It will be convenient subsequently to work with matrix or vector expressions, written in bold characters. For any matrix $\mathbf{M}$, let $\mathbf{M}^\top$ denote its transpose. Further, $\mathbf{I}^k$ is the $(k \times k)$ identity matrix, ι^k is the k dimensional column vector whose elements are all 1, and we recall the following definition of inequality between two k-dimensional vectors $\mathbf{m}$ and $\mathbf{n}$ with components m_i and n_i respectively: $\mathbf{m} \leqq \mathbf{n}$ if $m_i \leq n_i$, for all $i = 1,...,k$. The other basic relationships between vectors are: *i)* $\mathbf{m} = \mathbf{n}$ if $m_i = n_i$, for all $i = 1,...,k$, *ii)* $\mathbf{m} < \mathbf{n}$ if $m_i < n_i$, for all $i = 1,...,k$, *iii)* $\mathbf{m} \leq \mathbf{n}$ if $m_i \leq n_i$, for all $i = 1,...,k$, and $\mathbf{m} \neq \mathbf{n}$. We also need basic matrix operations, "+", "-" and "*", that refer to, respectively the addition, the subtraction and the multiplication.

Let us define:

$$\mathbf{Q} \equiv \begin{bmatrix} q_1 \\ q_2 \\ \vdots \\ q_k \end{bmatrix}, \mathbf{R} \equiv \begin{bmatrix} 0 & r_{12} & \cdots & r_{1k} \\ r_{21} & 0 & \cdots & r_{2k} \\ \cdots & \cdots & \ddots & \vdots \\ r_{k1} & r_{k2} & \cdots & 0 \end{bmatrix}, \mathbf{P} \equiv \begin{bmatrix} P_1 \\ P_2 \\ \vdots \\ P_k \end{bmatrix}, \beta \equiv \begin{bmatrix} \beta_1 \\ \beta_2 \\ \vdots \\ \beta_k \end{bmatrix}$$

$$\overline{\mathbf{P}} \equiv \begin{bmatrix} \overline{P}_1 \\ \overline{P}_2 \\ \vdots \\ \overline{P}_k \end{bmatrix}, \quad \underline{\mathbf{P}} \equiv \begin{bmatrix} \underline{P}_1 \\ \underline{P}_2 \\ \vdots \\ \underline{P}_k \end{bmatrix}, \quad \mathbf{X} \equiv \begin{bmatrix} x_1 \\ x_2 \\ \vdots \\ x_k \end{bmatrix}, \quad \overline{\mathbf{X}} \equiv \begin{bmatrix} \bar{x}_1 \\ \bar{x}_2 \\ \vdots \\ \bar{x}_k \end{bmatrix}.$$

In matrix form, the system (2) reads as:

$$\mathbf{P} = \mathbf{Q} + \mathbf{X} + \mathbf{R} * \mathbf{P}. \quad (5)$$

Throughout this article, we will assume:

Assumption 1 (INV) *The matrix* $\mathbf{I}^k - \mathbf{R}$ *is invertible.*

Under Assumption (INV), the system (5) can be solved to give:

$$\mathbf{P} = \mathbf{\Lambda} * (\mathbf{Q} + \mathbf{X}), \quad (6)$$

where $\mathbf{\Lambda} \equiv \left[\mathbf{I}^k - \mathbf{R}\right]^{-1}$.

Let $\mathcal{P}(\mathbf{X}) \equiv \mathbf{\Lambda} * (\mathbf{Q} + \mathbf{X})$ refer to the affine mapping from efforts to probabilities. Survival probabilities without protection policies are therefore:

$$\underline{\mathbf{P}} = \mathcal{P}(0 * \iota^{\mathbf{k}}), \quad (7)$$

where $0 * \iota^{\mathbf{k}}$ is a vector made of k zeroes. Without ecological interactions, $\mathbf{\Lambda}$ is the identity matrix, $\underline{\mathbf{P}} = \mathbf{Q}$ and $\overline{\mathbf{P}} = \underline{\mathbf{P}} + \overline{\mathbf{X}} = \mathbf{Q} + \overline{\mathbf{X}}$.

Now we can plug (6) into (4) to get rid of probabilities, and express Noah's problem only in terms of efforts. Define the two *composite* functions, which here are mappings from the values taken by function $\mathcal{P}(\mathbf{X})$ to the set of real numbers:

$$W \circ \mathcal{P}(\mathbf{X}) \equiv W(\mathcal{P}(\mathbf{X})),$$

$$U \circ \mathcal{P}(\mathbf{X}) \equiv U(\mathcal{P}(\mathbf{X})).$$

Under Assumption (INV), to each vector $\mathbf{X}$ corresponds a unique vector $\mathbf{P} = \mathcal{P}(\mathbf{X})$. Therefore we can define Noah's problem with ecological interactions, the constrained maximization of a function of protection efforts $\mathbf{X}$:

$$\max_{\mathbf{X}} W \circ \mathcal{P}(\mathbf{X}) + U \circ \mathcal{P}(\mathbf{X}), \quad (8)$$

subject to:

$$\beta^\top * \mathbf{X} \leq B, \quad (9)$$

$$0 * \iota^{\mathbf{k}} \leqq \mathbf{X} \leqq \overline{\mathbf{X}}. \quad (10)$$

Results

Two questions arise: *i)* could anything general be said about the solution to the problem expressed by (8), (9), (10)? And *ii)*, taking a more practical stance, could we engineer a simple rule that approximates the general solution?

Noah's policy is extreme

Weitzman [13] showed that the solution to Noah's problem lies on the boundary of the efforts set. As the set of constraints is made of linear constraints, the boundary involves corners, e.g. $x_i = 0$ or $x_i = \bar{x}_i$, and possibly a segment between two corners, therefore with $x_i \in [0, \bar{x}_i]$ for at most one species. This can be defined as an *extreme policy*. In words, the optimal protection policy gives full protection to a subset of species, partial protection for at most one species, and exposes the remaining species to the risk of no protection.

But what if probabilities are interdependent? We show that when species interact, the optimal solution is also extreme.

Theorem 1 *The solution to Noah's Ark problem with ecological interactions, defined by (8), (9) and (10), is an extreme policy.*

Proof. The proof rests on two pieces of information:

i) Noahs' problem is to maximize a continuous function over a compact set, therefore by Weicstrass *extreme value theorem* there exists a solution.

ii) The Hessian matrix of $W \circ \mathcal{P}(\mathbf{X}) + U \circ \mathcal{P}(\mathbf{X})$ is *not* negative semi-definite, a statement we shall prove below.

Item ii) violates the necessary second order condition for interior solutions to Noah's problem and, in combination with item i), leads to conclude the existence of a solution on the boundary of the efforts set.

In order to prove item ii), because $U \circ \mathcal{P}(\mathbf{X})$ is linear, we just have to ensure that the Hessian matrix of $W \circ \mathcal{P}(\mathbf{X})$ is not negative semi-definite. Recall that $\mathcal{P}(\mathbf{X})$ is a k-dimensional vector with typical element $\mathcal{P}_h(\mathbf{X}), h = 1,...,k$, and let $J_{\mathcal{P}}(\mathbf{X})$ stand for the *Jacobian* matrix:

$$J_{\mathcal{P}}(\mathbf{X}) \equiv \begin{bmatrix} \frac{\partial \mathcal{P}_1(\mathbf{X})}{\partial x_1} & \frac{\partial \mathcal{P}_1(\mathbf{X})}{\partial x_2} & \cdots & \frac{\partial \mathcal{P}_1(\mathbf{X})}{\partial x_k} \\ \frac{\partial \mathcal{P}_2(\mathbf{X})}{\partial x_1} & \frac{\partial \mathcal{P}_2(\mathbf{X})}{\partial x_2} & \cdots & \frac{\partial \mathcal{P}_2(\mathbf{X})}{\partial x_k} \\ \vdots & \vdots & \ddots & \vdots \\ \frac{\partial \mathcal{P}_k(\mathbf{X})}{\partial x_1} & \frac{\partial \mathcal{P}_k(\mathbf{X})}{\partial x_2} & \cdots & \frac{\partial \mathcal{P}_k(\mathbf{X})}{\partial x_k} \end{bmatrix}.$$

Note that, since each function $\mathcal{P}_h(\mathbf{X})$ is linear, the Jacobian matrix is made of invariant numbers, so we need not mention the application point $\mathbf{X}$ and we can simply refer to the matrix $J_{\mathcal{P}}$.

Denote $\nabla^2 W \circ \mathcal{P}(\mathbf{X})$ the *Hessian* matrix of $W \circ \mathcal{P}(\mathbf{X})$, a $k * k$ matrix with typical elements $\partial^2 (W \circ \mathcal{P}) / (\partial x_i \partial x_j)$. From meticulous derivations of the composite function $W \circ \mathcal{P}(\mathbf{X})$, and after simplifications allowed by the linearity of the mapping $\mathcal{P}(\mathbf{X})$, one obtains:

$$\nabla^2 W \circ \mathcal{P}(\mathbf{X}) \equiv J_{\mathcal{P}}^{\top} * \nabla^2 W(\mathbf{P}) * J_{\mathcal{P}} .$$

If $\nabla^2 W \circ \mathcal{P}(\mathbf{X})$ is negative semi-definite, then for any nonzero vector $m \in \mathbb{R}^k$ we must have:

$$m^{\top} * \nabla^2 W \circ \mathcal{P}(\mathbf{X}) * m = m^{\top} * J_{\mathcal{P}}^{\top} * \nabla^2 W(\mathbf{P}) * J_{\mathcal{P}} * m \leq 0 .$$

Notice that $J_{\mathcal{P}} * m$ is simply a nonzero $(k * 1)$ vector, which we may simply call n. Hence we can rewrite the above inequality as:

$$n^{\top} * \nabla^2 W(\mathbf{P}) * n \leq 0 ,$$

which would mean that $\nabla^2 W(\mathbf{X})$ is negative semi-definite, a possibility that has been ruled out by assumption. ■

A ranking rule for interacting species

Theorem 1 is a qualitative result, that does not indicate which species should be granted protection and why. This brings us to our second question; it would be welcome to have an explicit and easy-to-use approximation of the general solution. Facing the same problem, this is the practical point of view adopted by [13], which he describes as "the main theme" of his paper (p. 1294). His formula (1) offers a ranking that is not really a solution to the original problem, but rather a first order approximation of an optimal policy. In order to achieve this, he replaces the objective function by its linear approximation. He then obtains a classical linear programming problem, whose solution is to assign grades R_i given by formula (1) to species (those grades depend on the model parameters) and order them in decreasing order of importance up to the point where the budget is exhausted. Those grades are exactly the practical ranking Noah is looking for.

We follow the same approach here, *i.e.* we linearize the objective function. The astute reader knows that, in general, such approximations can be seriously misleading [2] and should not be followed blindly. Nevertheless, as proven in Theorem 2 below, there is something special about Noah's problem that makes this practice appropriate here.

Let us denote:

$$D_i \equiv \left.\frac{\partial W}{\partial P_i}\right|_{\mathbf{P}=\underline{\mathbf{P}}}, \quad U_i \equiv \left.\frac{\partial U}{\partial P_i}\right|_{\mathbf{P}=\underline{\mathbf{P}}},$$

and define the two matrices:

$$\mathbf{A} \equiv \begin{bmatrix} D_1 + U_1 \\ D_2 + U_2 \\ \vdots \\ D_k + U_k \end{bmatrix}, \quad \Upsilon \equiv \mathbf{A}^{\top} * \Lambda.$$

From simple calculations, the linearized problem in matrix form turns out to be:

$$\max_{\mathbf{X}} \Upsilon * \mathbf{X} + \text{constant terms}, \tag{11}$$

subject to (9) and (10).

As can be observed in the above approximation of Noah's problem, the introduction of ecological interactions changes the "slope" of the objective function to be maximized, which is now $\Upsilon \equiv \mathbf{A}^{\top} * \Lambda$ instead of just $\mathbf{A}^{\top}$. The crux, from the point of view of the present note, is to transform the information about ecological interactions conveyed by matrix $\mathbf{R}$, into operational data *via* the matrix $\Lambda = \left[\mathbf{I}^k - \mathbf{R}\right]^{-1}$. Given that $\mathbf{I}^k - \mathbf{R}$ is invertible, the computation of the matrix Λ is easily made and if Λ_{ij} denotes a typical element of Λ, then Υ is a k-dimensional line vector of the type:

$$\Upsilon = [\alpha_1 , \alpha_2 , \ldots, \alpha_k] ,$$

where

$$\alpha_i \equiv \sum_{h=1}^{k} (D_h + U_h) \Lambda_{hi} .$$

We can now define the "benefit"-cost ratios $\overline{R}^i \equiv \alpha_i / \beta_i$, or with explicit reference to relevant information:

$$\overline{R}^i \equiv \frac{\Delta P_i}{C_i} \sum_{h=1}^{k} (D_h + U_h) \Lambda_{hi} , \quad i = 1,\ldots,k. \tag{12}$$

As it is well-known, the argmax to the linear programming problem (11) is to fully protect the species with the highest grade $\overline{R}^i$, then the species with the second highest grade, and so on and so forth, up to the point where the budget is exhausted. It means that there exists a threshold value $\overline{R}^*$ such that all species i with $\overline{R}^i < \overline{R}^*$ are not embarked in the Ark, whereas those with grade larger than the threshold are all fully protected, except for at most one species with grade exactly equal the cutoff value $\overline{R}^*$ that is only partially protected. Let us call $\mathbf{X}^W$ this policy, which can be described formally as follows:

$$\begin{cases} \text{if } \overline{R}^i < \overline{R}^*, & \text{species i is granted zero protection,} \quad \Rightarrow x_i = 0\,, \\ \text{if } \overline{R}^i > \overline{R}^*, & \text{species i is granted full protection,} \quad \Rightarrow x_i = \overline{x}_i\,, \\ \text{if } \overline{R}^i = \overline{R}^*, & \text{species i is granted full protection if there is enough budget,} \quad \Rightarrow x_i = \overline{x}_i\,, \\ & \text{otherwise the remaining budget funds its partial protection,} \quad \Rightarrow x_i \in]0, \overline{x}_i[\,. \end{cases} \tag{13}$$

As shown in Theorem 2 below, $\mathbf{X}^W$ is a first order approximation of the optimal solution to Noah's Ark problem with ecological interactions. Put differently, there is a sense in which expression (12) can be taken for the new practical formula sought to construct *in situ* conservation priorities. Observe that the number assigned to each species i does not depend merely on its own "benefits" but actually on overall "benefits" generated by species i on all the species, $\sum_{h=1}^{k}(D_h+U_h)\Lambda_{hi}$, via ecological interactions. Therefore, a species with a strong own interest can be overridden by another, endowed with a less direct interest, but whose importance is enhanced because of its ecological role. Of course, when there are no ecological interactions, $\boldsymbol{\Lambda}$ is the identity matrix, with $\Lambda_{ii}=1, \Lambda_{hi}=0, \forall h \neq i$, and (12) boils down to Weitzman's original system of grades for species i:

$$\overline{R}^i = R^i \equiv \frac{\Delta P_i}{C_i}(D_i+U_i)\,.$$

One can ask to what extent can we rely on formula (12) to build a hierarchy among species? Can a conservation policy be based on such an approximation? Baumol and Bushnell in [2] have famously attracted the attention on a number of potential flaws with linear approximations, two of them being important for the problem at hand: *i*) a linear approximation to a nonlinear program need not provide an answer better than a randomly chosen admissible answer, *ii*) only if the objective function behaves monotonically in every variable within the admissible region can we be assured that a linear approximation will yield results which represent an improvement over the point where the linearization is made. Clearly, Noah's objective function does not meet this last condition, for an increase of the effort x_i can improve the chances of species i at the expense of another species j (obviously so when i is a predator for j).

Still, we can prove the following Theorem which establishes a special interest to the use of a linear approximation in this decision problem:

Theorem 2 *Consider the Noah's Ark Problem with ecological interactions, defined by (8), (9) and (10), and call* $\mathbf{X}^*$ *its optimal solution. Then,*

i) *the approximation of* $\mathbf{X}^*$ *by* $\mathbf{X}^W$*, indicated in (13), offers an improvement compared to the absence of protection,*

ii) *the approximation error,* $W\circ\mathcal{P}(\mathbf{X}^*) - W\circ\mathcal{P}(\mathbf{X}^W)$*, is no larger than* $K*\left(\overline{\mathbf{X}}^\top \iota^k\right)^2$*, where* $K=\max\{\partial^2(W\circ\mathcal{P})/(\partial x_i \partial x_j)\}$.

Proof. Item i). The solution proposed in Theorem 2 is inspired from gradient methods used to find optimal solutions based on the property of iterative improvements, like the famous Frank-Wolfe algorithm.

A first step is to replace the objective function by its first order Taylor approximation $Z(\mathbf{X})$ computed at an admissible vector $\mathbf{X}^0$ (here at the zero protection vector $\mathbf{X}^0 = 0*\iota^k$). Let us note $\nabla W\circ\mathcal{P}$ the Gradient, a $k*1$ vector with typical elements $\partial(W\circ\mathcal{P})/\partial x_i$, which corresponds actually to the vector $\mathbf{Y}\equiv\mathbf{A}^\top*\boldsymbol{\Lambda}$ given in the text.

Using those notations:

$$Z(\mathbf{X}) \simeq W\circ\mathcal{P}(0*\iota^k) + \nabla W\circ\mathcal{P}(0*\iota^k)^\top*(\mathbf{X}-0*\iota^k).$$

A second step is to find $\mathbf{X}$ that maximizes $Z(\mathbf{X})$ subject to the relevant constraints. Since in $Z(\mathbf{X})$ only the term $\nabla W\circ\mathcal{P}(0*\iota^k)^\top\mathbf{X}$ varies, this step is equivalent to maximize (11) subject to (9) and (10). And the policy $\mathbf{X}^W$ presented in the Theorem 2 is exactly the maximizer of this linear programming problem.

By definition of $\mathbf{X}^W$, we must have:

$$Z(\mathbf{X}^W) \geq Z(0*\iota^k)\,.$$

$$\begin{aligned} \Leftrightarrow\ & \nabla W\circ\mathcal{P}(0*\iota^k)^\top*(\mathbf{X}^W-0*\iota^k) \\ & \geq \nabla W\circ\mathcal{P}(0*\iota^k)^\top*(0*\iota^k-0*\iota^k)=0, \end{aligned} \tag{14}$$

so the vector $\mathbf{X}^W - 0*\iota^k$ is an ascent direction for $W\circ\mathcal{P}$. Although this means that the approximation $Z(\mathbf{X})$ is non decreasing along this direction, it is not guaranteed that the non linear objective will behave similarly, *i.e.* we cannot yet conclude $W\circ\mathcal{P}(\mathbf{X}^W) \geq W\circ\mathcal{P}(0*\iota^k)$.

By convexity of function $W\circ\mathcal{P}$ we can write:

$$W\circ\mathcal{P}(\mathbf{X}^W) - W\circ\mathcal{P}(0*\iota^k) \geq \nabla W\circ\mathcal{P}(0*\iota^k)^\top*(\mathbf{X}^W-0*\iota^k),$$

and since we have established in (14):

$$\nabla W\circ\mathcal{P}(0*\iota^k)^\top*(\mathbf{X}^W-0*\iota^k) \geq 0,$$

we are led to conclude:

$$W\circ\mathcal{P}(\mathbf{X}^W) - W\circ\mathcal{P}(0*\iota^k) \geq 0.$$

Item ii). Recall that $\nabla^2 W\circ\mathcal{P}$ stands for the Hessian matrix of $W\circ\mathcal{P}$. Using Taylor expansions, one can write:

$$\begin{aligned} W\circ\mathcal{P}(\mathbf{X}^*) = {} & W\circ\mathcal{P}(0*\iota^k) \\ & + \nabla W\circ\mathcal{P}(0*\iota^k)^\top*\mathbf{X}^* + \frac{1}{2!}(\mathbf{X}^*)^\mathsf{T}*\nabla^2 W\circ\mathcal{P}(\mathbf{Z}^*)*\mathbf{X}^*\,, \end{aligned}$$

for some admissible vector $\mathbf{Z}^*$, and

$$W \circ \mathcal{P}(\mathbf{X}^W) = W \circ \mathcal{P}(0 * \iota^k) + \nabla W \circ \mathcal{P}(0 * \iota^k)^\top * \mathbf{X}^W + \frac{1}{2!}(\mathbf{X}^W)^\mathsf{T} * \nabla^2 W \circ \mathcal{P}(\mathbf{Z}^w) * \mathbf{X}^W,$$

for some admissible vector $\mathbf{Z}^w$. Therefore

$$W \circ \mathcal{P}(\mathbf{X}^*) - W \circ \mathcal{P}(\mathbf{X}^W) = \nabla W \circ \mathcal{P}(0 * \iota^k)^\top * (\mathbf{X}^* - \mathbf{X}^W) + \frac{1}{2!}(\mathbf{X}^*)^\mathsf{T} * \nabla^2 W \circ \mathcal{P}(\mathbf{Z}^*) * \mathbf{X}^* - \frac{1}{2!}(\mathbf{X}^W)^\mathsf{T} * \nabla^2 W \circ \mathcal{P}(\mathbf{Z}^{1w}) * \mathbf{X}^W .$$

But, by definition of $\mathbf{X}^W$

$$\nabla W \circ \mathcal{P}(0 * \iota^k)^\top * (\mathbf{X}^* - \mathbf{X}^W) \leq 0 ,$$

so

$$\begin{aligned} W \circ \mathcal{P}(\mathbf{X}^*) - W \circ \mathcal{P}(\mathbf{X}^W) &\leq \frac{1}{2!}(\mathbf{X}^*)^\mathsf{T} * \nabla^2 W \circ \mathcal{P}(\mathbf{Z}^*) * \mathbf{X}^* - \frac{1}{2!}(\mathbf{X}^W)^\mathsf{T} * \nabla^2 W \circ \mathcal{P}(\mathbf{Z}^w) * \mathbf{X}^W \\ &\leq \frac{K}{2!}\left[(\mathbf{X}^*)^\mathsf{T} * \iota^k\right]^2 - \frac{K}{2!}\left[(\mathbf{X}^W)^\mathsf{T} * \iota^k\right]^2 \\ &\leq K\left(\overline{\mathbf{X}}^\mathsf{T} * \iota^k\right)^2, \end{aligned}$$

where $K = \max\{\partial^2(W \circ \mathcal{P})/(\partial x_i \partial x_j)\}$. ■

The upper bound K for the approximation error mentioned in the above theorem if of course related to the non-linearity of $W \circ \mathcal{P}$, formally captured by the second order derivatives $\partial^2(W \circ \mathcal{P})/(\partial x_i \partial x_j)$. As a matter of interpretation, we can say that the stronger the curvature of W (the stronger preference for diversity if W is convex) the larger this upper bound.

A Two-Species Example: Illustration and Discussion

We close this note with an illustration using a simple two-species example. Let us first study to which extent the consideration of ecological interactions can alter priorities. Assume for simplicity that $|r_{12}| < 1, |r_{21}| < 1$. The system (2) becomes:

$$\begin{bmatrix} P_1 \\ P_2 \end{bmatrix} = \begin{bmatrix} q_1 \\ q_2 \end{bmatrix} + \begin{bmatrix} x_1 \\ x_2 \end{bmatrix} + \begin{bmatrix} 0 & r_{12} \\ r_{21} & 0 \end{bmatrix} \begin{bmatrix} P_1 \\ P_2 \end{bmatrix} .$$

Here the matrix $I^k - R$ is invertible since $r_{12} r_{21} \neq 1$. Solving the system of interactions:

$$P_1 = \frac{q_1 + r_{12} q_2 + x_1 + r_{12} x_2}{1 - r_{12} r_{21}} , \tag{15}$$

$$P_2 = \frac{q_2 + r_{21} q_1 + x_2 + r_{21} x_1}{1 - r_{21} r_{12}} . \tag{16}$$

The grades also can be easily computed. They are:

$$\overline{R}^1 = \frac{\Delta P_1}{C_1}\left[\frac{D_1 + U_1}{1 - r_{12} r_{21}} + \frac{r_{21}(D_2 + U_2)}{1 - r_{21} r_{12}}\right] ,$$

$$\overline{R}^2 = \frac{\Delta P_2}{C_2}\left[\frac{r_{12}(D_1 + U_1)}{1 - r_{12} r_{21}} + \frac{D_2 + U_2}{1 - r_{21} r_{12}}\right] .$$

To further simplify, imagine that $C_1 = C_2 = C$, $\Delta P_1 = \Delta P_2 = \Delta P$. If ecological interactions are erroneously ignored, formally Noah assigns zero values by mistake to the system of interactions: $r_{12} = r_{21} = 0$. Suppose, without loss of generality, that on this erroneous basis the first species ranks higher:

$$R^1 = (D_1 + U_1) > R^2 = (D_2 + U_2) .$$

In other words $D_1 + U_1 = k * (D_2 + U_2)$, for some $k > 1$.

Two questions arise. Could this ranking be reversed once interactions are properly taken into account? And, if the answer is affirmative, why?

When the ranking is reversed:

$$\overline{R}^1 < \overline{R}^2 ,$$

$$\Leftrightarrow$$

$$\frac{D_1 + U_1}{1 - r_{12} r_{21}} + \frac{r_{21}(D_2 + U_2)}{1 - r_{21} r_{12}} < \frac{r_{12}(D_1 + U_1)}{1 - r_{12} r_{21}} + \frac{D_2 + U_2}{1 - r_{21} r_{12}} .$$

Since $1 - r_{12} r_{21} > 0$, and using $D_1 + U_1 = k * (D_2 + U_2)$, the last inequality is equivalent to:

$$\Leftrightarrow k + r_{21} < k r_{12} + 1 ,$$

$$\Leftrightarrow k < \frac{1 - r_{21}}{1 - r_{12}} . \quad (\text{since } |r_{12}| < 1).$$

So, a ranking reversal occurs when:

$$1 < k < \frac{1 - r_{21}}{1 - r_{12}} . \tag{17}$$

In order to fix ideas, consider that k is arbitrarily close to one, *i.e.* the two species provide similar "benefits" and therefore a ranking reversal, if any, is due to the consideration of ecological interactions. Then note that for the above inequality to hold, necessarily $r_{12} > r_{21}$, which may occur in various interesting ecological configurations:

i) *Predation*: species 1, a predator, feeds on species 2, its prey. So $r_{21} < 0$ whereas $r_{12} > 0$. Giving conservation priority to the prey is the most effective way to enjoy the benefits of both species.

ii) *Mutualism*: for example plant-pollinator interactions, $r_{12}, r_{21} > 0$. The synergistic relation between those two species is best enhanced by promoting species 2, which has the largest collective marginal impact.

iii) *Competition*: two species have to share a common resource in the same living area that cannot fully support both populations, hence $r_{12}, r_{21} < 0$, so conservation efforts focus on species 2 because its marginal negative impact is lower.

Let us now examine the robustness of our results by specifying an expected diversity function. Denote G the number of genes jointly owned by the two species whereas M_i is the total number of genes owned by species i. Assume, as in [13] (expression (5)) that the expected (genetic) diversity function takes the following functional form:

$$\begin{aligned} W(P_1,P_2) &= P_1P_2(M_1+M_2-G)+P_1(1-P_2)M_1 \\ &\quad +P_2(1-P_1)M_2+(1-P_1)(1-P_2)0 \\ &= M_1P_1+M_2P_2-GP_1P_2. \end{aligned}$$

Considering relations (15) and (16) between efforts and probabilities, we obtain:

$$\begin{aligned} W \circ \mathcal{P}(x_1,x_2) &= M_1\mathcal{P}_1(x_1,x_2) \\ &\quad + M_2\mathcal{P}_2(x_1,x_2) - G\mathcal{P}_1(x_1,x_2)\mathcal{P}_2(x_1,x_2)\ . \end{aligned}$$

Two questions arise. Can we compare the true solution and the approximate solution? And can we estimate the error due to the approximation of the optimal solution? From Theorem 2, the upper bound on the error due to the approximation can be computed from the Hessian $\nabla^2 W \circ \mathcal{P}(\mathbf{X})$. In this two-species example, it is easy to derive the following formulae:

$$\partial^2(W\circ\mathcal{P})/(\partial x_1\partial x_2) = \partial^2(W\circ\mathcal{P})/(\partial x_2\partial x_1) = -\frac{(1+r_{12}r_{21})}{(1-r_{12}r_{21})^2}G\ ,$$

$$\begin{aligned} \partial^2(W\circ\mathcal{P})/(\partial x_1)^2 &= -\frac{2r_{21}}{(1-r_{12}r_{21})^2}G\ , \\ \partial^2(W\circ\mathcal{P})/(\partial x_2)^2 &= -\frac{2r_{12}}{(1-r_{12}r_{21})^2}G\ . \end{aligned}$$

So the upper bound K on the approximation error, indicated in Theorem 2, is:

$$K = (1-r_{12}r_{21})^{-2} * G * \max\{-2r_{21}, -(1+r_{12}r_{21}), -2r_{12}\}\ , \quad (18)$$

a value which depends only on the number of genes owned jointly by the two species, G, and on the ecological interaction terms, r_{ji}.

Of course, this is only an upper bound. In some cases, the approximation could also give the exact solution. To illustrate this, assume as before that $C_1 = C_2 = C$, $\Delta P_1 = \Delta P_2 = \Delta P$, that utilities are identical, $U_1 = U_2 = U = 0$, and the upper bounds on efforts are the same for the two species, $\bar{x}_1 = \bar{x}_2 = \bar{x}$. Assume also that the total budget can cover the protection cost of only one species, $B = \bar{x} * C/\Delta P$. Noah then has to choose among two extreme policies, the first one $(x_1 = 0; x_2 = \bar{x})$ that provides the following expected diversity:

$$\begin{aligned} W(0,\bar{x}) &= M_1\frac{q_1+r_{12}q_2+r_{12}\bar{x}}{1-r_{12}r_{21}} + M_2\frac{q_2+r_{21}q_1+\bar{x}}{1-r_{12}r_{21}} \\ &\quad - G\frac{(q_1+r_{12}q_2+r_{12}\bar{x})(q_2+r_{21}q_1+\bar{x})}{(1-r_{12}r_{21})^2}\ , \end{aligned}$$

and the second one $(x_1 = \bar{x}; x_2 = 0)$ with expected diversity:

$$\begin{aligned} W(\bar{x},0) &= M_1\frac{q_1+r_{12}q_2+\bar{x}}{1-r_{12}r_{21}} + M_2\frac{q_2+r_{21}q_1+r_{21}\bar{x}}{1-r_{12}r_{21}} \\ &\quad - G\frac{(q_1+r_{12}q_2+\bar{x})(q_2+r_{21}q_1+r_{21}\bar{x})}{(1-r_{12}r_{21})^2}\ . \end{aligned}$$

It is optimal to protect species 2 if:

$$\begin{gathered} W(0,\bar{x}) > W(\bar{x},0) \\ \Leftrightarrow \\ (1-r_{12}r_{21})[M_2(1-r_{21}) - M_1(1-r_{12})] > \\ G[(1-r_{21})(q_1-q_2+r_{12}q_2-r_{21}q_1)+(r_{12}-r_{21})\bar{x}]\ . \end{gathered}$$

In the particular case where $G = 0$, then $M_i = \partial W/\partial P_i = D_i$, and the above condition boils down to a very simple expression:

$$W(0,\bar{x}) > W(\bar{x},0) \Leftrightarrow \frac{M_1}{M_2} = \frac{D_1}{D_2} < \frac{1-r_{21}}{1-r_{12}}\ ,$$

a condition which is also necessary for the approximated solution to select species 2 (remember condition (17)). It comes as no surprise that the optimal solution and its approximation concide, since when $G = 0$ the upper bound on the approximation error is zero, as can be seen from expression (18).

Acknowledgments

Thanks are due to two anonymous referees of PLOS ONE for very helpful and kind comments, to referees of the FAERE working paper series and to the audience at the 2014 World Congress of Environmental Economics, Istanbul.

Author Contributions

Contributed reagents/materials/analysis tools: PC CF CM. Contributed to the writing of the manuscript: PC CF CM.

References

1. Baumgärtner S (2004) Optimal investment in multi-species protection: interacting species and ecosystem health. EcoHealth 1: 101–110.
2. Baumol WJ, Bushnell RC (1967) Error Produced by Linearization in Mathematical Programming. Econometrica 35(3/4): 447–471.
3. Burgman MA, Ferson S, Akcakaya HR (1993) Risk Assessment in Conservation Biology. London: Chapman and Hall. pp 44–83.

4. Elith J, Leathwick JR (2009) Species distribution models: ecological explanation and prediction across space and time. Annual Review of Ecology, Evolution, and Systematics 40: 677–697.
5. Eppink FV, van den Bergh JCJM (2007) Ecological theories and indicators in economic models of biodiversity loss and conservation: A critical review. Ecological Economics 61(2–3): 284–293.
6. Garnett ST, Crowley GM (2000) The Action Plan for Australian Birds 2000. Environment Australia, Canberra, Australia.
7. Guisan A, Thuiller W (2005) Predicting species distribution: offering more than simple habitat models. Ecology Letters 8: 993–1009.
8. Hauser CE, McCarthy MA (2009) Streamlining 'search and destroy': cost-effective surveillance for invasives species management. Ecology Letters 12: 638–692.
9. Joseph L N, Maloney RF, Possingham HP (2008) Optimal allocation of resources among threatened species: a project prioritization protocol. Conservation Biology 23(2): 328–338.
10. McCarthy MA, Thompson CJ, Garnett T (2008) Optimal investment in conservation of species. Journal of Applied Ecology 45: 1428–1435.
11. Simianer H (2008) Accounting for non-independence of extinction probabilities in the derivation of conservation priorities based on Weitzman's diversity concept. Conservation Genetics 9(1): 171–179.
12. Van der Heide MC, van den Bergh JCJM, van Ierland EC (2005) Extending Weitzman's economic ranking of biodiversity protection: combining ecological and genetic considerations. Ecological Economics 55: 218–223.
13. Weitzman ML (1998) The Noah's Ark Problem. Econometrica 66(6): 1279–1298.
14. Witting L, Tomiuk J, Loeschcke V (2000) Modelling the optimal conservation of interacting species. Ecological Modelling 125: 123–143.

PERMISSIONS

All chapters in this book were first published in PLOS ONE, by The Public Library of Science; hereby published with permission under the Creative Commons Attribution License or equivalent. Every chapter published in this book has been scrutinized by our experts. Their significance has been extensively debated. The topics covered herein carry significant findings which will fuel the growth of the discipline. They may even be implemented as practical applications or may be referred to as a beginning point for another development.

The contributors of this book come from diverse backgrounds, making this book a truly international effort. This book will bring forth new frontiers with its revolutionizing research information and detailed analysis of the nascent developments around the world.

We would like to thank all the contributing authors for lending their expertise to make the book truly unique. They have played a crucial role in the development of this book. Without their invaluable contributions this book wouldn't have been possible. They have made vital efforts to compile up to date information on the varied aspects of this subject to make this book a valuable addition to the collection of many professionals and students.

This book was conceptualized with the vision of imparting up-to-date information and advanced data in this field. To ensure the same, a matchless editorial board was set up. Every individual on the board went through rigorous rounds of assessment to prove their worth. After which they invested a large part of their time researching and compiling the most relevant data for our readers.

The editorial board has been involved in producing this book since its inception. They have spent rigorous hours researching and exploring the diverse topics which have resulted in the successful publishing of this book. They have passed on their knowledge of decades through this book. To expedite this challenging task, the publisher supported the team at every step. A small team of assistant editors was also appointed to further simplify the editing procedure and attain best results for the readers.

Apart from the editorial board, the designing team has also invested a significant amount of their time in understanding the subject and creating the most relevant covers. They scrutinized every image to scout for the most suitable representation of the subject and create an appropriate cover for the book.

The publishing team has been an ardent support to the editorial, designing and production team. Their endless efforts to recruit the best for this project, has resulted in the accomplishment of this book. They are a veteran in the field of academics and their pool of knowledge is as vast as their experience in printing. Their expertise and guidance has proved useful at every step. Their uncompromising quality standards have made this book an exceptional effort. Their encouragement from time to time has been an inspiration for everyone.

The publisher and the editorial board hope that this book will prove to be a valuable piece of knowledge for researchers, students, practitioners and scholars across the globe.

LIST OF CONTRIBUTORS

Steeves Buckland
School of Biological Sciences, University of Bristol, Bristol, United Kingdom
NERC Biomolecular Analysis Facility, Department of Animal and Plant Sciences, University of Sheffield, Sheffield, United Kingdom
National Parks & Conservation Service, Reduit, Mauritius

Nik C. Cole
School of Biological Sciences, University of Bristol, Bristol, United Kingdom
Durrell Wildlife Conservation Trust, Jersey, Channel Islands
Mauritian Wildlife Foundation, Vacoas, Mauritius

Jim J. Groombridge
Durrell Institute of Conservation and Ecology, School of Anthropology and Conservation, University of Kent, Kent, United Kingdom

Clemens Küpper, Terry Burke and Deborah A. Dawson,
NERC Biomolecular Analysis Facility, Department of Animal and Plant Sciences, University of Sheffield, Sheffield, United Kingdom

Laura E. Gallagher
National Parks & Conservation Service, Reduit, Mauritius

Stephen Harris
School of Biological Sciences, University of Bristol, Bristol, United Kingdom

Alexander M. Brown, Simon J. Allen and Lars Bejder
Murdoch University Cetacean Research Unit, School of Veterinary and Life Sciences, Murdoch University, Perth, Western Australia, Australia

Anna M. Kopps
Centre for Ecology and Conservation, University of Exeter, Penryn, Cornwall, United Kingdom
Marine Evolution and Conservation, Centre for Ecological and Evolutionary Studies, University of Groningen, Groningen, The Netherlands

Bethan Littleford-Colquhoun
Centre for Ecology and Conservation, University of Exeter, Penryn, Cornwall, United Kingdom

Guido J. Parra
Cetacean Ecology, Behaviour and Evolution Lab, School of Biological Sciences, Flinders University, Adelaide, South Australia, Australia
South Australian Research and Development Institute, Adelaide, South Australia, Australia

Daniele Cagnazzi
Marine Ecology Research Centre, School of Environment, Science and Engineering, Southern Cross University, Lismore, New South Wales, Australia

Deborah Thiele
Fenner School of Environment & Society, Australian National University, Canberra, Australian Capital Territory, Australia

Carol Palmer
Marine Ecosystems, Flora and Fauna Division, Department of Land Resource Management, Palmerston, Northern Territory, Australia
Research Institute for the Environment and Livelihoods, Charles Darwin University, Darwin, Northern Territory, Australia

Celine H. Frère
GeneCology Research Centre, University of the Sunshine Coast, Maroochydore DC, Queensland, Australia

Takeshi Yuhara
Department of Environmental Science, Graduate School of Science, Toho University, Funabashi, Chiba, Japan

Masako Kawane
KYOUSEI Science Center for Life and Nature, Nara Women's University, Nara, Japan

Toshio Furota
Faculty of Sciences, Tokyo Bay Ecosystem Research Center, Toho University, Funabashi, Chiba, Japan

Stèphanie Manel
Aix Marseille Université, IRD, LPED UMR 151, Marseille, France
Centre de coopération internationale en recherche agronomique pour le développement, UMR AMAP, Montpellier, France

Thomas L. P. Couvreur
Institut de Recherche pour le Développement, UMR DIADE, Montpellier, France

Département des Sciences Biologiques, Laboratoire de Botanique Systématique et d'Ecologie, Université de Yaoundé I, Ecole Normale Supé rieure, Yaoundé, Cameroon

François Munoz
Université Montpellier 2, UMR AMAP, Montpellier, France

Pierre Couteron
Institut de Recherche Pour le Développement, UMR AMAP, Montpellier, France

Olivier J. Hardy
Université Libre de Bruxelles, Faculté des Sciences, Evolutionary Biology and Ecology Brussels, Belgium

Bonaventure Sonké
Département des Sciences Biologiques, Laboratoire de Botanique Systématique et d'Ecologie, Université de Yaoundé I, Ecole Normale Supé rieure, Yaoundé, Cameroon

Logan Volkmann
Department of Biological Sciences, Simon Fraser University, Burnaby, British Columbia, Canada

Iain Martyn
Interdisciplinary Research in the Mathematical and Computational Sciences (IRMACS) Centre, Simon Fraser University, Burnaby, British Columbia, Canada

Vincent Moulton
School of Computing Sciences, University of East Anglia, Norwich, United Kingdom

Andreas Spillner
Department of Mathematics and Computer Science, University of Greifswald, Greifswald, Germany

Arne O. Mooers
Department of Biological Sciences, Simon Fraser University, Burnaby, British Columbia, Canada
Interdisciplinary Research in the Mathematical and Computational Sciences (IRMACS) Centre, Simon Fraser University, Burnaby, British Columbia, Canada

Chunping Liu
State Key Laboratory of Forest Genetics and Breeding, Northeast Forestry University, Harbin, Heilongjiang Province, China
Laboratory of Forest Ecosystem Studies, Department of Ecosystem Studies, Graduate School of Agriculture and Life Science, The University of Tokyo, Tokyo, Japan

Yoshiaki Tsuda
Program in Plant Ecology and Evolution, Department of Ecology and Genetics, Evolutionary Biology Centre, Uppsala University, Uppsala, Sweden

Hailong Shen
School of Forestry, Northeast Forestry University, Harbin, China
State Key Laboratory of Forest Genetics and Breeding, Northeast Forestry University, Harbin, Heilongjiang Province, China

Lijiang Hu
State Key Laboratory of Forest Genetics and Breeding, Northeast Forestry University, Harbin, Heilongjiang Province, China

Yoko Saito and Yuji Ide
Laboratory of Forest Ecosystem Studies, Department of Ecosystem Studies, Graduate School of Agriculture and Life Science, The University of Tokyo, Tokyo, Japan

Joanne Russell, Ian K. Dawson, Allan Booth and Robbie Waugh
Cell and Molecular Sciences, The James Hutton Institute, Invergowrie, Scotland, United Kingdom

Maarten van Zonneveld
Regional Office for the Americas, Bioversity International, Cali, Colombia

Brian Steffenson
Department of Plant Pathology, University of Minnesota, Saint Paul, Minnesota, United States of America

Tak Fung
National University of Singapore, Department of Biological Sciences, Singapore, Singapore
Queen's University Belfast, School of Biological Sciences, Belfast, Northern Ireland, United Kingdom

Kevin Keenan
Queen's University Belfast, Institute for Global Food Security, School of Biological Sciences, Belfast, Northern Ireland, United Kingdom

Karel Mokany
Ecosystem Sciences, Commonwealth Scientific and Industrial Research Organisation, Canberra, ACT, Australia

David A. Westcott and Andrew J. Ford
Ecosystem Sciences, Commonwealth Scientific and Industrial Research Organisation, Atherton, QLD, Australia

Soumya Prasad
Centre for Ecological Sciences, Indian Institute of Science, Bangalore, India

Daniel J. Metcalfe
Ecosystem Sciences, Commonwealth Scientific and Industrial Research Organisation, Dutton Park, QLD, Australia

Zhe Wang, Huabo Liu, Jiao Gao, Zhengdong Zhang, Yingyue Li and Rongling Wu, Xiaoming Pang
National Engineering Laboratory for Tree Breeding, Key Laboratory of Genetics and Breeding in Forest Trees and Ornamental Plants, Ministry of Education, Center for Computational Biology, College of Biological Sciences and Biotechnology, Beijing Forestry University, Beijing, China

Ming Kang
South China Botanical Garden, Chinese Academy of Sciences, Guangzhou, China

Rey C. Thomas Jr. and Mudjekeewis D. Santos
Genetic Fingerprinting Laboratory, National Fisheries Research and Development Institute, Quezon City, Metro Manila, Philippines

Demian A. Willette
Department of Ecology and Evolutionary Biology, University of California Los Angeles, Los Angeles, California, United States of America

Kent E. Carpenter
Department of Biological Sciences, Old Dominion University, Norfolk, Virginia, United States of America

Olivia Charrier, Pierre Dupont, André Pornon and Nathalie Escaravage
University Toulouse III Paul Sabatier, Lab Evolution & Diversite Biologique EDB, F-31062 Toulouse, France
CNRS, EDB, UMR 5174, F-31062 Toulouse, France

Steven J. A. Kimble
Department of Forestry and Natural Resources, Purdue University, West Lafayette, Indiana, United States of America

O. E. Rhodes Jr.
Savannah River Ecology Laboratory, Aiken, South Carolina, United States of America

Rod N. Williams
Department of Forestry and Natural Resources, Purdue University, West Lafayette, Indiana, United States of America

Benito A. González
Laboratorio de Ecología de Vida Silvestre, Facultad de Ciencias Forestales y de la Conservación de la Naturaleza, Universidad de Chile, Santiago, Chile

Pablo Orozco-terWengel
School of Bioscience, Cardiff University, Cardiff, Glamorgan, Wales, United Kingdom

Rainer von Borries
Facultad de Medicina, Universidad de Chile, Santiago, Chile

Warren E. Johnson
Smithsonian Conservation Biology Institute, Smithsonian Institution, Front Royal, Virginia, United States of America

William L. Franklin
Department of Natural Resource Ecology and Management, Iowa State University, Ames, Iowa, United States of America

Juan C. Marín
Laboratorio de Genómica y Biodiversidad, Facultad de Ciencias, Universidad del Bío-Bío, Chillán, Chile

Vanesse Labeyrie, Monique Deu, Caroline Calatayud, Marylène Buiron, Jean-Christophe Glaszmann and Christian Leclerc
UMR AGAP, CIRAD, Montpellier, France

Adeline Barnaud
UMR DIADE, IRD, Montpellier, France

Peterson Wambugu
National Genebank of Kenya, KARI, Nairobi, Kenya

Stèphanie Manel
UMR LPED, Universite´ Aix-Marseille/IRD, Marseille, France
UMR AMAP, CIRAD, Montpellier, France

Helen Senn, Jennifer Kaden and Rob Ogden
WildGenes Laboratory, Royal Zoological Society of Scotland, Edinburgh, United Kingdom

Lisa Banfield
Conservation Department, Al Ain Zoo, Al Ain, Abu Dhabi, United Arab Emirates

Tim Wacher
Conservation Programmes, Zoologicial Society of London, Regents Park, London, United Kingdom

John Newby
Sahara Conservation Fund, L'Isle, Switzerland

Thomas Rabeil
Sahara Conservation Fund, Niamey, Niger

Andrew C. Kitchener
Department of Natural Sciences, National Museums Scotland, Chambers Street, Edinburgh, United Kingdom
Institute of Geography, School of Geosciences, University of Edinburgh, Drummond Street, Edinburgh, United Kingdom

Teresa Abaigar
Estación Experimental de Zonas Àridas, Consejo Superior de Investigaciones Científicas (CSIC), Almería, Spain

Teresa Luísa Silva
Estación Experimental de Zonas Àridas, Consejo Superior de Investigaciones Científicas (CSIC), Almería, Spain
CIBIO/InBIO, Centro de Investigção em Biodiversidade e Recursos Genéticos da Universidade do Porto, Vairão, Portugal
Departamento de Biologia da, Faculdade de Cieˆncias da Universidade do Porto, Porto, Portugal

Mike Maunder
College of Arts and Sciences, Florida International University, Miami, Florida, United States of America

Marco Mucciarelli
Department of Life Sciences and Systems Biology, University of Torino, Torino, Italy

Diana Ferrazzini and Piero Belletti
Department of Agricultural, Forest and Food Science, University of Torino, Grugliasco (Torino), Italy

Talatu Tende
A.P. Leventis Ornithological Research Institute, Department of Zoology, University of Jos, Plateau State, Nigeria
Department of Biology, Lund University, Lund, Sweden

Bengt Hansson and Staffan Bensch
Department of Biology, Lund University, Lund, Sweden

Ulf Ottosson
A.P. Leventis Ornithological Research Institute, Department of Zoology, University of Jos, Plateau State, Nigeria

Mikael Åkesson
Grimsö Wildlife Research Station, Department of Ecology, Swedish University of Agricultural Sciences, Riddarhyttan, Sweden

Gregory Carrier, Matthieu Garnier, Gaë l Bougaran, Jean-Paul Cadoret and Bruno Saint-Jean
IFREMER-PBA, Nantes, France

Loïc Le Cunff
UMT Geno-VigneH, Montpellier, France

Ian Probert and Colomban De Vargas
CNRS-UPMC, UMR 7144, Station Biologique de Roscoff, Roscoff, France

Erwan Corre
CNRS-UPMC, ABiMS, Station Biologique de Roscoff, Roscoff, France

Manish Roorkiwal, Hari D. Upadhyaya, Abhishek Rathore and Rajeev K. Varshney
International Crops Research Institute for the Semi-Arid Tropics (ICRISAT), Hyderabad, Andhra Pradesh, India

Eric J. von Wettberg and Emily Warschefsky
Department of Biological Sciences, Florida International University, Miami, Florida, United States of America
Center for Tropical Plant Conservation, Fairchild Tropical Botanic Garden, Miami, Florida, United States of America

Lara Awad and Rachid Cheddadi
University of Montpellier II, Institute of Evolutionary Sciences, CNRS UMR 5554, Montpellier, France

Bruno Fady and Anne Roig
INRA, UR629, Ecologie des Forêts Mèditerranèennes (URFM), Domaine St Paul, Avignon, France

Carla Khater
Center for Remote Sensing, National Council for Scientific Research-Lebanon, Bir Hassan, Beirut, Lebanon

Daniel Galiano
Programa de Pós-Graduação em Biologia Animal, Universidade Federal do Rio Grande do Sul, Porto Alegre, RS, Brazil

Jorge Bernardo-Silva
Departamento de Genética, Universidade Federal do Rio Grande do Sul, Porto Alegre, RS, Brazil

Thales R. O. de Freitas
Programa de Pós-Graduação em Biologia Animal, Universidade Federal do Rio Grande do Sul, Porto Alegre, RS, Brazil
Departamento de Genética, Universidade Federal do Rio Grande do Sul, Porto Alegre, RS, Brazil

Pierre Courtois, Charles Figuieres and Chloé Mulier
INRA-LAMETA, Montpellier, France

Index